Multi-Sensor Technology for Target Tracking, Positioning and Navigation

Multi-Sensor Technology for Target Tracking, Positioning and Navigation

Guest Editors

Bingbing Gao
Gaoge Hu

MDPI

Basel • Beijing • Wuhan • Barcelona • Belgrade • Novi Sad • Cluj • Manchester

Guest Editors
Bingbing Gao
School of Automation
Northwestern Polytechnical
University
Xi'an
China

Gaoge Hu
School of Automation
Northwestern Polytechnical
University
Xi'an
China

Editorial Office
MDPI AG
Grosspeteranlage 5
4052 Basel, Switzerland

This is a reprint of the Special Issue, published open access by the journal *Sensors* (ISSN 1424-8220), freely accessible at: https://www.mdpi.com/journal/sensors/special_issues/XSYORC9QLL.

For citation purposes, cite each article independently as indicated on the article page online and as indicated below:

Lastname, A.A.; Lastname, B.B. Article Title. *Journal Name* **Year**, *Volume Number*, Page Range.

ISBN 978-3-7258-4979-6 (Hbk)
ISBN 978-3-7258-4980-2 (PDF)
https://doi.org/10.3390/books978-3-7258-4980-2

Contents

About the Editors . . . . **vii**

Xianbing Xiao, Xingzhong Xiong, Fanqin Meng and Zhen Chen
Multi-Scale Feature Interactive Fusion Network for RGBT Tracking
Reprinted from: *Sensors* **2023**, *23*, 3410, https://doi.org/10.3390/s23073410 . . . . **1**

Yukun Wang, Xiuli Ning and Xiang Xu
An Improved In-Motion Coarse Alignment Method for SINS/GPS Integration with Initial Velocity Error Suppression
Reprinted from: *Sensors* **2023**, *23*, 3662, https://doi.org/10.3390/s23073662 . . . . **18**

Yang Luo, Xiqing Guo, Mingtao Dong and Jin Yu
Learning Modality Complementary Features with Mixed Attention Mechanism for RGB-T Tracking
Reprinted from: *Sensors* **2023**, *23*, 6609, https://doi.org/10.3390/s23146609 . . . . **31**

Fanrui Luo, Zhenyu Liu, Fengshan Zou, Mingmin Liu, Yang Cheng and Xiaoyu Li
Robust Localization of Industrial Park UGV and Prior Map Maintenance
Reprinted from: *Sensors* **2023**, *23*, 6987, https://doi.org/10.3390/s23156987 . . . . **50**

Ghina El Natour, Guillaume Bresson and Remi Trichet
Multi-Sensors System and Deep Learning Models for Object Tracking
Reprinted from: *Sensors* **2023**, *23*, 7804, https://doi.org/10.3390/s23187804 . . . . **74**

Linfeng Li, Jian Wang, Zhiming Chen and Teng Yu
Application of Adaptive Robust Kalman Filter Base on MCC for SINS/GPS Integrated Navigation
Reprinted from: *Sensors* **2023**, *23*, 8131, https://doi.org/10.3390/s23198131 . . . . **91**

Shan Huang, Zihan Song, Hyung-Rim Choi, Jae-Joong Kim, Do-Myung Park and Byung-Kwon Park
ConGPS: A Smart Container Positioning System Using Inertial Sensor and Electronic Map with Infrequent GPS
Reprinted from: *Sensors* **2023**, *23*, 9198, https://doi.org/10.3390/s23229198 . . . . **105**

Li Xue, Yongmin Zhong and Yulan Han
Constrained Cubature Particle Filter for Vehicle Navigation
Reprinted from: *Sensors* **2024**, *24*, 1228, https://doi.org/10.3390/s24041228 . . . . **118**

Jonathan Flores, Ivan Gonzalez-Hernandez, Sergio Salazar, Rogelio Lozano and Christian Reyes
Improving Optical Flow Sensor Using a Gimbal for Quadrotor Navigation in GPS-Denied Environment
Reprinted from: *Sensors* **2024**, *24*, 2183, https://doi.org/10.3390/s24072183 . . . . **134**

Tae-Ki Jung and Gyu-In Jee
VA-LOAM: Visual Assist LiDAR Odometry and Mapping for Accurate Autonomous Navigation
Reprinted from: *Sensors* **2024**, *24*, 3831, https://doi.org/10.3390/s24123831 . . . . **148**

Pan Xiong, Gang Bian, Qiang Liu, Shaohua Jin and Xiaodong Yin
Multiscale Bayes Adaptive Threshold Wavelet Transform Geomagnetic Basemap Denoising Taking Residual Constraints into Account
Reprinted from: *Sensors* **2024**, *24*, 3847, https://doi.org/10.3390/s24123847 . . . . **163**

Zhaohui Gao, Hua Zong, Yongmin Zhong and Guangle Gao
Limited Memory-Based Random-Weighted Kalman Filter
Reprinted from: *Sensors* **2024**, *24*, 3850, https://doi.org/10.3390/s24123850 . . . . **178**

Dingji Luo, Yucan Huang, Xuchao Huang, Mingda Miao and Xueshan Gao
W-VSLAM: A Visual Mapping Algorithm for Indoor Inspection Robots
Reprinted from: *Sensors* **2024**, *24*, 5662, https://doi.org/10.3390/s24175662 . . . . **199**

Kai Cui, Zhaohui Liu, Junfeng Han, Yuke Ma, Peng Liu and Bingbing Gao
Event-Driven Maximum Correntropy Filter Based on Cauchy Kernel for Spatial Orientation Using Gyros/Star Sensor Integration
Reprinted from: *Sensors* **2024**, *24*, 7164, https://doi.org/10.3390/s24227164 . . . . **225**

Wei Wang, Bin Jing, Xiaoru Yu, Yan Sun, Liping Yang and Chunliang Wang
YOLO-OD: Obstacle Detection for Visually Impaired Navigation Assistance
Reprinted from: *Sensors* **2024**, *24*, 7621, https://doi.org/10.3390/s24237621 . . . . **246**

Marko Periša, Petra Teskera, Ivan Cvitić and Ivan Grgurević
Empowering People with Disabilities in Smart Homes Using Predictive Informing
Reprinted from: *Sensors* **2025**, *25*, 284, https://doi.org/10.3390/s25010284 . . . . **261**

Bei Chen, Jiaxin Yan, Zebo Zhou, Rui Lai and Jiejian Lin
Autonomous Mission Planning for Fixed-Wing Unmanned Aerial Vehicles in Multiscenario Reconnaissance
Reprinted from: *Sensors* **2025**, *25*, 1176, https://doi.org/10.3390/s25041176 . . . . **281**

Lihua Wu, Yu Huang and Xintong Chen
A Lossless Scalar Calibration Algorithm Used for Tri-Axial Magnetometer Cross Array and Its Effectiveness Validation
Reprinted from: *Sensors* **2025**, *25*, 2164, https://doi.org/10.3390/s25072164 . . . . **297**

Cili Zuo, Demin Xie, Lianghong Wu, Xiaolong Tang and Hongqiang Zhang
An Improved Adaptive Monte Carlo Localization Algorithm Integrated with a Virtual Motion Model
Reprinted from: *Sensors* **2025**, *25*, 2471, https://doi.org/10.3390/s25082471 . . . . **313**

Hongge Mao and Xiaojun Yang
Three-Dimensional Extended Target Tracking and Shape Learning Based on Double Fourier Series and Expectation Maximization
Reprinted from: *Sensors* **2025**, *25*, 4671, https://doi.org/10.3390/s25154671 . . . . **334**

About the Editors

Bingbing Gao

Bingbing Gao received a Ph.D. degree in Navigation, Guidance, and Control from Northwestern Polytechnical University, Xi'an, China, in 2018. He is currently an Associate Professor with the School of Automation at Northwestern Polytechnical University. His research interests include nonlinear filtering, cooperative navigation, and inertial navigation. He has been named one of the top 2% of scientists in the world by Stanford University and Elsevier.

Gaoge Hu

Gaoge Hu received a Ph.D. degree in Navigation, Guidance, and Control from Northwestern Polytechnical University, Xi'an, China, in 2016. From 2016 to 2017, he held a role as a system engineer with Huawei Technologies Co Ltd. He is currently an Associate Professor with the School of Automation at Northwest Polytechnical University. His research interests include inertial-based navigation, state estimation, and multi-sensor data fusion.

sensors

MDPI

Article

Multi-Scale Feature Interactive Fusion Network for RGBT Tracking

Xianbing Xiao [1], Xingzhong Xiong [2,*], Fanqin Meng [2] and Zhen Chen [1]

1 School of Automation and Information Engineering, Sichuan University of Science and Engineering, Yibin 644000, China
2 Artificial Intelligence Key Laboratory of Sichuan Province, Sichuan University of Science and Engineering, Yibin 644000, China
* Correspondence: xzxiong@suse.edu.cn

Abstract: The fusion tracking of RGB and thermal infrared image (RGBT) is paid wide attention to due to their complementary advantages. Currently, most algorithms obtain modality weights through attention mechanisms to integrate multi-modalities information. They do not fully exploit the multi-scale information and ignore the rich contextual information among features, which limits the tracking performance to some extent. To solve this problem, this work proposes a new multi-scale feature interactive fusion network (MSIFNet) for RGBT tracking. Specifically, we use different convolution branches for multi-scale feature extraction and aggregate them through the feature selection module adaptively. At the same time, a Transformer interactive fusion module is proposed to build long-distance dependencies and enhance semantic representation further. Finally, a global feature fusion module is designed to adjust the global information adaptively. Numerous experiments on publicly available GTOT, RGBT234, and LasHeR datasets show that our algorithm outperforms the current mainstream tracking algorithms.

Keywords: RGBT tracking; multi-scale feature; information interaction; transformer; attention mechanism

1. Introduction

Visual object tracking [1,2] is widely used in autonomous driving, medical diagnostics, traffic monitoring, and other fields. Visual tracking aims at estimating the position and scale of the object in subsequent frames, given its state in the first frame. Although many RGB-based tracking algorithms have achieved excellent results, the performance of these trackers still needs to be further improved in some challenging scenarios. For example, the poor imaging quality of RGB in severe weather conditions usually leads to tracking failures. Therefore, RGBT tracking, which can make use of the advantages of RGB and infrared, has received extensive attention. Specifically, the infrared image is formed by thermal radiation with the benefits of insensitivity to light and intense penetration [3]. RGB images have rich texture and color information. Therefore, RGBT tracking can better handle complex scenarios due to its complementary advantages. Complementary advantages are shown in Figure 1.

To obtain more accurate and robust target attribute information and compensate for the information uncertainty of a single modality in target tracking, it is necessary to fuse the data collected by multi-sensors. The existing fusion tracking methods can be divided into pixel-level, feature-level, and decision-level. Pixel-level fusion tracking means that heterogeneous images are fused first, then target tracking is carried out based on the rich images [4,5]. Although rich information is preserved, it will bring greater computational costs during tracking. In feature-level fusion tracking, the features of RGB and infrared images are first extracted and then fused according to the designed fusion rules to obtain the fused feature and finally use the fused feature to perform tracking [6–8]. Decision-level fusion tracking is first performed in individual modalities to obtain tracking results or response maps. The results or response maps are then fused to get the final tracking result [9,10]. The computational cost is relatively low compared to pixel-level and feature-level fusion tracking methods.

(a)

(b)

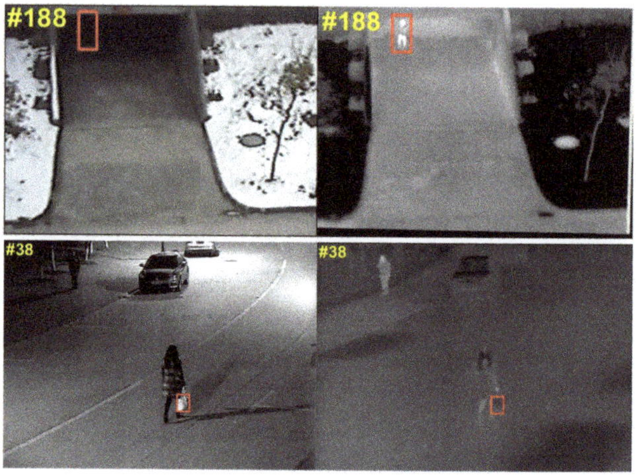

Figure 1. Illustration of the advantages of RGB and infrared modalities. (**a**) The advantage of thermal infrared modality over RGB modality. (**b**) The advantage of RGB modality over thermal infrared modality.

Although RGBT tracking has made significant progress, the mainstream methods can be roughly divided into two categories, namely the combined fusion tracking method and the discriminant fusion tracking method. The combined fusion tracking method aims to mine all valuable information of different modalities and combine them to achieve a richer representation. Li et al. [11] proposed a multi-adapter network to exploit modality-shared and modality-specific features, composed of a generic adapter, a modality adapter, and an instance adapter. Li et al. [12] proposed a challenge-aware network, summarizing existing datasets into several challenge attributes and training challenge branches one by one, which effectively enhanced the feature discrimination ability of poor modality. Although combined fusion is an effective fusion method, it tends to introduce noise and redundant information. However, the discriminant fusion tracking method aims to obtain discriminant information and achieve an effective fusion of information. To eliminate redundant information, Zhu et al. [13] proposed a fusion method based on the pruning strategy, using global average pooling and weighted random selection operations to score each channel and select high-quality channels for tracking. Zhang et al. [6] used the attention mechanism to calculate the reliability weights of multi-layer features to fuse multi-level deep features. Xu et al. [14] proposed a multi-modalities cross-layer bilinear pooling network, which uses channel and spatial attention mechanisms to predict the reliable weight of each position. However, these algorithms do not fully mine multi-scale information, limiting tracking performance when the target scale varies greatly. Moreover, they ignore the rich contextual information between features, and the interactive fusion of features can achieve more robust representations.

In response to the above problems, inspired by inception modules [15–17], we proposed a multi-scale feature interactive fusion network for RGBT tracking. Specifically, we add multiple branches to each layer for multi-scale feature extraction. In order to achieve effective fusion, we design a feature selection module that uses the channel-aware mechanism to calculate reliability weights for each branch and adaptively aggregate features from multiple branches. At the same time, the encoder and decoder of the Transformer are used to achieve self-enhancement and interaction enhancement. In addition, we designed a global feature fusion module to balance the contribution of different modalities in different scenarios, which adaptively adjusts global features in spatial and channel dimensions.

Finally, three fully connected layers are utilized for instance learning and target state estimation.

To sum up, the main contributions of this work are as follows:

i. We propose a new multi-scale feature interactive fusion network (MSIFNet) to implement robust RGBT tracking. The network can improve the recognition ability of targets of different sizes by fully exploiting multi-scale information, thus improving tracking accuracy and robustness;
ii. We design a feature selection module that adaptively selects multi-scale features for fusion by the channel-aware mechanism while effectively suppressing noise and redundant information brought by multiple branches;
iii. We propose a Transformer interactive fusion module to further enhance the aggregated feature and modality-specific features. It improves long-distance feature association and enhances semantic representation by exploring rich contextual information between features;
iv. We design a global feature fusion module, which adaptively adjusts the global information in spatial and channel dimensions, respectively, to integrate the global information more effectively.

2. Related Work

Visual tracking is a basic computer vision task, and many excellent algorithms have been proposed. This chapter will introduce relevant work from the following two aspects: (1) RGB tracker and (2) RGBT tracker.

(1) RGB trackers

Based on the correlation filter tracking algorithm [18,19], it has attracted ascendant attention because of its real-time speed, but handcrafted features limit the recognition ability. In response to this problem, siamFC [20] adopts the Siamese structure to introduce deep learning into tracking. With the application of Siamese networks in the tracking field, RGB tracking algorithms have developed rapidly. Additionally, a large number of algorithms based on Siamese networks have emerged, for example, siamRPN [21], SiamRPN++ [22] based on anchor, and siamFC++ [23], siamBAN [24], SiamCAR [25] based on the anchor-free mechanism. To optimize the tracking algorithm, Danelljan et al. [26] used off-line training IoU predictor and online training classifier for target state estimation, and they performed classification and regression tasks simultaneously in the tracking process to achieve robust tracking performance. To take advantage of the powerful feature expression ability of the Transformer, Wang et al. [27] used the Transformer to construct temporal information at different moments to model global dependencies better. Mayer et al. [28] aimed at the optimization-based tracking method that limits the expressive ability of the tracking network, used the Transformer to capture global relationships, and learned a more robust target prediction model.

(2) RGBT trackers

Traditional-based methods: Traditional RGBT tracking methods are mainly divided based on sparse representation and graph models. Since sparse representation can suppress noise and errors, early methods mainly focus on sparse representation. For example, Wu et al. [29] applied sparse representation to RGBT tracking for the first time, which integrated image patches of different modalities into a one-dimensional vector and carried out sparse representation in the target template. Li et al. [30] designed a fusion method based on Bayesian filtering, which considers intra-modality and inter-modality constraints for cross-modality sparse representation. Lan et al. [31] designed a modality-correlation-aware sparse representation model, adaptively selecting representative particles via low-rank and sparse regularization to handle appearance variations. The graph-based method can suppress the influence of noise background and has also received certain attention. To consider the synergism and heterogeneity between modalities, Li et al. [32] designed the cross-modality sorting graph model, introducing the cross-modality soft consistency to integrate multi-modalities information effectively. To eliminate background clutter, Shen et al. [33] proposed a cooperative low-rank graph model, which decomposes the input features into low-rank feature parts and sparse noise parts and uses the cooperative graph learning algorithm to renew dynamically. However, these works utilize handcrafted features for tracking, which limits the performance to handle various challenges.

Deep learning-based methods: Deep learning is known for its robust feature expressiveness, which can model the appearance of objects better than hand-crafted methods. Xu et al. [34] first applied deep learning to the field of RGBT tracking. Subsequently, deep RGBT trackers have dominated. DAPNet [13] performs fusion at different feature levels using the same aggregation network. DAFNet [35] also adopts a similar strategy to DAPNet for feature fusion. CMPP [36] utilizes the attention mechanism to model correlations between heterogeneous data. Considering the importance of time information in the video sequence, the time context information is correlated to achieve more effective information inheritance. CAT [12], ADRNet figure [37], and APFNet [38] all model robust appearance representations by training multiple challenge-aware branches independently. However, the implementation details are different, CAT [12] adaptively aggregates multiple challenge branches and then adds them to the backbone feature learning process. ADRNet designs an attribute-driven residual branch that models different challenge attributes and aggregates them through residual connections to obtain a powerful target representation. In APFNet, different challenge branches are aggregated by SKNet [39] adaptively and use Transformers to enhance features.

The above trackers perform well, but these networks do not explore multi-scale information, limiting tracking performance when the target scale varies greatly. To solve the above problems, we design a new multi-scale feature interactive fusion network to handle the RGBT tracking task.

3. Methods

This section will introduce the proposed multi-scale feature interactive fusion network. First, we outline the entire architecture, and then we present the structure of the individual module.

3.1. Network Architecture

As shown in Figure 2, our network uses a symmetrical parallel structure to mine the potential information of two modalities. Specifically, we used the VGG-M [40] as the backbone feature extraction network. To trade off speed and accuracy, we used only the first 3 layers, with convolution kernel sizes of 7×7, 5×5, and 3×3. Following the first and second layer convolution are the Relu function, local response normalization, and max-pooling. After the third layer convolution, only the Relu function is used.

Different convolution kernels have various receptive fields, and different receptive fields can better adapt to targets of different sizes. Based on this knowledge, we add 5×5 and 3×3 convolution branches in the first layer to extract multi-scale features and add 3×3 and 1×1 convolution branches in the second layer. Considering the third layer of the backbone network is a small convolution kernel, only a 1×1 convolution branch is added to the third layer. For feature maps of different sizes, we use max-pooling to fix feature maps to the exact resolution for better fusion. To better integrate multi-scale features, a feature selection module was designed to adaptively activate features from different branches for aggregation while eliminating noise and redundancy. Then, the aggregated feature and modality-specific features are fed into the Transformer interactive fusion module for interaction enhancement. After fusion, the enhanced feature is added to backbone features for subsequent feature extraction. Furthermore, we designed a gated network to block noise propagation. Finally, the global feature fusion module is used to further balance the contributions of different modalities in different scenarios, and three fully connected layers and a softmax layer are used to predict the position of the target.

In MANet [11], modality-shared and modality-specific features are helpful for modeling target information. To take full advantage of shared and specific features, our backbone network does not share parameters to extract modality-specific features, while other pairs of convolutions (e.g., 2 convolution branches of 5×5 in the first layer) share parameters to extract shared information.

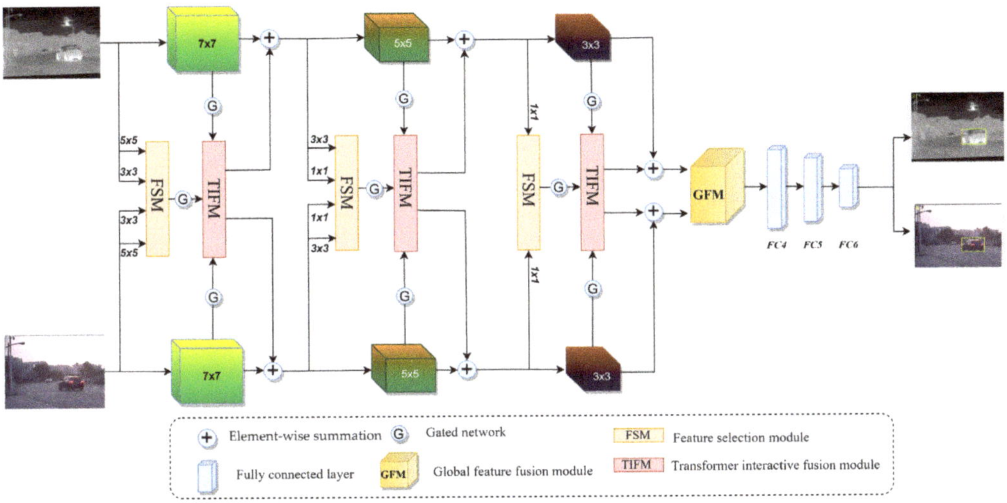

Figure 2. The overview of the proposed MSIFNet framework.

3.2. Feature Selection Module

There is redundant information in the feature maps of multiple branches. The reliability of features should be considered before fusion. Inspired by SKNet [39], we designed the feature selection module (FSM) to get more feature clues by deploying a small number of parameters. The FSM can adaptively activate valuable features for fusion while effectively suppressing noise and redundant information. Specifically, the FSM first performs simple element-wise summation, then uses the global average pooling operation (GAP) to obtain the channel vector. Then, feed the channel vector into 2 convolutions of 1×1 and a softmax function to obtain the channel weights. Finally, the original features are weighted with channel weights. The FSM adaptively activates features from different branches, combining them to focus the network on beneficial information. The specific situation of feature selection is shown in Figure 3. The whole process is summarized as follows:

$$X^a = \begin{cases} X^i_{5\times5} \oplus X^v_{5\times5} \oplus X^i_{3\times3} \oplus X^v_{3\times3}, & l = 1 \\ X^i_{1\times1} \oplus X^v_{1\times1} \oplus X^i_{3\times3} \oplus X^v_{3\times3}, & l = 2 \\ X^i_{1\times1} \oplus X^v_{1\times1}, & l = 3 \end{cases} \tag{1}$$

$$w_i = \phi\langle\delta\langle f(GAP(X^a))\rangle\rangle \tag{2}$$

$$X^{sel} = \begin{cases} w^i_{5\times5} \odot X^i_{5\times5} + w^v_{5\times5} \odot X^v_{5\times5} + w^i_{3\times3} \odot X^i_{3\times3} + w^v_{3\times3} \odot X^v_{3\times3}, & l = 1 \\ w^i_{3\times3} \odot X^i_{3\times3} + w^v_{3\times3} \odot X^v_{3\times3} + w^i_{1\times1} \odot X^i_{1\times1} + w^v_{1\times1} \odot X^v_{1\times1}, & l = 2 \\ w^i_{1\times1} \odot X^i_{1\times1} + w^v_{1\times1} \odot X^v_{1\times1}, & l = 3 \end{cases} \tag{3}$$

where l represents the l-th layer. f denotes 2 convolution operations of 1×1 inlaid with the ReLU function, expressed as $f = f_2(\xi(f_1))$. f_1 and f_2 represent 1×1 convolution operations. ξ and δ denote the ReLU function and softmax function, respectively. ϕ represents the reshape operation. w_i represents the generated weight. $\odot$ and $\oplus$ denote the element-wise product and element-wise summation.

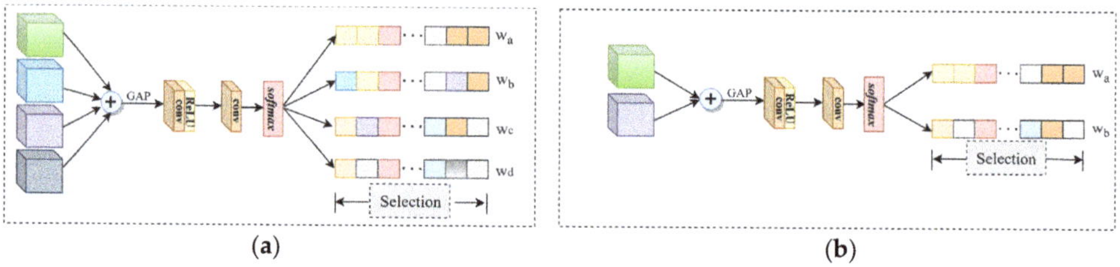

Figure 3. Specific structure of the FSM. The structure of the FSM of the first and second layers is shown in (**a**), and the third layer is shown in (**b**).

3.3. Transformer Interactive Fusion Module

The superior performance of the Transformer in computer vision proves the importance of modeling global dependencies. Inspired by [27,41], we present a Transformer interaction fusion module (TIFM) to explore the rich contextual information between features. The input of TIFM includes the modality-specific features and the aggregated feature (the output of the FSM). It should be noted that the aggregated feature and modality-specific features contain similar information (such as background information), and reducing redundant information can improve fusion efficiency. To solve this problem, a 1×1 convolution layer and sigmoid activation function are used to form the gated network to adjust the input adaptively.

As shown in Figure 4, to reduce the complexity of the model, we remove the position encoding and feed-forward network of the original Transformer. Each input feature is linearly transformed to produce a query matrix Q, a key matrix K, and a value matrix V. Through Q, K generates attention weights to modulate V. Q is then added to the output of Attention as a residual. Attention part can be expressed as follows:

$$Attention(Q,K,V) = softmax\left(\frac{QK^T}{\sqrt{C}}\right)V \tag{4}$$

where C is the dimension of key matrix to normalize attention. As we all know, the product of Q and K transpose can be regarded as the inner product of the corresponding vector so that the dependencies between any two elements of the global context can be modeled. Therefore, the encoders highlight the critical information by modeling its element dependencies. The decoders learn long-range dependencies among features to further enhance semantic information.

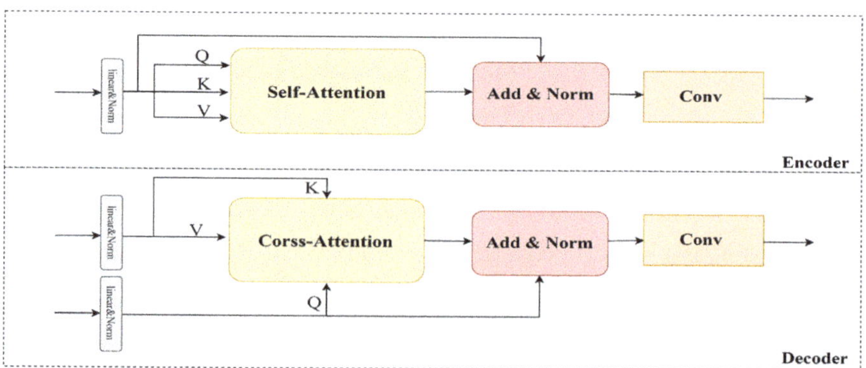

Figure 4. Specific structure of the encoder and decoder.

As shown in Figure 5, TIFM includes two encoders and two decoders. The encoders perform self-enhancement, and the decoders realize the interactive enhancement of the aggregated features (X^{sel}) and modality-specific features (X^{vis} and X^{inf}). Then, the element-wise summation operation combines the multiple enhancement features into a unified feature, and an encoder is connected to further enhance the semantic information. Specifically, we use the aggregated feature as the query for all features, resulting in a self-enhancement module (*SE*) and two feature interactive enhancement modules (*IE*). The whole process can be expressed as follows:

$$X^{\ominus} = SE\left(SE\left(X^{sel}, X^{sel}\right) + IE\left(X^{sel}, X^{vis}\right) + IE\left(X^{sel}, X^{inf}\right)\right) \tag{5}$$

where $X^{\ominus}$ represents the interactive enhanced features. Where $X^{sel}, X^{vis}, X^{inf} \in \mathbb{R}^{C\times H\times W}$ are the three inputs of TIFM. C, H, and W represent the number of channels, height, and width of the feature matrix.

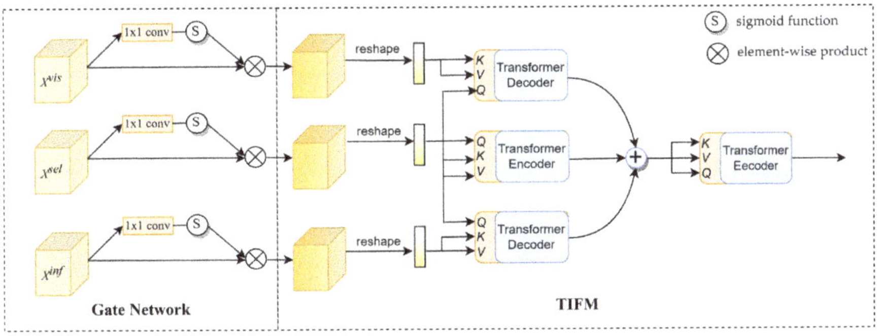

Figure 5. Specific structure of the Gate Network and TIFM.

3.4. Global Feature Fusion Module

In RGBT tracking, many algorithms aggregate the output features in the way of concatenating, which ignores the different contributions of different modalities in various scenarios. To weigh the contribution of the two modalities, we designed the global feature fusion module (GFM). The GFM aggregates multi-modalities features at the channel and spatial levels. As shown in Figure 6, it includes two branches, namely the channel integration branch and the spatial integration branch.

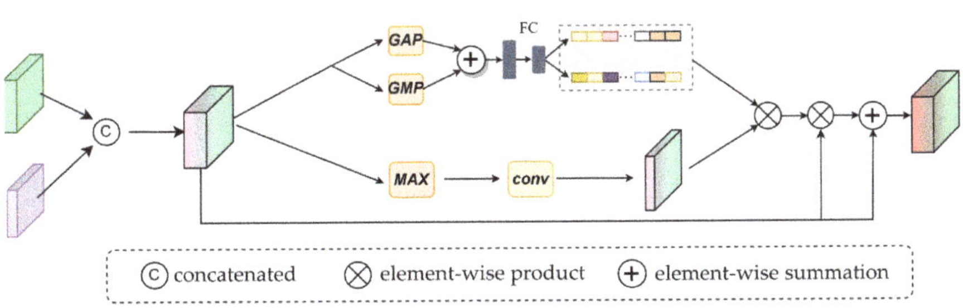

Figure 6. Specific structure of the GFM.

A channel integration branch similar to FSM is designed to obtain the channel weights of the two modalities. The difference from FSM is that the two input features are concatenated here, and GAP and global maximum pooling (GMP) are used to obtain global information and salient information. Finally, two fully connected layers and a softmax function to transform the channel vector nonlinearly obtain the channel weight w_c.

$$w_c = \phi_2(\phi_1(GAP(P^a) + GMP(P^a)) \tag{6}$$

$$P^a = cat(x_i, x_v) \tag{7}$$

where ϕ_1, ϕ_2 denote the fully connected layer. P^a represents the concatenated feature. x_i and x_v represent the output of the third layer.

At the same time, we design a spatial integration branch, which uses the max function to obtain the maximum value of the corresponding channel dimension of each spatial position and uses a 3 × 3 convolution smoothing feature to get the spatial weight w_s. w_s evaluates the contribution of each spatial location while highlighting candidates and suppressing distractions.

$$w_s = conv_{3\times3}(max(P^a)) \tag{8}$$

After the channel and spatial weights are obtained, they are fixed between 0 and 1 through softmax function σ. Additionally, we multiply them to get a 3D feature matrix with the same size as the original feature, then multiply the 3D feature with P^a. Finally, the element-wise summation of the P^a is carried out to get the optimized feature P. The mathematical expression is given as follows:

$$P = P^a \otimes [\sigma(w_s) \otimes \sigma(w_c)] + P^a \tag{9}$$

4. Experiments

To verify the effectiveness of our proposed method, we conducted many representative experiments. The experimental environment is configured as follows: NVIDIA GeForce RTX3090 GPU server, PyTorch 1.12, and Python 3.8.

4.1. Implementation Details

4.1.1. Network Training

In this work, we train MSIFNet in multiple steps, and stochastic gradient descent (SGD) algorithm is adopted to optimize MSIFNet. During training, we use softmax cross-entropy loss function for binary classification. The weight decay and momentum are set to 0.0005 and 0.9, respectively. The first step is to train the backbone network. The backbone network is initialized by the pre-trained VGG-M model on ImageNet [42], and the specific-modality branch and the full connection layer are trained with 100 epoch iterations. The learning rate of the convolution layer and the full connection layer is 0.0005 and 0.001, respectively. In the second step, we loaded the training parameters of the first step, randomly initialized the parameters of the multi-scale branches and FSM, and conducted the training of 500 epoch iterations. The learning rate was consistent with that of the first step. In the third stage, we load the training parameters of the second step and randomly initialize TIFM and GFM to train 1000 epoch iterations, and the learning rate of TIFM is 0.001, and the others are 0.0005. In each iteration, eight frames of images are randomly obtained from the video sequence. Gaussian samples are performed on these 8 frames to get 256 positive and 768 negative samples, of which 32 positive samples and 96 negative samples are generated for each frame image. It is considered a positive sample when the ratio of overlap rate (IoU) between the sample and the ground truth is [0.7, 1], while a sample in the range of [0, 0.5] is considered a negative sample. Notably, we train MSIFNet with the GTOT [43] dataset and test it on RGBT234 [44] and LasHeR [45] datasets. When testing on GTOT, we randomly selected 50 video sequences on RGBT234 for training.

4.1.2. Online Tracking

Like MDNet [46], we freeze all network parameters except Fc4-Fc6 and randomly initialize a new FC6 branch during tracking. Then, Gaussian sampling is performed on the target bounding box in the first frame, and 500 positive samples and 5000 negative samples are obtained as training sets for 50 epoch iterations of fine-tuning FC4, FC5, and FC6. The learning rate for Fc4 and Fc5 is 0.0001, and for Fc6 is 0.001. To make the tracking results more accurate, we update the parameters of the three fully connected layers in the short- and long-term and use the bounding box regression technique to solve the target scale change problem. Given the unreliability of subsequent frames, we only collected 1000 samples in the first frame to train the regression factor. For the t frame, the network uses the tracking result of the t-1 frame to sample 256 candidate samples and send them to the network to predict the target state. We select the five candidate samples with the highest score and take their average as the tracking result of the current frame. The tracking is successful when the result score exceeds 0, and the results are adjusted using a bounding box regressor for more accurate positioning. The network repeats until the entire sequence is tracked.

4.2. Result Comparisons

4.2.1. Datasets and Evaluation Metrics

(1) GTOT dataset: the GTOT dataset contains 50 pairs of RGBT video sequences collected in different scenarios and conditions, aligned spatially and temporally, totaling about 15K frames. According to the target state, it is divided into 7 challenge attributes to analyze the performance of the tracker under different conditions.

(2) RGBT234 dataset: it is a large dataset after the RGBT210 [47] dataset, adding 34 video sequences based on RGBT210. It includes 234 pairs of highly aligned RGBT video sequences and 12 challenge attributes for approximately 234K frames. It provides more accurate annotations and considers the challenges posed by various environments.

(3) LasHeR dataset: the LasHeR dataset is a more comprehensive and extensive RGBT dataset containing 1224 pairs of aligned video sequences and 19 attribute annotations. Among them, 245 sequences were selected as the test dataset, and the rest were used for training.

In the above datasets, precision rate (PR) and success rate (SR) are commonly used to evaluate tracker performance. Specifically, PR represents the percentage of frames for which the Euclidean distance between the tracking result and the ground truth is below the set threshold. Since the target of the GTOT is smaller, the threshold is set to 5 pixels. We set the threshold to 20 pixels for RGBT234 and LasHeR datasets. SR measures the percentage of frames where the overlap ratio between the tracking result and the ground truth is greater than the set threshold. Normalized PR (NPR) [48] is also commonly used to evaluate the LasHeR dataset.

4.2.2. Evaluation of GTOT Dataset

Overall Comparison: To verify the effectiveness of MSIFNet on the GTOT dataset, we compared the tracking results with some advanced RGBT tracking algorithms, including M^5L [49], CAT [12], APFNet [38], MANet [11], HDINet [50], DAFNet [35], MACNet [51], SGT [47], ECO [52], MDNet+RGBT, RT-MDNet [53], etc. The results of the PR and SR evaluations are shown in Figure 7. Specifically, MSIFNet gains 10.4% in PR and SR compared to the baseline tracker MDNet + RGBT. Compared to the recent APFNet, MSIFNet is 0.4% higher on SR and only 0.1% lower on PR. Compared to HDINet and CAT trackers, our trackers have a performance improvement of 1.6%/2.3% and 1.5%/2.4% on PR/SR, respectively. In summary, compared with some advanced RGBT algorithms, the effectiveness of our algorithm is proven.

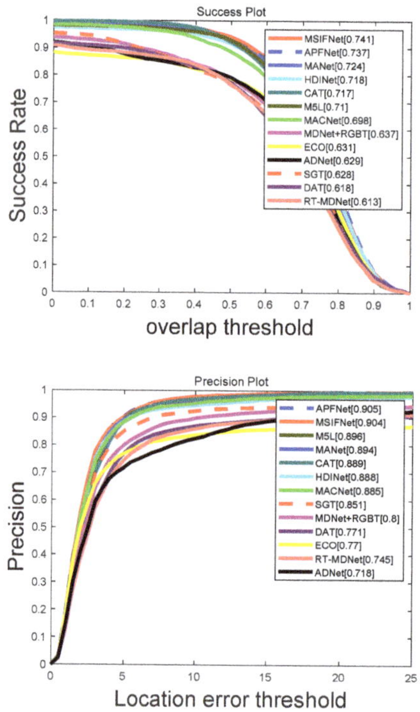

Figure 7. Evaluation curves for PR/SR metrics on GTOT.

Attribute-based performance: To evaluate the specific performance of our algorithm in addressing challenges, we compared the results of MSIFNet with those of other state-of-the-art RGBT trackers on various challenge attributes. The GTOT dataset contains 7 challenge attribute labels, including occlusion (OCC), large scale variation (LSV), a small object (SO), fast motion (FM), deformation (DEF), low illumination (LI), and thermal crossover (TC). The evaluation results are presented in Table 1. The best result is indicated in red, and the second and third are green and blue, respectively. The comparative results show that the overall performance of MSIFNet is optimal, especially in solving challenges such as OCC, LSV, FM, TC, and SO, which fully demonstrates the effectiveness of our method. In particular, MSIFNet ranks first for performance in handling large-scale changes and small object challenges, which proves the effectiveness of our proposed design of multi-scale information mining. MSIFNet is first place in both PR and SR on OCC, demonstrating that the GFM has a solid ability to highlight important information about the target and suppress the background.

Table 1. Attribute-based PR/SR scores (%) on the GTOT dataset.

	SGT	MDNet + RGBT	MANet	CAT	APFNet	MSIFNet
OCC	81.0/56.7	82.9/64.1	88.2/69.6	89.9/69.2	90.3/71.3	90.9/72.2
LSV	84.2/54.7	77.0/57.3	86.9/70.6	85.0/67.9	87.7/71.2	88.0/72.0
FM	79.9/55.9	80.5/59.8	87.9/69.4	83.9/65.4	86.5/68.4	88.2/70.2
LI	88.4/65.1	79.5/64.3	91.4/73.6	89.2/72.3	91.4/74.8	91.2/74.5
TC	84.8/61.5	79.5/60.9	88.9/70.2	89.9/71.0	90.4/71.6	91.4/72.9
SO	91.7/61.8	87.0/66.2	93.2/70.0	94.7/69.9	94.3/71.3	95.9/72.3
DEF	91.9/73.3	81.6/68.8	92.3/75.2	92.5/75.5	94.6/78.0	92.8/77.5
ALL	85.1/62.8	80.0/63.7	89.4/72.4	88.9/71.7	90.5/73.7	90.4/74.1

4.2.3. Evaluation on RGBT234 Dataset

Overall Comparison: As shown in Figure 8, a comprehensive evaluation was performed on the RGBT234 dataset. The algorithm in this paper also achieves advanced performance. Compared with JMMAC [54], the proposed algorithm improves PR by 2.7%, while SR is only 0.3% lower. In addition, compared to CAT, M^5L and HDINet, MSIFNet has a performance improvement of 1.3%/0.9%, 2.2%/2.8%, and 3.4%/1.1% on PR/SR, respectively. However, compared to high-performance CMPP [36], MSIFNet has a certain gap. The reason is that the proposed algorithm only uses the information of the current frame to model the target appearance, while CMPP enhances the representation of the current frame with the help of a large amount of historical information.

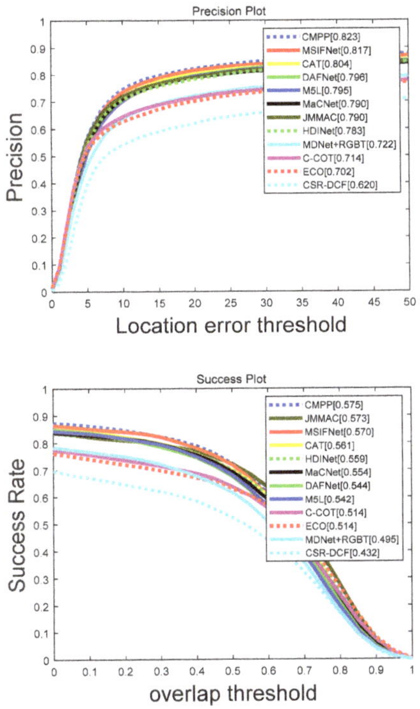

Figure 8. Evaluation curves for PR/SR metrics on RGBT234.

Attribute-based performance: At the same time, we also compared MSIFNet with 5 excellent trackers on 12 challenge attributes on the RGBT234 dataset. Challenging attributes include scale variation (SV), partial occlusion (PO), deformation (DEF), background clutter (BC), low resolution (LR), low illumination (LI), camera movement (CM), fast motion (FM), heavy occlusion (HO), no occlusion (NO), motion blur (MB), and thermal crossover (TC). As can be seen from Table 2, MSIFNet performs well in the challenges of HO, SV, LI, LR, TC, MB, DEF, CM, and FM. Notably, MSIFNet ranked in the top three across all challenges. The PR/SR metric on SV is 82.0%/57.6% compared to state-of-the-art RGBT trackers, proving that MSIFNet can adapt well to scale changes and mine multi-scale clues well. Moreover, the performance of our algorithm on LI is also excellent, which proves that the FSM and TIFM can achieve modality information interaction and fusion well. The FSM and TIFM make the RGB modality have the infrared modality information, which can cope with the challenge of low illumination well.

Table 2. Attribute-based PR/SR scores (%) on the RGBT234 dataset.

	MDNet + RGBT	DAFNet	M^{5}L	CAT	HDINet	MSIFNet
NO	86.2/61.1	90.0/63.6	93.1/64.6	93.2/66.8	88.4/65.1	92.6/66.0
PO	76.1/51.8	85.9/58.8	86.3/58.9	85.1/59.3	84.9/60.4	84.7/59.5
HO	61.9/42.1	68.6/45.9	66.5/45.0	70.0/48.0	67.1/47.3	73.8/50.5
LI	67.0/45.5	81.2/54.2	82.1/54.7	81.0/54.7	77.7/53.2	83.2/55.7
LR	75.9/51.5	81.8/53.8	82.3/53.5	82.0/53.9	80.1/54.5	85.1/56.5
TC	75.6/51.7	81.1/58.3	82.1/56.4	80.3/57.7	77.2/57.5	84.4/59.0
DEF	66.8/47.3	74.1/51.6	73.6/51.1	76.2/54.1	76.2/56.5	74.7/53.5
FM	58.6/36.3	74.0/46.5	72.8/46.5	73.1/47.0	71.7/47.5	74.7/47.5
SV	73.5/50.5	79.1/54.4	79.6/54.2	79.7/56.6	77.5/55.8	82.0/57.6
MB	65.4/46.3	70.8/50.0	73.8/52.8	68.3/49.0	70.8/52.6	75.8/54.4
CM	64.0/45.4	72.3/50.6	75.2/52.9	75.2/52.7	69.7/51.4	77.3/54.9
BC	64.4/43.2	79.1/49.3	75.0/47.7	81.1/51.9	71.1/47.8	80.5/52.3
ALL	72.2/49.5	79.6/54.4	79.5/54.2	80.4/56.1	78.3/55.9	81.7/57.0

4.2.4. Evaluation of LasHeR Dataset

To further demonstrate the robustness of MSIFNet to different datasets, we evaluate the performance of the tracker on LasHeR. The comparison results of NPR and SR indicators are shown in Figure 9. Compared to 11 RGBT algorithms, MSIFNet obtained the best results in NPR and SR. Specifically, MSIFNet outperformed the best-performing MaCNet on NPR/SR, up 0.8% and 0.2%, respectively. In addition, MSIFNet has a 3.3%/3.8% and 2.4%/3.8% improvement in NPR and SR compared to CAT and MANet++ [55]. These results demonstrate the robustness of MSIFNet to different datasets.

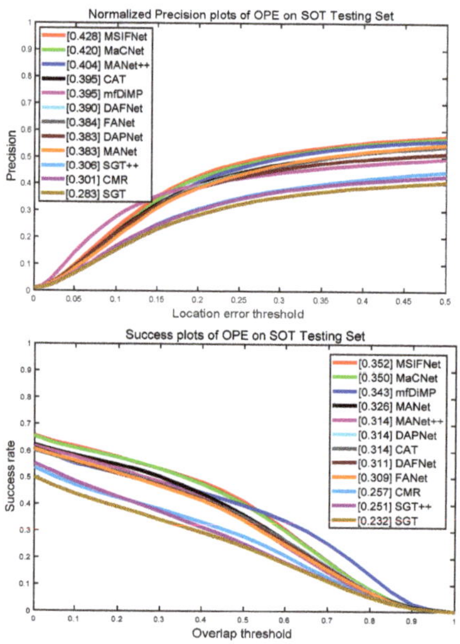

Figure 9. Evaluation curves for NPR/SR metrics on LasHeR.

4.3. *Qualitative Analysis*

In Figure 10, we perform a qualitative analysis of our algorithm. We compared MSIFNet with four advanced RGBT algorithms on six pairs of video sequences, namely children4, mancross1, diamond, Yellowcar, single3, and Night2. In the case of the first children4 sequence target moving the fast and low resolution, the target bounding box of all other trackers gradually drifts, and the accuracy of MSIFNet is barely affected. In the mancross1 sequence and diamond sequence, there are challenging attributes such as background clutter, scale variation, and heavy occlusion. Other algorithms keep the target in the lost state after heavy occlusion, but our method can track the object in a good state after occlusion. Meanwhile, in the Yellowcar sequence, our tracker can still maintain almost the same size as the ground truth box as the target scale changes. Our tracker can also robustly track the target for the single3 and Night2 sequences with a small object, fast motion, low illumination, background clutter, and partial occlusion attributes. It is clear that MSIFNet can locate the target accurately during the tracking process and is capable enough to cope with various challenges.

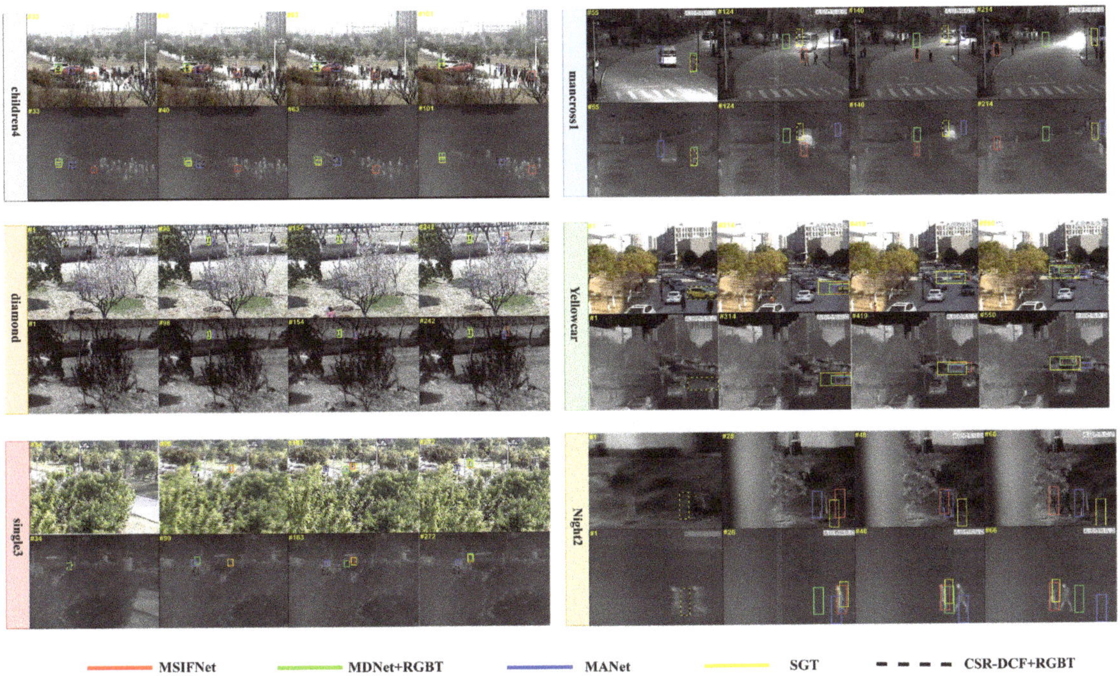

Figure 10. Qualitative comparison with four excellent RGBT trackers.

4.4. Ablation Study

To verify the validity of the proposed modules, ablation experiments were performed on the RGBT234 dataset. We compared the tracking performance of the following methods: (1) MSIFNet-FSM, which removes the FSM and performs simple addition operations for multi-scale features; (2) MSIFNet-TIFM, which removes the TIFM and adds the output of FSM directly to the backbone features; (3) MSIFNet-GFM, which removes the GFM and directly concatenates the output features of the third layer; and (4) MSIFNet-Gate, which removes the Gated network and inputs the TIFM for interactive fusion without gating. As seen from Table 3, MSIFNet-FSM has the lowest performance, mainly because a large amount of redundant information is inevitably introduced when aggregating data from multiple branches, which proves the necessity of introducing aggregation modules into multi-scale branches. The performance of MSIFNet is significantly better than that of MSIFNet-TIFM, which verifies that the proposed TIFM can effectively enhance semantic information by learning long-distance dependencies. The accuracy of MSIFNet-GFM and MSIFNet-Gate is somewhat reduced in MSIFNet, which also demonstrates the effectiveness of the GFM and Gate network.

Table 3. The PR/SR scores of our method were compared with variants on the RGBT234 datasets to verify the validity of the proposed modules.

		MSIFNet-FSM	MSIFNet-TIFM	MSIFNet-GFM	MSIFNet-Gate	MSIFNet
RGBT234	PR	0.799	0.806	0.813	0.811	**0.817**
	SR	0.557	0.561	0.564	0.565	**0.570**

We compared the performance of the following variants to verify the effectiveness of the multi-scale features extracted from each layer. (1) MSIFNet-L1 means deleting the 5×5 and 3×3 branches of the first layer and directly fusing the backbone features while leaving the other layers unchanged. (2) MSIFNet-L2 means that the 3×3 and 1×1 branches of the second layer are deleted, while others remain unchanged. (3) MSIFNet-L3 means to delete the convolution branch of 1×1 in the third layer. The results of the comparison are shown in Table 4. Experimental results show that the performance of MSIFNet is better than that of MSIFNet-L1, MSIFNet-L2, and MSIFNet-L3, which proves the effectiveness of multi-scale features in each layer.

Table 4. The PR/SR scores of our method were compared with variants on the RGBT234 datasets to verify the validity of the multi-scale features extracted from each layer.

		MSIFNet-L1	MSIFNet-L2	MSIFNet-L3	MSIFNet
RGBT234	PR	0.801	0.806	0.814	**0.817**
	SR	0.563	0.563	0.568	**0.570**

5. Conclusions

In this paper, we propose a multi-scale feature interaction fusion network (MSIFNet), which can mine the multi-scale information of RGB and infrared images to better identify targets of different sizes and further improve tracking performance. In particular, we use a feature selection module to adaptively select features from multiple branches, and the Transformer interactive fusion module is used to mine complementary information between features and enhance semantic representation. Moreover, a global feature fusion module is designed to adjust the global information in spatial and channel dimensions, respectively. Numerous experiments on publicly available GTOT, RGBT234, and LasHeR datasets have shown that our MSIFNet has advanced performance in handling various challenges. In the future, we will further explore multi-modalities fusion mechanisms to achieve more robust feature representation.

Author Contributions: Conceptualization, X.X. (Xingzhong Xiong); methodology, X.X. (Xianbing Xiao); investigation, X.X. (Xianbing Xiao); writing—original draft preparation, X.X. (Xianbing Xiao); writing—review and editing, X.X. (Xingzhong Xiong), F.M. and Z.C.; supervision, X.X. (Xingzhong Xiong), F.M. and Z.C.; project administration, X.X. (Xingzhong Xiong) and F.M.; funding acquisition, X.X. (Xingzhong Xiong) and F.M. All authors have read and agreed to the published version of the manuscript.

Funding: This work was supported by the Science and Technology Department of Sichuan Province (2023NSFSC1987) and in part by the Postgraduate Innovation Fund Project of Sichuan University of Science and Engineering under grant Y2022147 and the Artificial Intelligence Key Laboratory of Sichuan Province (grant number 2019RZJ04).

Institutional Review Board Statement: Not applicable.

Informed Consent Statement: Not applicable.

Data Availability Statement: Public datasets.

Conflicts of Interest: The authors declare that they have no conflict of interest.

References

1. Shen, Y.; Liu, Z.; Zhang, G. PAC interaction inspection using real-time contact point tracking. *IEEE Trans. Instrum. Meas.* **2018**, *68*, 4051–4064. [CrossRef]
2. Mehmood, K.; Jalil, A.; Ali, A.; Khan, B.; Murad, M.; Khan, W.U.; He, Y. Context-aware and occlusion handling mechanism for online visual object tracking. *Electronics* **2020**, *10*, 43. [CrossRef]
3. Gade, R.; Moeslund, T.B. Thermal cameras and applications: A survey. *Mach. Vis. Appl.* **2014**, *25*, 245–262. [CrossRef]
4. Schnelle, S.R.; Chan, A.L. Enhanced target tracking through infrared-visible image fusion. In Proceedings of the 14th International Conference on Information Fusion, Chicago, IL, USA, 5–8 July 2011; IEEE: New York, NY, USA, 2011; pp. 1–8.
5. Chan, A.L.; Schnelle, S.R.J.O.E. Fusing concurrent visible and infrared videos for improved tracking performance. *Opt. Eng.* **2013**, *52*, 017004. [CrossRef]
6. Zhang, X.; Ye, P.; Peng, S.; Liu, J.; Xiao, G. DSiamMFT: An RGB-T fusion tracking method via dynamic Siamese networks using multi-layer feature fusion. *Signal Process. Image Commun.* **2020**, *84*, 115756. [CrossRef]
7. Xia, W.; Zhou, D.; Cao, J.; Liu, Y.; Hou, R. CIRNet: An improved RGBT tracking via cross-modality interaction and re-identification. *Neurocomputing* **2022**, *493*, 327–339. [CrossRef]
8. Lu, A.; Qian, C.; Li, C.; Tang, J.; Wang, L. Duality-gated mutual condition network for RGBT tracking. In *IEEE Transactions on Neural Networks and Learning Systems*; IEEE: New York, NY, USA, 2022.
9. He, F.; Chen, M.; Chen, X.; Han, J.; Bai, L. SiamDL: Siamese Dual-Level Fusion Attention Network for RGBT Tracking. *SSRN*, 2022; *submitted*. [CrossRef]
10. Wang, Y.; Wei, X.; Tang, X.; Wu, J.; Fang, J. Response map evaluation for RGBT tracking. *Neural Comput. Appl.* **2022**, *34*, 5757–5769. [CrossRef]
11. Long Li, C.; Lu, A.; Hua Zheng, A.; Tu, Z.; Tang, J. Multi-adapter RGBT trackin. In Proceedings of the IEEE/CVF International Conference on Computer Vision Workshops, Seoul, Republic of Korea, 27–28 October 2019.
12. Li, C.; Liu, L.; Lu, A.; Ji, Q.; Tang, J. Challenge-aware RGBT tracking. In *Part XXII 16, Proceedings of the Computer Vision–ECCV 2020: 16th European Conference, Glasgow, UK, 23–28 August 2020*; Springer: Berlin/Heidelberg, Germany, 2020; pp. 222–237.
13. Zhu, Y.; Li, C.; Luo, B.; Tang, J.; Wang, X. Dense feature aggregation and pruning for RGBT tracking. In Proceedings of the 27th ACM International Conference on Multimedia, Nice, France, 21–25 October 2019; pp. 465–472.
14. Xu, Q.; Mei, Y.; Liu, J.; Li, C. Multimodal cross-layer bilinear pooling for RGBT tracking. *IEEE Trans. Multimedia* **2021**, *24*, 567–580. [CrossRef]
15. Szegedy, C.; Liu, W.; Jia, Y.; Sermanet, P.; Reed, S.; Anguelov, D.; Erhan, D.; Vanhoucke, V.; Rabinovich, A. Going deeper with convolutions. In Proceedings of the IEEE Conference on Computer Vision and Pattern Recognition, Boston, MA, USA, 7–12 June 2015; pp. 1–9.
16. Szegedy, C.; Vanhoucke, V.; Ioffe, S.; Shlens, J.; Wojna, Z. Rethinking the inception architecture for computer vision. In Proceedings of the IEEE Conference on Computer Vision and Pattern Recognition, Lasvegas, NV, USA, 26 June–1 July 2016; pp. 2818–2826.
17. Szegedy, C.; Ioffe, S.; Vanhoucke, V.; Alemi, A. Inception-v4, inception-resnet and the impact of residual connections on learning. In Proceedings of the AAAI Conference on Artificial Intelligence, San Francisco, CA, USA, 4–9 February 2017.
18. Henriques, J.F.; Caseiro, R.; Martins, P.; Batista, J. High-speed tracking with kernelized correlation filters. *IEEE Trans. Pattern Anal. Mach. Intell.* **2014**, *37*, 583–596. [CrossRef]
19. Danelljan, M.; Hager, G.; Shahbaz Khan, F.; Felsberg, M. Learning spatially regularized correlation filters for visual tracking. In Proceedings of the IEEE International Conference on Computer Vision, Santiago, Chile, 7–13 December 2015; pp. 4310–4318.
20. Bertinetto, L.; Valmadre, J.; Henriques, J.F.; Vedaldi, A.; Torr, P.H. Fully-convolutional siamese networks for object tracking. In *Part II 14, Proceedings of the Computer Vision–ECCV 2016 Workshops, Amsterdam, The Netherlands, 8–10, 15–16 October 2016*; Springer: Berlin/Heidelberg, Germany, 2016; pp. 850–865.
21. Li, B.; Yan, J.; Wu, W.; Zhu, Z.; Hu, X. High performance visual tracking with siamese region proposal network. In Proceedings of the IEEE Conference on Computer Vision and Pattern Recognition, Salt Lake City, UT, USA, 18–23 June 2018; pp. 8971–8980.
22. Li, B.; Wu, W.; Wang, Q.; Zhang, F.; Xing, J.; Yan, J. Siamrpn++: Evolution of siamese visual tracking with very deep networks. In Proceedings of the IEEE/CVF Conference on Computer Vision and Pattern Recognition, Long Beach, CA, USA, 16–17 June 2019; pp. 4282–4291.
23. Xu, Y.; Wang, Z.; Li, Z.; Yuan, Y.; Yu, G. Siamfc++: Towards robust and accurate visual tracking with target estimation guidelines. In Proceedings of the AAAI Conference on Artificial Intelligence, New York, NY, USA, 7–12 February 2020; pp. 12549–12556.
24. Chen, Z.; Zhong, B.; Li, G.; Zhang, S.; Ji, R. Siamese box adaptive network for visual tracking. In Proceedings of the IEEE/CVF Conference on Computer Vision and Pattern Recognition, Seattle, WA, USA, 13–19 June 2020; pp. 6668–6677.
25. Guo, D.; Wang, J.; Cui, Y.; Wang, Z.; Chen, S. SiamCAR: Siamese fully convolutional classification and regression for visual tracking. In Proceedings of the IEEE/CVF Conference on Computer Vision and Pattern Recognition, Seattle, WA, USA, 13–19 June 2020; pp. 6269–6277.
26. Danelljan, M.; Bhat, G.; Khan, F.S.; Felsberg, M. Atom: Accurate tracking by overlap maximization. In Proceedings of the IEEE/CVF Conference on Computer Vision and Pattern Recognition, Long Beach, CA, USA, 15–20 June 2019; pp. 4660–4669.
27. Wang, N.; Zhou, W.; Wang, J.; Li, H. Transformer meets tracker: Exploiting temporal context for robust visual tracking. In Proceedings of the IEEE/CVF Conference on Computer Vision and Pattern Recognition, Nashville, TN, USA, 19–25 June 2021; pp. 1571–1580.

28. Mayer, C.; Danelljan, M.; Bhat, G.; Paul, M.; Paudel, D.P.; Yu, F.; Van Gool, L. Transforming model prediction for tracking. In Proceedings of the IEEE/CVF Conference on Computer Vision and Pattern Recognition, New Orleans, LA, USA, 18–24 June 2022; pp. 8731–8740.
29. Wu, Y.; Blasch, E.; Chen, G.; Bai, L.; Ling, H. Multiple source data fusion via sparse representation for robust visual tracking. In Proceedings of the 14th International Conference on Information Fusion, Chicago, IL, USA, 5–8 July 2011; IEEE: New York, NY, USA, 2011; pp. 1–8.
30. Li, L.; Li, C.; Tu, Z.; Tang, J. A fusion approach to grayscale-thermal tracking with cross-modal sparse representation. In *Proceedings of the Image and Graphics Technologies and Applications: 13th Conference on Image and Graphics Technologies and Applications, IGTA 2018, Beijing, China, 8–10 April 2018*; Revised Selected Papers 13; Springer: Berlin/Heidelberg, Germany, 2018; pp. 494–505.
31. Lan, X.; Ye, M.; Zhang, S.; Zhou, H.; Yuen, P.C. Modality-correlation-aware sparse representation for RGB-infrared object tracking. *Pattern Recognit. Lett.* **2020**, *130*, 12–20. [CrossRef]
32. Li, C.; Zhu, C.; Huang, Y.; Tang, J.; Wang, L. Cross-modal ranking with soft consistency and noisy labels for robust RGB-T tracking. In Proceedings of the European Conference on Computer Vision (ECCV), Munich, Germany, 8–14 September 2018; pp. 808–823.
33. Shen, L.; Wang, X.; Liu, L.; Hou, B.; Jian, Y.; Tang, J.; Luo, B. RGBT tracking based on cooperative low-rank graph model. *Neurocomputing* **2022**, *492*, 370–381. [CrossRef]
34. Xu, N.; Xiao, G.; Zhang, X.; Bavirisetti, D.P. Relative object tracking algorithm based on convolutional neural network for visible and infrared video sequences. In Proceedings of the 4th International Conference on Virtual Reality, Hong Kong, China, 24–26 February 2018; pp. 44–49.
35. Gao, Y.; Li, C.; Zhu, Y.; Tang, J.; He, T.; Wang, F. Deep adaptive fusion network for high performance RGBT tracking. In Proceedings of the IEEE/CVF International Conference on Computer Vision Workshops, Seoul, Republic of Korea, 27–28 October 2019.
36. Wang, C.; Xu, C.; Cui, Z.; Zhou, L.; Zhang, T.; Zhang, X.; Yang, J. Cross-modal pattern-propagation for RGB-T tracking. In Proceedings of the IEEE/CVF Conference on Computer Vision and Pattern Recognition, Seattle, WA, USA, 13–19 June 2020; pp. 7064–7073.
37. Zhang, P.; Wang, D.; Lu, H.; Yang, X. Learning adaptive attribute-driven representation for real-time RGB-T tracking. *Int. J. Comput. Vis.* **2021**, *129*, 2714–2729. [CrossRef]
38. Xiao, Y.; Yang, M.; Li, C.; Liu, L.; Tang, J. Attribute-based progressive fusion network for rgbt tracking. In Proceedings of the AAAI Conference on Artificial Intelligence, Washington, DC, USA, 7–14 February 2022; pp. 2831–2838.
39. Li, X.; Wang, W.; Hu, X.; Yang, J. Selective kernel networks. In Proceedings of the IEEE/CVF Conference on Computer Vision and Pattern Recognition, Long Beach, CA, USA, 16–20 June 2019; pp. 510–519.
40. Chatfield, K.; Simonyan, K.; Vedaldi, A.; Zisserman, A. Return of the devil in the details: Delving deep into convolutional nets. In Proceedings of the British Machine Vision Conference 2014, Nottinghamshire, UK, 1–5 September 2014.
41. Chen, X.; Yan, B.; Zhu, J.; Wang, D.; Yang, X.; Lu, H. Transformer tracking. In Proceedings of the IEEE/CVF Conference on Computer Vision and Pattern Recognition, Nashville, TN, USA, 20–25 June 2021; pp. 8126–8135.
42. Krizhevsky, A.; Sutskever, I.; Hinton, G.E. Imagenet classification with deep convolutional neural networks. *Commun. ACM* **2017**, *60*, 84–90. [CrossRef]
43. Li, C.; Cheng, H.; Hu, S.; Liu, X.; Tang, J.; Lin, L. Learning collaborative sparse representation for grayscale-thermal tracking. *IEEE Trans. Image Process.* **2016**, *25*, 5743–5756. [CrossRef]
44. Li, C.; Liang, X.; Lu, Y.; Zhao, N.; Tang, J. RGB-T object tracking: Benchmark and baseline. *Pattern Recognit.* **2019**, *96*, 106977. [CrossRef]
45. Li, C.; Xue, W.; Jia, Y.; Qu, Z.; Luo, B.; Tang, J.; Sun, D. LasHeR: A large-scale high-diversity benchmark for RGBT tracking. *IEEE Trans. Image Process.* **2021**, *31*, 392–404. [CrossRef] [PubMed]
46. Nam, H.; Han, B. Learning multi-domain convolutional neural networks for visual tracking. In Proceedings of the IEEE Conference on Computer Vision and Pattern Recognition, Las Vegas, NV, USA, 26 June–1 July 2016; pp. 4293–4302.
47. Li, C.; Zhao, N.; Lu, Y.; Zhu, C.; Tang, J. Weighted sparse representation regularized graph learning for RGB-T object tracking. In Proceedings of the 25th ACM International Conference on Multimedia, New York, NY, USA, 23–27 October 2017; pp. 1856–1864.
48. Muller, M.; Bibi, A.; Giancola, S.; Alsubaihi, S.; Ghanem, B. Trackingnet: A large-scale dataset and benchmark for object tracking in the wild. In Proceedings of the European Conference on Computer Vision (ECCV), Munich, Germany, 8–14 September 2018; pp. 300–317.
49. Tu, Z.; Lin, C.; Zhao, W.; Li, C.; Tang, J. M 5 l: Multi-modal multi-margin metric learning for RGBT tracking. *IEEE Trans. Image Process.* **2021**, *31*, 85–98. [CrossRef]
50. Mei, J.; Zhou, D.; Cao, J.; Nie, R.; Guo, Y. Hdinet: Hierarchical dual-sensor interaction network for rgbt tracking. *IEEE Sens. J.* **2021**, *21*, 16915–16926. [CrossRef]
51. Zhang, H.; Zhang, L.; Zhuo, L.; Zhang, J. Object tracking in RGB-T videos using modal-aware attention network and competitive learning. *Sensors* **2020**, *20*, 393. [CrossRef]
52. Danelljan, M.; Bhat, G.; Shahbaz Khan, F.; Felsberg, M. Eco: Efficient convolution operators for tracking. In Proceedings of the IEEE Conference on Computer Vision and Pattern Recognition, Honolulu, HI, USA, 21–26 July 2017; pp. 6638–6646.
53. Jung, I.; Son, J.; Baek, M.; Han, B. Real-time mdnet. In Proceedings of the European Conference on Computer Vision (ECCV), Munich, Germany, 8–14 September 2018; pp. 83–98.

54. Zhang, P.; Zhao, J.; Bo, C.; Wang, D.; Lu, H.; Yang, X. Jointly modeling motion and appearance cues for robust RGB-T tracking. *IEEE Trans. Image Process.* **2021**, *30*, 3335–3347. [CrossRef] [PubMed]
55. Lu, A.; Li, C.; Yan, Y.; Tang, J.; Luo, B. RGBT tracking via multi-adapter network with hierarchical divergence loss. *IEEE Trans. Image Process.* **2021**, *30*, 5613–5625. [CrossRef] [PubMed]

sensors

MDPI

Article

An Improved In-Motion Coarse Alignment Method for SINS/GPS Integration with Initial Velocity Error Suppression

Yukun Wang [1], Xiuli Ning [2] and Xiang Xu [3,*]

1 School of Mechatronical Engineering, Beijing Institute of Technology, Beijing 100081, China; wyk521215@163.com
2 China National Institute of Standardization, Beijing 100191, China; nxl_warm0908@163.com
3 School of Automation, Nanjing University of Science and Technology, Nanjing 210094, China
* Correspondence: hsianghsu@163.com

Abstract: The integrated system with the strapdown inertial navigation system (SINS) and the global positioning system (GPS) is the most popular navigation mode. It has been used in many navigation fields. Before the integrated system works properly, it must determine the initial attitude for SINS. In SINS/GPS-integrated systems, the navigational velocity can be used to carry out the initial alignment when the system is installed in the in-motion vehicle. However, the initial velocity errors are not considered in the current popular in-motion alignment methods for SINS/GPS integration. It is well-known that the initial velocity errors must exist when the initial velocity is obtained from the GPS outputs. In this paper, an improved method was proposed to solve this problem. By analyzing the original observation vectors in the in-motion coarse alignment method, an average operation was used to construct the intermediate vectors, and the new observation vector can be calculated by subtracting the intermediate vector from the original observation vector. Then, the initial velocity errors can be eliminated from the newly constructed observation vector. Thus, the interferences of the initial velocity errors for the initial alignment process can be suppressed. The simulation and field tests are designed to verify the performance of the proposed method. The tests results showed that the proposed method can obtain the higher accurate results than the current methods when the initial velocity is considered. Additionally, the results of the proposed method were similar to the current methods when the initial velocity errors were not considered. This shows that the initial velocity errors were eliminated effectively by the proposed method, and the alignment accuracy were not decreased.

Keywords: initial velocity errors; average operation for observation vectors; in-motion coarse alignment; SINS; GPS

1. Introduction

Navigation system is important for many applications, such as autonomous underwater vehicle (AUV), land vehicles, military systems and so on [1–3]. In many navigation systems, the strapdown inertial navigation system has the self-contained properties, and high sampling rates. Thus, SINS was found to be widely used for many applications [4,5]. However, the accumulated errors, which is caused by the measured errors of the inertial sensors, will be contained in the SINS outputs. Then, the positioning accuracy of SINS degrades. To address this problem, the SINS-based integrated navigation system was investigated [6,7]. Then, the accumulated errors can be corrected by the external measurements. In these integrated navigation systems, the SINS/GPS-integrated system is one of the most popular categories [8,9].

Before SINS/GPS works in full accuracy, the initial alignment process must be carried on for SINS. Many researchers were devoted to initial alignment research [10–22]. Firstly, the self-alignment methods in static or swaying base were investigated [10–12]. In [10], the attitude determination method with the Kalman filter was investigated. The convergence rate of the coarse alignment process was improved. In [11], the alignment errors were analyzed by an analytical calculation method when the SINS was on the stationary base. In [12], the parameter identification method was proposed for improving the vector observations. The random noises of the inertial sensors were weakened. These methods can be divided into the self-contained alignment process, because the alignment process can be finished only by the inertial sensors. The external equipment was not needed. Moreover, the self-alignment process can obtain the high accuracy during a short time, because the alignment process was not corrupted by the external interferences. However, the self-contained methods cannot be applicable to the in-motion situations, because the vehicle's motion will interfere the extraction of the Earth gravitational apparent motion. Thus, the applied range of the self-contained alignment methods was limited.

To extend the applied range, the in-motion alignment methods were devised [13–17]. In [13,14], the in-motion coarse alignment method for SINS/GPS integration was proposed, the detailed deduction of the observation vectors construction was described. With the SINS/GPS integration, the alignment results were high accuracy, which can be found in [13,14]. However, the outliers, which are contained in the GPS outputs, were ignored. Thus, the methods could not obtain the high accuracy when there were outliers in the GPS outputs. In [15], the position loci method was used to implement the in-motion alignment process. Based on the position loci, the outliers of GPS outputs were suppressed and the robustness of the in-motion alignment method was improved. These methods were based on the inertial frame theory. In [16,17], the in-motion alignment methods for SINS/DVL (Doppler Velocity Log) and SINS/OD (Odometer) systems were investigated, the alignment principles were also based on the inertial frame theory. Moreover, these in-motion alignment methods used the observation vectors, which were based on the chain rule of the direction cosine matrix (DCM). The alignment process was transformed to the attitude determination problem. However, it is noted that almost all aforementioned in-motion methods, which used the inertial frame theory, used the initial velocity of the vehicle to construct the observed vectors. Additionally, the initial velocity must be obtained from the external equipment since SINS worked in the initial alignment stage. However, the initial velocity errors were not concerned. Thus, there was a flaw in the proposed methods.

In this paper, we design an improved method for in-motion coarse alignment process, where the initial velocity errors are concerned. The intermediate observed vectors are introduced to construct the new observed vectors. Then, according to the vector subtraction operations, the initial velocity errors are eliminated from the new observed vectors. Based on the new observed vectors, the performance of the in-motion coarse alignment method is improved. In this paper, the proposed method is used for the SINS/GPS system. Additionally, we think it can be extended for other integration systems, such as SINS/DVL and SINS/OD integration.

The rest of this paper is organized as follows. Section 2 explores the related works of the current in-motion coarse alignment for SINS/GPS integration. Then, the existing problems are analyzed. In Section 3, the construction for the new observation vectors is investigated. The simulation and field tests are designed to verify the performance of the proposed method in Section 4. Finally, the conclusions are drawn in Section 5.

2. Related Works and Problem Statements

In this section, the previous works will be reviewed and the observations vectors will be given. Moreover, the initial velocity errors will be introduced.

2.1. Related Works

The most popular methods for in-motion coarse alignment process are based on the inertial frame theory [13–15]. Thus, based on the inertial frame theory, the chain rule of the DCM can be given by:

$$C_b^n(t) = C_{b(t)}^{n(t)} = C_{n(0)}^{n(t)} C_{b(0)}^{n(0)} C_{b(t)}^{b(0)} = C_{n(0)}^{n(t)} C_b^n(0) C_{b(t)}^{b(0)} \tag{1}$$

Additionally, the attitude rates of $C_{b(t)}^{b(0)}$ and $C_{n(0)}^{n(t)}$ are given by:

$$\begin{cases} \dot{C}_{b(t)}^{b(0)} = C_{b(t)}^{b(0)} \left[\omega_{ib}^b \times\right] \\ \dot{C}_{n(t)}^{n(0)} = C_{n(t)}^{n(0)} \left[\omega_{in}^n \times\right] \end{cases} \tag{2}$$

where b denotes the body frame (b-frame), it is the right-forward-up (RFU) orthogonal frame in this paper; n denotes the navigation frame (n-frame), it is the east-north-up (ENU) orthogonal frame in this paper; $b(0)$ and $n(0)$ denote the b-frame and n-frame at the beginning of the alignment process, they are the non-rotating frame during the whole alignment procedure. In the alignment process, the angular velocity ω_{ib}^b can be obtained instead by the gyroscopes' outputs. Additionally, the ω_{in}^n can be calculated by the velocity and position measurements of the GPS.

According to (1) and (2), only the DCM $C_b^n(0)$ is unknown. If $C_b^n(0)$ can be determined, the initial alignment can be finished. Referring to [13], the initial alignment process was transformed as the attitude determination problem with the observation vectors:

$$C_b^n(0)\boldsymbol{\alpha}_v = \boldsymbol{\beta}_v \tag{3}$$

where

$$\begin{cases} \boldsymbol{\alpha}_v = \int_0^t C_{b(t)}^{b(0)} \boldsymbol{f}^b dt \\ \boldsymbol{\beta}_v = C_{n(t)}^{n(0)} \boldsymbol{v}^n - \boldsymbol{v}^n(0) + \int_0^t C_{n(t)}^{n(0)} \boldsymbol{\omega}_{ie}^n \times \boldsymbol{v}^n dt - \int_0^t C_{n(t)}^{n(0)} \boldsymbol{g}^n dt \end{cases} \tag{4}$$

where $\boldsymbol{f}^b$ denotes the specific force, it can be measured by the accelerometers; $\boldsymbol{\omega}_{ie}^n$ denotes the Earth rate in n-frame; $\boldsymbol{g}^n$ denotes the gravity in n-frame. It is noted that the reference vector $\boldsymbol{\alpha}_v$ can be calculated by the outputs of accelerometers and gyroscopes in the in-motion coarse alignment process. Additionally, the observed vector $\boldsymbol{\beta}_v$ can be calculated by the outputs of the GPS.

Using the observation vectors, the $\boldsymbol{K}$ matrix can be constructed by:

$$\boldsymbol{K} = \sum([\boldsymbol{\beta}_v \otimes] - [\boldsymbol{\alpha}_v \odot])^T([\boldsymbol{\beta}_v \otimes] - [\boldsymbol{\alpha}_v \odot]) \tag{5}$$

where

$$\begin{cases} [\boldsymbol{\beta}_v \otimes] = \begin{bmatrix} \boldsymbol{0} & -\boldsymbol{\beta}_v^T \\ \boldsymbol{\beta}_v & [\boldsymbol{\beta}_v \times] \end{bmatrix} \\ [\boldsymbol{\alpha}_v \odot] = \begin{bmatrix} \boldsymbol{0} & -\boldsymbol{\alpha}_v^T \\ \boldsymbol{\alpha}_v & -[\boldsymbol{\alpha}_v \times] \end{bmatrix} \end{cases} \tag{6}$$

According to Equations (5) and (6), the attitude quaternion can be extracted from the matrix $\boldsymbol{K}$ [15]. Thus, the DCM $C_b^n(0)$ can be obtained from the attitude quaternion. Then, the initial alignment process can be finished.

2.2. Problem Statement

Based on the aforementioned related works, which are shown in [13–15], the in-motion coarse alignment methods with SINS/GPS integration are implemented. Based on these methods, many researchers were devoted to extend the in-motion coarse alignment method to many other fields, such as underwater navigation vehicle [16,18], land vehicle and so on [19]. Additionally, the robust methods were proposed to address the outliers from the external auxiliary equipment, such as GPS, DVL and odometers [20–22]. However, from the observed vector β_v, it can be found that there exists an initial velocity $v^n(0)$ in the observation vector. When SINS is under in-motion coarse alignment process, the initial velocity can be obtained from the GPS outputs. Due to the specifications' performance of the GPS, this initial velocity must be corrupted by the noises even outliers. If the initial velocity contains the errors, the alignment accuracy, which is based on the accuracy of the observation vector, will degrade. Moreover, in contrast to the real-time measurement errors of GPS, the initial velocity errors are the constant value during the whole alignment process. It is hard to use the outlier's detection method to suppress them. Therefore, how to suppress the initial velocity errors is worth focusing on. We will give the simple effective method to solve this problem. The detailed method can be found in the next section.

3. The New Observation Vectors

According to the GPS measurements, the initial velocity from the GPS outputs can be modelled as:

$$\tilde{v}^n(0) = v^n(0) + \delta v_0^n \tag{7}$$

where $\tilde{v}^n(0)$ denotes the measured initial velocity from the GPS; $v^n(0)$ denotes the truth initial velocity; δv_0^n denotes the initial velocity error. Substituting (7) into (4), we can obtain the calculated observed vector as:

$$\begin{aligned} \tilde{\beta}_v &= C_{n(t)}^{n(0)}\tilde{v}^n - \tilde{v}^n(0) + \int_0^t C_{n(t)}^{n(0)}\omega_{ie}^n \times v^n dt - \int_0^t C_{n(t)}^{n(0)} g^n dt \\ &= C_{n(t)}^{n(0)}\tilde{v}^n - v^n(0) - \delta v_0^n + \int_0^t C_{n(t)}^{n(0)}\omega_{ie}^n \times v^n dt - \int_0^t C_{n(t)}^{n(0)} g^n dt \\ &= \breve{\beta}_v - \delta v_0^n \end{aligned} \tag{8}$$

where $\tilde{v}^n$ denotes the real-time measured velocity from GPS; and $\breve{\beta}_v = \beta_v + C_{n(t)}^{n(0)}\delta v^n$. It is noted that $\tilde{v}^n$ also contain the noises, but it is not a constant value during the whole alignment process, and it can be suppressed by the robust methods [22]. In this study, we only focused on the initial velocity errors.

From [13,22], the following equation was obtained:

$$\breve{\beta}_v = C_b^n(0)\tilde{\alpha}_v \tag{9}$$

where

$$\tilde{\alpha}_v = \int_0^t \tilde{C}_{b(t)}^{b(0)}\tilde{f}^b dt \tag{10}$$

Additionally, $\tilde{C}_{b(t)}^{b(0)}$ and $\tilde{f}^b$ are obtained from the gyroscopes and accelerometers. It is noted that the upper tilde means there are errors in $\tilde{C}_{b(t)}^{b(0)}$ and $\tilde{f}^b$ with the inertial sensors.

From (8), using the discrete form of the observation vectors, the averaged observed vectors can be calculated by:

$$\begin{aligned} \bar{\beta}_{v,M} &= \frac{1}{M}\sum_{k=1}^{M}\tilde{\beta}_{v,k} \\ &= \frac{1}{M}\sum_{k=1}^{M}\left(\breve{\beta}_{v,k} - \delta v_0^n\right) \\ &= \frac{1}{M}\sum_{k=1}^{M}\breve{\beta}_{v,k} - \delta v_0^n \end{aligned} \tag{11}$$

where the subscript k denotes the vectors at time instant k; The M denotes the current time instant M.

Using (7), the new observed vector can be calculated by the vector subtraction:

$$\begin{aligned}\hat{\boldsymbol{\beta}}_{v,M} &= \tilde{\boldsymbol{\beta}}_{v,M} - \bar{\boldsymbol{\beta}}_{v,M} \\ &= \tilde{\boldsymbol{\beta}}_{v,M} - \tfrac{1}{M}\sum_{k=1}^{M}\tilde{\boldsymbol{\beta}}_{v,k} \\ &= \breve{\boldsymbol{\beta}}_{v,M} - \delta \boldsymbol{v}_0^n - \left(\tfrac{1}{M}\sum_{k=1}^{M}\breve{\boldsymbol{\beta}}_{v,k} - \delta \boldsymbol{v}_0^n\right) \\ &= \breve{\boldsymbol{\beta}}_{v,M} - \tfrac{1}{M}\sum_{k=1}^{M}\breve{\boldsymbol{\beta}}_{v,k}\end{aligned} \tag{12}$$

Based on (9), it was:

$$\hat{\boldsymbol{\beta}}_{v,M} = \boldsymbol{C}_b^n(0)\hat{\boldsymbol{\alpha}}_{v,M} \tag{13}$$

where

$$\hat{\boldsymbol{\alpha}}_{v,M} = \tilde{\boldsymbol{\alpha}}_{v,M} - \frac{1}{M}\sum_{k=1}^{M}\tilde{\boldsymbol{\alpha}}_{v,k} \tag{14}$$

From the above deduction, it can be found that the initial velocity errors $\delta \boldsymbol{v}_0^n$ are eliminated from the new observed vectors $\hat{\boldsymbol{\beta}}_{v,M}$. Thus, using the new observed vector $\hat{\boldsymbol{\beta}}_{v,M}$, the alignment results will be more accurate than the previous method, which the initial velocity errors $\delta \boldsymbol{v}_0^n$ are not eliminated. Using the new observation vectors $\hat{\boldsymbol{\beta}}_{v,M}$ and $\hat{\boldsymbol{\alpha}}_{v,M}$, the initial attitude $\boldsymbol{C}_b^n(0)$ can be determined. Then, the initial alignment can be finished with (1).

The algorithm flowchart of the proposed method is summarized in Algorithm 1. From Algorithm 1, it can be divided into three parts. Firstly, the initial parameters are needed to set. The initial parameters contain $\boldsymbol{K}$ matrix, DCM, and vector observations. Secondly, the outputs of the inertial sensors and the GPS are obtained. Using the outputs information, the alignment process can be carried out. Thirdly, the alignment process is carried out. Additionally, the alignment results are obtained at real-time.

Algorithm 1. Initial alignment method for SINS/GPS integration

Initialization: $\boldsymbol{K}_0 = \boldsymbol{0}, \boldsymbol{C}_b^{b0}(0) = \boldsymbol{C}_n^{n0}(0) = \boldsymbol{I}_3, \tilde{\boldsymbol{\alpha}}_{v,0} = \tilde{\boldsymbol{\beta}}_{v,0} = \boldsymbol{0},$
Inputs: $\left\{\tilde{f}^b\right\}_{k=1}^{N}, \left\{\tilde{\omega}_{ib}^b\right\}_{k=1}^{N}, \left\{\tilde{v}^n\right\}_{k=1}^{N}$
for: $k = 1, 2, 3 \ldots$ **do**
    Calculate $\boldsymbol{C}_{n(k)}^{n0}$ and $\boldsymbol{C}_{b(k)}^{b0}$ using (2)
    if GPS outputs are available
        Calculate the observation vectors $\tilde{\boldsymbol{\alpha}}_{v,k}$ and $\tilde{\boldsymbol{\beta}}_{v,k}$ by (4)
        Calculate the averaged observation vector according to (7)
        $\bar{\boldsymbol{\beta}}_{v,M} = \frac{1}{M}\sum_{k=1}^{M}\tilde{\boldsymbol{\beta}}_{v,k}$
        Calculate the new observed vector by (8)
        $\hat{\boldsymbol{\beta}}_{v,M} = \tilde{\boldsymbol{\beta}}_{v,M} - \bar{\boldsymbol{\beta}}_{v,M}$
        Calculate the averaged reference vector according to (10)
        $\hat{\boldsymbol{\alpha}}_{v,M} = \tilde{\boldsymbol{\alpha}}_{v,M} - \frac{1}{M}\sum_{k=1}^{M}\tilde{\boldsymbol{\alpha}}_{v,k}$
        Construct the matrix
        $\boldsymbol{K}_M = \boldsymbol{K}_{M-1} + \left(\left[\hat{\boldsymbol{\beta}}_{v,M}\otimes\right] - [\hat{\boldsymbol{\alpha}}_{v,M}\odot]\right)^T\left(\left[\hat{\boldsymbol{\beta}}_{v,M}\otimes\right] - [\hat{\boldsymbol{\alpha}}_{v,M}\odot]\right)$
        Extract $\boldsymbol{q}_{n0}^{b0}$ from $\boldsymbol{K}_M$, and transform $\boldsymbol{q}_{b0}^{n0}$ to $\boldsymbol{C}_{n0}^{b0}$
    end if
        Calculate the current attitude according to
        $\boldsymbol{C}_b^n(k) = \boldsymbol{C}_{n0}^{n(k)}\boldsymbol{C}_{b0}^{n0}\boldsymbol{C}_{b(k)}^{b0}$
        $k = k + 1$
    end for

4. Simulation and Field Test Results

To verify the performance of the proposed method, the simulation and field tests were designed in this section. The alignment results are shown in the next two subsections. The current popular methods, which were proposed in [13,15], were designed as the comparative methods and four different initial velocity errors are considered.

Scheme 1: The current popular method [13] with the initial velocity errors [5 5 5] m/s.
Scheme 2: The current popular method [13] with the initial velocity errors [1 1 1] m/s.
Scheme 3: The current popular method [13] with the initial velocity errors [0.1 0.1 0.1] m/s.
Scheme 4: The current popular method [15] with the initial velocity errors [5 5 5] m/s.
Scheme 5: The current popular method [13] without the initial velocity errors.
Scheme 6: the proposed method with the initial velocity error [5 5 5] m/s.

4.1. Simulation Test

In the simulation tests, the Zigzag trajectory was designed for verification. The trajectory and the movement states of the vehicle are shown in Figures 1 and 2. The initial position of the vehicle was set as L = 32° N, λ = 118° E, where L and λ denote the latitude and longitude, respectively.

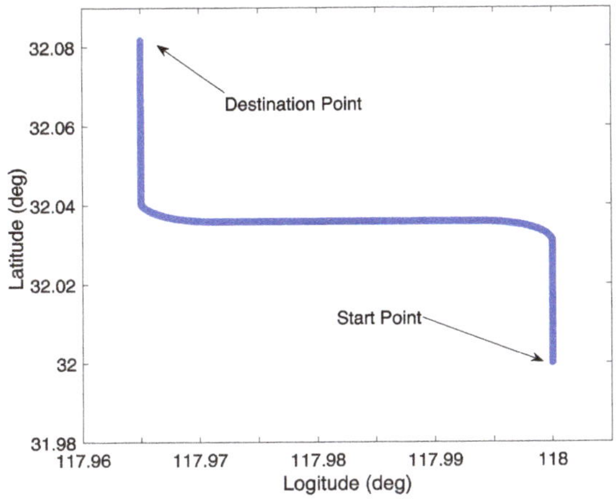

Figure 1. The in-motion trajectory of the vehicle.

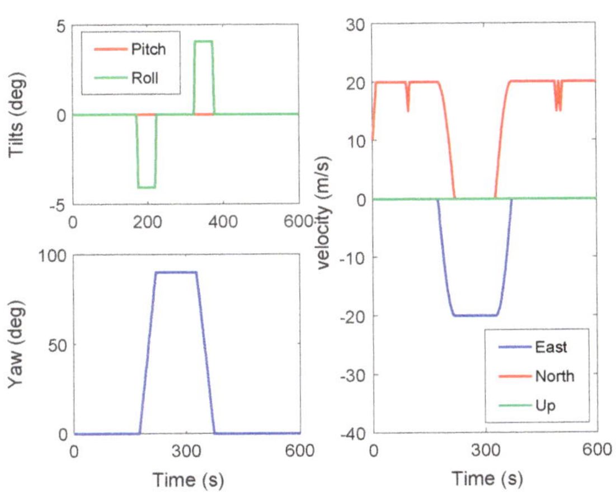

Figure 2. The in-motion status of the vehicle.

The bias and the angular rate random walk of the gyroscopes were set as 0.01°/h and 0.005°/$\sqrt{\text{h}}$. The bias and the velocity random walk of the accelerometers were set as 100 μg and 50 μg/$\sqrt{\text{Hz}}$. The sampling rate of inertial sensors was set as 200 Hz. The velocity noise of GPS was set as 0.1 m/s and the position noises of GPS was set as 10 m. The noises of GPS were correspondent to Gaussian distribution. The sampling rate of the GPS was set as 1 Hz.

The alignment results are shown in Figures 3–5. In Figure 3, the pitch errors are presented; it was found that the alignment errors of Schemes 1 and 4 were worse than other schemes. This is because the initial velocity errors in Schemes 1 and 4 were set as m/s. The alignment errors of Scheme 2 were less than the same one of Schemes 1 and 4. This is because the initial velocity errors of Scheme 2 were set as [1 1 1] m/s. However, with the same initial velocity errors [5 5 5] m/s, the proposed method, which is Scheme 6, can obtain the high accuracy alignment results of the pitch. From Figure 3, it was found that the alignment errors of the proposed method were around 0.005°. The alignment results were close to Scheme 5, in which the initial velocity errors were not set. Additionally, the alignment results of Scheme 3 were also close to Scheme 5 because the initial velocity was set as [0.1 0.1 0.1] m/s, which is similar to the noises of the GPS outputs. These results showed that the different initial velocity errors will produce different alignment results. The larger initial velocity errors were contained in the GPS outputs and the large alignment errors will be contained in the alignment results of the traditional methods.

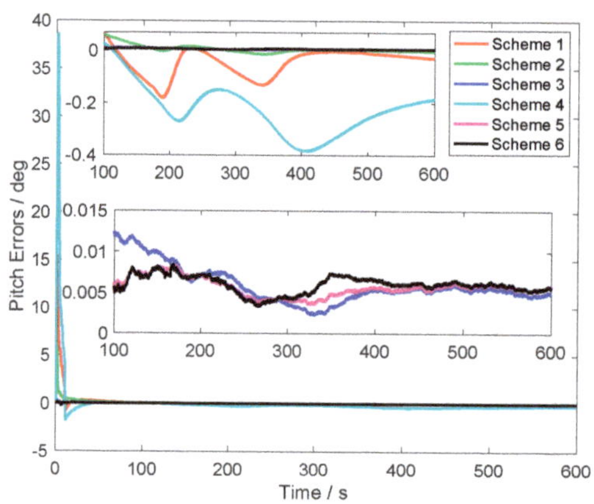

Figure 3. The pitch errors.

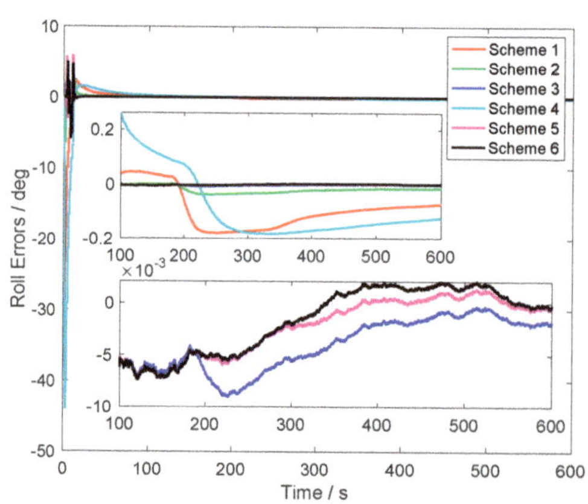

Figure 4. The roll errors.

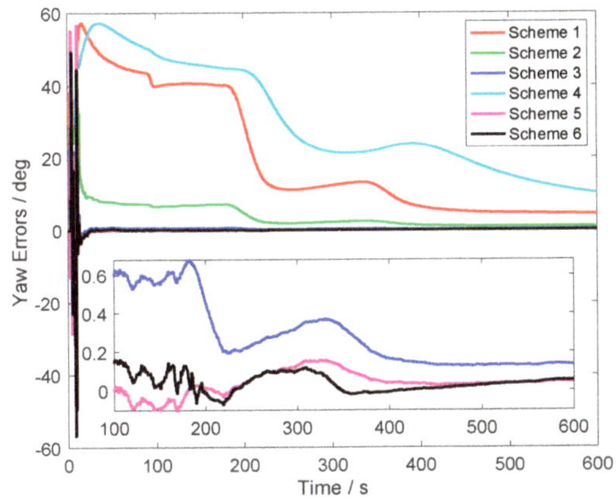

Figure 5. The yaw errors.

In Figure 4, similar results were found, the roll errors of Scheme 5 were less than 0.005° after 200 s. The roll errors of Scheme 1 and Scheme 4 were larger than 0.1°. Although the alignment errors of Scheme 2 were less than the same one of Scheme 1 and Scheme 4, they were still larger than the proposed method. When alignment process lasts 200 s, compared with the alignment errors of Scheme 3, it was found that the roll errors of Scheme 6, which was the proposed method, were closer to the alignment results of Scheme 5. This conclusion shows that the proposed method could suppress the initial velocity errors effectively. It is noted that the roll errors of Scheme 6 had a similar convergence rate with the errors of Scheme 5. This conclusion showed that the vector subtraction operations did not weaken the characteristics of the vector observations.

In Figure 5, the yaw errors are shown. From Schemes 1, 2 and 4, it was found that the yaw accuracy was degraded by the initial velocity errors. In Scheme 1, the errors were larger than 5° when the alignment process lasted for 300 s. In Scheme 2, the errors were smaller than the errors of Scheme 1. This is because the initial velocity errors were small in Scheme 2. However, the errors were still larger than the errors of the proposed method. Although the alignment errors of Scheme 3 were less than 0.5°, they were still larger than the alignment errors of Scheme 5, which were not corrupted by the initial velocity errors. At the end of alignment process, the yaw errors were around 0.1° of Schemes 5 and 6. From the enlarged figure in Figure 5, it was found that the curve of Scheme 6, which was the proposed method, was consistent with the results of Scheme 5. This conclusion showed that the initial velocity errors were eliminated by the proposed method effectively. Moreover, the convergence rate of Scheme 6 was not degraded.

From Figures 3–5, it was found that the curves of the alignment errors of Schemes 5 and 6 were coincidental. These results show that the proposed method can eliminate the initial velocity error effectively.

4.2. Field Test

To show the performance of the proposed method in practical systems, the field test was designed. The experimental vehicle and equipment are shown in Figure 6. The navigational computer was produced by our team. It was combined with a PC104 board. The CPU (central processing unit) can operate up to 500 MHz. The GPS receiver was produced by the NovAtel, the BESTVEL and BESTPOS logs were used to output the velocity and position of the vehicle, the sampling rate of the GPS was set as 1 Hz. The reference system, which is named CPT7, was produced by NovAtel. The accuracies of the pitch, roll and yaw of CPT7 were 0.01°, 0.01° and 0.1°, respectively. The specifications of inertial measurement unit (IMU), which were combined with triaxial accelerometers and gyroscopes, are shown in Table 1.

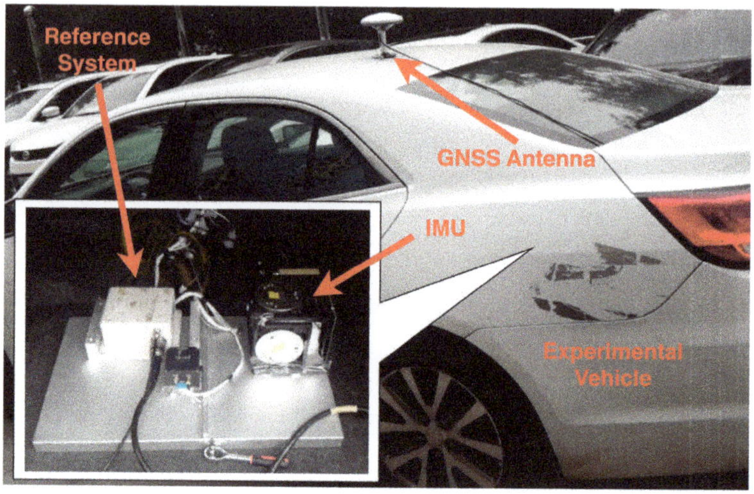

Figure 6. The experimental vehicle and equipment.

Table 1. The specification of the inertial sensors.

Error Items	Gyroscopes (x-, y-, z-Axes)	Accelerometers (x-, y-, z-Axes)
Bias	$\leq(0.05, 0.05, 0.05)^\circ/\mathrm{h}$	$\leq(1000, 1000, 1000)$ μg
Random Walk Coefficients	$\leq(0.005, 0.005, 0.005)^\circ/\sqrt{\mathrm{h}}$	$\leq(50, 50, 50)$ $\mu\mathrm{g}/\sqrt{\mathrm{Hz}}$

It was found that the specification of the inertial sensors was not a determined value; this is because the errors of the inertial sensors were time-varying. The moving trajectory and the states of the vehicle are shown in Figures 7 and 8. The averaged moving velocity was around 20 m/s. Additionally, the alignment process was carried out when the vehicle was moving at any time. The alignment results are shown in Figures 9–11.

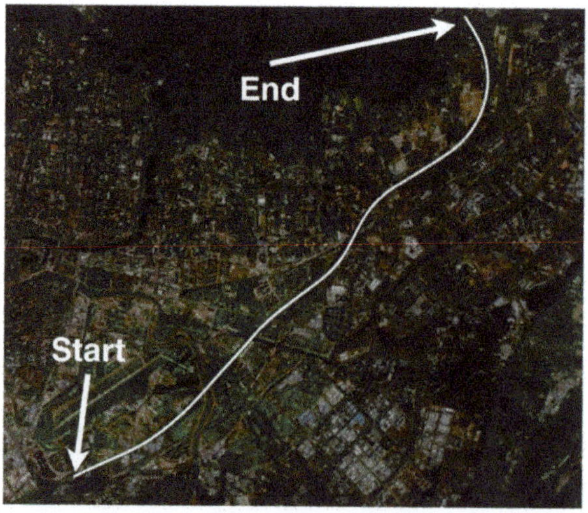

Figure 7. The in-motion trajectory of the vehicle.

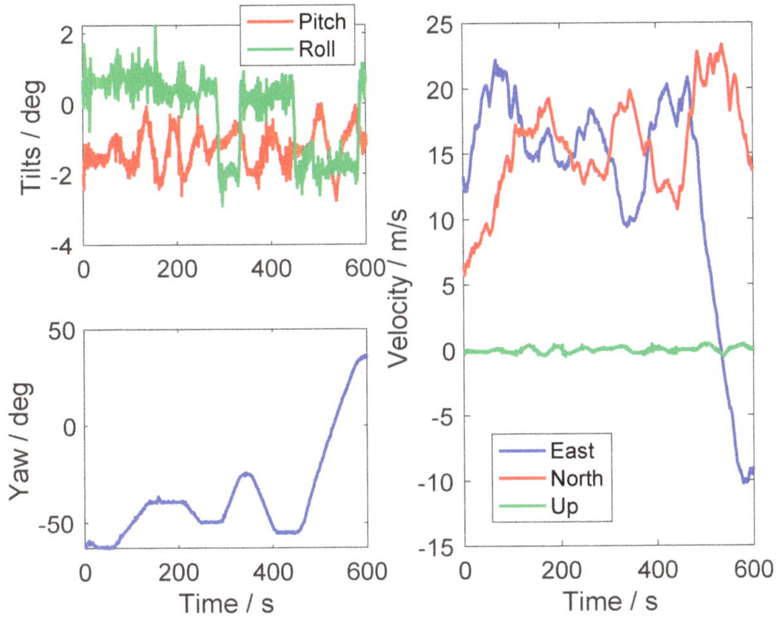

Figure 8. The in-motion status of the vehicle.

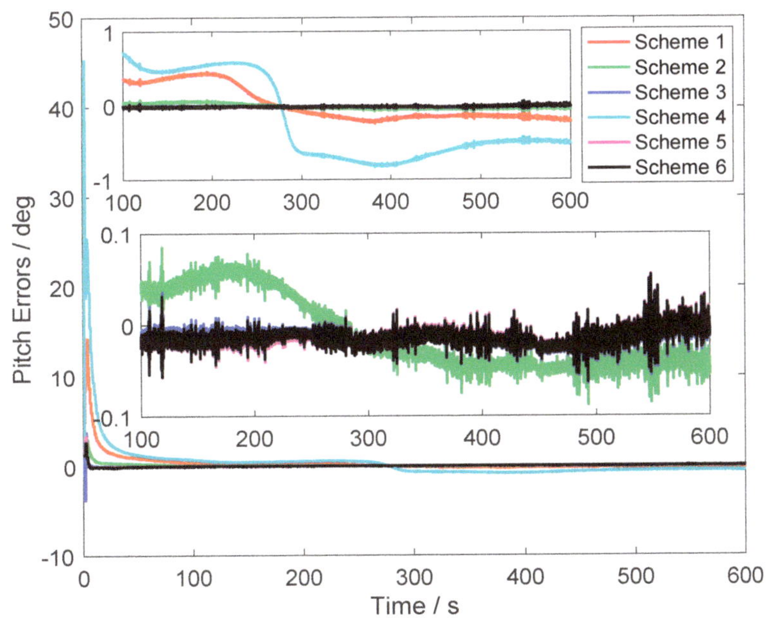

Figure 9. The pitch errors.

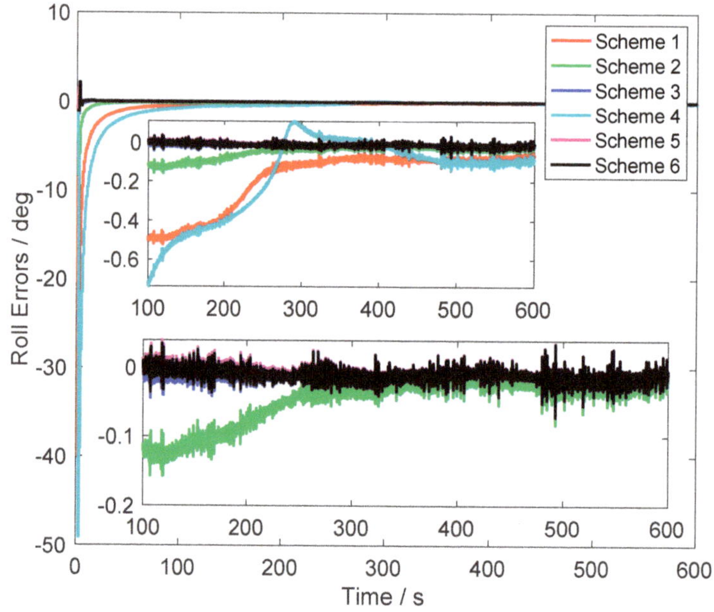

Figure 10. The roll errors.

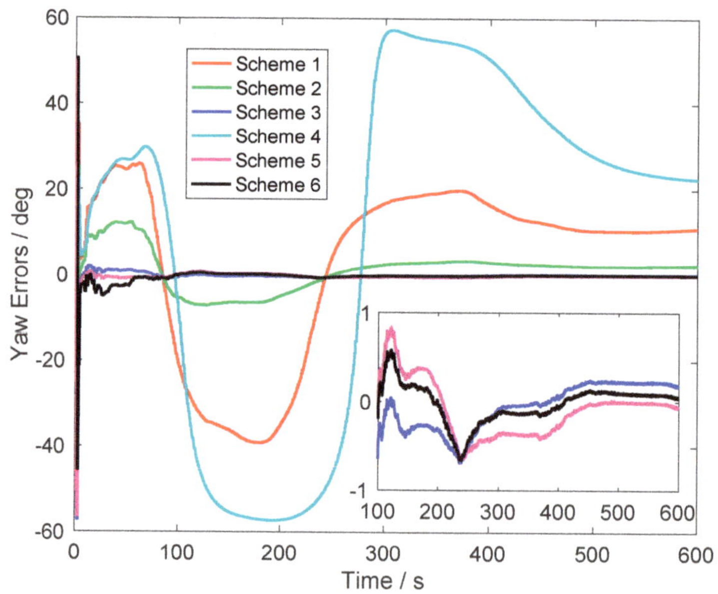

Figure 11. The roll errors.

In Figure 9, the pitch errors are shown. Due to the big initial velocity errors, it was found that the alignment errors of Schemes 1 and 4 were larger than 0.1°. The results were unsatisfactory. The pitch errors of Scheme 2 were around 0.05° when the alignment time lasted 150 s. However, it was also larger than the proposed method, which was shown as Scheme 6. The errors of Schemes 3, 5 and 6 were closer to each other. In Scheme 3, the initial velocity errors were set as [0.1 0.1 0.1] m/s. Additionally, there were no initial velocity errors in Scheme 5. However, the proposed method, which was Scheme 6, contained the initial velocity errors, which were [5 5 5] m/s. These results showed that the initial velocity errors were suppressed by the proposed method. The alignment errors of pitch of Scheme 6 were less than 0.01°.

In Figure 10, the roll errors are shown. The errors of Schemes 1 and 4 were also fluctuant. This was because the initial velocity errors were large, and there was no effective method to suppress them. The errors of Schemes 1 and 4 were around 0.1° when the alignment process lasted 300 s. The errors of Scheme 2 were fewer than the errors of Schemes 1 and 4. This was because the initial velocity errors were small. However, they were still worse than the errors of Scheme 6. The errors of Schemes 3, 5 and 6 were similar, i.e., they were less than 0.01°. In Scheme 3, the initial velocity errors were 0.1 m/s. Additionally, in Scheme 5, the initial velocity errors were 0 m/s. However, in Scheme 6, which was the proposed method, the initial velocity errors were 5 m/s. Compared with the results of Schemes 3 and 5, it was found that the large initial velocity errors in Scheme 6 were suppressed.

In Figure 11, the yaw errors are shown. The errors of Schemes 1 and 4 were large and were caused by the relatively large initial velocity errors. Moreover, the alignment errors of Scheme 2 were larger than 2° when the alignment time lasted 300 s. The other three methods, which were Schemes 3, 5 and 6, had the similar alignment results. However, it is noted that the initial velocity errors in Schemes 3 and 5 were small. Moreover, the alignment errors of Scheme 6 were closer to Scheme 5 than Scheme 3. It was shown that although the yaw errors of Scheme 3 could converge towards less than 0.5°, they were also affected by the initial velocity errors. Thus, the alignment errors were a little bigger than the proposed method. After 300 s, the yaw errors of Scheme 6 were less than 0.2°, and 5 m/s initial velocity errors were contained in Scheme 6. The field test also verified the performance of the proposed method.

5. Conclusions

In this paper, an improved method for in-motion coarse alignment process of SINS/GPS integration was proposed. The initial velocity errors, which were contained in the observed vectors, were considered. If the initial velocity errors were contained in the observed vectors, the alignment error would have been large. To address this problem, the average operation was used to construct the intermediate observed vectors. Then, the new observed vectors were constructed by the intermediate observed vectors with the vector subtraction operations. In the new observed vectors, the initial velocity errors were eliminated effectively. Thus, the alignment accuracy was improved. Moreover, the characteristics of the vector observations were reserved. The simulation and field tests were designed to verify the performance of the proposed method. The results showed that the proposed method could suppress the influences of initial velocity errors on the initial alignment procedure. The proposed method can also be used in other in-motion coarse alignment processes.

Author Contributions: Conceptualization, Y.W. and X.X.; methodology, X.X.; software, X.N.; validation, Y.W. All authors have read and agreed to the published version of the manuscript.

Funding: This research was funded by National Natural Science Foundation of China, grant number 52271346 and 61803278.

Institutional Review Board Statement: Not applicable.

Informed Consent Statement: Not applicable.

Data Availability Statement: Not applicable.

Acknowledgments: The authors would like to thank Zhonghu Hao from the School of Mechatronic Engineering, Beijing Institute of Technology, China, for providing the revision suggestion.

Conflicts of Interest: The authors declare no conflict of interest.

References

1. Groves, P.D. *Principles of GNSS, Inertial, and Multi-Sensor Integrated Navigation Systems*; Artech House: London, UK, 2013.
2. Wu, Y.; Pan, X. Velocity/position integration formula part II: Application to strapdown inertial navigation computation. *IEEE Trans. Aerosp. Electron. Syst.* **2013**, *49*, 1024–1034. [CrossRef]
3. Gao, W.; Ben, Y.; Zhang, X.; Li, Q.; Yu, F. Rapid fine strapdown INS alignment method under marine mooring condition. *IEEE Trans. Aerosp. Electron. Syst.* **2011**, *47*, 2887–2896. [CrossRef]
4. Qin, F.; He, H.; Xu, J. Phase Modulation-Based SINS Damping Method for Autonomous Vehicles. *IEEE Sens. J.* **2018**, *18*, 2483–2493.
5. Yao, Y.; Xu, X.; Li, Y.; Liu, Y.; Sun, J.; Tong, J. Transverse Navigation under the Ellipsoidal Earth Model and its Performance in both Polar and Non-polar areas. *J. Navig.* **2016**, *69*, 335–352. [CrossRef]
6. Wang, B.; Yu, L.; Deng, Z.; Fu, M. A particle filter-based matching algorithm with gravity sample vector for underwater gravity aided navigation. *IEEE/ASME Trans. Mechatron.* **2016**, *21*, 1399–1408. [CrossRef]
7. Shen, C.; Bai, Z.; Cao, H.; Xu, K.; Wang, C.; Zhang, H.; Wang, D.; Tang, J.; Liu, J. Optical flow sensor/INS/magnetometer integrated navigation system for MAV in GPS-denied environment. *J. Sens.* **2016**, 1–10. [CrossRef]
8. Hong, S.; Lee, M.H.; Kwon, S.H.; Chun, H.H. A car test for the estimation of GPS/INS alignment errors. *IEEE Trans. Intell. Transp. Syst.* **2004**, *5*, 208–218. [CrossRef]
9. Hong, S.; Lee, M.H.; Chun, H.; Kwon, S.; Speyer, J.L. Experimental study on the estimation of lever arm in GPS/INS. *IEEE Trans. Veh. Technol.* **2006**, *55*, 431–448. [CrossRef]
10. Xu, X.; Xu, X.; Zhang, T.; Li, Y.; Tong, J. A Kalman Filter for SINS Self-Alignment Based on Vector Observation. *Sensors* **2017**, *17*, 264. [CrossRef]
11. Silva, F.O.; Hemerly, E.M.; Waldemar Filho, C.L. Error analysis of analytical coarse alignment formulations for stationary SINS. *IEEE Trans. Aerosp. Electron. Syst.* **2016**, *52*, 1777–1796. [CrossRef]
12. Xu, X.; Xu, X.; Zhang, T.; Li, Y.; Zhou, F. Improved Kalman filter for SINS coarse alignment based on parameter identification. *Zhongguo Guanxing Jishu Xuebao/J. Chin. Inert. Technol.* **2016**, *24*, 320–324 and 329.
13. Wu, Y.; Pan, X. Velocity/Position integration formula part I: Application to in-flight coarse alignment. *IEEE Trans. Aerosp. Electron. Syst.* **2013**, *49*, 1006–1023. [CrossRef]
14. Zhang, T.; Zhu, Y.; Xu, X.; Wang, J.; Li, Y. In-motion coarse alignment based on the vector observation for SINS. *IEEE Trans. Instrum. Meas.* **2018**, *68*, 3740–3750. [CrossRef]
15. Xu, X.; Xu, D.; Zhang, T.; Zhao, H. In-Motion Coarse Alignment Method for SINS/GPS Using Position Loci. *IEEE Sens. J.* **2019**, *19*, 3930–3938. [CrossRef]
16. Xu, X.; Guo, Z.; Yao, Y.; Zhang, T. Robust Initial Alignment for SINS/DVL Based on Reconstructed Observation Vectors. *IEEE/ASME Trans. Mechatron.* **2020**, *25*, 1659–1667. [CrossRef]
17. Huang, Y.; Zhang, Y.; Wang, X. Kalman-Filtering-Based In-Motion Coarse Alignment for Odometer-Aided SINS. *IEEE Trans. Instrum. Meas.* **2017**, *66*, 3364–3377. [CrossRef]
18. Xu, X.; Sun, Y.; Gui, J.; Yao, Y.; Zhang, T. A Fast Robust In-Motion Alignment Method for SINS With DVL Aided. *IEEE Trans. Veh. Technol.* **2020**, *69*, 3816–3827. [CrossRef]
19. Li, J.; Gao, W.; Zhang, Y. Gravitational apparent motion-based SINS self-alignment method for underwater vehicles. *IEEE Trans. Veh. Technol.* **2018**, *67*, 11402–11410. [CrossRef]
20. Xu, X.; Gui, J.; Sun, Y.; Yao, Y.; Zhang, T. A robust in-motion alignment method with inertial sensors and Doppler velocity log. *IEEE Trans. Instrum. Meas.* **2020**, *70*, 1–13. [CrossRef]
21. Sun, Y.; Yang, G.; Cai, Q.; Wen, Z. A robust in-motion attitude alignment method for odometer-aided strapdown inertial navigation system. *Rev. Sci. Instrum.* **2020**, *91*, 125006. [CrossRef]
22. Xu, X.; Sun, Y.; Yao, Y.; Zhang, T. A Robust In-Motion Optimization-Based Alignment for SINS/GPS Integration. *IEEE Trans. Intell. Transp. Syst.* **2021**, *23*, 4362–4372. [CrossRef]

sensors

MDPI

Article

Learning Modality Complementary Features with Mixed Attention Mechanism for RGB-T Tracking

Yang Luo [1,2], Xiqing Guo [1,2,*], Mingtao Dong [3] and Jin Yu [1,2]

1 Aerospace Information Research Institute, Chinese Academy of Sciences, Beijing 100094, China; luoyang211@mails.ucas.ac.cn (Y.L.); jinyu@aoe.ac.cn (J.Y.)
2 University of Chinese Academy of Sciences, Beijing 100040, China
3 Institute of Image Recognition and Machine Intelligence, Northeastern University, Shenyang 110167, China; 2100729@stu.neu.edu.cn
* Correspondence: guoxq100036@aircas.ac.cn

Abstract: RGB-T tracking involves the use of images from both visible and thermal modalities. The primary objective is to adaptively leverage the relatively dominant modality in varying conditions to achieve more robust tracking compared to single-modality tracking. An RGB-T tracker based on a mixed-attention mechanism to achieve a complementary fusion of modalities (referred to as MACFT) is proposed in this paper. In the feature extraction stage, we utilize different transformer backbone branches to extract specific and shared information from different modalities. By performing mixed-attention operations in the backbone to enable information interaction and self-enhancement between the template and search images, a robust feature representation is constructed that better understands the high-level semantic features of the target. Then, in the feature fusion stage, a modality shared-specific feature interaction structure was designed based on a mixed-attention mechanism, effectively suppressing low-quality modality noise while enhancing the information from the dominant modality. Evaluation on multiple RGB-T public datasets demonstrates that our proposed tracker outperforms other RGB-T trackers on general evaluation metrics while also being able to adapt to long-term tracking scenarios.

Keywords: multi-modality adaptive fusion; mixed-attention mechanism; RGB-T tracking

1. Introduction

Visual object tracking is an essential branch of computer vision that has recently made significant progress and has been widely applied in various industries, including autonomous driving, intelligent security, and robotics [1]. The current research on visual object tracking predominantly focuses on visible modality (RGB tracking). This is primarily due to the rich color and texture information that RGB images provide, making target feature extraction more feasible as well as enabling the use of relatively low-cost imaging devices based on visible modality [2].

However, researchers have identified the limitations of RGB tracking in extreme environments such as rain, fog, and low-light conditions. Thermal infrared imaging, on the other hand, relies on the target's thermal radiation and is less affected by intensity variations in illumination, making it possible to penetrate rain and fog. It provides a complementary and extended modality to RGB tracking [3]. This complementarity has led researchers to progressively concentrate more on object tracking based on the fusion of these two modalities to create trackers with higher accuracy and robustness [4]. As shown in Figure 1, the complementary fusion of visible images and thermal images can provide more stable features for target-tracking tasks to overcome environmental challenges that cannot be dealt with in a single modality.

Figure 1. Illustration of the complementary information between multi-modal images.

The object tracking method that combines visible and thermal modalities is known as RGB-T tracking. The key point in RGB-T tracking is to achieve complementary information fusion between the visible and thermal modalities. Early RGB-T tracking methods mostly employed CNN for feature modeling. Some methods [5,6] simply concatenated the features extracted from the visible and thermal modalities and fed them into a fully connected network to obtain the target position. However, this simple approach introduced excessive background noise, leading to a decrease in the network's recognition capability. Other methods [1,7–15] first generated candidate boxes (RoIs) from the search frame and then performed feature fusion on the RoIs from different modalities using mechanisms such as gating, attention, and specific data attribute annotations. Finally, binary classification of foreground/background was conducted on the fused features, and the bounding box was regressed. One significant drawback of such methods is that the sizes and ratios of RoI regions are limited and local in nature. They cannot adapt flexibly to variations in the target's appearance, as illustrated in Figure 2. Additionally, these methods fail to incorporate sufficient background information for feature learning. This limitation results in inadequate modeling of the global environmental context in the search framework due to the potential insufficient feature interaction between RoIs from different modalities. Consequently, the mutual enhancement and complementary effects between the two modalities are restricted.

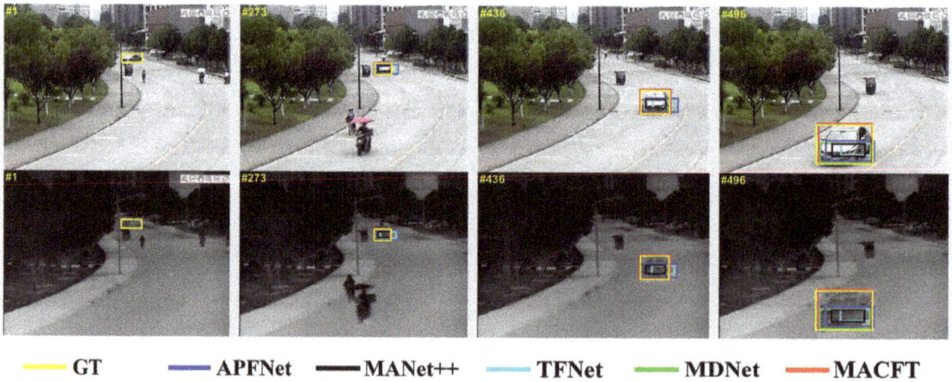

Figure 2. Comparing the model proposed in this paper with other advanced RoI-based RGB-T tracking models when the target undergoes large deformation (the yellow box represents ground truth). We can observe that our MACFT model accurately extracts the features of the target and, due to not being restricted by RoI's scale, can generate bounding boxes that are closer to the ground truth.

To enhance the adaptive fusion capability of the algorithm for complementary information from different modalities, we have designed a tracking architecture based on a combination of multiple attention mechanisms (referred to as MACFT). This architecture can learn the complementary features between visible and thermal modalities. The core of the architecture is the modality shared-specific feature interaction (SSFI) module. The SSFI module performs cross-attention and mixed-attention operations on the shared and specific features from both the visible and thermal modalities. The cross-attention operation facilitates the interaction of information between modalities, while also merging the shared and specific features to suppress the weights of low-quality modalities. The mixed-attention operation is used to enhance the fused modality information and further remove redundant noise.

Furthermore, in the feature extraction stage, considering that the effective receptive field of CNN may grow much more slowly in practice than theoretically due to the continuous downsampling process [16], there is still a significant amount of local information retained even after multiple convolutional layers. To achieve long-range modeling of target pixels, we employ the Vision Transformer [17] architecture, which has been widely used in recent years for Single-Object Tracking (SOT), as the backbone of our network. We extend it to a dual-branch Siamese architecture. Additionally, although visible and thermal images have different imaging wavelengths, they still share many correlations, such as object edges, contour information, and fine-grained texture details [9]. Therefore, apart from the dual-branch backbone used to extract features from visible and thermal images separately, we introduce a shared-parameter backbone. By introducing a KL divergence loss function as a constraint, we maintain the consistency between modalities and learn shared features across modalities. Finally, we insert the SSFI module after the backbone to perform feature fusion and achieve cross-modal interaction.

After cross-fusing features from the modality-specific feature branch and the modality-shared feature branch, we generate the target bounding box with only a simple structured corner point predictor. After extensive experiments on several public RGB-T tracking datasets, we verify that the proposed method is effective and outperforms most of the current state-of-the-art RGB-T tracking models.

The main contributions of this paper are summarized as follows. Firstly, a high-performance RGB-T tracking method has been proposed, which utilizes a deep transformer-based backbone to extract both the specific features and shared features from visible and thermal modality images. This method demonstrates excellent performance in challenging scenarios such as significant target deformations and occlusions. Secondly, a novel modality feature fusion structure based on a combination of multiple attention mechanisms is designed, and through the cross-input and fusion of cross-modal shared/specific information, the complementary features between modalities can be fully interacted. It can adaptively suppress low-quality modalities and enhance the features of dominant modalities for tracking. Thirdly, as a versatile fusion strategy, the module we designed can be easily extended to other multi-modal scenarios.

2. Related Work

2.1. RGB Tracking Methods

Visual object tracking based on RGB images has been developed for quite some time, but it was not until recent years when deep neural networks were successfully applied to computer vision tasks that models with high tracking accuracy emerged. They can be broadly classified into two categories, namely, discriminative models and generative models.

Discriminative models are trained using the whole image labeled with the target location, aiming to obtain a classifier that can distinguish between foreground and background. They often need to generate multiple proposal regions during the tracking process before feeding the classifier through the forward propagation to obtain positive and negative samples. These samples are further refined through post-processing methods such as the cosine window penalty, and the best samples are retained. Typical discriminative trackers include MDNet [8], ATOM [18], DiMP [6], etc. Although these models are known for their high accuracy, the excessive online processing required for them consumes many computational resources, which results in slow operation; indeed, the RGB-T tracking methods extended from the above-mentioned approaches also face the challenge of not being able to meet real-time requirements, such as DMCNet [12], MaCNet [3], APFNet [1], etc. To address this drawback, Jung et al. [19] modified the forward propagation based on MDNet with the ROI Align [20], thereby increasing its speed.

Contrary to the discriminative model, the generative model solely relies on offline training and has no online update step. The final prediction of the target is obtained by computing the joint probability density of both the template and search image. Typical generative models such as the Siamese network-based tracker [21–23]. The generative model's major strength is its small computational effort and fast speed, but its tracking accuracy is often inferior to that of discriminative models due to insufficient information interaction between the template and the search images and the lack of an online update step.

In recent years, the field of computer vision has witnessed the gradual replacement of CNN as the backbone architecture with the transformer. In the domain of visual object tracking, Chen et al. [24] designed self-contextual augmentation (ECA) and cross-feature augmentation (CFA) modules in their proposed model TransT that differ from the traditional transformer structure to achieve a better fusion of features between templates and search regions. Yan et al. [25] proposed a method based on the transformer codec structure to achieve tracking directly by bounding box prediction, in which the image is fed into a simple fully convolutional network for feature extraction and then directly fed into the transformer prediction head for bounding box prediction without pre-generating proposal regions. Cui et al. [26] proposed an asymmetric attention mechanism for information interaction between the template and search images, which saves computational resources and provides the possibility to update the template online during the tracking process. Furthermore, attention mechanisms have also been widely utilized in other fields. For instance, Zhou et al. [27] and Zheng et al. [28] applied attention mechanisms in tool condition monitoring and bearing defect detection.

To improve the speed of our tracker, the model proposed in this paper is based on the idea of the generative model, whereby the process of tracking relies entirely on offline training. Inspired by the work of Chen et al. [29], the full interaction between the target template and the search image is implemented within the transformer backbone, which also gives our tracker a very high accuracy.

2.2. RGB-T Tracking Methods

RGB-T tracking, based on the fusion of visible and thermal modalities, is a branch of single-object tracking that focuses on effectively utilizing complementary information from multiple modalities to achieve a more stable and robust tracking performance. In 2018, Wang et al. [30] introduced correlation filters into the field of RGB-T tracking, using average correlation peak energy to compute weights. In 2021, Zhang et al. [31] proposed a novel robust RGB-T tracking framework called JMMAC. JMMAC is based on the well-known ECO algorithm [32] in single-modality tracking and takes motion principles into full consideration to design an adaptive structure that allows the model to switch between appearance modeling and motion modeling.

In recent years, more high-performance RGB-T trackers have been proposed because of the introduction of deep neural networks. Li et al. [10] developed a multi-adapter convolutional network for RGB-T tracking by following MDNet's multi-domain learning framework that performs feature extraction for modal-generic features, modal-specific features, and instance features. However, performing forward propagation too many times leads to slow running speed of the model, which cannot meet the real-time requirements. Zhu et al. [14] also used MDNet, but through a 1×1 convolutional kernel + Relu + LRN (local response normalization) to form a feature aggregation module to fuse the features of two modalities and randomly reject the feature channels using the channel dropout technique, which effectively eliminates redundant information and noise while preventing model overfitting. Zhang et al. [5] first implemented the task of RGB-T fusion tracking based on a Siamese network combined with multi-modal features. In this method, the modality weight calculation is designed according to the size of the predicted target position deviation, which improves the reliability of the model. The disadvantage of this method is that only the concatenation method is used for the fusion of modes, which fails to effectively suppress low-quality mode noise. Zhu et al. [7] proposed a trident architecture to integrate the fused modality features and two modality-specific features, thus achieving robust target representation. Zhang et al. [33] introduced a complementary perception module for multi-modal feature fusion, which reduces the modality discrepancy between single-modal features to enhance the discriminability of fused features. This is done to fully utilize training data and address various challenges such as illumination variations, occlusion, thermal crossover, and fast motion. Zhu et al. [11] designed a feature fusion structure based on the attention mechanism proposed in SKNet [34] during the modality fusion stage, enabling different neurons of the model to have adaptive receptive fields. Zhang et al. [35] built upon the mfDiMP [36] framework and unified multiple modality fusion strategies (including pixel-level fusion, feature-level fusion, and decision-level fusion) into a hierarchical fusion framework. They also designed a self-attention-based module to model the confidence of modalities and exploit modal information in a non-local manner.

With the trend of the transformer architecture surpassing CNN in performance in recent years, many researchers in the field of RGB-T object tracking have attempted to incorporate the core idea of the self-attention mechanism into their algorithm models. One typical example is APFNet proposed by Xiao et al. [1]. Building upon the challenges posed by RGB-T-specific attributes, APFNet adopts specific strategies for fusion and incorporates three independent and separate encoders (based on the self-attention mechanism) and decoders (based on the cross-attention mechanism) in each fusion branch. This enables self-enhancement within modalities and interaction enhancement between modalities. The limitation of this method is that it still relies on CNN to extract general features and performs attention operations in the high-level semantic space after downsampling multiple times but fails to consider more fine image information. Zhu et al. [37] introduced ViPT, which is based on a unified ViT backbone for feature extraction. By leveraging the idea of visual prompt tuning [38], ViPT adds the thermal modality information to the tokens for learning, while preserving the original feature structure of the transformer. This approach significantly reduces the number of trainable parameters, but the price is insufficient fusion, making the model perform poorly in scenes such as illumination changes.

3. Method

In this section, our proposed RGB-T tracking model MACFT is described in detail, as shown in Figure 3. The whole model consists of three parts, which comprise the modality complementary feature extraction network based on the transformer backbone, the modality shared-specific feature interaction module based on the mixed-attention mechanism, and the target localization regression network; each of these components will be discussed more specifically later. The overall information pipeline of the model can be summarized as follows: after the images from the two modalities are sent to different feature extraction backbones, the modal-shared features and modal-specific features will be obtained, respectively, and then these features will be crossed in pairs and fused by the SSFI module (generating a group of shared features of visible mode and specific features of thermal mode, and a group of shared features of thermal mode and specific features of visible mode). The fused features are then fed into the target localization network to generate the position of the target. Additionally, the explanation of symbols used in this section will be shown in Abbreviations section.

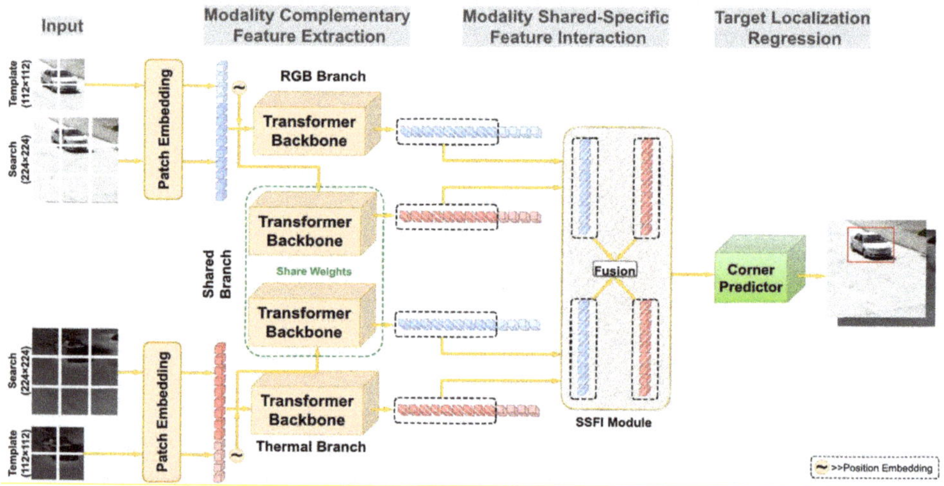

Figure 3. An overview of the MACFT model, which is divided into three parts, which are used for modal-specific/shared feature extraction, information fusion between modalities, and bounding box regression.

3.1. *Transformer-Based Complementary Feature Extraction Backbone*

In our proposed model, the feature extraction part is divided into four branches, namely, the visible modal-specific feature extraction branch, the thermal modal-specific feature extraction branch, and two modal-shared feature extraction branches. ViT [17] served as the base backbone for all four branches, initiated with parameters from the CLIP [39] pre-trained model. Furthermore, the template image is concatenated with the search image in the same dimension to achieve information interaction. The process is detailed as follows.

3.1.1. Model-Specific Feature Extraction Branch

First, given a video sequence, we select the first frame as the reference frame and crop the template image $\mathbf{Z} \in \mathbb{R}^{H_z \times W_z \times 3}$ according to the labeled bounding box, where $H_z \times W_z$ is the size of the template image, and accordingly, the subsequent frames in the video sequence are used as the search image with the size set to $\mathbf{X} \in \mathbb{R}^{H_x \times W_x \times 3}$. To accommodate the generic ViT backbone, we set the search image size $H_x \times W_x$ to 224×224 and the template image size $H_z \times W_z$ to 112×112.

Before using the transformer backbone for feature extraction, the input image is first serialized, i.e., patch embedded, by dividing the 2D image into N blocks of size $D = P^2 \times C$ (P is the size of patch; here, $P = 16$ since the backbone used is ViT-B-16, and C is the number of channels of the image (here $C = 3$), before projecting the patch onto the space with dimension D through a linear transformation. For our input template image and search image, the dimensions of the image after serialization will be transformed into $\mathbf{Z}_p \in \mathbb{R}^{N_z \times D}$ and $\mathbf{X}_p \in \mathbb{R}^{N_x \times D}$, where $N_z = H_z W_z / P^2$ and $N_x = H_x W_x / P^2$.

For the transformer backbone, there is an additional step in the input process which is to add positional encoding to help the model distinguish the order of the input sequence. For the search image sequence, the pre-trained positional encoding $p_s = p_{ptr}$ can be used directly because its size is consistent with the pre-trained ViT backbone input. However, for the template image, its size is not aligned with the ViT input and cannot accommodate the pre-trained positional encoding, so we set up a learnable positional encoding structure that consists of a two-layer fully connected network, which is added to the template image sequence to obtain the positional encoding vector p_z of the template image.

After adding the position encoding vector, we obtain the final input vectors $z \in \mathbb{R}^{N_z \times D^*}$ and $x \in \mathbb{R}^{N_x \times D^*}$, and subsequently, we concatenate z, x along the first dimension and feed them into the transformer backbone network for learning. Let the concatenated vector be r:

$$r = Concat(z, x, dim = 0). \tag{1}$$

Let the output of layer n in the backbone network be r^n; then we have:

$$\begin{aligned} [r^*] &= [r^{n-1}] + MA([r^{n-1}]), \\ [r^n] &= [r^*] + FFN([r^*]). \end{aligned} \tag{2}$$

where MA denotes mixed-attention operation and FFN denotes feed forward network. The calculation within MA can be expressed as follows:

$$MA([r^n]) = softmax\left(\begin{bmatrix} a(z^n, z^n), a(z^n, x^n) \\ a(x^n, z^n), a(x^n, x^n) \end{bmatrix}\right)\left(\begin{bmatrix} z^n W_V \\ x^n W_V \end{bmatrix}\right) \tag{3}$$

where $a(x, y) = (xW_Q)(yW_K)^T / \sqrt{d}$.

It is in the cross-attention operations such as $a(z^n, x^n)$ and $a(x^n, z^n)$ that the model learns the relationship between the template image and the search image, and in the self-attention operations such as $a(z^n, z^n)$ and $a(x^n, x^n)$ that this relationship is self-enhanced. We construct two identical and independent feature extraction branches for RGB and thermal images based on the above method for extracting features unique between different modalities. Let the feature vectors output from the two branches be:

$$\mathbf{R}_v \in \mathbb{R}^{(N_z+N_x) \times D^*},\ \mathbf{R}_t \in \mathbb{R}^{(N_z+N_x) \times D^*}. \tag{4}$$

3.1.2. Modal-Shared Feature Extraction Branch

To better construct a complementary feature representation between the visible and thermal modalities, we add two modal-shared feature extraction backbones to the model (again using ViT as the backbone with the same input structure as the modal-specific feature branch) for learning features shared by both modalities. Further, to be able to constrain the backbone of this branch to ensure that its output features are consistent across the two modalities, we introduce a Kullback–Leibler (KL) divergence loss $\mathcal{L}_{div}$. The formula is expressed as follows:

$$\mathcal{L}_{div} = KL(\mathbf{G}_v \| \mathbf{G}_t) = \frac{1}{N_z + N_x} \sum_{n=1}^{N_z+N_x} (g_{vn} log(g_{vn} - g_{tn})) \tag{5}$$

where $\mathbf{G}_v \in \mathbb{R}^{(N_z+N_x) \times D^*}$, $\mathbf{G}_t \in \mathbb{R}^{(N_z+N_x) \times D^*}$ denote the output vectors (after *softmax*) of the visible modality input and thermal modality input through the shared feature branch backbone, respectively, while g_{vn}, g_{tn} denote the nth term of the first dimension in $\mathbf{G}_v, \mathbf{G}_t$. KL divergence is based on the information entropy principle, and it can accurately measure the information lost when approximating one distribution with another, so it can be used to calculate the similarity between two distributions.

3.2. Modality Shared-Specific Feature Interaction Module

After obtaining the features from the four branches (the visible branch, the thermal branch, and the shared branches), we design a modality shared-specific feature interaction module based on the mixed-attention mechanism, with the aim of learning an adaptive discriminative weight for features from different modalities, thus enhancing high-quality modality while suppressing low-quality modality.

As shown in Figure 4, we use two types of attention mechanisms in the SSFI module: the cross-attention module (CAM) and the mixed-attention module (MAM).

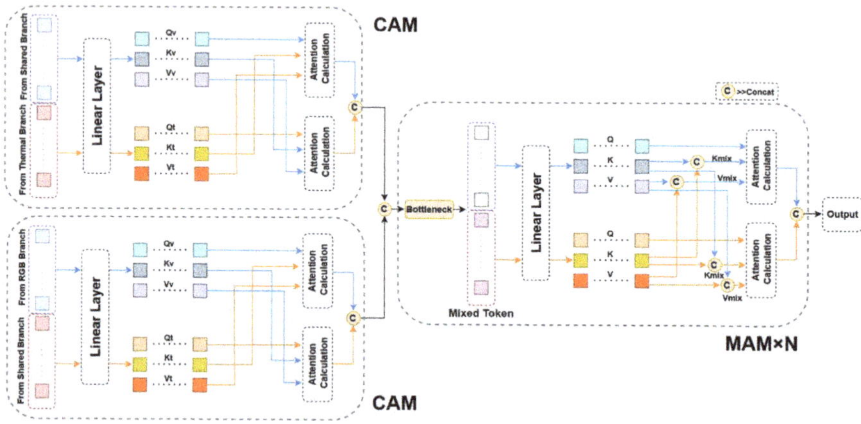

Figure 4. Schematic diagram of the SSFI module. It consists of 2 CAM modules and N = 6 MAM modules. The CAM modules utilize cross-attention to aggregate features from the visible specific/thermal shared branch and the thermal specific/visible shared branch. The MAM modules further enhance the aggregated features by introducing mixed-attention.

Given a vector consisting of visible features connected with thermal features $x = Concat(x_v, x_t)$, we first map the input into three equally shaped weight matrices called *query*, *key*, and *value* by a linear layer. $q_v,\ k_v,\ v_v$ are used below to denote the visible part and $q_t,\ k_t,\ v_t$ denote the thermal part. From (3), the calculation of MAM can be described as:

$$MAM([x]) = softmax\left(\begin{bmatrix} a(x_v, x_v), a(x_v, x_t) \\ a(x_t, x_v), a(x_t, x_t) \end{bmatrix}\right)\left(\begin{bmatrix} x_v v_v \\ x_t v_t \end{bmatrix}\right) \tag{6}$$

where $a(x_v, x_t) = (x_v q_v)(x_t k_t)^T / \sqrt{d}$, $a(x_v, x_v)$, $a(x_t, x_t)$, $a(x_t, x_v)$ in the same way.

The CAM module, on the other hand, eliminates the self-attention component and retains only the cross-attention, which can be described as:

$$CAM([x]) = \begin{bmatrix} softmax(a(x_v, x_t)) * x_t v_t \\ softmax(a(x_t, x_v)) * x_v v_v \end{bmatrix} \tag{7}$$

In the SSFI module, we connect the outputs from the visible modal-specific feature branch with the modal-shared feature branch (thermal part), and from the thermal modal-specific feature branch with the modal-shared feature branch (visible part), but only the part with the same size as the search image is selected and the template part is discarded:

$$\begin{aligned} \mathbf{R}_v, \mathbf{R}_t \in \mathbb{R}^{(N_z+N_x)\times C} \rightarrow \mathbf{R}_v^x, \mathbf{R}_t^x \in \mathbb{R}^{N_x \times C}, \\ \mathbf{G}_v, \mathbf{G}_t \in \mathbb{R}^{(N_z+N_x)\times C} \rightarrow \mathbf{G}_v^x, \mathbf{G}_t^x \in \mathbb{R}^{N_x \times C}. \end{aligned} \tag{8}$$

$$\begin{aligned} \mathbf{H}_{vt} &= Concat(\mathbf{R}_v^x, \mathbf{G}_t^x), \\ \mathbf{H}_{tv} &= Concat(\mathbf{R}_t^x, \mathbf{G}_v^x). \end{aligned} \tag{9}$$

Subsequently, $\mathbf{H}_{vt}, \mathbf{H}_{tv}$ are fed into the CAM module for modal information interaction:

$$\begin{aligned} \mathfrak{C}_{tv} &= CAM(\mathbf{H}_{tv}), \\ \mathfrak{C}_{vt} &= CAM(\mathbf{H}_{vt}). \end{aligned} \tag{10}$$

Only cross-attention is used here for the fusion of features from different branches due to the fact that sufficient self-enhancement has been performed in the backbone for the features from each modality, and there is no need to add extra self-attention modules. Additionally, removing the self-attention operation can also reduce the model's parameter count.

However, in the following MAM modules, we introduce self-attention operation because the initial interaction mapped features from different modalities into a new space, necessitating further enhancement of feature representation with the self-attention mechanism. These points will be supported in Chapter 4 in the subsection on ablation studies.

Finally, we connect the output of the CAM module and reduce the dimension and send it to the MAM module to further enhance the information interaction between the modalities:

$$\mathtt{C}_{mix} = DR\ (Concat\ (\mathtt{C}_{tv}, \mathtt{C}_{vt})). \tag{11}$$

$$\mathtt{M}_{mix} = \text{MAM}\ (\mathtt{C}_{tv}, \mathtt{C}_{vt}). \tag{12}$$

$$Output = DR\ (\mathtt{M}_{mix}). \tag{13}$$

3.3. *Target Localization Regression Network*

After the SSFI module, we use a lightweight corner predictor to localize the target. First, the features are reshaped into 2D; then, for the top left and bottom right corners of the bounding box, the features are downsampled using 5 convolutional layers, and finally the corner coordinates are calculated using the *softArgmax* layer and are reverse mapped to the coordinates corresponding to the original image according to the scale of the image crop.

3.4. *Training Method*

3.4.1. Training in Stages

We employ a staged approach for model training, which comprises three stages. The first two stages involve backbone training, requiring the modality fusion network to be removed while retaining only the corner predictor.

In the first stage, we freeze the modal-shared feature branch and load the pre-trained model to initialize the backbone of the modal-specific feature branches. We then train the backbone of the corresponding branches using RGB and thermal image data and save the model parameters.

In the second stage, the modal-specific feature branches are frozen, and the modal-shared branch is unfrozen. During the forward pass, RGB and thermal images are repeatedly used to update the parameters of the backbone, and the model parameters are saved.

In particular, during the training of the feature extraction backbone in the first two stages, in order to reduce the number of trainable parameters of the network and considering the shared characteristics of the low-level features between images, we freeze the first eight multi-head attention modules of the ViT backbone and only fine-tune the last four.

In the third stage, all feature extraction backbone branches are frozen, and the models trained in the first two stages are loaded for initializing the backbone parameters. The modality fusion network (SSFI) is added for training and the model parameters are saved.

3.4.2. Loss Function

The model proposed in this paper uses a total of three loss functions in the offline training phase, which are l_1 loss, generalized GioU loss [40], and KL divergence loss, and the loss functions are combined in the first and third stages of model training in the following manner:

$$\mathcal{L} = \lambda_{\text{giou}}\, \mathcal{L}_{giou}\left(b_i, b_i^*\right) + \lambda_{\mathcal{L}_1} \mathcal{L}_1\left(b_i, b_i^*\right). \tag{14}$$

where $\mathcal{L}_1$ denotes the l_1 loss, $\mathcal{L}_{giou}$ denotes the generalized GIoU loss, b_i denotes the target bounding boxes predicted by the model on the search image, b_i^* denotes the true value of the target bounding boxes, and λ_{giou} and $\lambda_{\mathcal{L}_1}$ are the weight parameters of the loss.

In the second stage of model training, we additionally introduce the KL divergence loss, and the loss function is combined in the following way:

$$\mathcal{L} = min\big[\lambda_{\text{giou}}\,\mathcal{L}_{giou}\big(b_v, b_v^*\big) + \lambda_{\mathcal{L}_1}\mathcal{L}_1\big(b_v, b_v^*\big),\\ \lambda_{\text{giou}}\,\mathcal{L}_{giou}\big(b_t, b_t^*\big) + \lambda_{\mathcal{L}_1}\mathcal{L}_1\big(b_t, b_t^*\big)\big]\\ +\lambda_{KL}\mathcal{L}_{div}(b_v, b_t). \tag{15}$$

where $\mathcal{L}_{div}$ denotes KL divergence loss, λ_{KL} is its weight, and b_v, b_t represent the target bounding boxes obtained from the RGB image and thermal image after modal-shared feature branches and corner predictors, respectively.

3.5. Inference

The RGB-T tracking model proposed in this paper does not contain any additional post-processing operations in the inference stage, and consists of only a few steps of image sampling, forward pass, and coordinate transformation, so MACFT can run in real time at a high speed.

4. Experiments

This section focuses on the experimental setup and procedure, evaluated on three major RGB-T public datasets (RGBT234 [41], LasHeR [42], and VTUAV [35]) and compared with current state-of-the-art models.

4.1. Implementation Details

4.1.1. Experimental Platform Setup

MACFT training was performed in Python 3.8.13, Pytorch 1.7.1, and CUDA 11.7. All experiments were conducted on a workstation equipped with two RTX 3090 GPUs, a Core i7 12700k CPU, and 32 GB of RAM.

4.1.2. Parameter Details

The first stage of training trains visible and thermal modal-specific feature branches for 40 epochs, which was in total an 80-epoch training: 40 epochs of training for the second stage and 80 epochs of training for the third stage. Each epoch consists of 12,000 image samples, with a batch size of 64. All models are learned using the AdamW optimizer [43], with the weight decay set to 10^{-4} and the initial learning rate set to 10^{-5} for the backbone and 10^{-4} for the rest.

In particular, when testing the LasHeR dataset, the model is trained using LasHeR's own training set, while when testing the RGBT234 dataset and VTUAV dataset, the model is trained using the full LasHeR dataset. In all training stages, similar to our baseline model, λ_{giou} and $\lambda_{\mathcal{L}_1}$ are set to 2 and 5, respectively, whereas in the second stage of training, λ_{KL} is set to a large number (in this case, 800) with the purpose of suppressing the optimizer's focus on other loss terms, thereby facilitating the learning of modal-shared features.

4.2. Public Dataset Evaluation

4.2.1. Evaluation Metrics

In this paper, we use the two most commonly used metrics to evaluate the performance of the tracking algorithm, which are the precision rate (PR) and the success rate (SR). The precision rate is calculated as the center location error (CLE) between the predicted bounding box and the ground truth, which is eventually expressed as the percentage of frames where the CLE is within a threshold (usually set to 20 pixels), and by varying the size of the threshold, a precision rate curve can be obtained. The success rate is calculated as the percentage of frames whose intersection over union (IoU) between the predicted bounding box and the ground truth is greater than the threshold value, and by setting different thresholds, a success rate curve can be obtained.

4.2.2. Evaluation on RGBT234 Dataset

The RGBT234 dataset contains a total of 234 RGB-T video sequences with a total of about 116.7 K frames, as well as 12 different challenge attributes, no division between the training and test sets, and the images of the two modalities are not precisely aligned. As shown in Table 1, we evaluate 10 different RGB-T tracking algorithms (APFNet [1], TFNet [7], DAFNet [13], MANet++ [9], MANet [10], SiamCDA [33], ViPT [37], JMMAC [31], MaCNet [3], DMCNet [12]) on the RGBT234 dataset and compare them with the algorithm proposed in this paper. The results show that our MACFT model achieves a precision rate of 85.7% and a success rate of 62.2% on the RGBT234 dataset, which is 5.7% and 4% better than the baseline, respectively.

Table 1. Evaluation results on the RGBT234 dataset. Red, blue, and green represent the first, second, and third place, respectively.

	APFNet	TFNet	DAFNet	DMCNet	MANet++	SiamCDA	ViPT	JMMAC	MaCNet	MACFT
PR (↑)	0.827	0.806	0.796	0.839	0.795	0.760	0.835	0.790	0.790	0.857
SR (↑)	0.579	0.560	0.544	0.593	0.559	0.569	0.617	0.573	0.554	0.622

4.2.3. Evaluation on LasHeR Dataset

The LasHeR dataset is the largest RGB-T tracking dataset with accurate annotation and alignment, with a total of about 734.8 K frames, 19 challenge attributes, divided training and testing sets, and a higher tracking difficulty compared to the RGBT234 dataset. To further evaluate the effectiveness of the model proposed in this paper, as shown in Figure 5, we compare its results evaluated on the LasHeR dataset with 10 RGB-T tracking algorithms (APFNet [1], DMCNet [12], MaCNet [3], mfDiMP [36], MANet [10], MANet++ [9], DAPNet [14], DAFNet [13], TransT [24], ViPT [37]). Our MACFT model achieves a 65.3% precision rate and a 51.4% success rate on the LasHeR dataset, which is a 7.5% and 5% improvement, respectively, compared to the baseline.

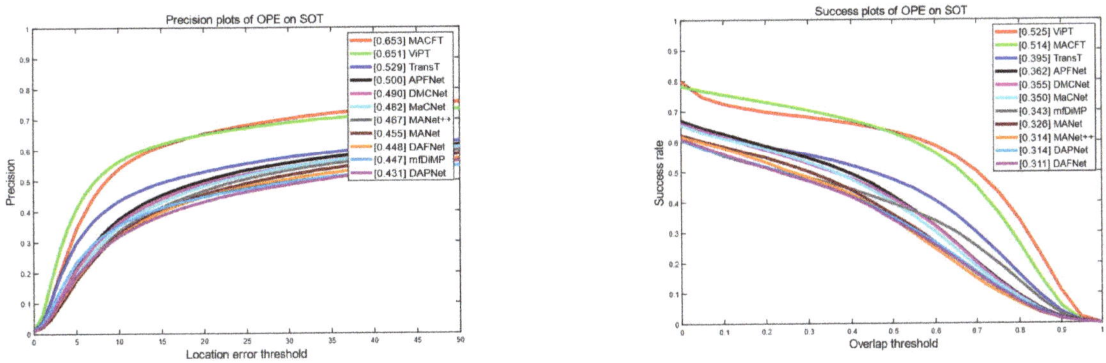

(**a**) (**b**)

Figure 5. Evaluation results on the LasHeR dataset; (**a**) is the precision rate curve, (**b**) is the success rate curve.

4.2.4. Evaluation on VTUAV Dataset

The VTUAV dataset is a large-scale RGB-T dataset designed for unmanned aerial vehicle (UAV) tracking, and it is currently the highest-resolution RGB-T dataset that considers both short-term and long-term tracking scenarios. However, the disadvantage of VTUAV is that the visible and thermal modalities are not aligned, and each frame in the sequence is not precisely manually annotated. Nevertheless, from another perspective, such data are more in line with the images obtained in real-world applications.

We evaluated our approach on both short-term and long-term tracking subsets of the VTUAV dataset and compared it with state-of-the-art RGB-T tracking algorithms. The results are shown in Tables 2 and 3. In particular, HMFT-LT is a model optimized specifically for long-term tracking (using GlobalTrack [44] as the global detector and RT-MDNet [19] as the tracking selector). When RT-MDNet recognizes that the target does not exist, GlobalTrack is selected to search for the target in the entire image.

Table 2. Quantitative comparison against state-of-the-art RGB-T trackers on the VTUAV short-term subset. Red, blue, and green represent the first, second, and third place, respectively.

Trackers	PR	SR	FPS
MACFT	80.1%	66.8%	31.7
HMFT [35]	75.8%	62.7%	30.2
ADRNet [4]	62.2%	46.6%	25.0
mfDiMP [36]	67.3%	55.4%	28.0
DAFNet [13]	62.0%	45.8%	21.0
FSRPN [45]	65.3%	54.4%	30.3

Table 3. Quantitative comparison against state-of-the-art RGB-T trackers on the VTUAV long-term subset. Red, blue, and green represent the first, second, and third place, respectively.

Trackers	PR	SR	FPS
MACFT	54.1%	46.7%	31.7
HMFT-LT [35]	53.6%	46.1%	8.1
HMFT [35]	41.4%	35.5%	30.2
ADRNet [4]	17.5%	25.3%	25.0
mfDiMP [36]	31.5%	27.2%	28.0
DAFNet [13]	25.3%	18.8%	21.0
FSRPN [45]	36.6%	31.4%	30.3

In the short-term subset evaluation, MACFT achieves a precision rate of 80.1% and a success rate of 66.8%, far surpassing the second-ranked tracker. In the long-term subset evaluation, MACFT achieves a precision rate of 54.1% and a success rate of 46.7%. It can be observed that even without specific optimization for long-term tracking tasks, MACFT still achieves state-of-the-art tracking performance on the long-term subset, which also demonstrates its potential for long-term tracking tasks.

4.2.5. Evaluation Based on Challenge Attributes

To better evaluate the effectiveness of our proposed MACFT model, we evaluated it on the RGBT234 dataset based on 12 labeled challenge attributes, which are: No Occlusion (NO), Partial Occlusion (PO), Heavy Occlusion (HO), Low Illumination (LI), Low Resolution (LR), Thermal Crossover (TC), Object Deformation (DEF), Fast Motion (FM), Scale Variation (SV), Motion Blur (MB), Camera Movement (CM), and Background Clutter (BC).

As shown in Table 4, our proposed model (MACFT) outperforms other state-of-the-art tracking models (DMCNet [12], APFNet [1], TFNet [7], MaCNet [3], DAFNet [13], ViPT [37], JMMAC [31], respectively) in most challenging attributes, especially in scenarios involving object occlusion and deformation, which further validates the effectiveness of the tracking model.

Table 4. Precision rate and success rate (PR/SR) of MACFT and other trackers under different challenge attribute subsets on the RGBT234 dataset. Red, blue, and green represent the first, second, and third place, respectively.

	DMCNet	ViPT	APFNet	TFNet	MaCNet	DAFNet	JMMAC	MACFT
NO	93.7/70.6	94.5/74.2	96.0/70.5	92.1/69.0	94.0/68.4	83.6/60.0	91.6/70.2	97.3/73.0
PO	89.9/63.0	84.4/63.3	85.8/60.4	80.8/56.7	81.1/56.5	80.1/55.9	79.9/59.4	85.7/63.9
HO	71.5/51.2	77.1/57.2	75.0/51.6	68.7/46.6	71.9/48.5	65.6/42.8	67.8/47.6	77.8/56.1
LI	85.3/59.2	79.1/57.2	86.8/58.5	81.5/54.8	81.4/54.0	82.5/55.1	88.4/61.8	82.7/58.3
LR	83.2/58.1	84.6/61.3	83.8/57.3	81.0/53.8	77.9/53.1	77.6/51.9	75.3/51.4	84.2/58.7
TC	87.0/61.9	88.9/66.2	84.4/59.1	80.1/57.2	81.5/59.0	81.8/57.5	72.7/50.4	82.7/60.7
DEF	75.4/54.9	81.3/63.4	79.4/56.8	71.4/50.9	73.5/51.4	65.1/44.2	68.0/51.2	83.4/63.8
FM	81.0/53.6	89.7/64.8	86.6/55.6	76.1/47.4	80.9/49.2	65.6/41.5	68.0/36.6	85.0/59.6
SC	82.3/59.8	82.6/62.9	83.7/58.5	77.6/56.0	78.3/55.7	73.7/50.9	81.0/60.6	86.1/64.1
MB	79.4/60.3	86.8/67.7	80.1/60.2	71.2/53.3	75.8/57.4	62.9/46.1	74.3/56.7	86.9/65.5
CM	81.2/59.4	85.4/65.0	81.0/59.1	76.0/54.1	75.9/54.2	68.1/47.0	74.2/55.0	88.4/65.2
BC	82.1/54.7	83.7/59.1	81.0/54.1	80.2/50.7	79.6/49.0	78.3/45.7	63.8/44.3	85.0/57.9
ALL	83.9/59.3	83.5/61.7	82.7/57.9	80.6/56.0	79.0/55.4	79.6/54.4	79.0/57.3	85.7/62.2

4.2.6. Qualitative Evaluation

Our results for the tracking performance of MACFT on several challenging video sequences are presented in Figure 6. These sequences included object deformation, camera movement, object occlusion, low illumination, high illumination, and other challenges. To test our method, we compared it with several advanced RGB-T trackers. This comparison indicates that MACFT had a higher ability to utilize the modality complementary information while excluding the interference of low-quality modality. Furthermore, MACFT had a more accurate understanding of the high-level semantic features of the target, making it more robust in tracking.

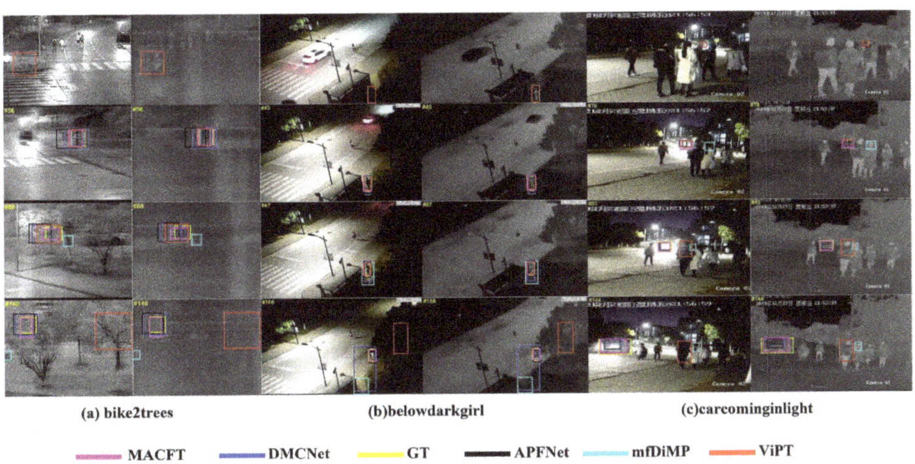

Figure 6. A visual example of the effectiveness of our proposed tracking method. (**a**–**c**) are three challenging sequences selected from the LasHeR dataset, and also show the tracking results of other advanced RGB-T trackers for comparison.

4.2.7. Comprehensive Evaluation of Efficiency and Effectiveness

Our method runs at 33.3 FPS (50.2 FPS for a lighter version with modality-shared feature branches removed) on an RTX 3090 GPU and can run in real time. Figure 7 gives a visual display of the comprehensive comparison of MACFT's success rate and running speed on the RGBT234 dataset with other advanced RGB-T tracking methods (the compared methods include TFNet [7], APFNet [1], MANet++ [9], MDNet [8], JMMAC [31], HMFT [35], SiamCDA [33], and DAFNet [13]).

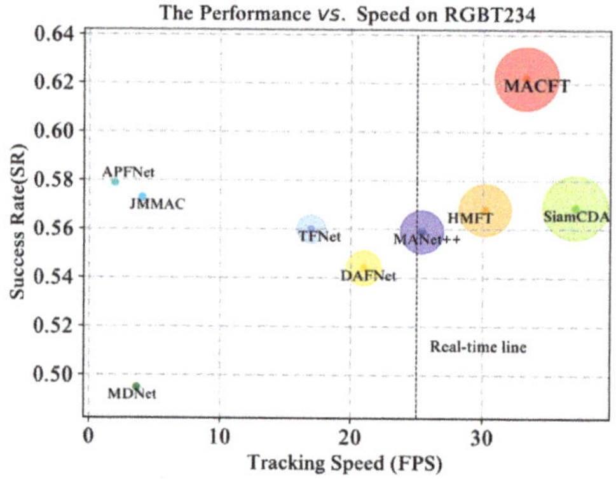

Figure 7. Comprehensive comparison of tracking success rate and speed on the RGBT234 dataset. Our MACFT model achieves the highest tracking success rate under the premise of ensuring real-time running (33.3 FPS on average).

4.3. *Visualization*

To prove that the fusion structure we designed can suppress the noise well from low-quality modality while enhancing the feature expression of the dominant modality, we selected specific sequences for visualization. The results are shown in Figure 8.

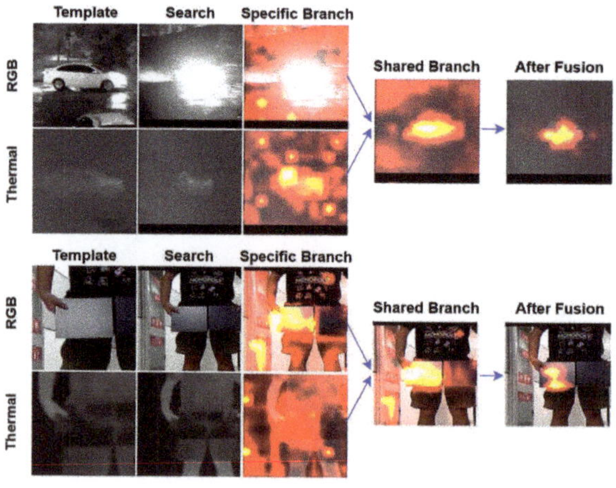

Template
Search
Specific Branch
Shared Branch
After Fusion
RGB
Thermal

Figure 8. Visualization of the attention weights of different branches when one of the modalities is of low quality.

The visualization method involves computing the attention weights by multiplying the query weight matrix with the key weight matrix at different network positions. Then, the average is taken along the token dimension to obtain a 14×14 (search size/patch size) attention weights map, which represents the regions of the image that the network focuses on at the current position.

We selected the car light sequence (thermal modality is dominant) from the RGBT234 dataset and the left mirror sequence (visible modality is dominant) from the LasHeR dataset for visualization. Among the attention weights maps, we can clearly observe that when facing a lack of target information in a certain modality, the network struggles to concentrate its attention entirely on the target location after feature extraction by the backbone. However, after incorporating our designed fusion structure, the noise introduced from the low-quality modality image is effectively suppressed, while the features from the dominant modality image are further enhanced.

4.4. Ablation Study

4.4.1. Model Pruning Experiments

To verify the validity of our model structure, specific analyses were performed for each component, and we set up several variants of the MACFT model (prefix B for baseline and DM for dual-modality), namely (1) MACFT (B-T): uses the baseline model + thermal modality, (2) MACFT (B-RGB): uses the baseline model + visible modality, (3) MACFT (DM): uses 2 modal-specific branches but removes all fusion modules based on attention mechanism, only concatenating the features from the two modalities and sending them to a fully connected layer for fusion, (4) MACFT (DM + CAM): uses 2 modal-specific branches (removing modal-shared feature branch) and 6 cascaded CAM modules for modality fusion (5) MACFT (DM + MAM): uses 2 modal-specific branches (removing modal-shared feature branch) and 6 cascaded MAM modules for modality fusion, (6) MACFT (DM + CAM + COM): uses all three branches and uses CAM modules in early feature fusion, but in late feature fusion, uses CAM modules instead of all MAM modules, (7) MACFT (w/o-FT): uses the complete model of MACFT but removes the finetune operation of the backbone.

The performance of the above control groups on the two RGB-T datasets is shown in detail in Table 5. The experimental results show that each component in MACFT improves the tracking performance to different degrees, and also indicates that the designed modal-shared branch and modality fusion network are effective.

Table 5. Evaluation results of MACFT and its variants on the RGBT234 dataset and the LasHeR dataset.

Trackers	LasHeR		RGBT234	
	PR (↓)	SR (↓)	PR (↓)	SR (↓)
MACFT (B-T)	42.8% (−22.5%)	33.2% (−18.2%)	67.4% (−18.3%)	48.8% (13.4%)
MACFT (B-RGB)	57.8% (−7.5%)	46.4% (−5.0%)	80.0% (−5.7%)	58.2% (−4.0%)
MACFT (DM)	62.2% (−3.1%)	48.7% (−2.7%)	83.7% (−2.0%)	58.9% (−3.3%)
MACFT (DM + CAM)	63.8% (−1.5%)	50.3% (−1.1%)	83.7% (−2.0%)	60.7% (−1.5%)
MACFT (DM + MAM)	63.9% (−1.4%)	50.2% (−1.2%)	83.9% (−1.8%)	60.7% (−1.5%)
MACFT (DM + CAM + COM)	64.3% (−1.0%)	50.7% (−0.7%)	85.3% (−0.4%)	61.6% (−0.6%)
MACFT (w/o-FT)	64.3% (−1.0%)	50.8% (−0.6%)	84.0% (−1.7%)	60.9% (−1.3%)
MACFT	65.3%	51.4%	85.7%	62.2%

In particular, comparison between MACFT (DM + MAM) and MACFT (DM + CAM) confirms our view that since the features from different modalities are sufficiently self-enhanced in the backbone, it does not make sense to simply add a self-attention operation after the backbone without introducing the modal-shared branch. However, after introducing modal-shared feature branches and cross-fusing them with modal-specific branches, the feature distribution is changed so there is room for learning with the introduction of the self-attention operation.

4.4.2. Parameter Tuning Experiments

To find the optimal number of MAM modules in series under the current volume of datasets, we evaluated the precision rate and success rate on the LasHeR test set and conducted experiments on the number of MAM modules in series from 4 to 8, respectively, and the results are shown in Table 6. The best results are obtained with 6 MAM modules in series under the current volume of training datasets and using fewer MAM modules is not enough to build the relationship of complementary information interactions between modalities, while using more MAM modules makes the model parameters not adjusted properly under the same epoch of training due to the limitation of the volume of data.

Table 6. The relationship between the number of MAM modules and the tracking performance; the evaluation results in the table are all tested on the LasHeR dataset.

Number of MAM Blocks	PR	SR
4	64.80%	51.00%
5	64.90%	51.20%
6	65.30%	51.40%
7	65.00%	51.20%
8	64.80%	51.10%

4.4.3. Parameter Size and Inference Speed

As an object tracking algorithm, the inference speed of MACFT is crucial as it directly affects the practical usability of the algorithm. Table 7 shows the inference speed of MACFT and its variants on an RTX 3090 GPU, as well as the success rate of tracking on the LasHeR dataset, and the number of trainable parameters for these models.

Table 7. Inference speed, success rate, and corresponding model parameters using MACFT and its variants on LasHeR (* represents the maximum number of trainable parameters in three-stage training, excluding frozen network layers).

Trackers	Speed (FPS)	Params *	SR
MACFT	33.3	59.6 M	65.3%
MACFT (DM + CAM + COM)	35.5	59.6 M	64.3%
MACFT (DM + CAM)	50.5	45.4 M	63.8%

It can be observed that the inclusion of the modal-shared branch has a significant impact on the inference speed. If the requirement for tracking performance is not very high but the demand for inference speed is high in practical applications, a light version without the modal-shared branch can be considered.

5. Conclusions

In this paper, we propose an RGB-T tracking method, MACFT, based on a mixed-attention mechanism. Four separate branches are established to extract modal-specific features and modal-shared features, and information interaction between the template image and the search image is introduced in each layer of the backbone. In the modality fusion stage, both cross-attention and mixed-attention are used to fuse the image features from different branches, which effectively reduces the noise from low-quality modality and enhances the feature information of the dominant modality. After conducting extensive experiments, we demonstrate that MACFT can understand the high-level semantic features of the target well, thus enabling it to cope with multiple challenges and achieve robust tracking.

However, our work still has some limitations, such as insufficient discriminability of target instances, large parameter count, and complex training procedures. In the future, we will address these issues by exploring various approaches, including but not limited to designing lightweight fusion modules, employing knowledge distillation to reduce parameter count, and integrating modal feature fusion into the feature extraction process to simplify training steps. At the same time, we will also explore ways to make the model compatible with more other modalities.

Author Contributions: Conceptualization, Y.L., X.G., M.D. and J.Y.; methodology, Y.L. and M.D.; software, Y.L.; validation, Y.L.; formal analysis, X.G., Y.L. and M.D.; investigation, Y.L.; resources, X.G. and J.Y.; data curation, Y.L.; writing—original draft preparation, Y.L.; writing—review and editing, X.G. and J.Y. All authors have read and agreed to the published version of the manuscript.

Funding: This work was funded by The National Key Research and Development Program of China, grant Nos. 2022YFB3901800, 2022YFB3901805.

Institutional Review Board Statement: Not applicable.

Informed Consent Statement: Not applicable.

Data Availability Statement: The raw results of our experiments could be available at: https://drive.google.com/file/d/1Fing9GjH6i5RTvRXuNRZ5fWt03DSE_ZO/view?usp=drive_link (accessed on 4 July 2023).

Conflicts of Interest: The authors declare no conflict of interest.

Abbreviations

Symbols	**Explanations**
Concat	Means to connect two tensors in the same dimension.
W_Q, W_K, W_V	Refers to the three weight matrices used by the self-attention mechanism in transformer, namely: query, key, and value.
d	Refers to the dim of W_Q and W_Q
softmax	Refers to the normalized exponential function; the function is to make the range of each element mapped between (0, 1).
KL	Refer to the Kullback–Leibler divergence calculation.
DR	Refers to operations that use linear layer dimensionality reduction.
softArgmax	Refers to finding the index corresponding to the maximum value through *softmax*.

References

1. Xiao, Y.; Yang, M.; Li, C.; Liu, L.; Tang, J. Attribute-Based Progressive Fusion Network for RGBT Tracking. *Proc. AAAI Conf. Artif. Intell.* **2022**, *36*, 2831–2838. [CrossRef]
2. Tang, Z.; Xu, T.; Wu, X.-J. A Survey for Deep RGBT Tracking. *arXiv* **2022**, arXiv:2201.09296.
3. Zhang, H.; Zhang, L.; Zhuo, L.; Zhang, J. Object Tracking in RGB-T Videos Using Modal-Aware Attention Network and Competitive Learning. *Sensors* **2020**, *20*, 393. [CrossRef]
4. Zhang, P.; Wang, D.; Lu, H.; Yang, X. Learning Adaptive Attribute-Driven Representation for Real-Time RGB-T Tracking. *Int. J. Comput. Vis.* **2021**, *129*, 2714–2729. [CrossRef]
5. Zhang, X.; Ye, P.; Peng, S.; Liu, J.; Gong, K.; Xiao, G. SiamFT: An RGB-Infrared Fusion Tracking Method via Fully Convolutional Siamese Networks. *IEEE Access* **2019**, *7*, 122122–122133. [CrossRef]
6. Bhat, G.; Danelljan, M.; Van Gool, L.; Timofte, R. Learning Discriminative Model Prediction for Tracking. In Proceedings of the 2019 IEEE/CVF International Conference on Computer Vision (ICCV), Seoul, Republic of Korea, 27 October–2 November 2019; IEEE: Piscataway, NJ, USA, 2019; pp. 6181–6190.
7. Zhu, Y.; Li, C.; Tang, J.; Luo, B.; Wang, L. RGBT Tracking by Trident Fusion Network. *IEEE Trans. Circuits Syst. Video Technol.* **2022**, *32*, 579–592. [CrossRef]
8. Nam, H.; Han, B. Learning Multi-domain Convolutional Neural Networks for Visual Tracking. In Proceedings of the 2016 IEEE Conference on Computer Vision and Pattern Recognition (CVPR), Las Vegas, NV, USA, 27–30 June 2016; IEEE: Piscataway, NJ, USA, 2016; pp. 4293–4302.
9. Lu, A.; Li, C.; Yan, Y.; Tang, J.; Luo, B. RGBT Tracking via Multi-Adapter Network with Hierarchical Divergence Loss. *IEEE Trans. Image Process.* **2021**, *30*, 5613–5625. [CrossRef] [PubMed]
10. Long Li, C.; Lu, A.; Hua Zheng, A.; Tu, Z.; Tang, J. Multi-adapter RGBT tracking. In Proceedings of the IEEE/CVF International Conference on Computer Vision Workshops, Seoul, Republic of Korea, 27 October–2 November 2019.
11. Zhu, Y.; Li, C.; Tang, J.; Luo, B. Quality-Aware Feature Aggregation Network for Robust RGBT Tracking. *IEEE Trans. Intell. Veh.* **2021**, *6*, 121–130. [CrossRef]
12. Lu, A.; Qian, C.; Li, C.; Tang, J.; Wang, L. Duality-Gated Mutual Condition Network for RGBT Tracking. *IEEE Trans. Neural Netw. Learn. Syst.* **2022**, 1–14. [CrossRef] [PubMed]
13. Gao, Y.; Li, C.; Zhu, Y.; Tang, J.; He, T.; Wang, F. Deep Adaptive Fusion Network for High Performance RGBT Tracking. In Proceedings of the 2019 IEEE/CVF International Conference on Computer Vision Workshop (ICCVW), Seoul, Republic of Korea, 27–28 October 2019; IEEE: Piscataway, NJ, USA, 2019; pp. 91–99.

14. Zhu, Y.; Li, C.; Luo, B.; Tang, J.; Wang, X. Dense Feature Aggregation and Pruning for RGBT Tracking. In Proceedings of the 27th ACM International Conference on Multimedia, Nice, France, 21–25 October 2019; pp. 465–472.
15. Li, C.; Liu, L.; Lu, A.; Ji, Q.; Tang, J. Challenge-Aware RGBT Tracking. In *Computer Vision–ECCV 2020: 16th European Conference, Glasgow, UK, August 23–28 2020, Proceedings, Part XXII 16*; Springer: Berlin/Heidelberg, Germany, 2020; pp. 222–237.
16. Luo, W.; Li, Y.; Urtasun, R.; Zemel, R. Understanding the Effective Receptive Field in Deep Convolutional Neural Networks. *Adv. Neural Inf. Process. Syst.* **2016**, *29*, 4905–4913.
17. Dosovitskiy, A.; Beyer, L.; Kolesnikov, A.; Weissenborn, D.; Zhai, X.; Unterthiner, T.; Dehghani, M.; Minderer, M.; Heigold, G.; Gelly, S.; et al. An Image is Worth 16x16 Words: Transformers for Image Recognition at Scale. *arXiv* **2010**, arXiv:2010.11929.
18. Danelljan, M.; Bhat, G.; Khan, F.S.; Felsberg, M. Atom: Accurate tracking by overlap maximization. In Proceedings of the IEEE/CVF Conference on Computer Vision and Pattern Recognition, Long Beach, CA, USA, 15–20 June 2019; pp. 4660–4669.
19. Jung, I.; Son, J.; Baek, M.; Han, B. Real-Time MDNet. In Proceedings of the European Conference on Computer Vision (ECCV), Munich, Germany, 8–14 September 2018; pp. 83–98.
20. He, K.; Gkioxari, G.; Dollár, P.; Girshick, R. Mask R-CNN. In Proceedings of the IEEE International Conference on Computer Vision, Venice, Italy, 22–29 October 2017; pp. 2961–2969.
21. Li, B.; Yan, J.; Wu, W.; Zhu, Z.; Hu, X. High Performance Visual Tracking with Siamese Region Proposal Network. In Proceedings of the 2018 IEEE/CVF Conference on Computer Vision and Pattern Recognition, Salt Lake City, UT, USA, 18–22 June 2018; IEEE: Piscataway, NJ, USA, 2018; pp. 8971–8980. [CrossRef]
22. Bertinetto, L.; Valmadre, J.; Henriques, J.F.; Vedaldi, A.; Torr, P.H.S. Fully-Convolutional Siamese Networks for Object Tracking. In *Computer Vision—ECCV 2016 Workshops*; Hua, G., Jégou, H., Eds.; Lecture Notes in Computer Science; Springer International Publishing: Cham, Switzerland, 2016; Volume 9914, pp. 850–865. [CrossRef]
23. Li, B.; Wu, W.; Wang, Q.; Zhang, F.; Xing, J.; Yan, J. SiamRPN++: Evolution of Siamese Visual Tracking with Very Deep Networks. In Proceedings of the 2019 IEEE/CVF Conference on Computer Vision and Pattern Recognition (CVPR), Long Beach, CA, USA, 15–20 June 2019; IEEE: Piscataway, NJ, USA, 2019; pp. 4277–4286. [CrossRef]
24. Chen, X.; Yan, B.; Zhu, J.; Wang, D.; Yang, X.; Lu, H. Transformer Tracking. In Proceedings of the 2021 IEEE/CVF Conference on Computer Vision and Pattern Recognition (CVPR), Nashville, TN, USA, 19–25 June 2021; IEEE: Piscataway, NJ, USA, 2021; pp. 8122–8131. [CrossRef]
25. Yan, B.; Peng, H.; Fu, J.; Wang, D.; Lu, H. Learning Spatio-Temporal Transformer for Visual Tracking. In Proceedings of the 2021 IEEE/CVF International Conference on Computer Vision (ICCV), Montreal, QC, Canada, 11–17 October 2021; IEEE: Piscataway, NJ, USA, 2021; pp. 10428–10437. [CrossRef]
26. Cui, Y.; Jiang, C.; Wang, L.; Wu, G. MixFormer: End-to-End Tracking with Iterative Mixed Attention. In Proceedings of the 2022 IEEE/CVF Conference on Computer Vision and Pattern Recognition (CVPR), New Orleans, LA, USA, 18–24 June 2022; IEEE: Piscataway, NJ, USA, 2022; pp. 13598–13608. [CrossRef]
27. Zhou, Y.; Kumar, A.; Parkash, C.; Vashishtha, G.; Tang, H.; Xiang, J. A novel entropy-based sparsity measure for prognosis of bearing defects and development of a sparsogram to select sensitive filtering band of an axial piston pump. *Measurement* **2022**, *203*, 111997. [CrossRef]
28. Zheng, G.; Chen, W.; Qian, Q.; Kumar, A.; Sun, W.; Zhou, Y. TCM in milling processes based on attention mechanism-combined long short-term memory using a sound sensor under different working conditions. *Int. J. Hydromechatron.* **2022**, *5*, 243–259. [CrossRef]
29. Chen, B.; Li, P.; Bai, L.; Qiao, L.; Shen, Q.; Li, B.; Gan, W.; Wu, W.; Ouyang, W. Backbone is All Your Need: A Simplified Architecture for Visual Object Tracking. In *Computer Vision–ECCV 2022: 17th European Conference, Tel Aviv, Israel, October 23–27 2022, Proceedings, Part XXII*; Springer: Berlin/Heidelberg, Germany, 2022; pp. 375–392.
30. Wang, Y.; Li, C.; Tang, J. Learning soft-consistent correlation filters for RGB-T object tracking. In *Chinese Conference on Pattern Recognition and Computer Vision (PRCV)*; Springer: Berlin/Heidelberg, Germany, 2018; pp. 295–306.
31. Zhang, P.; Zhao, J.; Bo, C.; Wang, D.; Lu, H.; Yang, X. Jointly Modeling Motion and Appearance Cues for Robust RGB-T Tracking. *IEEE Trans. Image Process.* **2021**, *30*, 3335–3347. [CrossRef] [PubMed]
32. Danelljan, M.; Bhat, G.; Khan, F.S.; Felsberg, M. ECO: Efficient Convolution Operators for Tracking. In Proceedings of the 2017 IEEE Conference on Computer Vision and Pattern Recognition (CVPR), Honolulu, HI, USA, 21–26 July 2017; IEEE: Piscataway, NJ, USA, 2017; pp. 6931–6939. [CrossRef]
33. Zhang, T.; Liu, X.; Zhang, Q.; Han, J. SiamCDA: Complementarity-and distractor-aware RGB-T tracking based on Siamese network. *IEEE Trans. Circuits Syst. Video Technol.* **2021**, *32*, 1403–1417. [CrossRef]
34. Li, X.; Wang, W.; Hu, X.; Yang, J. Selective kernel networks. In Proceedings of the IEEE/CVF Conference on Computer Vision and Pattern Recognition, Long Beach, CA, USA, 15–20 June 2019; pp. 510–519.
35. Zhang, P.; Zhao, J.; Wang, D.; Lu, H.; Ruan, X. Visible-Thermal UAV Tracking: A Large-Scale Benchmark and New Baseline. In Proceedings of the 2022 IEEE/CVF Conference on Computer Vision and Pattern Recognition (CVPR), New Orleans, LA, USA, 18–24 June 2022; IEEE: Piscataway, NJ, USA, 2022; pp. 8876–8885. [CrossRef]
36. Zhang, L.; Danelljan, M.; Gonzalez-Garcia, A.; van de Weijer, J.; Shahbaz Khan, F. Multi-Modal Fusion for End-to-End RGB-T Tracking. In Proceedings of the 2019 IEEE/CVF International Conference on Computer Vision Workshop (ICCVW), Seoul, Republic of Korea, 27–28 October 2019; IEEE: Piscataway, NJ, USA, 2019; pp. 2252–2261. [CrossRef]

37. Zhu, J.; Lai, S.; Chen, X.; Wang, D.; Lu, H. Visual prompt multi-modal tracking. In Proceedings of the IEEE/CVF Conference on Computer Vision and Pattern Recognition, Vancouver, CAN, 18–22 June 2023; pp. 9516–9526.
38. Jia, M.; Tang, L.; Chen, B.-C.; Cardie, C.; Belongie, S.; Hariharan, B.; Lim, S.-N. Visual prompt tuning. In Proceedings of the European Conference on Computer Vision, Tel Aviv, Israel, 23–27 October 2022; Springer: Berlin/Heidelberg, Germany, 2022; pp. 709–727.
39. Radford, A.; Kim, J.W.; Hallacy, C.; Ramesh, A.; Goh, G.; Agarwal, S.; Sastry, G.; Askell, A.; Mishkin, P.; Clark, J. Learning transferable visual models from natural language supervision. In Proceedings of the International Conference on Machine Learning, Virtual Event. 18–24 July 2021; pp. 8748–8763.
40. Carion, N.; Massa, F.; Synnaeve, G.; Usunier, N.; Kirillov, A.; Zagoruyko, S. End-to-end object detection with transformers. In *Computer Vision–ECCV 2020: 16th European Conference, Glasgow, UK, August 23–28, 2020, Proceedings, Part I 16*; Springer: Berlin/Heidelberg, Germany, 2020; pp. 213–229.
41. Li, C.; Liang, X.; Lu, Y.; Zhao, N.; Tang, J. RGB-T Object Tracking: Benchmark and Baseline. *Pattern Recognit.* **2019**, *96*, 106977. [CrossRef]
42. Li, C.; Xue, W.; Jia, Y.; Qu, Z.; Luo, B.; Tang, J.; Sun, D. LasHeR: A Large-scale High-diversity Benchmark for RGBT Tracking. *IEEE Trans. Image Process.* **2022**, *31*, 392–404. [CrossRef] [PubMed]
43. Loshchilov, I.; Hutter, F. Decoupled weight decay regularization. *arXiv* **2017**, arXiv:1711.05101.
44. Huang, L.; Zhao, X.; Huang, K. GlobalTrack: A Simple and Strong Baseline for Long-Term Tracking. *Proc. AAAI Conf. Artif. Intell.* **2020**, *34*, 11037–11044. [CrossRef]
45. Kristan, M.; Matas, J.; Leonardis, A.; Felsberg, M.; Pflugfelder, R.; Kamarainen, J.-K.; Cehovin Zajc, L.; Drbohlav, O.; Lukezic, A.; Berg, A.; et al. The Seventh Visual Object Tracking VOT2019 Challenge Results. In Proceedings of the 2019 IEEE/CVF International Conference on Computer Vision Workshop (ICCVW), Seoul, Republic of Korea, 27–28 October 2019; IEEE: Piscataway, NJ, USA, 2019; pp. 2206–2241.

sensors

MDPI

Article

Robust Localization of Industrial Park UGV and Prior Map Maintenance

Fanrui Luo [1], Zhenyu Liu [1,*], Fengshan Zou [2], Mingmin Liu [2], Yang Cheng [1] and Xiaoyu Li [1]

1 School of Information Science and Engineering, Shenyang University of Technology, Shenyang 110870, China; 202120679@smail.sut.edu.cn (F.L.); tanxiaojian@smail.sut.edu.cn (Y.C.); lxy_4210@smail.sut.edu.cn (X.L.)
2 SIASUN Robot & Automation Co., Ltd., Shenyang 110169, China; zoufengshan@siasun.com (F.Z.); liumingmin@siasun.com (M.L.)
* Correspondence: liuzhenyu@sut.edu.cn

Abstract: The precise localization of unmanned ground vehicles (UGVs) in industrial parks without prior GPS measurements presents a significant challenge. Simultaneous localization and mapping (SLAM) techniques can address this challenge by capturing environmental features, using sensors for real-time UGV localization. In order to increase the real-time localization accuracy and efficiency of UGVs, and to improve the robustness of UGVs' odometry within industrial parks—thereby addressing issues related to UGVs' motion control discontinuity and odometry drift—this paper proposes a tightly coupled LiDAR-IMU odometry method based on FAST-LIO2, integrating ground constraints and a novel feature extraction method. Additionally, a novel maintenance method of prior maps is proposed. The front-end module acquires the prior pose of the UGV by combining the detection and correction of relocation with point cloud registration. Then, the proposed maintenance method of prior maps is used to hierarchically and partitionally segregate and perform the real-time maintenance of the prior maps. At the back-end, real-time localization is achieved by the proposed tightly coupled LiDAR-IMU odometry that incorporates ground constraints. Furthermore, a feature extraction method based on the bidirectional-projection plane slope difference filter is proposed, enabling efficient and accurate point cloud feature extraction for edge, planar and ground points. Finally, the proposed method is evaluated, using self-collected datasets from industrial parks and the KITTI dataset. Our experimental results demonstrate that, compared to FAST-LIO2 and FAST-LIO2 with the curvature feature extraction method, the proposed method improved the odometry accuracy by 30.19% and 48.24% on the KITTI dataset. The efficiency of odometry was improved by 56.72% and 40.06%. When leveraging prior maps, the UGV achieved centimeter-level localization accuracy. The localization accuracy of the proposed method was improved by 46.367% compared to FAST-LIO2 on self-collected datasets, and the located efficiency was improved by 32.33%. The z-axis-located accuracy of the proposed method reached millimeter-level accuracy. The proposed prior map maintenance method reduced RAM usage by 64% compared to traditional methods.

Keywords: LiDAR-IMU SLAM; feature extraction; prior map maintenance; relocation; unmanned ground vehicle

1. Introduction

With the rapid development of science and technology, the construction of intelligent and digital industrial parks has entered a transformative phase [1,2]. Ensuring the stable operation of UGVs within industrial parks and guaranteeing their accurate arrival at designated positions are key challenges that must be addressed. The accuracy and efficiency of pose estimation serve as fundamental prerequisites for achieving the reliable arrival of UGVs at the designated position [3–5]. Simultaneous localization and mapping (SLAM) techniques can use external information obtained by sensors to perform real-time 6DOF state estimation, effectively resolving the problem of precise localization in the absence of prior GPS measurements, thus providing a crucial foundation for UGV operations within industrial parks [6]. SLAM technology can be classified into visual and LiDAR SLAM. Visual SLAM, which relies on camera-based scene feature extraction, is susceptible to lighting conditions and environmental textures, making it unsuitable for real-time UGV localization [7]. On the other hand, LiDAR SLAM compares the emission signals of laser beams with the received echo signals to obtain point cloud information from the scene, which is less affected by environmental factors [8]. Therefore, LiDAR SLAM is better suited to real-time UGV localization within industrial parks.

1.1. Odometry Based on LiDAR-IMU

Currently, LiDAR SLAM can be classified into loosely coupled and tightly coupled methods, based on the approaches used for LiDAR-IMU fusion [9]. Loosely coupled methods involve the direct integration of IMU pre-integration results into the front-end of the LiDAR odometry. These pre-integration results are utilized as initial estimates for LiDAR scanning registration, and they serve as prior values for distortion compensation. The IMU pre-integration results are not incorporated into the joint optimization process. Zhang et al. [10] proposed LiDAR odometry and mapping (LOAM), which is a widely recognized and classic loosely coupled method for fusing LiDAR and IMU data. LOAM has been extensively evaluated on the KITTI dataset [11], demonstrating outstanding performance, and serving as a milestone in the field. However, ensuring its robustness in complex environments is challenging, due to its heavy reliance on stable edge and planar features present in the environment. Shan et al. [12] proposed a lightweight and ground-optimized version of LOAM (LeGO-LOAM), based on the original LOAM. LeGO-LOAM adopts a two-step L-M optimization approach on separated ground points and on points representing edges and planes in space. This method conducts separate three-degrees-of-freedom pose estimations and combines them to achieve a robust six-degrees-of-freedom pose estimation. The aforementioned method can enhance pose estimation accuracy while concurrently reducing system computational complexity. However, its heavy reliance on ground features poses challenges to achieving robust pose estimation in environments characterized by complex ground features. On this basis, reference [13] proposed Livox LOAM specifically designed for solid-state LiDAR systems, and reference [14] proposed M-LOAM for simultaneous localization and mapping utilizing multiple LiDARs. These methods are classified as loosely coupled LiDAR-IMU methods.

Unlike loosely coupled methods, tightly coupled methods typically incorporate the original feature points from LiDAR into IMU data, and they integrate the IMU pre-integration results into joint optimization. Tightly coupled methods offer the advantage of mitigating drift by incorporating residual calculations of point-to-edge and point-to-plane distances, especially in scenarios where the quality of point cloud scene features deteriorates. Tightly coupled methods can be primarily categorized into optimization-based methods and filtering-based methods. Ye et al. [15] proposed a tightly coupled LiDAR-IMU method known as tightly coupled 3D LiDAR industrial odometry and mapping (LIOM). The LIOM method mitigates IMU and LiDAR errors by optimizing the sliding window and incorporating rotation-constrained refinement. It achieves reliable pose estimation in scenarios with reduced feature availability, via the tightly coupled approach. However, this method suffers from high computational complexity and significant time consumption. Qin et al. [16] proposed a lightweight tightly coupled method known as robocentric LiDAR industrial state estimator for robust and efficient navigation (LINS). This method employs an iterated-error-state Kalman filter (ESKF) to achieve the tight fusion of LiDAR and IMU measurements, resulting in enhanced efficiency without compromising accuracy. Shan et al. [17] proposed a tightly coupled LiDAR inertial odometry via the smoothing and mapping (LIO-SAM) method, based on factor graph optimization. This method introduces a sliding window of key frames from the LiDAR data, effectively reducing the computational complexity. The factor graph incorporates IMU pre-integration factors, LiDAR odometry factors, loop closure factors, and GPS factors for joint optimization, leading to improved precision. Xu et al. [18,19] proposed FAST-LIO and FAST-LIO2, which utilize the iterative extended Kalman filter (IEKF) for tightly coupled LiDAR-IMU fusion. FAST-LIO introduces a novel Kalman gain calculation formula inspired by the matrix inverse lemma. This formula transforms the dimension of the inverse matrix from the measurement dimension to the state dimension. Motion distortion is compensated for using a backward propagation method. FAST-LIO2 significantly reduces computational burden and enables high-frequency odometry output at 50 Hz. To enhance computational efficiency, FAST-LIO2 introduces an incremental k-d tree for map maintenance, leading to an accelerated K-nearest neighbor (KNN) search speed and improved computational efficiency.

1.2. The Problems in the Real-Time Localization of UGV in Industrial Parks

To fulfill the demands of high-precision and high-frequency odometry for path planning and motion control of UGVs within industrial parks, LiDAR SLAM encounters the following novel challenges:

1. Given the intricate and expansive nature of industrial parks, traditional algorithms encounter challenges in mitigating odometry drift in extensive environments and over long distances. The utilization of low-frequency odometry with inadequate real-time performance can result in errors and delays in UGV motion control. Consequently, traditional methods struggle to fulfill the high-frequency and real-time localization demands of UGVs [20–23].
2. By leveraging prior maps for real-time localization, the localization accuracy of UGVs within industrial parks can be significantly enhanced. However, the utilization of prior maps for large-scale industrial parks presents challenges, as it necessitates substantial RAM occupancy. Consequently, performing a KNN search on UGVs with limited RAM capacity becomes arduous and impractical [24].
3. The edge and planar features in industrial park environments exhibit relative prominence; however, they are often accompanied by numerous unstable features. Traditional methods encounter challenges in efficiently and accurately extracting high-quality feature points, as well as in concurrently extracting ground points in a targeted manner [25–28].

Reference [29] proposed utilizing prior maps to compute the residuals of feature point clouds, enabling the generation of highly precise and high-frequency poses that fulfill the odometry accuracy demands of UGVs. Many existing methods maintain a prior map using Octree or kd-tree structures. While the speed of KNN search may satisfy the requirements, the associated RAM usage should not be overlooked, especially when dealing with sizable prior maps. The limited RAM capacity of UGV host systems often poses a challenge in meeting these requirements. References [30–32] employed calculations of structural tensors, point feature histograms, and viewpoint feature histograms to extract feature points. While these methods demonstrate acceptable performance in feature extraction, their extraction speed is inadequate to meet the requirements of high-frequency odometry. The curvature feature extraction method categorizes feature points into planar points and edge points. However, this method exhibits poor performance in accurately extracting stable ground points as the characteristics of ground points on LiDAR scanning lines differ from planar points. Hence, further improvements are necessary to enhance the speed and accuracy.

1.3. New Contributions

In summary, to fulfill the real-time localization requirements of UGVs in industrial parks, this study presents a tightly coupled LiDAR-IMU odometry method that leverages prior maps. The key contributions of this research can be outlined as follows:

1. A method is proposed for extracting high-quality feature points using a bidirectional projection plane slope difference filter. This method not only reduces the feature extraction speed but also enhances the quality of feature points. Moreover, it enables the separate extraction of ground points and planar points.

2. A maintenance method is proposed for large prior maps, which integrate three-layer voxels with an ikd-tree structure. This method effectively reduces RAM consumption and enables the real-time localization of UGVs using the large prior maps.
3. A back-end optimization model incorporating ground constraints is proposed. This method assigns adaptive weights to observation error equations of ground and planar points using the proposed pseudo occupancy method. This method addresses the issue of inadequate lateral or longitudinal constraints that may arise due to a limited number of points on the plane or ground.

2. Overall System Framework

This paper presents a real-time localization method for UGVs using prior maps. The method involves searching for the nearest neighbor points in the observation error equation to calculate the residual within the prior maps. This approach significantly enhances localization accuracy while maintaining processing speed. The overall design, as illustrated in Figure 1, is divided into four parts:

1. Data processing for LiDAR and IMU: the initial pose estimation of the current frame is obtained through the preintegration of the IMU measurements. Subsequently, high-quality planar and ground points are extracted using a bidirectional projection plane slope difference filter. The IMU pre-integration results are propagated backwards to compensate for the motion distortion of LiDAR points in the current frame.
2. Tightly coupled LiDAR-IMU odometry: the error equations for prediction and observation with ground constraints in the iterated extended Kalman filter (IEKF) are formulated using the points information and the pre-integration results from the IMU. As part of the estimation process, the pose is cyclically and iteratively updated.
3. Relocation: the initial frame captured by the LiDAR is detected and corrected using prior historical keyframes for relocation. The initial frame is registered with the successfully detected historical keyframe, enabling the acquisition of accurate UGV position information within a prior map.
4. Prior map maintenance: through the real-time maintenance of the three-layer voxels comprising the overall ROM voxels, nearest-neighbor RAM voxels, and ikd-tree voxels, the prior maps can be efficiently maintained, enabling a fast KNN search while ensuring lightweight operations.

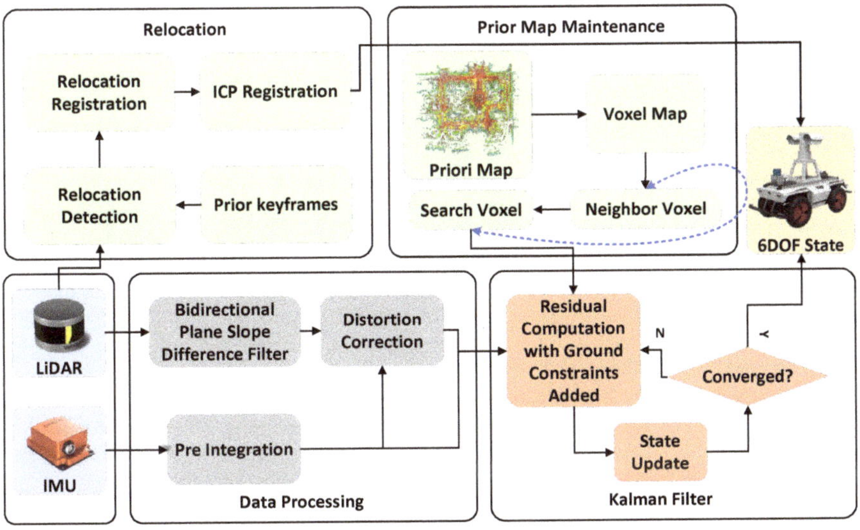

Figure 1. Overall system framework.

3. Tightly Coupled LiDAR-IMU Odometry

3.1. Feature Extraction Based on Bidirectional Projection Plane Slope Difference Filter

The current most popular curvature feature extraction method involves calculating the relationship between each point and neighboring points on the local plane along the current LiDAR scanning line. However, this method suffers from relatively slow extraction speed and limited accuracy. Moreover, the curvature feature extraction method roughly classifies feature points into planar and edge points, but fails to account for the distinctive characteristics of ground points on the LiDAR scanning lines, leading to poor performance in extracting stable ground points and potential height value drift. Additionally, the curvature method exhibits a limited ability to filter out noisy and unstable points, resulting in potential lateral drift. Considering the scanning characteristics of 3D LiDAR, each scanning line is captured at a specific angle. When a line scans a plane, the points do not lie on a straight line in three-dimensional space. This observation makes feature extraction via projection planes more reliable. To address these limitations, this paper proposes a feature extraction method based on the bidirectional projection plane slope difference filter. This method only requires calculating the relationship between a point on the current line and its preceding and subsequent points. By mapping the calculation from three-dimensional space to two planes, the proposed method can extract high-quality ground, planar, and edge points simultaneously. Additionally, special processing is applied to lines in the proximity of the ground, enhancing both extraction accuracy and speed. The main steps of the proposed method are as follows.

(1) Near-end ground feature point extraction: the point cloud of the current frame is denoted as P. Leveraging the characteristics of 3D LiDAR, it is observed that LiDAR points from lower scanning lines tend to contain more stable ground features. This method establishes the lowest LiDAR scanning line P_{min} as the reference line for near-end ground feature constraints. The line is divided into n segments based on the sector area, and the average distance from the LiDAR center to the LiDAR points in each segment is calculated. The segment P_{ground} with the highest average distance is selected as the constraint for ground feature points. The average height z_{ground} of these selected points represents an approximate estimation of the ground plane height in the near-end region. The lines ranging from P_{min} to P_h, which are in proximity to the lowest scanning line, are identified as the near-end ground lines. On the other hand, the lines from P_h to P_{max} are classified as non-near-end ground lines. To mitigate the influence of noise and obstacles in the vicinity of the ground plane, points with heights, z_s^i, of the point i in the LiDAR scanning s-line falling outside the range $[z_{ground} - \mathrm{u}, z_{ground} + \mathrm{u}]$ within the near-end ground lines are filtered out, where u is the floating range threshold of the near-end ground height. The value of u is set according to the strictness of the near-end ground point filtering. This process facilitates the extraction of high-quality near-end ground points, thus enhancing the reliability of ground feature constraints for pose estimation.

(2) Far-end feature point extraction: this method involves the projection of points from each LiDAR scanning line, excluding those within the near-end ground lines of the current frame, onto the l-plane comprising the x-axis and y-axis of the LiDAR coordinate system. Simultaneously, the points are projected onto the h-plane, which consists of the z-axis and the angle θ within the plane formed by the x-axis and y-axis. Subsequently, the method applies a slope difference filter to each LiDAR scanning line on the two projection planes. The calculations performed in three-dimensional space are projected onto these two two-dimensional planes via the bidirectional projection plane slope filter. By evaluating the slope difference of the same LiDAR point on the two projection planes, the smoothness of the point on the local plane can be determined. This process allows for the efficient extraction of high-quality ground points, planar points, and edge points. The slope and slope difference of points in the l-plane and h-plane can be defined as follows:

$$K_i^l = \frac{\rho_{i+1} - \rho_i}{\rho_i \Delta\theta_i} \tag{1}$$

$$d_i^l = K_i^l - K_{i-1}^l \tag{2}$$

$$K_i^h = \frac{z_{i+1} - z_i}{\Delta\theta_i} \tag{3}$$

$$d_i^h = \arccos\left(\frac{\Delta\theta_i \Delta\theta_{i-1} + (\Delta\theta_i K_i^h)(\Delta\theta_{i-1} K_{i-1}^h)}{\sqrt{(\Delta\theta_i K_i^h)^2 + (\Delta\theta_i)^2}\sqrt{(\Delta\theta_{i-1} K_{i-1}^h)^2 + (\Delta\theta_{i-1})^2}}\right) \tag{4}$$

$$\Delta\theta_i = \theta_{i+1} - \theta_i \tag{5}$$

In the formula, K_i^l and d_i^l are the slope and slope difference of point i on the l-plane. K_i^h and d_i^h are the slopes and slope difference of point i on the h-plane. ρ_i is the distance between point i on the l-plane and the LiDAR center. z_i is the height value of point i. θ_i is the angle value of point i on the l-plane. $\Delta\theta_i$ is the difference in angle between point $i+1$ and point i on the l-plane. As shown in Figure 2, the extraction effect of a point is shown. The near-end ground points, high-quality planar points and high-quality ground points are red, and the corner points are green.

(a) (b)

Figure 2. The results of feature extraction. (**a**) is the feature extraction effect of the near-ground end. (**b**) is the feature extraction effect of the far end.

The slope difference filter formulas enable the evaluation of whether the current point, along with its preceding and succeeding points on the same line, can be fitted into a straight line in the l-plane or h-plane. A close approximation to a straight line in the l-plane indicates that the point lies on a non-ground plane within three-dimensional space. The projection of ground points onto the l-plane exhibits a circular arc shape. Conversely, an approach to a straight line in the h-plane suggests that there is no significant change in the z-axis height between the points. When the slope difference on the l-plane approaches zero and the slope difference on the h-plane is less than the threshold value for a jump on the h-plane, it indicates that the point is a stable planar point. The constraint on the l-plane serves to filter out noise and unstable points. When the slope difference on the l-plane approaches zero and the slope difference on the h-plane is greater than the jump threshold, it suggests that the angle between the laser beam of the current point and the plane is small, indicating that the point is located at the far end of the plane. These points are considered unstable planar far-end points that need to be filtered out. When the slope difference on the l-plane cannot approach zero and the slope difference on the h-plane does not approach zero but is less than the jump threshold, it indicates that the point is a corner point. The constraint on the h-plane serves to filter out the far jump points. When the slope difference on the l-plane cannot approach zero and the slope difference on the h-plane is greater than the jump threshold, it suggests that the point transitions from one plane to another far plane. Lastly, when the slope difference on the l-plane cannot approach zero and the slope difference on the h-plane strictly approaches zero, it indicates that the point is a stable ground point.

Algorithm 1 outlines the feature extraction method in this paper. d_i^l is the slope difference on the l-plane of the point i in the current frame. d_i^h is the slope difference on the h-plane of the point i in the current frame. d_{lm} is the threshold at which the slope difference of the point on the l-plane approaches zero. d_{hn} is the threshold at which the slope difference of the point on the h-plane strictly approaches zero. d_{hm} is the threshold at which the slope difference of a point on the h-plane does not jump. $\overline{P}_p$ is a high-quality planar point, $\overline{P}_d$ is a high-quality ground point, and $\overline{P}_d$ is a high-quality edge point.

Algorithm 1: Feature extraction

Input : P from current frame point cloud
Output: $\overline{P}_p$,$\overline{P}_d$,$\widetilde{P}$

for (P^i *from every points of* P) **do**
  put point p^i in the corresponding P_s according to LiDAR line s;
end
Divide P_{min} into n equal parts according to the fan shaped area;
Put the maximum part of the average distance from the LiDAR to the point in P_{ground};
Put the average height of the P_{ground} in z_{ground};
for (*each point* P_s^i *from* P_{min} *to* P_h) **do**
  if ($z_s^i > z_{ground} - u$ & $z_s^i < z_{ground} + u$) **then**
    Put P_s^i into $\overline{P}_d$;
  end
end
for (*each point* P_s^i *from* P_h *to* P_{max}) **do**
  for (*each point* P_s^i *from* P_s) **do**
    if ($i = 0$) **then**
      Calculate the slope and slope difference k_i^l, k_i^h of the projection planes l and h based on (1),(2),(3),(4),(5);
    end
    if ($i \neq 0$) **then**
      Calculate the slope and slope difference k_i^l, k_i^h, d_i^l, d_i^h of the projection planes l and h based on (1),(2),(3),(4),(5);
      if ($d_i^l < dlm$ & $d_i^h < dhm$) **then**
        Put P_s^i into $\overline{P}_p$;
      end
      if ($d_i^l > dlm$ & $d_i^h > dhn$ & $d_i^h < dhm$) **then**
        Put P_s^i into $\widetilde{P}$;
      end
      if ($d_i^l > dlm$ & $d_i^h < dhn$) **then**
        Put P_s^i into $\overline{P}_d$;
      end
    end
  end
end

3.2. *IMU Preintegration and Compensation of Point Cloud Distortion*

The IMU preintegration and compensation of point cloud distortion are performed using discretized forward and backward propagation. The forward propagation is conducted over the time interval Δt_f between consecutive IMU frames. It begins with the IMU measurement taken after the LiDAR point timestamp at the start of the current frame and ends with the IMU measurement taken before the LiDAR point timestamp at the end of the current frame. On the other hand, backward propagation compensates for point cloud distortion in the current frame based on the time difference Δt_l between adjacent LiDAR scanning points.

(1) Forward propagation: in the continuous model, the position, velocity and attitude of the IMU can be defined as follows:

$$v_b^G = v_a^G + \int_{t\epsilon[a,b]} R_t^G(a_t - b_{at} - n_{at}) + g^G dt \tag{6}$$

$$p_b^G = p_a^G + \int_{t\epsilon[a,b]} v_t + R_t^G(a_t - b_{at} - n_{at})t + g^G t dt \tag{7}$$

$$R_b^G = R_a^G exp(\int_{t\epsilon[a,b]} \omega_t - b_{\omega t} - n_{\omega t} dt) \tag{8}$$

In the equation, v_a^G and v_b^G are the velocities of IMU at time a and b in the world coordinate system. p_a^G and p_b^G are the positions of IMU at time a and b in the world coordinate system. R_a^G, R_b^G and R_t^G are the attitude of the IMU at times a, b, and t in the world coordinate system. a_t is the acceleration measured at time t in the IMU coordinate system. v_t is the speed at time t in the IMU coordinate system. ω_t is the angular velocity measured at time t in the IMU coordinate system. g^G is the gravitational component in the world coordinate system. b_{at} and n_{at} are acceleration bias and noise at time t. $b_{\omega t}$ and $n_{\omega t}$ are the angular velocity bias and noise at time t. After removing the unknown noise and offsetting, discretizing and simplifying it, the IMU forward propagation model is defined as follows:

$$v_{k+1} = v_k + (R_k^G a_k + g^G)\Delta t_f \tag{9}$$

$$p_{k+1} = p_k + v_k \Delta t_f \tag{10}$$

$$R_{k+1} = R_k exp(\omega_k \Delta t_f) \tag{11}$$

$$v_0 = \bar{v}_i \tag{12}$$

$$p_0 = \bar{p}_i \tag{13}$$

$$R_0 = \bar{R}_i \tag{14}$$

In the formula, v_k, v_{k+1} and v_0 are the speeds of the current frame IMU at times k, $k+1$, and 0 in the world coordinate system. p_k, p_{k+1} and p_0 are the positions of the current frame IMU at times k, $k+1$, and 0 in the world coordinate system. R_k, R_{k+1} and R_0 are the attitude of the IMU at times k, $k+1$ and 0 in the world coordinate system. a_k is the accelerometer measurement at time k in the IMU coordinate system of the current frame. ω_k is the angular velocity measurement at time k in the IMU coordinate system of the current frame. $\bar{v}_i$, $\bar{p}_i$ and $\bar{R}_i$ are the true value of velocity, position and rotation matrix of the last frame after the Kalman filter. When the time $k+1$ is the time of the last LiDAR point in the current frame, v_{k+1}, p_{k+1} and R_{k+1} are the $\hat{v}_{i+1}$, $\hat{p}_{i+1}$ and $\hat{R}_{i+1}$ entering the Kalman filter at the end of the current frame $i+1$.

(2) Backward propagation: similar to the discretized forward propagation, the backward propagation propagates the IMU measurements in the opposite time direction -Δt_l to compensate for the distortion of points in the current frame. This process occurs at the frequency of LiDAR scanning points. Since the IMU measurement frequency is lower than the LiDAR scanning frequency, there are multiple LiDAR points between two consecutive IMU measurements. Consequently, the most recent IMU measurement prior to the times-

tamp of the current LiDAR point is utilized as the input value for backward propagation. The IMU backward propagation model is defined as follows:

$$v_b = v_{b+1} - \Delta t_l (R_{b+1}^G a_k + g^G) \quad (15)$$

$$p_b = p_{b+1} - \Delta t_l v_{b+1} \quad (16)$$

$$R_b = log((\omega_k \Delta t_l)^T R_{b+1}) \quad (17)$$

v_b, p_b and R_b represent the velocity, position, and attitude of the b-th point in the current LiDAR scan frame in the world coordinate system. a_k and ω_k are the acceleration and angular velocity obtained from the closest IMU measurement taken prior to the timestamp of the $b+1$-th point in the current LiDAR scan frame. When time $b+1$ is the time of the last LiDAR point in the current frame, v_{b+1}, p_{b+1} and R_{b+1} are the initial inputs $\hat{v}_{i+1}$, $\hat{p}_{i+1}$ and $\hat{R}_{i+1}$ for the backward propagation in the current frame $i+1$.

3.3. Back-End Optimization Based on Iterated Extended Kalman Filter

3.3.1. Error State Estimation

To facilitate error calculation and state update, the state of UGV is defined as $[R^{G^T}, p^{G^T}, v^{G^T}, b_{\omega}^T, b_a^T, g^{G^T}]^T$, and the error ε between the estimated state and the actual state is defined as $[\varepsilon^{R^T}, \varepsilon^{p^T}, \varepsilon^{v^T}, \varepsilon^{b\omega^T}, \varepsilon^{ba^T}, \varepsilon^{g^T}]^T$. The integrated state can be expressed as $[(R^G exp[\varepsilon^R])^T, (p^G + \varepsilon^p)^T, (v^G + \varepsilon^v)^T, (b_\omega + \varepsilon^{b_\omega})^T, (b_a +, \varepsilon^{b_a})^T, (g^G + \varepsilon^g)^T]^T$. In the formula, ε^R represents the error in the state rotation matrix. ε^p is the error in the state position. ε^v is the error in the state velocity. ε^{b_ω} is the error in the angular velocity bias. ε^{b_a} is the error in the accelerometer bias error.

(1) Predicted error state: the predicted error is defined as the discrepancy between the real state and the propagated state. By utilizing the propagated formula of the real state and the propagated formula of the propagated state, the propagated formula of the predicted error state and the covariance propagation formula are defined as follows:

$$\hat{\varepsilon}_{i+1} = F_{\hat{\varepsilon}_i} \Delta t_f \hat{\varepsilon}_i + \hat{\varepsilon}_i \quad (18)$$

$$\hat{P}_{i+1} = F_{\hat{\varepsilon}_i} \hat{P}_i F_{\hat{\varepsilon}_i}^T + F_{\omega_i} Q F_{\omega_i}^T \quad (19)$$

In the formula, $\hat{\varepsilon}_i$ is the predicted error state at time i. ω_i is the Gaussian white noise at time i. $F_{\hat{\varepsilon}_i}$ is the error state propagated matrix at time i. F_{ω_i} is the Jacobian matrix of the Gaussian white noise at time i. $\hat{P}_i$ and Q are the covariance of $\hat{\varepsilon}_i$ and ω_i at time i. $F_{\hat{\varepsilon}_i}$ and F_{ω_i} are more concretely derived in [18].

After obtaining the estimation of the propagated error state, the predicted error state of the iterated update is defined as follows:

$$\tilde{\varepsilon}_k = \hat{\varepsilon}_k^a + h_a \tilde{\varepsilon}_k^a \quad (20)$$

$$P_k^{a+1} = (h_a)^{-1} P_k^a (h_a)^{-T} \quad (21)$$

In the equation, $\tilde{\varepsilon}_k$ is the error between the true state of the frame k and the propagated state of frame k. $\hat{\varepsilon}_k^a$ is the error between the a-th update state of the frame k and the propagated state of the frame k. P_k^a is the covariance of the a-th state update of the frame k. h_a is the Jacobian matrix of $\tilde{\varepsilon}_k$ relative to $\tilde{\varepsilon}_k^a$.

(2) Observed error state: the LiDAR point cloud of the current frame is projected onto the global coordinate system using IMU preintegration. The residual distance between the

projected points of the current frame and the nearest local plane of the prior map is defined as the observed state error. The observed error state is defined as follows:

$$0 = z_j = S_J[(R_k M_j + P_k) - M_j^G] \approx z_j^a + H_j^k \bar{\varepsilon}_k^a \tag{22}$$

$$z_j^a = S_J[(R_k^a M_j + P_k^a) - M_j^G] \tag{23}$$

In the formula, z_j^a is the residual distance between point j of the a-th update state of frame k in state $\hat{x}_k^a$ and the nearest local plane in the prior map. z_j is the residual distance between point j of the current frame k in state x_k and the nearest local plane in the prior map. S_J is the normal vector of the local plane fitted by the nearest neighbor points of point j in the current frame k within the prior map. M_j is the coordinate of point j in the coordinate system of the current frame k. R_k^a and P_k^a are the rotation matrix and translation vector of state $\hat{x}_k^a$. R_k and P_k are the rotation matrix and translation vector of state x_k. M_j^G is the nearest-neighbor point of point j in the prior map of the current frame k. x_k and $\hat{x}_k^a$ represent the true state of frame k and the a-th update state of frame k. $\bar{\varepsilon}_k^a$ is the error between the true state and the a-th update state of frame k. H_k^a is the Jacobian matrix of z_j relative to $\bar{\varepsilon}_k^a$. The nearest neighbor points are obtained through KNN search in the ikd-tree.

3.3.2. Iterated State Update with Ground Constraints Added

Combining the predicted error state in (24) and the observed error state in (27) yields the maximum posterior estimate (MAP). To address the limitations posed by a limited number of points on the ground or plane, a new MAP formula is proposed, which incorporates ground constraints and utilizes a feature extraction method to accurately divide high-quality ground and planar points. The proposed method assigns adaptive weights to the observed error equations of plane and ground points. Specifically, a higher weight is assigned to the observed error equation of ground points when there are more planar points, and a higher weight is assigned to the observed error equation of planar points when there are more ground points. By employing this new formula, the insufficiency of lateral or longitudinal constraints caused by a scarcity of points on the ground or plane can be reduced. The new formula is defined as follows:

$$\min_{\bar{\varepsilon}_k^a}(\|\hat{\varepsilon}_k^a + h_a \bar{\varepsilon}_k^a\|^2_{P_k^{a^{-1}}} + \frac{m+n}{2n}\Sigma_{j=1}^{n}\left\|z_j^a + H_j^a \bar{\varepsilon}_k^a\right\|^2_{U_k^{a^{-1}}} + \frac{m+n}{2m}\Sigma_{j=n+1}^{n+m}\left\|z_j^a + H_j^a \bar{\varepsilon}_k^a\right\|^2_{U_k^{a^{-1}}}) \tag{24}$$

In the equation, $\|x\|_p^2 = x^T P x$. U_k^a is the covariance of the observed noise for the a-th state update in the frame k. n is the number of planar points in current frame k. m is the number of ground points in current frame k. According to references [33,34], the Kalman gain K_k^a of the k-th frame and the update error state $\bar{\varepsilon}_a^{a+1}$ of the a-th iteration in the frame k after dimensions are reduced are defined as follows:

$$K_k^a = (H U_k^a H + P^{-1})^{-1} H^T U_k^{a^{-1}} \tag{25}$$

$$\bar{\varepsilon}_a^{a+1} = -K_k^a z_k^a - (I - K_k^a H) h_a^{-1} \hat{\varepsilon}_k^a \tag{26}$$

If n is greater than m, $H = [H_1^a, \cdots, H_n^{a^T}, H_{n+1}^{a^T}, \cdots, H_{n+m}^{a^T}, \cdots, H_{2n}^{a^T}]^T$, $R = diag(R_1, \cdots, R_n, R_{n+1}, \cdots, R_{n+m}, \cdots, R_{2n})$, $z^a = [z_1^{a^T}, \cdots, z_n^a, z_{n+1}^{a^T}, \cdots, z_{n+m}^{a^T}, \cdots, z_{2n}^{a^T}]^T$. $[H_{n+m}^{a^T}, \cdots, H_{2n}^{a^T}]$, $[R_{n+m}, \cdots, R_{2n}]$, $[z_{n+m}^{a^T}, \cdots, z_{2n}^{a^T}]$ is the pseudo occupancies of ground constraints reusing high-quality ground points. If m is greater than n, $H = [H_1^a, \cdots, H_n^{a^T}, H_{n+1}^{a^T}, \cdots, H_m^{a^T}, \cdots, H_{2m}^{a^T}]^T$, $R = diag(R_1, \cdots, R_n, R_{n+1}, \cdots, R_m, \cdots, R_{2m})$, $z^a = [z_1^{a^T}, \cdots, z_n^a, z_{n+1}^{a^T}, \cdots, z_m^{a^T}, \cdots, z_{2m}^{a^T}]^T$. $[H_{n+1}^{a^T}, \cdots, H_m^{a^T}]$, $[R_{n+1}, \cdots, R_m]$, $[z_{n+1}^{a^T}, \cdots, z_m^{a^T}]$ is the pseudo occupancies of planar constraints reusing high-quality planar points. By combining the states of $\hat{x}_k^a$ and $\bar{\xi}_a^{a+1}$, the updated state $\hat{x}_k^{a+1}$ of time $a+1$ can be obtained. When the error between $\hat{x}_k^{a+1}$ and $\hat{x}_k^a$ is small enough, the iteration will be terminated. The final state $\bar{x}_k$ and covariance matrix $\bar{P}_k$ of the frame k are defined as follows:

$$\bar{x}_k = \hat{x}_k^{a+1} \tag{27}$$

$$\bar{P}_k = (I - KH)P \tag{28}$$

4. Relocation

To obtain the prior pose of UGV in the prior map, a relocation method is devised by combining detection and correction with the iterative closest point (ICP) method. This method is integrated into the front-end odometry to initialize the UGV's pose. The proposed method involves extracting keyframes from significant positions within the prior map, forming prior keyframes. Following the method described in reference [35], we generate descriptor matrices for the current frame and historical keyframes. By employing a two-stage matching approach, candidate relocalization keyframes for the current frame are identified through the column vectors formed using ring key encoding information. Rotation and translation offsets are calculated through the row summation of these column vectors. Subsequently, the descriptor of the current frame is matched with the descriptor matrix of the candidate relocalization keyframes assigned with the computed offsets, resulting in the identification of the relocalization keyframe. A rough correction is then applied to the UGV's current attitude using the calculated angle offset. Finally, an accurate UGV pose within the prior map is obtained by performing ICP point cloud registration on the corrected initial frame and the relocation keyframe. To address the limitations associated with traditional point-to-point, point-to-plane, and linear least-squares optimization methods for point-to-plane ICP [36–38], which involve multiple iterations, time consumption, and are prone to overfitting, the generalized ICP (G-ICP) [39] is employed for point cloud registration. This method reduces the number of iterations while improving matching accuracy.

5. Prior Map Maintenance

The observed equation requires identifying the nearest neighboring points in the prior map to perform residual fitting. To achieve a fast and stable KNN search on UGVs with limited RAM and CPU capabilities, the prior map is maintained by combining the proposed three-layer voxels and the ikd-tree. This method ensures efficient and reliable execution, as shown in Figure 3.

First, the prior map is voxelized based on the position of the world coordinate system and stored in ROM accordingly. The voxel files are named $[sign(x), 100|x| + |y|, sign(y)]$ according to the voxel positions in a prior map coordinate system, forming the first-layer ROM voxels. By retrieving file names containing position information, the voxels near the current position, obtained from the relocation, are extracted from the system ROM into the corresponding voxel container in RAM, forming the second-layer RAM voxels. The point clouds within the nearest nine RAM voxel containers to the current position are loaded into the ikd-tree, forming the third-layer ikd-tree voxels. This method reduces the number of point clouds in RAM and the size of the ikd-tree, effectively minimizing RAM consumption, and enabling UGV to perform real-time localization using a large prior map. Once the initialization of prior map maintenance is completed, the map is continuously updated in real time. Real-time updates are performed on the second and third layers of voxels using the 6DOF state estimation output from the odometry. When the LiDAR center moves to another voxel range, this method removes the point clouds outside the ikd-tree voxel range of the LiDAR center in the ikd-tree, and inserts the point clouds of the newly added ikd-tree voxels from the RAM voxels into the ikd-tree. Simultaneously, the RAM containers outside the RAM voxel range of the LiDAR center at the corresponding position are emptied. The newly added RAM voxels are extracted from the corresponding ROM voxel into the corresponding RAM voxel container at the respective position. To ensure independent operation and avoid interference between the map maintenance modules of the second and third layers, as well as to prevent any impact on the real-time KNN search of the odometry, this method runs the RAM voxel maintenance, ikd-tree voxel maintenance, and odometry in three separate threads. This setup improves the system's real-time performance. Additionally, by setting RAM voxels and ikd-tree voxels separately, the real-time performance of inserting voxel point clouds into the ikd-tree is enhanced, thereby accelerating the operating cycle of the odometry. The process of prior map maintenance is summarized in Algorithm 2.

Algorithm 2: Map maintenance

Input : G from prior map points, $\overline{X}$ from relocation result, X from location information of odometer

Voxelize points G according to its position with voxel edge length l;
Create a RAM voxel point cloud container V_{xy}. V_{xy} is established to store the point clouds within the range of x-axis [x*l, (x + 1)*l] and y-axis [yl, (y + 1)*l];
put the corresponding voxel points into the corresponding voxel container V_{xy};
Put the point cloud in each voxel container v into the ROM based on location information $[sign(x), 100|x| + |y|, sign(y)]$;
Empty point clouds in all voxel containers V_{xy};
Based on the location information $[sign(x), 100|x| + |y|, sign(y)]$, put 7×7 ROM voxels centered around $\overline{X}$ into the corresponding 7×7 voxel containe V_{xy};
Construct an ikd-tree from the points in the 3×3 voxel container vs. centered on $\overline{X}$;
Threadspawn(RAM maintenance,X);
Threadspawn(ikd-tree maintenance,X);
Function RAM maintenance (X)
    while (*the main program thread did not end*) **do**
        if (*X is not within the center voxel*) **then**
            Clear non empty V_{xy} outside of 7×7 V_{xy} centered on X;
            Add points of corresponding ROM voxels to the empty V_{xy} in 7×7 V_{xy} centered on X;
        end
    end
end
Function ikd-tree maintenance (X)
    while (*the main program thread did not end*) **do**
        if (*X is not within the center voxel*) **then**
            Delete points in ikd-tree outside the range of 3×3 V_{xy} centered on X;
            Insert points into ikd-tree from 3×3 V_{xy} ranges centered on X that have not been inserted in to ikd-tree ;
        end
    end
end

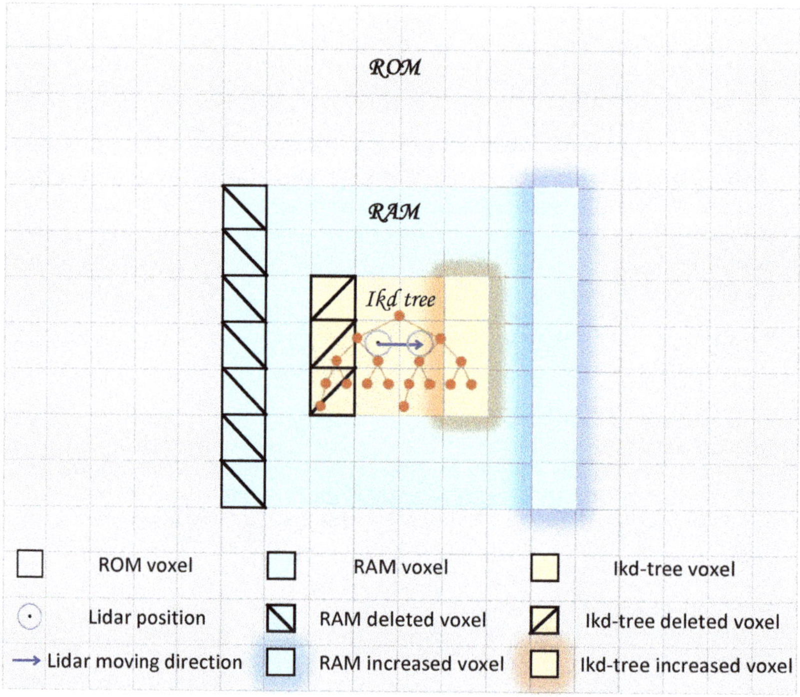

Figure 3. Map maintenance diagram.

6. Experiment and Analysis

In this experimental section, we describe our ablation experiments, which we conducted to analyze the accuracy and efficiency of each proposed component separately and compare them with advanced SLAM methods. Adhering to the principles of fairness and impartiality, all comparative experiments were conducted in the same environment, and the experimental data were both accurate and reliable. The experimental hardware platform consisted of a laptop, a wheeled UGV, a Velodyne VLP-16 LiDAR, and an Xsens MTI-100 IMU, as depicted in Figure 4. The Velodyne VLP-16 LiDAR mounted on the experimental platform had a measurement range of 100 m, a vertical viewing angle ranging from +15° to −15°, a horizontal viewing angle of 360°, and vertical and horizontal resolutions of 2° and 0.1° to 0.4°, respectively. It offered a measurement accuracy of ±3 cm and operated at a frequency of 20 Hz. The Xsens MTI-100 IMU incorporated a built-in gyroscope, accelerometer, and magnetometer, with a sampling frequency of up to 10 KHz. The laptop utilized in the experiments featured an Intel (R) Core (TM) i7-12700H processor, operating at a frequency of 2.3 GHz, with 16 GB of RAM. The software system ran on Ubuntu 18.04 and employed ROS Melodic [40] as the middleware. The algorithm was implemented in C++.

To evaluate the performance of each module, experiments were conducted using self-collected indoor and outdoor closed-loop datasets from an industrial park, as well as publicly available KITTI datasets. The self-collected datasets were acquired within a standardized industrial park environment. The outdoor dataset encompasses notable structures within the industrial park, including industrial factories, vegetation, parking lots, office buildings, and roads. The trajectory length of this dataset is approximately 308 m. The indoor dataset consists of distinctive structures such as columns, beams, walls, and industrial equipment, with a trajectory length of approximately 202 m. It is important to note that both the indoor and outdoor datasets were meticulously designed to commence and conclude at the same position during data collection. To bolster the credibility of the experiments, the odometry performance was also assessed using the publicly available

KITTI dataset. This dataset was generated by equipping a car with various sensors and driving along well-structured roads. The car's setup included a Velodyne HDL-64E LiDAR, Oxts R3003 IMU, two grayscale cameras, two color cameras, four optical lenses, and a GPS navigation system.

Figure 4. The UGV for collecting indoor and outdoor datasets.

To assess the accuracy and efficiency of the proposed method, trajectory maps for both the proposed method and other existing methods are generated. Comparative experiments are conducted by calculating the time consumption of each module. The corresponding errors, including trajectory error, absolute pose error, and closed-loop error, are computed. Additionally, to evaluate the global consistency of each method's trajectory, the root mean square error (RMSE) is calculated for each trajectory. Moreover, time consumption curves for each component are plotted to facilitate a more intuitive analysis of the method's efficiency.

7. Analysis of Mapping Results without Prior Maps

To evaluate the efficiency and accuracy of the proposed feature extraction method and odometry method, a horizontal comparison was performed. The proposed method was compared with FAST-LIO2 and FAST-LIO2 combined with the curvature feature extraction method. To ensure reliability and credibility, the experiments were conducted using the KITTI dataset. The evaluation tool EVO [41] was employed to visualize the experimental trajectories and ground truth trajectory, and to compute the absolute pose error for accuracy analysis. Furthermore, the number of feature points per frame and the time consumption curves of each component were plotted to analyze the efficiency.

7.1. Analysis of Mapping Accuracy without Prior Maps

Experiments were conducted on the KITTI dataset using three different methods: FAST-LIO2, FAST-LIO2 combined with curvature feature extraction, and the proposed odometry method combined with the bidirectional projection plane slope difference filter feature extraction. The results and trajectories obtained from these experiments are presented in Figures 5 and 6, respectively.

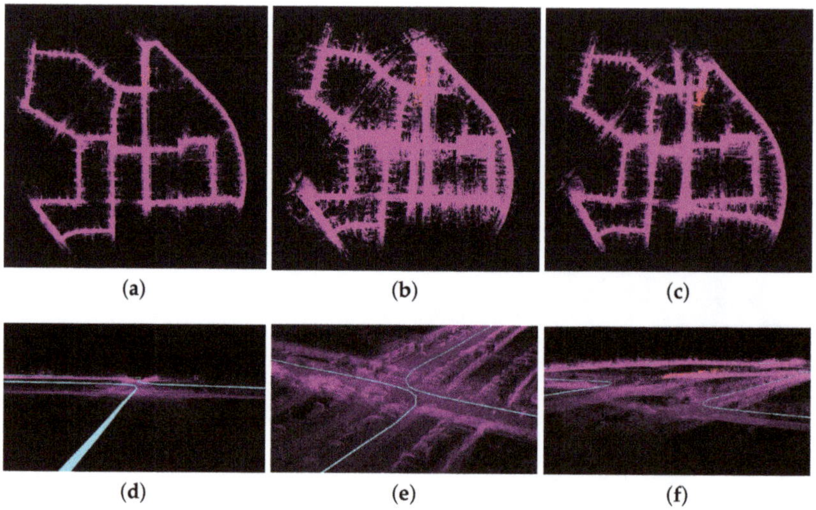

Figure 5. The results of three methods on the KITTI dataset. (**a**) is the result of the combination of proposed odometry method and bidirectional projection plane slope difference filter method. (**b**) is the result of FAST-LIO2. (**c**) is the result of the combination of FAST-LIO2 and the curvature method. (**d–f**) are the corresponding local results.

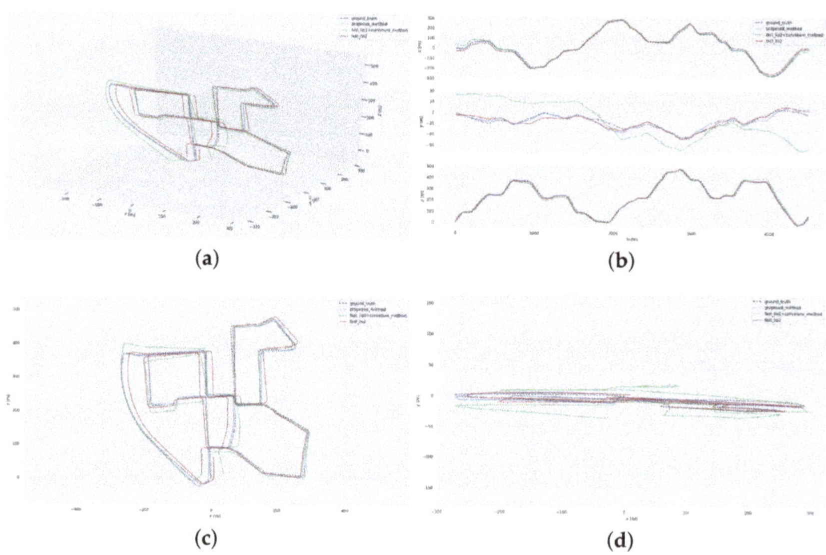

Figure 6. The trajectory map generated by EVO. (**a**) is a 3D trajectory. (**b**) is a three-axis trajectory. (**c**) is a vertical planar trajectory. (**d**) is a horizontal planar trajectory. The black trajectory represents the ground truth trajectory. The blue trajectory corresponds to the trajectory obtained using the proposed method. The green trajectory corresponds to the trajectory obtained using FAST-LIO2 combined with curvature-based feature extraction. The red trajectory corresponds to the trajectory obtained using FAST-LIO2.

Figure 5 depicts the top view and local details of the running results obtained from the three methods. The top view (a, b, c) and the local detail view (d, e, f) of Figure 5 provide a visual comparison of the mapping effects of the three methods. Both FSAT-

LIO2 and FAST-LIO2 combined with the curvature feature extraction method exhibit significant drift, particularly at turns and certain positions where feature degradation occurs in space. Notably, the FAST-LIO2 combined with the curvature feature extraction method demonstrates severe vertical height drift. In contrast, the proposed odometry method, combined with the proposed feature extraction method, effectively suppresses drift during turns, positions of feature degradation, and vertical heights. Furthermore, it exhibits minimal translocation at revisited positions. The aforementioned effect can be attributed to the curvature feature extraction method's inability to accurately extract stable ground points, leading to insufficient constraints on vertical height and significant vertical height drift. Moreover, the curvature extraction method fails to filter out noise and unstable points during plane feature extraction, resulting in insufficient lateral constraints and some degree of drift in the lateral direction. The method that does not perform feature extraction utilizes a large number of original points for residual calculation. While its vertical height constraints are superior to the curvature method, it fails to filter out unstable points and noise within the point clouds. When the proportion of noise and unstable points cannot be ignored, it affects the effectiveness of residual fitting, resulting in significant drift. In contrast, the proposed method strengthens ground constraints by filtering ground points near the ground and extracting high-quality ground and planar points from the far end using the bidirectional projection plane slope difference filter. It incorporates ground constraints into the back-end optimization formulation, imposing strict constraints on both vertical height and horizontal plane, which yields outstanding mapping results. Figure 6 presents the 3D trajectory, horizontal planar trajectory, vertical planar trajectory, and three-axis trajectory generated by the EVO for the three methods. Comparing the accuracy of odometry among the three methods, the proposed odometry method combined with the proposed feature extraction method demonstrates the trajectory that most closely aligns with the true trajectory. Additionally, it exhibits the best closed-loop effect at the end. In contrast, FSAT-LIO2 and FAST-LIO2 combined with the curvature feature extraction method result in significant translocation at the end. Furthermore, FAST-LIO2 combined with the curvature feature extraction method exhibits severe vertical height drift. The results depicted in Figure 6 validate the accuracy and efficiency of the proposed method, confirming the advantages and feasibility of both the proposed feature extraction method and odometry method.

To assess the global consistency of the three methods and quantitatively evaluate their errors, we calculate the absolute pose error (APE) between their respective trajectories and the ground truth trajectory. The calculated APE values are presented in Table 1.

Table 1. Absolute pose error (APE) of the three methods.

Methods/APE	Max (m)	Mean (m)	Median (m)	Min (m)	Rmse (m)	Std (m)
Proposed method	25.645	11.785	11.232	1.089	12.787	4.968
FAST-LIO2	33.429	16.553	16.991	1.222	18.319	7.847
FAST-LIO2 + curvature	48.196	22.783	22.108	6.580	24.707	9.559

When testing on the KITTI dataset, the root mean square error (RMSE) of the absolute pose error obtained using the proposed odometry method combined with the proposed feature extraction method exhibits a 48.24% increase compared to FAST-LIO2 combined with the curvature feature extraction method. Furthermore, the RMSE of the proposed method is 30.19% higher than that of FAST-LIO2. These results demonstrate that the proposed odometry method combined with the proposed feature extraction method achieves superior performance.

7.2. Analysis of Mapping Efficiency without Prior Maps

This paper presents a comprehensive analysis of three methods by plotting the number of feature points, the time consumption for feature extraction, and the time consumption for odometry in each frame. Mean values are calculated to provide a summary of the results. Through a visual comparison of feature point counts and time consumption across different method components, the impact of the proposed feature extraction method and odometry method on time consumption and computational complexity is analyzed. The curves depicting the number of feature points and time consumption for each component are illustrated in Figure 7. Furthermore, Table 2 presents the average number of feature points across all frames and the average time consumption for each component.

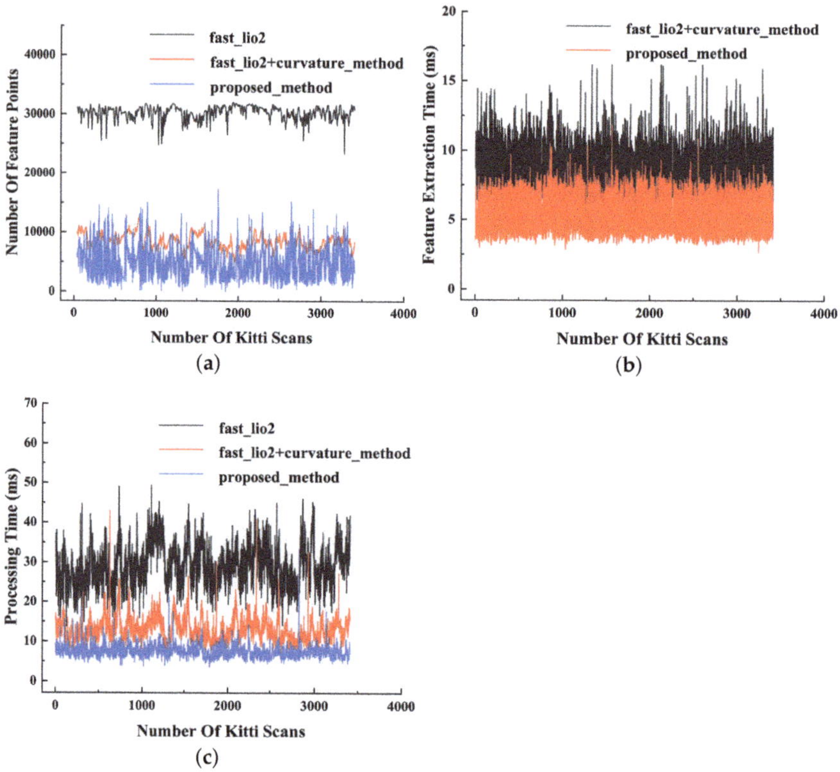

Figure 7. The number of feature points and the curve of time consumption for each part of the KITTI dataset. (**a**) is the curve of the number of feature points of each frame. (**b**) is the curve of the feature extraction time of each frame. (**c**) is the curve of the odometry processing time of each frame.

Table 2. The mean number of feature points and the time consumption means of each part on the KITTI dataset.

Method	Proposed Method	FAST-LIO2 + Curvature	FAST-LIO2
Number of feature points	5155	8549	30321
Feature extraction time (ms)	5.423	8.483	0
Odometer processing time (ms)	7.553	13.164	29.980
Total time(ms)	12.976	21.647	29.980

The feature point count, feature extraction time, and odometry time of the three methods per frame are compared in Figure 7a–c. The average time consumption and average number of feature points are presented in Table 2. To analyze the improvement in odometry efficiency offered by the proposed method, a comparison is made between the time consumption and calculated amount of the three methods. The proposed feature extraction method reduces the average number of feature points extracted per frame by 39.70% compared to the curvature feature extraction method. Furthermore, it achieves an 83.01% reduction in feature points compared to the method without feature extraction. Consequently, the proposed method achieves higher odometry accuracy with fewer feature points. In terms of time consumption, the proposed feature extraction method reduces the average feature extraction time by 36.07% compared to the curvature feature extraction method. When the proposed feature extraction method is combined with the proposed odometry method, it reduces the average odometry time by 42.62% compared to FAST-LIO2 combined with the curvature feature extraction method. Additionally, the proposed method reduces the average odometry time by 74.81% compared to FAST-LIO2. This reduction in time consumption is due to the proposed method's projection of 3D space onto a 2D plane, which requires less calculation for feature extraction compared to 3D space feature extraction methods. Overall, the proposed method reduces the overall time consumption per frame by 40.06% compared to FAST-LIO2 combined with the curvature method and achieves a 56.72% reduction compared to FAST-LIO2 alone. These findings demonstrate that the proposed method significantly improves the efficiency of the overall system, and therefore the real-time requirements for odometry processing are met.

8. Analysis of Localization Results Using Prior Maps

To assess the influence of prior maps on the UGV's localization accuracy and the impact of the proposed prior map maintenance method on KNN search speed and RAM consumption, closed-loop experiments are conducted on self-collected datasets acquired from indoor and outdoor environments. In the case of utilizing prior maps, horizontal comparisons are performed using different methods. The located accuracy of each method is evaluated by analyzing the closed-loop errors along the three axes and the overall performance. The HTOP [42] is employed to calculate the average RAM consumption of the proposed map maintenance method and the alternative method, as well as to determine the average KNN search speed for both methods. Furthermore, this paper presents the time consumption means of individual components of odometry to analyze the efficiency and practicality of the proposed method.

8.1. Analysis of Localization Accuracy Using Prior Maps

Closed-loop experiments were conducted on self-collected indoor and outdoor datasets using FAST-LIO2 combined with the proposed prior map maintenance method, and proposed odometry method combined with the proposed prior map maintenance method and proposed feature extraction method. When utilizing the prior map for localization, the observed angle deviation of the closed-loop position in the outdoor results, as well as the angle and height deviation of the closed-loop position in the indoor results, are attributed to discrepancies between the prior map coordinate system and the odometry coordinate system. However, the computed closed-loop error values are accurate. Figure 8 illustrates the indoor and outdoor results, along with the results of the local closed-loop position. The closed-loop errors for all three axes and the overall errors are presented in Table 3.

Table 3. Indoor and outdoor closed-loop error of different methods.

Method/Closed-Loop Error	Δx(m)	Δy(m)	Δz (m)	Total (m)
FAST-LIO2 + Prior map (Outdoor)	0.018	0.092	0.025	0.097
Proposed method + Prior map method (Outdoor)	0.030	0.032	0.0003	0.044
FAST-LIO2 + Prior map (Indoor)	0.018	0.022	0.004	0.029
Proposed method + Prior map method (Indoor)	0.017	0.012	0.0004	0.021

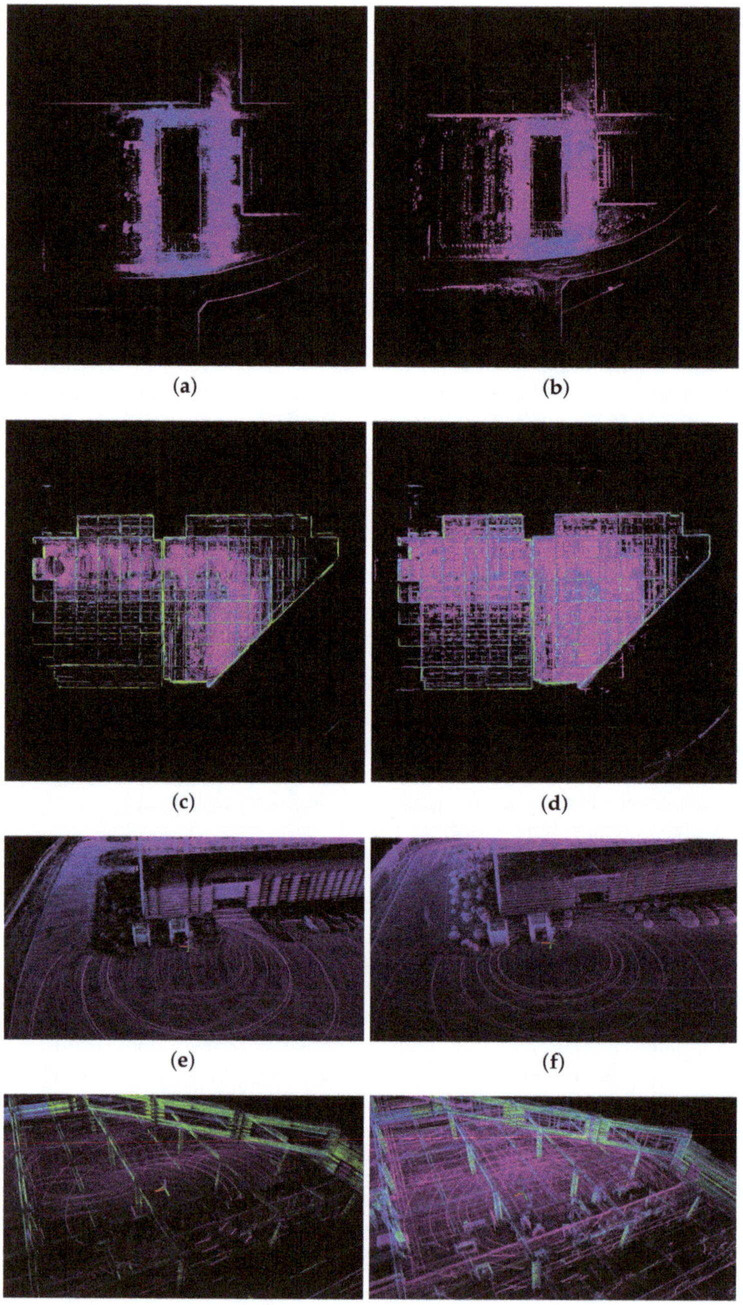

Figure 8. The results of combining different methods with prior maps on indoor and outdoor self-collected datasets. (**a**,**c**,**e**,**g**) are the overall and local figures of the indoor and outdoor results of the proposed odometry method combined with the proposed prior map maintenance method and proposed feature extraction method. (**b**,**d**,**f**,**h**) are the overall and local figures of the indoor and outdoor results of FAST-LIO2 combined with the proposed prior map maintenance method.

The comparative analysis of the intuitive mapping effect and closed-loop error between the two methods is presented in the outdoor and indoor results, as well as in Figure 8 and Table 3. In the outdoor test scenario, when FAST-LIO2 is combined with the proposed prior map maintenance method, it is evident that the initial and final positions overlap, showcasing an outstanding closed-loop effect. The total closed-loop error measures 0.097 m, corresponding to a proportion of total mileage of 0.0315%. This achievement demonstrates a centimeter-level localization accuracy, validating the effectiveness of utilizing prior maps for localization. Moreover, when the proposed odometry method is combined with the proposed prior map maintenance method and the proposed feature extraction method, the initial and final positions also exhibit overlapping characteristics. The utilization of the proposed feature extraction method facilitates the filtration of unstable point clouds such as grass and cars, resulting in the extraction of high-quality ground points. Furthermore, it is observed that the extracted planar points exhibit clearer lines and boundaries compared to the other method, indicating a high quality of extracted planar points. The total closed-loop error is reduced to 0.044 m, with a proportion of total mileage of 0.0143%. This achievement signifies improved localization accuracy, with a smaller number of feature points. Remarkably, the localization accuracy along the Z-axis reaches the millimeter level. In the indoor test scenario, the overall closed-loop error for the combination of FAST-LIO2 and the proposed prior map maintenance method is measured at 0.029 m, with a proportion of total mileage at 0.0144%. The total closed-loop error for the proposed odometry method, combined with the proposed prior map maintenance method and the proposed feature extraction method, is reduced to 0.021 m, corresponding to a proportion of total mileage of 0.0104%. In both outdoor and indoor tests, the proposed odometry method, combined with the proposed prior map maintenance method and the proposed feature extraction method, achieves an improvement of 54.639% and 38.095% in closed-loop accuracy compared to the combination of FAST-LIO2 and the proposed prior map maintenance method. These findings highlight that localization using prior maps meets the required accuracy standards for UGVs in industrial parks. Additionally, the proposed method significantly enhances the localization accuracy of UGVs within industrial park environments.

8.2. Analysis of Localization Efficiency Using Prior Maps

To evaluate the localization efficiency of the proposed UGV located method in industrial parks, this paper computes the average time consumption for each segment of both indoor and outdoor datasets. The results for the average time consumption of each segment are presented in Table 4.

Table 4. The time consumption means of each part on the indoor and outdoor datasets.

Method	Scene	FAST-LIO2 + Proposed Prior Map Maintenance Method	Proposed Method + Proposed Prior Map Maintenance Method
Feature extraction time (ms)	Outdoor	0	0.963
	Indoor	0	1.345
Odometer processing time (ms)	Outdoor	10.561	6.426
	Indoor	12.634	6.964
Total time (ms)	Outdoor	10.561	7.389
	Indoor	12.634	8.309

The average processing time for the proposed combination of FAST-LIO2 with the prior map maintenance method is 11.598 ms, and for the proposed odometry method with the prior map maintenance method and feature extraction method is 7.849 ms. The total processing speed of the proposed odometry method, when combined with the prior map maintenance method and feature extraction method, exhibits a 32.33% improvement compared to the combination of FAST-LIO2 with the prior map maintenance method.

It can be observed that the proposed odometry method, in conjunction with the prior map maintenance method and feature extraction method, satisfies the requirements for UGV odometry processing speed in industrial parks, while ensuring sufficient localization accuracy. Therefore, it is more suitable for real-time UGV operations in industrial park environments.

8.3. Analysis of Effectiveness of Map Maintenance

To assess the practicality and validity of the map maintenance module, we employ the proposed map maintenance method alongside the approach of incorporating all point clouds from prior maps into the ikd-tree. This enables us to evaluate the associated RAM usage and KNN search speed. The computed results are presented in Table 5.

Table 5. Test results of map maintenance method.

Method	Average RAM Consumption (GB)	Average RAM Consumption Percentage	KNN Search Average Time (ms)	Number of Points in Ikd-Tree	Map Range in Ikd-Tree (m)	Prior Map Range (m)
Proposed map maintenance method	15	98%	0.00316	2,335,000	240× 240	1320× 1320
ikd-tree	5.2	34%	0.00348	12,450,000	1320× 1320	1320× 1320

The prior map area utilized is 1320 m $\times$ 1320 m, consisting of approximately 12.45 million point clouds. The same number of point clouds, 12.45 million, are added to the ikd-tree. On average, the RAM usage amounts to 15 GB, occupying 98% of the total RAM. It is evident that the method of including all prior map point clouds in the ikd-tree exceeds the host's RAM capacity, resulting in lag and data loss. Consequently, the system becomes unable to run when handling larger prior maps of industrial parks. In contrast, the proposed map maintenance method significantly reduces RAM consumption on the host. Specifically, the number of point clouds in the ikd-tree is reduced to 2.335 million, resulting in an average RAM usage of 5.2 GB, occupying 34% of the total RAM. This observation demonstrates the effectiveness of the proposed map maintenance method in reducing RAM consumption, thereby enabling the loading and maintenance of prior maps for large industrial parks. Since the search speed of the ikd-tree relies on the number of layers within the tree structure, and the relationship between the number of layers and the number of point clouds follows a power-of-two pattern, the KNN search speed of the proposed map maintenance method is only 10% higher than that of the method that includes all point clouds in the ikd-tree.

9. Conclusions

This paper presents a method for real-time UGV localization in industrial parks utilizing prior maps. This paper presents a novel feature point extraction method, improves the back-end optimization function of FAST-LIO2, and introduces a new prior map maintenance method. To enhance the quality and efficiency of feature extraction, a novel feature extraction method based on the bidirectional projection plane slope difference filter is proposed. This method improves extraction speed while maintaining extraction accuracy. It enables the separate extraction of ground and planar points, as well as edge points, using a unified method. The back-end optimization incorporates ground constraints through the proposed pseudo occupancy method, enhancing the vertical and horizontal constraint effects of ground and planar points. Experimental results on the KITTI dataset demonstrate that the proposed method significantly improves feature extraction and odometry accuracy and efficiency. To enable KNN search using prior maps with limited RAM on UGV, a novel prior map maintenance method is introduced, combining three-layer voxels and the ikd-tree. This method achieves efficient RAM consumption during prior map maintenance.

The experimental results confirm its effectiveness in reducing RAM consumption. The proposed method achieves centimeter-level localization accuracy, meeting the requirements for accurate and efficient UGV localization when utilizing prior maps.

In future work, we will focus on the regular updating of the prior map. When there are changes in the scene information, we will incorporate newly acquired point clouds of added objects and remove point clouds of disappeared objects from the previous prior map. The prior map will undergo long-term maintenance to ensure its relevance and accuracy. Furthermore, in order to improve the accuracy of relocation detection, the generation method of the descriptor matrix will be enhanced. The original descriptor matrix only contained the highest point information of the loop key, and will be improved to more comprehensively represent the point cloud information within the loop key, thereby achieving more precise relocation detection and reducing false positives and false negatives.

Author Contributions: Conceptualization, F.L. and X.L.; methodology, F.L.; software, F.L.; validation, F.L., Y.C. and M.L.; formal analysis, F.L. and Z.L.; investigation, F.L.; resources, Z.L., F.Z. and M.L.; data curation, F.L.; writing—original draft preparation, F.L.; writing—review and editing, F.L.; visualization, F.L.; supervision, F.Z.; project administration, Z.L.; funding acquisition, Z.L. All authors have read and agreed to the published version of the manuscript.

Funding: This work was partially supported by Shenyang University of Technology: 2023JH2/101300225.

Institutional Review Board Statement: Not applicable.

Informed Consent Statement: Not applicable.

Data Availability Statement: The data presented in this study is not publicly available at this time but may be obtained from the authors upon reasonable request.

Conflicts of Interest: The authors declare no conflict of interest.

References

1. Pan, Y.; Wu, N.; Qu, T.; Li, P.; Zhang, K.; Guo, H. Digital-twin-driven production logistics synchronization system for vehicle routing problems with pick-up and delivery in industrial park. *Int. J. Comput. Integr. Manuf.* **2021**, *34*, 814–828. [CrossRef]
2. Nath, S.V.; Dunkin, A.; Chowdhary, M.; Patel, N. *Industrial Digital Transformation: Accelerate Digital Transformation with Business Optimization, AI, and Industry 4.0*; Packt Publishing Ltd.: Birmingham, UK, 2020.
3. Sivaneri, V.O.; Gross, J.N. UGV-to-UAV cooperative ranging for robust navigation in GNSS-challenged environments. *Aerosp. Sci. Technol.* **2017**, *71*, 245–255. [CrossRef]
4. Guérin, F.; Guinand, F.; Brethé, J.F.; Pelvillain, H. UAV-UGV cooperation for objects transportation in an industrial area. In Proceedings of the 2015 IEEE International Conference on Industrial Technology (ICIT), Seville, Spain, 17–19 March 2015; pp. 547–552.
5. Garzón, M.; Valente, J.; Zapata, D.; Barrientos, A. An aerial–ground robotic system for navigation and obstacle mapping in large outdoor areas. *Sensors* **2013**, *13*, 1247–1267. [CrossRef]
6. Taheri, H.; Xia, Z.C. SLAM; definition and evolution. *Eng. Appl. Artif. Intell.* **2021**, *97*, 104032. [CrossRef]
7. Macario Barros, A.; Michel, M.; Moline, Y.; Corre, G.; Carrel, F. A comprehensive survey of visual slam algorithms. *Robotics* **2022**, *11*, 24. [CrossRef]
8. Huang, L. Review on LiDAR-based SLAM techniques. In Proceedings of the 2021 International Conference on Signal Processing and Machine Learning (CONF-SPML), Stanford, CA, USA, 14 November 2021; pp. 163–168.
9. Xu, X.; Zhang, L.; Yang, J.; Cao, C.; Wang, W.; Ran, Y.; Tan, Z.; Luo, M. A review of multi-sensor fusion slam systems based on 3D LIDAR. *Remote Sens.* **2022**, *14*, 2835. [CrossRef]
10. Zhang, J.; Singh, S. LOAM: Lidar odometry and mapping in real-time. In Proceedings of the Robotics: Science and Systems, Berkeley, CA, USA, 12–16 July 2014; Volume 2, pp. 1–9.
11. Geiger, A.; Lenz, P.; Urtasun, R. Are we ready for autonomous driving? the kitti vision benchmark suite. In Proceedings of the 2012 IEEE Conference On computer Vision and Pattern Recognition, Providence, RI, USA, 16–21 June 2012; pp. 3354–3361.
12. Shan, T.; Englot, B. Lego-loam: Lightweight and ground-optimized lidar odometry and mapping on variable terrain. In Proceedings of the 2018 IEEE/RSJ International Conference on Intelligent Robots and Systems (IROS), Madrid, Spain, 1–5 October 2018; pp. 4758–4765.
13. Lin, J.; Zhang, F. Loam livox: A fast, robust, high-precision LiDAR odometry and mapping package for LiDARs of small FoV. In Proceedings of the 2020 IEEE International Conference on Robotics and Automation (ICRA), Paris, France, 31 May–31 August 2020; pp. 3126–3131.

14. Jiao, J.; Ye, H.; Zhu, Y.; Liu, M. Robust odometry and mapping for multi-lidar systems with online extrinsic calibration. *IEEE Trans. Robot.* **2021**, *38*, 351–371. [CrossRef]
15. Ye, H.; Chen, Y.; Liu, M. Tightly coupled 3d lidar inertial odometry and mapping. In Proceedings of the 2019 International Conference on Robotics and Automation (ICRA), Montreal, QC, Canada, 20–24 May 2019; pp. 3144–3150.
16. Qin, C.; Ye, H.; Pranata, C.E.; Han, J.; Zhang, S.; Liu, M. Lins: A lidar-inertial state estimator for robust and efficient navigation. In Proceedings of the 2020 IEEE International Conference on Robotics and Automation (ICRA), Paris, France, 31 May–31 August 2020; pp. 8899–8906.
17. Shan, T.; Englot, B.; Meyers, D.; Wang, W.; Ratti, C.; Rus, D. Lio-sam: Tightly coupled lidar inertial odometry via smoothing and mapping. In Proceedings of the 2020 IEEE/RSJ International Conference on Intelligent Robots and Systems (IROS), Las Vegas, NV, USA, 24 October 2020–24 January 2021; pp. 5135–5142.
18. Xu, W.; Zhang, F. Fast-lio: A fast, robust lidar-inertial odometry package by tightly coupled iterated kalman filter. *IEEE Robot. Autom. Lett.* **2021**, *6*, 3317–3324. [CrossRef]
19. Xu, W.; Cai, Y.; He, D.; Lin, J.; Zhang, F. Fast-lio2: Fast direct lidar-inertial odometry. *IEEE Trans. Robot.* **2022**, *38*, 2053–2073. [CrossRef]
20. Wu, D.; Zhong, X.; Peng, X.; Hu, H.; Liu, Q. Multimodal Information Fusion for High-Robustness and Low-Drift State Estimation of UGVs in Diverse Scenes. *IEEE Trans. Instrum. Meas.* **2022**, *71*, 1–15. [CrossRef]
21. Asadi, K.; Suresh, A.K.; Ender, A.; Gotad, S.; Maniyar, S.; Anand, S.; Noghabaei, M.; Han, K.; Lobaton, E.; Wu, T. An integrated UGV-UAV system for construction site data collection. *Autom. Constr.* **2020**, *112*, 103068. [CrossRef]
22. Liu, K.; Ou, H. A light-weight lidar-inertial slam system with high efficiency and loop closure detection capacity. In Proceedings of the 2022 International Conference on Advanced Robotics and Mechatronics (ICARM), Guilin, China, 9–11 July 2022; pp. 284–289.
23. Liu, Z.; Chen, H.; Di, H.; Tao, Y.; Gong, J.; Xiong, G.; Qi, J. Real-time 6d lidar slam in large scale natural terrains for ugv. In Proceedings of the 2018 IEEE Intelligent Vehicles Symposium (IV), Changshu, China, 26–30 June 2018; pp. 662–667.
24. Hamieh, I.; Myers, R.; Rahman, T. Construction of autonomous driving maps employing lidar odometry. In Proceedings of the 2019 IEEE Canadian Conference of Electrical and Computer Engineering (CCECE), Edmonton, AB, Canada, 5–8 May 2019; pp. 1–4.
25. Alharthy, A.; Bethel, J. Heuristic filtering and 3D feature extraction from LIDAR data. *Int. Arch. Photogramm. Remote Sens. Spat. Inf. Sci.* **2002**, *34*, 29–34.
26. Guo, S.; Rong, Z.; Wang, S.; Wu, Y. A LiDAR SLAM with PCA-based feature extraction and two-stage matching. *IEEE Trans. Instrum. Meas.* **2022**, *71*, 1–11. [CrossRef]
27. Li, Y.; Olson, E.B. Extracting general-purpose features from LIDAR data. In Proceedings of the 2010 IEEE International Conference on Robotics and Automation, Anchorage, AK, USA, 3–7 May 2010; pp. 1388–1393.
28. Serafin, J.; Olson, E.; Grisetti, G. Fast and robust 3d feature extraction from sparse point clouds. In Proceedings of the 2016 IEEE/RSJ International Conference on Intelligent Robots and Systems (IROS), Daejeon, Korea, 9–14 October 2016; pp. 4105–4112.
29. Egger, P.; Borges, P.V.; Catt, G.; Pfrunder, A.; Siegwart, R.; Dubé, R. Posemap: Lifelong, multi-environment 3d lidar localization. In Proceedings of the 2018 IEEE/RSJ International Conference on Intelligent Robots and Systems (IROS), Madrid, Spain, 1–5 October 2018; pp. 3430–3437.
30. Li, Y.; Olson, E.B. Structure tensors for general purpose LIDAR feature extraction. In Proceedings of the 2011 IEEE International Conference on Robotics and Automation, Shanghai, China, 9–13 May 2011; pp. 1869–1874.
31. Rusu, R.B.; Marton, Z.C.; Blodow, N.; Beetz, M. Learning informative point classes for the acquisition of object model maps. In Proceedings of the 2008 10th International Conference on Control, Automation, Robotics and Vision, Hanoi, Vietnam, 17–20 December 2008; pp. 643–650.
32. Rusu, R.B.; Bradski, G.; Thibaux, R.; Hsu, J. Fast 3d recognition and pose using the viewpoint feature histogram. In Proceedings of the 2010 IEEE/RSJ International Conference on Intelligent Robots and Systems, Taipei, Taiwan, 18–22 October 2010; pp. 2155–2162.
33. Raitoharju, M.; Piché, R. On computational complexity reduction methods for Kalman filter extensions. *IEEE Aerosp. Electron. Syst. Mag.* **2019**, *34*, 2–19. [CrossRef]
34. Higham, N.J. *Accuracy and Stability of Numerical Algorithms*; SIAM: Philadelphia, PA, USA, 2002.
35. Kim, G.; Kim, A. Scan context: Egocentric spatial descriptor for place recognition within 3d point cloud map. In Proceedings of the 2018 IEEE/RSJ International Conference on Intelligent Robots and Systems (IROS), Madrid, Spain, 1–5 October 2018; pp. 4802–4809.
36. Besl, P.J.; McKay, N.D. Method for registration of 3-D shapes. In *Proceedings of the Sensor Fusion IV: Control Paradigms and Data Structures*; SPIE: Bellingham, WA, USA, 1992; Volume 1611, pp. 586–606.
37. Chen, Y.; Medioni, G. Object modelling by registration of multiple range images. *Image Vis. Comput.* **1992**, *10*, 145–155. [CrossRef]
38. Low, K.L. Linear least-squares optimization for point-to-plane icp surface registration. *Chapel Hill Univ. North Carol.* **2004**, *4*, 1–3.
39. Segal, A.; Haehnel, D.; Thrun, S. Generalized-icp. In Proceedings of the Robotics: Science and Systems, Seattle, WA, 28 June–1 July 2009; Volume 2, p. 435.

40. Będkowski, J.; Pełka, M.; Majek, K.; Fitri, T.; Naruniec, J. Open source robotic 3D mapping framework with ROS-robot operating system, PCL-point cloud library and cloud compare. In Proceedings of the 2015 International Conference on Electrical Engineering and Informatics (ICEEI), Denpasar, Indonesia, 10–11 August 2015; pp. 644–649.
41. Sanfourche, M.; Vittori, V.; Le Besnerais, G. eVO: A realtime embedded stereo odometry for MAV applications. In Proceedings of the 2013 IEEE/RSJ International Conference on Intelligent Robots and Systems, Tokyo, Japan, 3–7 November 2013; pp. 2107–2114.
42. Muhammad, H. Htop—An Interactive Process Viewer for Linux. 2015. Available online: http://hisham.hm/htop/ (accessed on 22 May 2015).

sensors

MDPI

Article

Multi-Sensors System and Deep Learning Models for Object Tracking

Ghina El Natour [1,*], Guillaume Bresson [2] and Remi Trichet [1]

[1] Continental, 1 Av. Paul Ourliac, 31100 Toulouse, France; remi.trichet@gmail.com
[2] Vedecom, 23 bis All. des Marronniers, 78000 Versailles, France; gubresson@gmail.com
* Correspondence: ghina_natour@hotmail.com

Abstract: Autonomous navigation relies on the crucial aspect of perceiving the environment to ensure the safe navigation of an autonomous platform, taking into consideration surrounding objects and their potential movements. Consequently, a fundamental requirement arises to accurately track and predict these objects' trajectories. Three deep recurrent network architectures were defined to achieve this, fine-tuning their weights to optimize the tracking process. The effectiveness of this proposed pipeline has been assessed, with diverse tracking scenarios demonstrated in both sub-urban and highway environments. The evaluations have yielded promising results, affirming the potential of this approach in enhancing autonomous navigation capabilities.

Keywords: multi-sensors system; tracking; recurrent neural networks; sensor fusion; metric learning

1. Introduction

Nowadays, autonomous vehicles (AVs) are gradually becoming part of our lives. Autonomously navigating robots have been developed since 1961 when James Adams first developed an autonomous lunar rover during his Ph.D. Over the years, various autonomous robots and aircraft were developed, and Tesla launched Autopilot in 2015. Nowadays, Waymo driverless taxi cars are employed in Arizona, USA.

However, the path to developing autonomous vehicles has not been without challenges. Back in 2010, manufacturers such as Honda, Tesla, and Toyota claimed that, by 2020, fully autonomous vehicles would be operational, and users would be sitting in the back seat. By 2023, it became evident that this ambitious objective faced multiple challenges, particularly concerning safety. Although recent studies [1] claim that cars equipped with ADAS (Advanced Driver Assistance Systems) technology experience a 23% reduction in crash frequency, fully automated vehicle safety remains a significant concern. On March 18, 2018, the first driverless deadly accident occurred when an Uber autonomous vehicle hit and killed a pedestrian walking by her bicycle in Arizona, USA. According to the National Transportation Safety Board accident report, the near-range sensor was not correctly operational and failed to predict the object's movement because it was not at a crosswalk. Following this incident, multiple AV accidents have been reported, serving as a reminder that more work should be brought to sensor technology and object tracking for autonomous vehicles.

Autonomous vehicles rely on an array of sensors, including cameras, radars, lidars, . . ., to perceive their surrounding environment. At each time step, noisy sensor detections are received and treated to identify the surrounding objects and their potential trajectories. In this paper, we are interested in the fusion of two hybrid sensors at the core of the most prevalent association: vision and radar. The complementarity of these sensors makes their fusion a natural choice.

The goal is to track objects in front of an AV. Figure 1 illustrates the sequential steps involved in the object-tracking process. Our approach consists of replacing classical algorithms that tackle association, prediction, and update problems (Figure 1a) with recurrent neural networks (RNN)-based algorithms. RNNs are chosen for their ability to learn the dynamic properties of moving objects, but the scarcity of annotated data poses a challenge for LSTM-based (Long-Short Term Memory) [2] tracking techniques. The problem is divided into multiple tasks to overcome this, and simple models are trained to solve those tasks, as illustrated in Figure 1b.

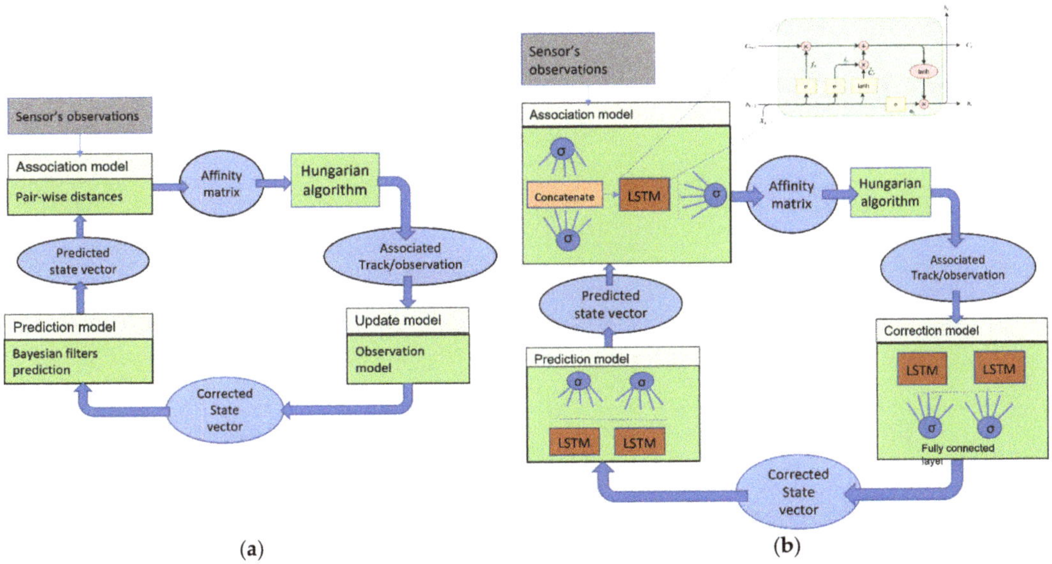

Figure 1. An illustration of the object-tracking process. (**a**) Bayesian-based approach. (**b**) Our approach consists of replacing classical algorithms to solve the association, prediction, and update problems by LSTM-based algorithms.

In the Bayesian filter methodology, the object state vector is updated based on previous estimations and the received observations. Authors in [3] and [4] showed that the Kalman filter works better on the sensor error than on the sensor output. Building on this idea, we incorporate a single layer of an LSTM-based model that learns the observation error model for each sensor. At each time stamp, we predict the observation error using this model to refine and correct the observations accordingly.

Furthermore, in the proposed association method, we want to learn an association metric that is more representative than Euclidian distance. Indeed, the Euclidian distance is generally applied to compute the association metric, but it is not fully representative of the object state because it only considers the object's position (2D coordinates). The learned association metric considers the object state vector, including (x,y) position, class type, and orientation. To address the problem of the time-varying number of objects, especially using neural networks involving a fixed input size, the association model evaluates one combination at a time, and then a classic Hungarian algorithm loops over all combinations to find the best correspondences.

Finally, we propose a new existence score computation aimed at managing the track's lifecycle. The existence score of a track is crafted to encompass its association scores with both sensor observations as well as the count of skipped frames. We want to take advantage of the association probability estimated using the association model because this information allows us to automatically account for whether a track is associated with one or two sensor observations. This approach serves the dual purpose of managing the initiation and termination of a track when it enters or exits the sensor's field of view. Secondly, to eliminate false positive tracks originating solely from sensor-only observations.

In this paper, we propose a novel LSTM-based tracking algorithm that addresses several challenges in object tracking for autonomous systems and that performs the following:

- Leverages the need for a large amount of annotated data by dividing the tracking problem into multiple tasks and utilizing single-layer models trained for the association, the prediction, and the update steps.
- Improve the accuracy of objects' positions by estimating the observation errors rather than direct position estimation, utilizing the updated model.
- Learning an optimized association metric by combining diverse cues such as object position, type, orientation, and trajectory into a unified model that estimates the association score for an observation/track pair.
- Manage tracks by computing a variable existence score that incorporates association scores, participating sensors, lifetime, and track disappearance.

State of the Art

In this paper, we are interested in the fusion of two hybrid sensors for object tracking that nest within the most common sensor association: vision and radar. The complementary nature of these sensors has inspired previous approaches [5]. Wang et al. [6] utilized radar detections to guide object search in images, while Yu et al. [7] presented a high-level fusion of camera and radar sensors, leveraging deep learning for object detection and Kalman filtering for tracking. Although recent research tends to favor the elimination of radar sensors in favor of relying solely on camera sensors for advanced and autonomous driving functions, as mentioned by the authors in their review [8], it is important to note that camera sensors still have significant limitations, particularly in the distance and velocity estimation. To address these limitations and ensure better precision and completeness of detections, the integration of range sensors remains necessary.

Joly et al. [9] conducted a comparative study of several nonlinear filters and found that the particle filter offers enhanced robustness compared with the Kalman filter, though the number of particles affects both performance and computational costs. One limitation of Bayesian filtering is the necessity to model the behavior of the objects being tracked, which may not accurately reflect the real movements of different object types, such as pedestrians, vehicles, or cyclists.

Since 2010, the rise of deep learning (DL) has led to groundbreaking advancements in autonomous vehicles and object tracking. RNNs successfully applied to language modeling by Wang [10] set the foundation for progress in this domain. In particular, LSTM networks have become the most studied type of RNNs due to their ability to overcome the vanishing gradient problem through multiple gates controlling the network's input/output status and memory. LSTMs' model-free nature allows them to model both linear and nonlinear dynamics, effectively capturing object motion patterns through trajectory analysis. Comparatively, Bayesian filters require substantial efforts in designing state transition and measurement models, making LSTMs a favorable alternative, as noted by Iter [11], Gu [12], and Altché [13]. Ondruska [14] demonstrated the versatility of RNNs in multi-object tracking by directly mapping raw radar sensor inputs to object tracks in the sensor space, eliminating the need for formal data association. Further research has explored LSTM's predictive capabilities for trajectory estimation of pedestrians, Alahi [15], and vehicles, Ma [16]. Park [17] introduced an LSTM encoder-decoder architecture designed for vehicle trajectory prediction. Their model takes as input the positional and velocity data of the ego vehicle as well as that of surrounding vehicles. The model's output consists of an occupancy grid map containing the k most probable trajectories for each vehicle. However, the beam width K is manually set to a value of 10, thereby introducing heightened complexity into their algorithm. In their work, Sengupta [18] introduced a sensor fusion approach that leveraged deep neural networks and LSTMs to integrate data from a monocular camera and a mmWave radar. Both sensors' object positions are extracted in 2D coordinates, with subsequent utilization of the LSTM-based model for predicting object trajectories. However, in their study, the sensor-only tracks (tracks detected by only one sensor) were not effectively managed, which led to an increased risk of false positives over time. Moreover, their fusion methodology primarily yielded a fused position comprising the lateral coordinates from the camera and the longitudinal coordinates from the radar rather than providing a refined representation of the sensor's observations. This discrepancy is addressed and rectified in our approach through the application of the LSTM correction model. Da Zhang [19] presented a significant breakthrough with the Deep RL Tracker (DRLT) algorithm, which combines Convolutional Neural Networks (CNNs), RNNs, and RL to achieve precise object tracking in videos. While showing good performance compared with state-of-the-art algorithms, DRLT requires the initial ground truth location for each object and leaves room for addressing the association problem.

Data association in multi-sensor-based tracking is a significant challenge, involving multiple objects and detections received at each time step, with varying numbers over time and potential false positives. Addressing this problem entails two key steps: metric learning and affinity matrix optimization. The affinity matrix contains similarity scores obtained by comparing cues, such as motion and appearance, between track/observation pairs. The goal is to achieve an optimal assignment of tracks to observations. The Hungarian algorithm, also called Global Nearest Neighbor (GNN) [20,21], is a well-known and widely used approach to solve the bipartite graph problem of affinity matrix optimization. Additionally, Multiple-Hypothesis Tracking (MHT) [22] and Joint Probability Data Association (JPDA) [23] are considered to outperform the GNN in clusters but become impractical for real-time settings as the number of objects increases.

Methods addressing the association metric can be categorized into two approaches. The first approach involves computing pair-wise distances between sensors' measurements and state predictions. The second approach utilizes deep learning algorithms, combining multiple cues such as appearance and dynamic cues. Milan [24] presented an online pedestrian tracking approach using RNN. They performed prediction, data association, and state updates within a unified network. The association vector was computed using an LSTM-based network, utilizing Euclidean distance between the predicted state and the observations as input. Similarly, Liu [25] designed an LSTM network to learn the measurement-to-track association probability from radar noisy measurements and existing tracks. The input to the LSTM network is also a pair-wise distance matrix generated from all measurements and state prediction of each target without explicit metric learning. Using the Euclidean distance as an association metric might not accurately reflect the track-to-object correspondence as it only considers coordinates parameters. Building on this, Sadeghian [26] utilized hierarchical RNNs to encode multiple cues, including appearance, motion, and interaction, for offline metric learning. Following a similar line, Wan [27] fused three features (appearance, motion, and velocity) into one LSTM network for metric learning. The Hungarian algorithm is employed to solve the global optimization problem. Luo [28] concatenated motion features to appearance features extracted by a CNN. However, in certain scenarios, the appearance of objects is not available; sensors such as radar or lidar provide very limited appearance information. Huang [29] introduced a straightforward tracking-by-detection framework, achieving good performance on the BDD100K MOT and MOTS datasets. Their main contribution is in the association step of objects bounding boxes across frames. However, they discard motion information, relying solely on appearance embeddings for object association. Kieritz in [30] proposed to perform joint detection and association in a unified model. The input cues for the association layer are spatial distance, object appearance, and detection and tracking score. The association score matrix is then optimized using the Hungarian algorithm. In their work [31], Yoon introduced an encoder-decoder model with bidirectional LSTM to calculate an association affinity matrix. This model takes non-appearance features as input, including bounding box coordinates and detection confidences. The optimization of the association matrix was achieved using the Hungarian algorithm. Similarly, we do not use appearance cues as input to our association model. Nevertheless, our model is considerably simpler and requires less data for training.

In this paper, we aim to provide new insights into sensor fusion for object tracking, seeking to overcome traditional Bayesian filter limitations and achieve improved performance and accuracy in various tracking scenarios. The remainder of this paper is structured as follows. Section 2 details the proposed multi-sensor system for object tracking. Section 3 evaluates the proposed system experimentally. Finally, Section 4 summarizes and concludes the paper.

2. Materials and Methods

2.1. The Sensors

Our study utilizes a multi-sensor system, including radar and a camera with overlapping fields of view, to track objects in front of an autonomous truck. Additionally, an RT range sensor serves as a ground truth reference. We collected a database in highway and peri-urban environments, with sensors triggered at 10 frames per second. The raw data are prepossessed, and objects are detected. The state vector of each sensor observation is provided, including its position, type, and orientation angle.

2.2. Problem Formalization

The sensor's observation set at time t is defined as follows: $O_t^k = \{o_1 \dots o_N\}_t^k$ where N is the number of observations of sensor k. Each observation is represented by a state vector containing the position, the velocity, the orientation, etc.

A tracked object is identified by a unique identifier, a state vector, and a trajectory which is composed of the state vectors from $t0$ to $t1$. The track set is defined as follows: $X_t = \{x_1, \dots, x_M\}_t$ where M is the number of simultaneous tracked objects.

The goal is to continuously track objects and accurately estimate their positions at each timestamp. For this, the future position of a track at the next timestamp needs to be predicted to associate it with the received observation. Therefore, each track is associated with at least one observation if it is still visible by the sensors. However, data association poses significant challenges due to the variable cardinality of objects and observations at each timestamp. Ambiguities in association can arise from occlusion or association problems (misses, mismatches, and false negatives), as illustrated in Figure 2.

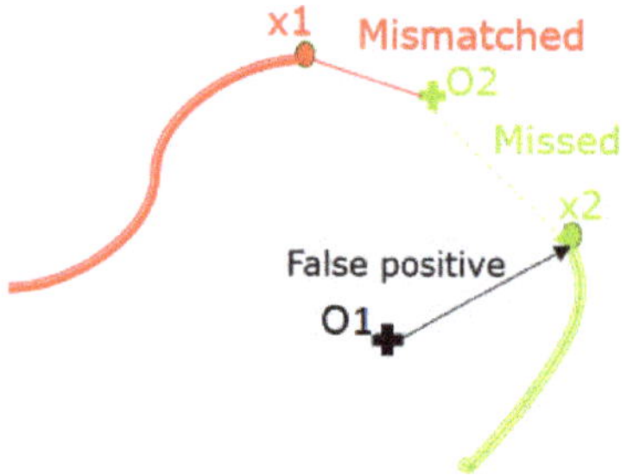

Figure 2. Illustration of association ambiguities. O stands for observations and x stands for the objects.

Once associated, the predictions are then corrected with the observations. We want to avoid the modeling step because the models may not reflect the correct dynamic of the objects or the correct sensor observations. Thus, we proposed three deep learning models to handle the prediction, association, and update steps for object tracking.

2.3. The Models

The proposed models are built upon LSTM recurrent neural networks. The inclusion of recurrent neural networks allows us to capture the intricate interdependencies present in sequences of temporal data, such as object trajectory sequences in our context. The term "recurrent" stems from the feedback loop that is passed through the iterations, enabling the network to retain and process information over time. To optimize the network's performance, we empirically set the hidden layer size to 200, ensuring a balance between complexity and efficiency.

2.3.1. Association Model

Our primary focus is on solving the data association between tracks and observations received at time t. Figure 3 illustrates an example of a scene at time t with two objects and three sensors' observations. To achieve this, we employ an association model that evaluates all possible combinations of object/observation associations. The model takes an object's state vector sequence from time t_0 to time $t-1$ and concatenates it with an observation obtained from one of the sensors at time t.

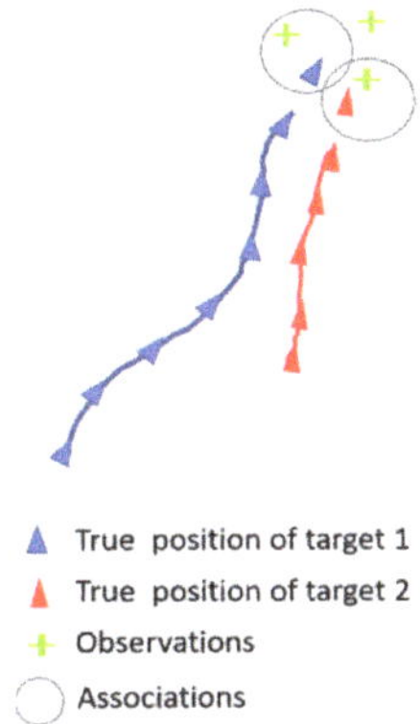

Figure 3. Scene illustration with 2 objects and 3 observations at time t. The circles depict the combinations to be evaluated.

An LSTM layer of hidden size set to 400 is followed by a fully connected dense layer with only one neuron and a sigmoid activation function, which makes it possible to output a probability of association in the range [0, 1], indicating the likelihood of an association. During training, we employ the logarithmic loss function (binary cross entropy) commonly used for binary classification problems. Hence, the LSTM-based feature extractor is learned by minimizing the binary cross entropy (BCE) loss, formulated as follows: $L_{BCE} = -(\sum_{n=i} y_i \log(f(s_i)) + (1-y_i)\log(1-f(s_i)))$, where s is the input, f is the activation sigmoid function, y is the correct classification, and i is the class to predict. The architecture of the association model is depicted in Figure 4 below.

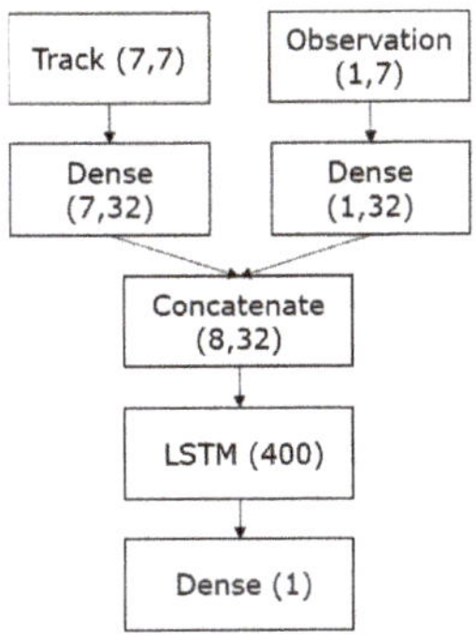

Figure 4. The association model architecture.

2.3.2. Correction Model

The goal of the correction model is to replace the update step of a Kalman filter, brought in to correct the Kalman predictions using the received observations and the sensor's observation models. Figure 5 illustrates the scene with two objects already associated with two observations out of three, and the observation positions are corrected (estimated positions using the correction model).

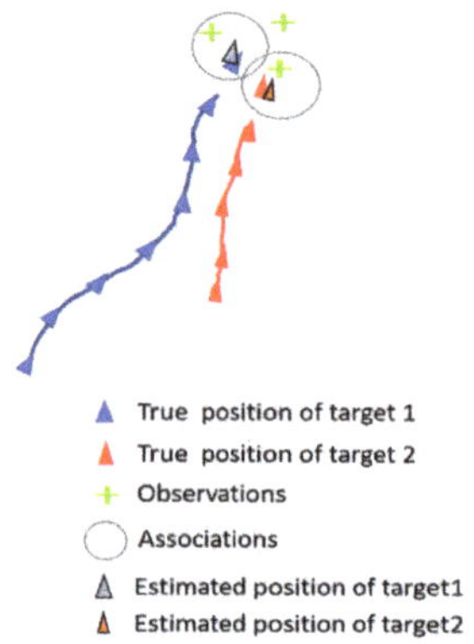

Figure 5. Scene illustration with two tracks and three observations already associated and the observations corrections at instant *t*.

The state vector of a moving object at a given time *t* depends on the previous state vector at $t-1$. The correction model is designed to predict the observation's errors in an instant, *t* having as input an observation error sequence shown in Figure 6. For the training, the observation error at each time instant is computed by comparing the sensors' observation to the ground truth positions.

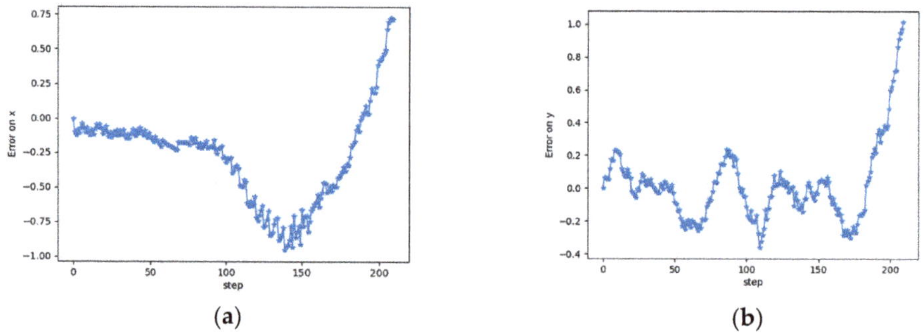

Figure 6. Observation error plots on x(m) (**a**) and y(m) (**b**) coordinates.

The correction model utilizes the observation's error sequence as input, which is then passed through two LSTM layers, each with a hidden layer size of 200. Subsequently, a fully connected dense layer is applied with a sigmoid activation function to predict the next observation's errors. The model architecture is depicted in Figure 7.

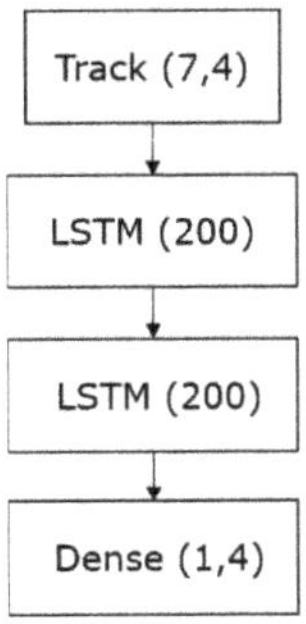

Figure 7. The update model architecture.

The learning is done by optimizing the MSE (Mean Squared Error) loss function for prediction:

$L_{MSE} = \frac{1}{N}\sum_{i=1}^{N} \left|\left(Y_t^i - \hat{Y}_t^i\right)\right|^2$, where Y_t^i is the correct observation error at t and $\hat{Y}_t^i$ is the estimated observation error at t.

2.3.3. Prediction Model

To ensure accurate track-to-observation association, we want to predict the object's position at the same time instant as the observation. For this purpose, the same correction model architecture is employed to predict the next position from a sequence of object trajectories. The input to the model is the track trajectory sequence starting from time t_0 up to the current instant t. The model then processes this trajectory sequence and produces the prediction of the object's state at time $t + 1$.

Similarly to the correction model, the MSE loss function is optimized for prediction: $L_{MSE} = \frac{1}{N}\sum_{i=1}^{N} \left|\left(Y_t^i - \hat{Y}_t^i\right)\right|^2$, where Y_t^i is the correct position at time t and $\hat{Y}_t^i$ is the estimated position at time t.

2.3.4. Models Training

We trained the three models using a database of 10 diverse sequences with a total running time of approximately 12 min. Trajectories were divided into training and testing sets, including cars, trucks, motorcycles, and pedestrians, with varying speeds between 0 and 80 kph. The ego vehicle speed varies from 20 to 80 kph.

We extracted 5000 trajectory sequences of length = 7 time steps from the ground truth objects trajectories. Additionally, the 8th state vector from the ground truth sequence is used as the output for the prediction model training. The computational complexity of the LSTM depends on the depth of the model and the sequence length, and because we chose simple models' architectures, the complexity of the network is roughly dependent on the sequence length. Thus, the sequence's length was chosen empirically in a way that it is not too short to have significant information for the model to learn and not too long, so it will not make the training too hard. The radar and camera observations errors corresponding to the trajectories were computed and used to train the update model. Finally, the correct association probabilities (0 or 1) are generated to train the association model.

The extracted data was scaled to a range of [0, 1] to ensure uniformity in the data with different scales, aligning with the sigmoid activation function of the units in the final layer responsible for the predictions. For optimization during training, we opted for the ADAM optimizer, known for its faster convergence and stability compared to other optimizers, according to this study in [32]. We chose a learning rate of 0.001 based on our experiments, which falls within the optimal range of 0.001 to 0.01 recommended in the previous study. The models were trained for 300 epochs, iteratively refining their performance and enabling them to make more accurate predictions.

2.4. Track Management

Track management plays a vital role in the tracking process as it handles object birth and death while dealing with association ambiguities. An existence score is computed for each tracked object based on its association scores with radar and camera observations to achieve this, along with the number of skipped detections.

The existence score (*ES*) is calculated as follows:

$$ES = RAC_i + CAC_i - SF \tag{1}$$

Here, RAC_i and CAC_i are the radar association cost and camera association cost for object i. The existence score increases when an object has a high association probability with an observation.

To mitigate false births caused by noisy detections, a validity threshold is set. When a new object is detected, its state is initially set to 'hidden' until it consecutively appears in the next iterations, at which point its state is changed to 'visible.' The number of skipped frames (*SF*) for an object is incremented by 1 if the track is not observed by any sensor (not associated with any observation) and by 0.5 if it is observed by only one of the sensors. Consequently, the existence score decreases when the object is not observed by one or both sensors.

If a track is no longer detected, its state is set to 'hidden,' and it is propagated into the future by predicting its next position to handle object occlusion. Finally, when the existence score falls below the death threshold, set to 0, the track is deleted. Figure 8 illustrates the various states a track can pass through during its lifetime, demonstrating the dynamic nature of the track management process.

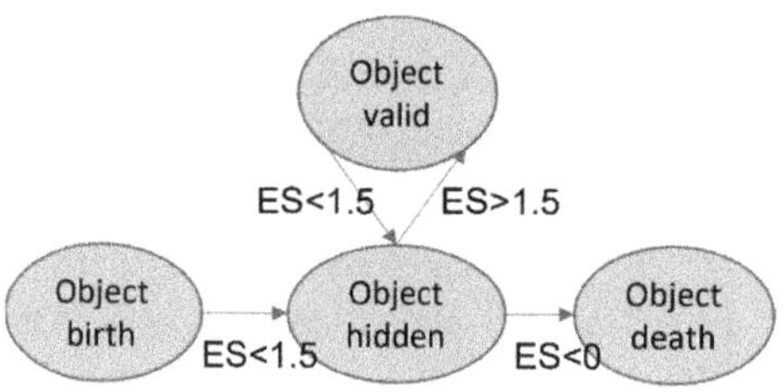

Figure 8. Illustration of an object's life state.

3. Evaluation and Results

To gain deeper insights into the performance of the proposed algorithms, we conducted evaluations of a multi-object tracker using various types of sequences available in the dataset. Specifically, we considered scenarios involving heading targets and lane changing in highway environments, as well as oncoming vehicles in peri-urban environments.

The truck utilized in the evaluation is equipped with an ARS430 radar sensor with a lateral field of view of $\pm 60°$. The range resolution is 2.23 m, while the azimuth resolution is $3.2°$ at $0°$ and increases for angles below $0°$. The radar provides a range accuracy of approximately ± 0.55 m and $\pm 0.1°$ on azimuth at $0°$ for the far scan. In addition, an MFC520 camera is used. Objects are detected within the camera's field of view and classified within a detection range of up to 150 m. The lateral accuracy of the MFC is approximately ± 0.5 m, while the longitudinal accuracy varies, being ± 0.5 m for near objects and higher for farther objects.

Figure 9 illustrates the mounting setup of the sensors (MFC520 and ARS430) on the truck, which is also equipped with an RT-range (real-time) system from OXTS used as a ground truth for object positions. The RT system measures the relative positions between an RT inertial system and other mobile targets equipped with RT-range systems. The accuracy for object range measurements is 0.03 m. The sensor observations are transformed into a common frame relative to the ego vehicle.

Figure 9. The setup of the sensors (MFC520 and ARS430) mounted on a truck for training and test collection data.

The evaluation database consists of 10 trajectories, corresponding to a running time of approximately 16 min. Both peri-urban and highway scenarios are included in the evaluation setup.

Unlike appearance-based methods, our proposed method does not rely on appearance cues; instead, we receive camera observations at the object level. As a result, comparing our method with appearance-based algorithms is challenging due to the black-box nature of the object detection algorithm employed for the MFC camera. As stated by Mouzaffari in [33], unlike object detection, which employs a standardized evaluation framework, the field of vehicle behavior prediction lacks a consistent benchmark. This absence of a benchmark hinders fair comparisons among various deep learning-based approaches and between deep learning-based methods and other techniques. To evaluate our approach, we compare the obtained results with those of an Extended Kalman-based tracking algorithm using the same database acquired by the same sensors. For the evaluation of the multi-object tracking algorithm, we use the following metrics:

- The RMSE (Root-Mean-Square Error) to evaluate object position estimation, calculated as the mean Euclidean distance over all predicted positions and real positions: RMSE = $\sqrt{\frac{1}{N}\sum_{i=1}^{N}\left(Y_t^i - \hat{Y}_t^i\right)^2}$, where Y *and* $\hat{Y}$ are the respective correct (RT ground truth) and predicted outputs.
- The MOTA (multi-object tracking accuracy) [34] to assess the association model and track management process: MOTA= $1 - \frac{\sum_b (m_t + fp_t + mme_t)}{\sum_t g_t}$, where t is the frame index, and m, fp, and mme denote the number of misses, false positives, and mismatches, respectively, over g, the total number of objects in the scene across all iterations b. MOTA is a widely used metric as it effectively combines the three sources of errors mentioned above.

3.1. Results

To assess the accuracy of the tracking system, we compared the RMS position errors obtained from different sources with the ground truth provided by the RT-range system. For this evaluation, we extracted 1000 subsequences of length 10 from the test sequences. RMS errors were computed at each time step across the 1000 subsequences, considering various test scenarios such as cars on the ego lane, cars changing lanes, and two cars on adjacent lanes. The resulting RMS errors are plotted in Figure 10.

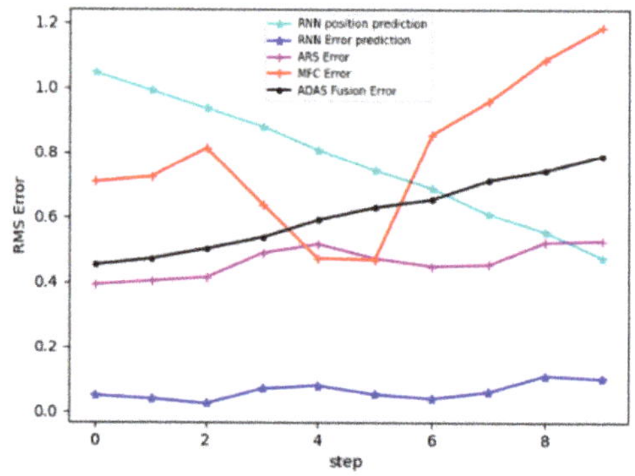

Figure 10. RMS errors (in m) graphs comparison on 10 timestamps of the sensors observations errors with the ADAS fusion error using Kalman filtering and the LSTM correction and prediction errors. The LSTM-corrected position error (in blue) is the lowest, with an average of 0.09 m.

The MFC camera exhibited an error of approximately 0.91 m for objects at distances between 20 m and 60 m from the ego vehicle. In contrast, the radar observations had an average error of 0.5 m, while the LSTM correction achieved an average error of 0.09 m. Comparing this with an Extended Kalman filter-based algorithm, which had an average error of 0.7 m, the LSTM correction outperformed it by a factor of approximately 7.

Regarding the LSTM prediction error, it was approximately 1 m at the beginning of the sequence and decreased to around 0.47 m by the end of the sequence. This behavior indicates that the LSTM improved its prediction accuracy with longer sequences, capturing richer dynamic properties of moving targets. It is worth noting that the LSTM update error estimation was lower than the state prediction error, indicating that the LSTM-based method fits the sensor error model better than the object's trajectory. To facilitate the training step, we divided the tracking problem into multiple stages and used simple model architectures for each step. Consequently, we did not require an extensive training dataset. Our proposed LSTM-based method is model-free, making it adaptable to various dynamic scenarios and sensor observations. However, it does require a training dataset with reliable ground truth for learning purposes. A comprehensive comparison between our method and the Kalman filter-based method is presented in Table 1, highlighting the advantages and efficacy of our LSTM-based approach.

Table 1. Comparison of our LSTM-based method with Kalman-based tracking.

	Kalman	**Our Proposed Algorithm**
Sensor model	Yes	No
Dynamic model	Yes	No
Needs learning	No	Yes
Estimation update accuracy	0.7 m	0.09
Computational complexity	Low	Moderate
Versatility	Restricted	Wide range of scenarios

Table 2 shows MOTA score equal to 95%. This score is derived from the combination of three types of errors: false positives (2.1%), misses (0.17%), and mismatches (2.6%). An important observation from the MOTA results is that the association algorithm tends to make more false associations than not associate observations at all. This behavior can be attributed to the limited occurrence of false positive observations compared with real positives in real-world scenarios, and this characteristic is reflected in the training data. Furthermore, as highlighted by Gioele [35], the MOTA score exhibits a significant correlation with the count of missed detections (false negatives). Notably, our algorithm has successfully minimized the number of false negatives, a result attributed to the inclusion of occluded tracks and sensor-only tracks within the hidden state.

Table 2. Association error and MOTA error.

Error Type	% Association Model	% Euclidean Distance
False Positives	2.12%	3.45%
Misses	0.17%	0.15%
Mismatches	2.60%	4.11%
MOTA	95.10%	92.30%

One notable advantage of our trained model is that it exhibits a decrease of approximately 1.5% in mismatches error when compared with the Euclidean distance algorithm. This improvement is attributed to the fact that the Euclidean distance algorithm relies solely on the track's coordinates, whereas our association model takes into account additional cues such as trajectories and object types, resulting in a more accurate association of observations. Additionally, contrarily to the association algorithm, the misses error for the Euclidean distance algorithm is affected by an empirically chosen distance threshold. This can impact the accuracy of the misses error, and the threshold may not adapt optimally to different scenarios.

Figure 11 illustrates the LSTM correction results of object trajectories. The figure includes the ground truth positions and sensor observations. In Figure 11a, the radar observations closely align with the RT range ground truth (depicted by green squares) trajectories. However, it is evident that the camera observations are less precise along the longitudinal axis. The LSTM-corrected positions (indicated by blue dots) demonstrate the closest match to the RT ground truth. Figure 11b presents a scenario where both radar and camera observations exhibit high observation errors. Despite this, the LSTM correction manages to provide positions that are significantly closer to the ground truth than the raw observations alone. This demonstrates the LSTM's capability to effectively learn and correct dynamic object behavior, specifically for cars in this scenario. Moreover, it indicates the potential for extending this approach to other object types, such as pedestrians and cyclists, by appropriately fitting the model to relevant data.

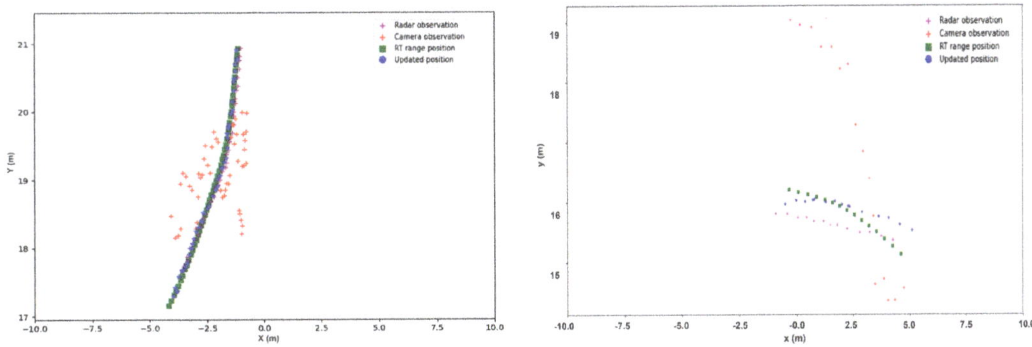

(a) (b)

Figure 11. LSTM corrections (blue dots) for object trajectories with sensor observations (red and magenta crosses) and RT range (green squares). In (**a**), radar has a better estimation of object position than camera observation, and LSTM-corrected position best aligns with the RT ground truth. In (**b**), both sensors have notable observation errors, but LSTM-corrected positions are much closer to the ground truth compared with relying solely on the raw observations.

In Figures 12 and 13, we present two tracking scenarios in highway and sub-urban environments. The figures use color annotations to indicate the visibility status of each object. When an object is 'hidden', its ID is written in magenta, and when it is 'visible', the ID is written in blue. In the first frame, the object ID annotations are in magenta for all three tracks because their existence scores do not satisfy the validity threshold, which we set empirically to 1.5. The existence score of an object increases if it is detected and tracked over time until it becomes higher than the validity threshold, at which point the ID color switches to blue.

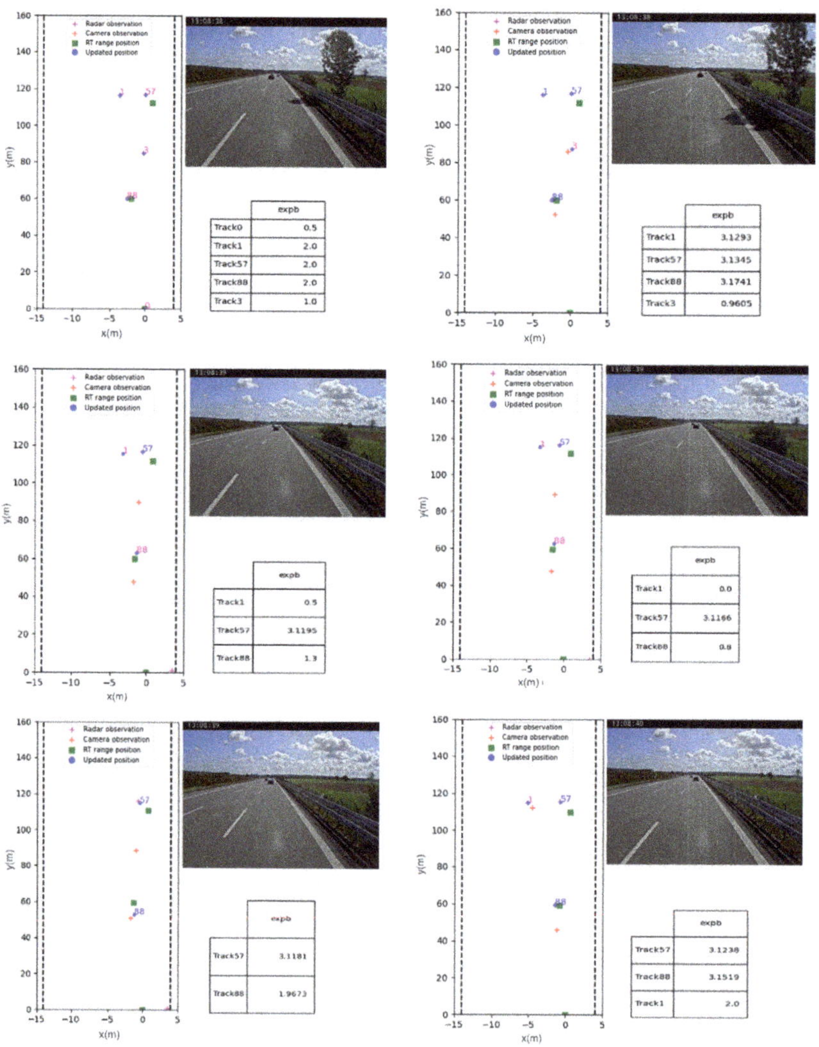

Figure 12. A highway tracking scenario with tracking results and table of existence scores. First, all the tracks are in magenta with 'hidden' status, then sensor-only tracks (Track (3) and Track (88) later on) and occluded track (Track (1) where both sensors observations were lost) have their status shifted to 'hidden'.

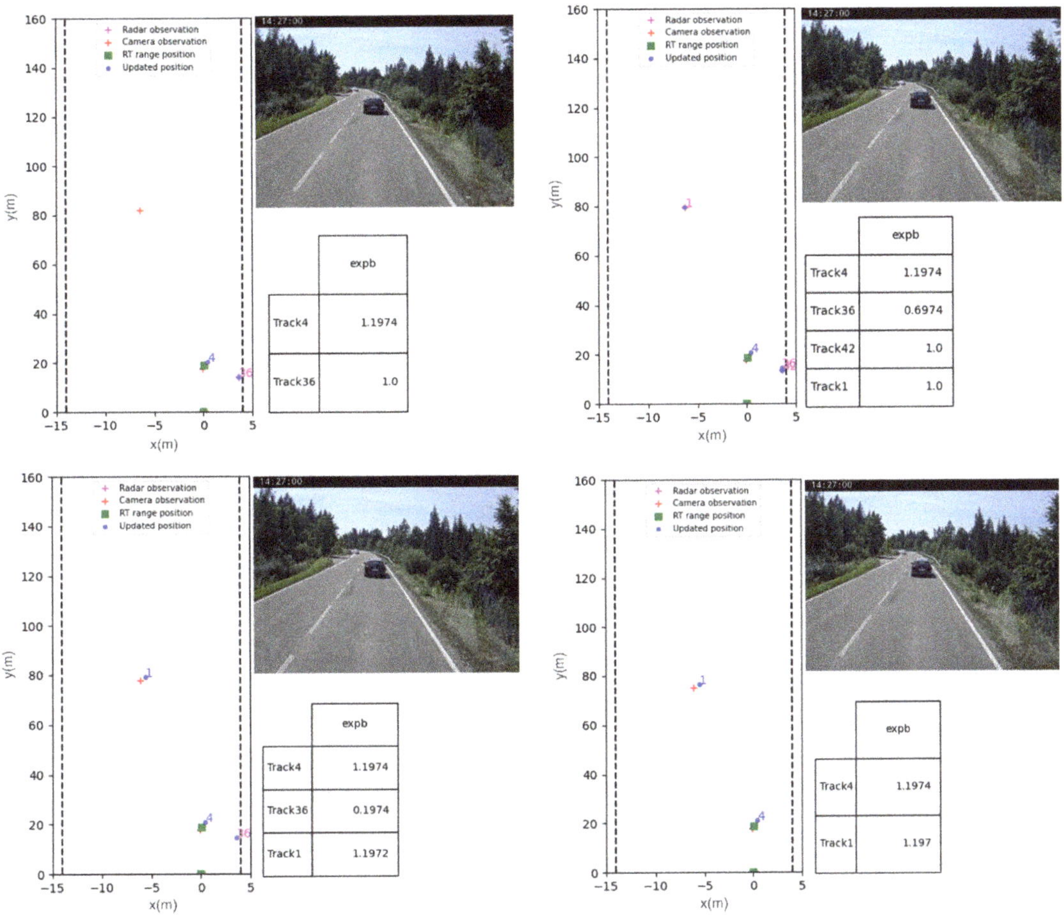

Figure 13. Frames extracted from peri-urban tracking scenario of two cars. Only the car in ego lane is equipped with an RT range ground truth system. Track (36) being a radar-only false positive, is being correctly deleted from the tracking scenario because of a decreasing existence score.

However, in Figure 12, Track (3) remains 'hidden' because it is a camera-only track and its existence score remains lower than the validity threshold. Meanwhile, Track (1) color turns magenta when it becomes unobservable by both the camera and the radar. This occurs due to occlusion by Track (88). As a result, its existence score continues to decrease until it reaches zero, leading to its deletion from the tracking scenario. A comparable situation is observed with Track (88). When not detected by the radar but only by the camera, its existence score continues to decrease. However, once the radar starts detecting Track (88) again, its status shifts to 'visible', as shown in Figure 12.

In Figure 13, two observed cars initialized with a 'hidden' state. In the subsequent frame, these objects are validated, and the tracking algorithm begins to track Tracks (1) and (4). It is important to note that only the heading vehicle is equipped with an RT-range system, while the second oncoming car is observed by the MFC (Multi-Function Camera). Additionally, in this figure, an example of how false positive detections are handled is presented. An initial existence $ES = 1.0$ score is assigned to Track (36) originating from a radar false positive detection. Then, the existence score gradually decreases because of unstable radar-only observations, ultimately leading to the deletion of Track (36).

3.2. Discussion

Compared to a classical Kalman filter, the LSTM correction model shows much better results in highway and peri-urban environments. The Kalman filter's limitation in modeling observation errors that increase with distance becomes apparent in challenging scenarios. Conversely, our model adeptly adapts to this behavior, making it particularly valuable for our Multi-Function Camera (MFC) when estimating the distances of detected objects, where errors grow with distance.

Additionally, radar detections are vulnerable to clutter from multipaths and reflections, leading to a higher rate of false positives. Our association management utilizes an intelligent existence score to address these sensor-related challenges, which accounts for association scores of a track with both the sensor's observations and for the number of skipped frames. This track management effectively reduces the influence of individual sensor observations, thus bolstering the overall tracking performance.

By breaking down the tracking process into manageable components and training individual models for each step, we were able to focus on specific aspects of the tracking problem. This approach allowed us to fine-tune the models for their respective tasks, optimizing their performance for specific challenges. Consequently, we achieved better results in scenarios characterized by complex dynamics, occlusions, and sensor-specific limitations.

Furthermore, the training of individual models demanded a smaller training dataset compared to a single monolithic model. As a result, we mitigated the burden of gathering and annotating a large amount of training data, which can be particularly challenging and expensive in the context of multi-object tracking. Instead, our approach allowed us to make efficient use of available data and yielded promising results even with a more limited dataset.

However, the selection and distribution of learning data play pivotal roles in DL-based approaches. It is useful that these datasets are thoughtfully curated to encompass a comprehensive spectrum of scenarios and dynamic sequences. Furthermore, in the context of correction models, it is important to acknowledge that these models are designed to predict errors specific to individual sensors. Consequently, a distinct model must be trained for each sensor to ensure accuracy and effectiveness. Therefore, a challenge emerges when the inherent characteristics of a sensor undergo modifications, as this necessitates an adjustment in the training approach.

4. Conclusions

Despite the significant progress in recent studies, multi-object tracking remains a challenging subject due to complexities introduced by object motion, interactions, and occlusions. Imperfect sensor detections, including false alarms, missed detections, and noisy readings, add further complications.

To address these challenges, we proposed an LSTM-based multi-object tracking solution, utilizing three trained models for data association, track update, and predictions. Our approach simplifies the training process, and the LSTM models with simple architectures were trained on a database collected from both highway and peri-urban environments.

Our results demonstrate that the LSTM models learned association metrics that outperformed the classically used Euclidean or Mahalanobis distances, providing more representative track-observation relationships.

For predicting sensor errors, the LSTM-based models showed promising results compared to the Kalman-updated estimations, accurately learning the sensors' observation models. However, we acknowledge that further improvements can be achieved in the prediction of the state vector and in predicting the trajectory of a track by using a larger training database and deeper models to learn object dynamics better. Furthermore, we proposed a track management process using an existence score that integrates the association scores. The existence score is variable and is computed at each iteration for each track in order to efficiently handle the death and birth of the tracks but also to reduce false positives and missed detections.

In future work, we aim to address object interactions, which are crucial in predicting future state vectors, especially in crowded scenes for autonomous vehicle applications. Considering interactions will enable better predictions of object movements based on their neighborhood and environment. One potential approach is to input multiple close tracks and their trajectories into the prediction model to make simultaneous predictions for all neighboring objects.

Overall, our LSTM-based approach shows promising results in multi-object tracking and lays the foundation for addressing more complex scenarios in the future.

Author Contributions: Conceptualization, G.E.N., G.B., R.T.; methodology, software, G.E.N.; formal analysis, investigation, resources, data curation, G.B., G.E.N., R.T.; writing—original draft preparation, G.E.N.; writing—review and editing, G.B.; supervision, project administration, funding acquisition G.B. All authors have read and agreed to the published version of the manuscript.

Funding: This research received no external funding.

Institutional Review Board Statement: Not applicable.

Informed Consent Statement: Not applicable.

Data Availability Statement: Data are contained within the article.

Acknowledgments: We acknowledge the invaluable resources provided by Institute Vedecom and by Continental. Their collaboration greatly enhanced the depth and breath of our research. We also extend our appreciation to the participants who generously shared their time and insights.

Conflicts of Interest: The authors declare no conflict of interest.

References

1. Spicer, R.; Vahabaghaie, A.; Bahouth, G.; Drees, L.; von Bülow, R.M.; Baur, P. Field effectiveness evaluation of advanced driver assistance systems. *Traffic Inj. Prev.* **2018**, *19*, S91–S95. [CrossRef] [PubMed]
2. Hochreiter, S.; Schmidhuber, J. Long short-term memory. *Neural Comput.* **1997**, *9*, 1735–1780. [CrossRef] [PubMed]
3. Noureldin, A.; El-Shafie, A.; Bayoumi, M. GPS/INS Integration Utilizing Dynamic Neural Networks for Vehicular Navigation. *Inf. Fusion* **2011**, *12*, 48–57. [CrossRef]
4. Wu, G.X.; Ding, Y.; Tahsin, T.; Atilla, I. Adaptive neural network and extended state observer-based non-singular terminal sliding modetracking control for an underactuated USV with unknown uncertainties. *Appl. Ocean Res.* **2023**, *135*, 103560. [CrossRef]
5. El Natour, G.; Ait-Aider, O.; Rouveure, R.; Berry, F.; Faure, P. Toward 3D reconstruction of outdoor scenes using an MMW radar and a monocular vision sensor. *Sensors* **2015**, *15*, 25937–25967. [CrossRef] [PubMed]
6. Wang, X.; Xu, L.; Sun, H.; Xin, J.; Zheng, N. Bionic vision inspired on-road obstacle detection and tracking using radar and visual information. In Proceedings of the 17th International IEEE Conference on Intelligent Transportation Systems (ITSC), Qingdao, China, 8–11 October 2014; IEEE: Piscataway, NJ, USA, 2014; pp. 39–44.
7. Yu, Z.; Bai, J.; Chen, S.; Huang, L.; Bi, X. Camera-Radar Data Fusion for Target Detection via Kalman Filter and Bayesian Estimation. *SAE Tech. Pap.* **2018**. [CrossRef]
8. Guo, S.; Wang, S.; Yang, Z.; Wang, L.; Zhang, H.; Guo, P.; Gao, Y.; Guo, J. A Review of Deep Learning-Based Visual Multi-Object Tracking Algorithms for Autonomous Driving. *Appl. Sci.* **2022**, *12*, 10741. [CrossRef]
9. Joly, C.; Betaille, D.; Peyret, F. Étude comparative des techniques de filtrage non-linéaire appliquées à la localisation 2D d'un véhicule en temps réel. *Trait. Du Signal* **2008**, *25*, 20.
10. Wang, H.; Bansal, M.; Gimpel, K.; McAllester, D. Machine comprehension with syntax, frames, and semantics. In Proceedings of the 53rd Annual Meeting of the Association for Computational Linguistics and the 7th International Joint Conference on Natural Language Processing (Volume 2: Short Papers), Beijing, China, 26–31 July 2015; pp. 700–706.
11. Iter, D.; Kuck, J.; Zhuang, P. Target Tracking with Kalman Filtering, knn and Lstms. 2016, pp. 1–7. Available online: https://cs229.stanford.edu/proj2016/report/IterKuckZhuang-TargetTrackingwithKalmanFilteringKNNandLSTMs-report.pdf (accessed on 2 August 2023).
12. Gu, J.; Yang, X.; De Mello, S.; Kautz, J. Dynamic facial analysis: From Bayesian filtering to recurrent neural network. In Proceedings of the 2017 IEEE Conference on Computer Vision and Pattern Recognition (CVPR), Honolulu, HI, USA, 21–26 July 2017; pp. 1548–1557.
13. Altché, F.; de La Fortelle, A. An LSTM network for highway trajectory prediction. In 2017 IEEE 20th International Conference on Intelligent Transportation Systems (ITSC), Yokohama, Japan, 16–19 October 2017; IEEE: Piscataway, NJ, USA, 2017; pp. 353–359.
14. Ondruska, P.; Posner, I. Deep tracking: Seeing beyond seeing using recurrent neural networks. In Proceedings of the AAAI Conference on Artificial Intelligence, Phoenix, AZ, USA, 12–17 February 2016.

15. Alahi, A.; Goel, K.; Ramanathan, V.; Robicquet, A.; Fei-Fei, L.; Savarese, S. Social lstm: Human trajectory prediction in crowded spaces. In Proceedings of the 2016 IEEE Conference on Computer Vision and Pattern Recognition (CVPR), Las Vegas, NV, USA, 27–30 June 2016; pp. 961–971.
16. Ma, Y.; Zhu, X.; Zhang, S.; Yang, R.; Wang, W.; Manocha, D. TrafficPredict: Trajectory Prediction for Heterogeneous Traffic-Agents. *Proc. Conf. AAAI Artif. Intell.* **2019**, *33*, 6120–6127. [CrossRef]
17. Park, S.H.; Kim, B.; Kang, C.M.; Chung, C.C.; Choi, J.W. Sequence-to-sequence prediction of vehicle trajectory via LSTM encoder-decoder architecture. In Proceedings of the 2018 IEEE Intelligent Vehicles Symposium (IV), Changshu, China, 26–30 June 2018; IEEE: Piscataway, NJ, USA, 2018; pp. 1672–1678.
18. Sengupta, A.; Jin, F.; Cao, S. A DNN-LSTM based target tracking approach using mmWave radar and camera sensor fusion. In Proceedings of the 2019 IEEE National Aerospace and Electronics Conference (NAECON), Dayton, OH, USA, 15–19 July 2019; IEEE: Piscataway, NJ, USA, 2019; pp. 688–693.
19. Zhang, D.; Maei, H.; Wang, X.; Wang, Y.-F. Deep reinforcement learning for visual object tracking in videos. *arXiv* **2017**, arXiv:1701.08936.
20. Kuhn, H.W. The Hungarian method for the assignment problem. *Nav. Res. Logist. Q.* **1955**, *2*, 83–97. [CrossRef]
21. Munkres, J. Algorithms for the assignment and transportation problems. *J. Soc. Ind. Appl. Math.* **1957**, *5*, 32–38. [CrossRef]
22. Reid, D. An algorithm for tracking multiple targets. *IEEE Trans. Autom. Control* **1979**, *24*, 843–854. [CrossRef]
23. Fortmann, T.; Bar-Shalom, Y.; Scheffe, M. Sonar tracking of multiple targets using joint probabilistic data association. *IEEE J. Ocean. Eng.* **1983**, *8*, 173–184. [CrossRef]
24. Milan, A.; Rezatofighi, S.H.; Dick, A.; Reid, I.; Schindler, K. Online multi-target tracking using recurrent neural networks. In Proceedings of the AAAI Conference on Artificial Intelligence, San Francisco, CA, USA, 4–9 February 2017.
25. Liu, H.; Zhang, H.; Mertz, C. DeepDA: LSTM-based deep data association network for multi-targets tracking in clutter. In Proceedings of the 2019 22th International Conference on Information Fusion (FUSION), Ottawa, ON, Canada, 2–5 July 2019; IEEE: Piscataway, NJ, USA, 2019; pp. 1–8.
26. Sadeghian, A.; Alahi, A.; Savarese, S. Tracking The Untrackable: Learning To Track Multiple Cues with Long-Term Dependencies. In Proceedings of the 2017 IEEE International Conference on Computer Vision (ICCV), Venice, Italy, 22–29 October 2017; pp. 300–311.
27. Wan, X.; Wang, J.; Zhou, S. An online and flexible multi-object tracking framework using long short-term memory. In Proceedings of the 2018 IEEE/CVF Conference on Computer Vision and Pattern Recognition Workshops (CVPRW), Salt Lake City, UT, USA, 18–22 June 2018; pp. 1230–1238.
28. Luo, W.; Sun, P.; Zhong, F.; Liu, W.; Zhang, T.; Wang, Y. End-to-end active object tracking via reinforcement learning. In Proceedings of the 35th International Conference on Machine Learning, PMLR, Stockholm, Sweden, 10–15 July 2018; pp. 3286–3295.
29. Huang, K.; Lertniphonphan, K.; Chen, F.; Li, J.; Wang, Z. Multi-Object Tracking by Self-Supervised Learning Appearance Model. In Proceedings of the IEEE/CVF Conference on Computer Vision and Pattern Recognition, Vancouver, BC, Canada, 18–22 June 2023; pp. 3162–3168.
30. Kieritz, H.; Hübner, W.; Arens, M. Joint detection and online multi-object tracking. In Proceedings of the 2018 IEEE/CVF Conference on Computer Vision and Pattern Recognition Workshops (CVPRW), Salt Lake City, UT, USA, 18–22 June 2018; pp. 1459–1467.
31. Kwangjin, Y.; Yong, K.D.; Young-Chul, Y.; Jeon, M. Data association for multi-object tracking via deep neural networks. *Sensors* **2019**, *19*, 559.
32. Xiang, Y.; Alahi, A.; Savarese, S. Learning to track: Online multi-object tracking by decision making. In Proceedings of the IEEE International Conference on Computer Vision, Santiago, Chile, 7–13 December 2015; pp. 4705–4713.
33. Mozaffari, S.; Al-Jarrah, O.Y.; Dianati, M.; Jennings, P.; Mouzakitis, A. Deep learning-based vehicle behavior prediction for autonomous driving applications: A review. *IEEE Trans. Intell. Transp. Syst.* **2020**, *23*, 33–47. [CrossRef]
34. Rainer, S.; Keni, B.; Rachel, B. The CLEAR 2006 evaluation. In *Multimodal Technologies for Perception of Humans: First International Evaluation Workshop on Classification of Events, Activities and Relationships, CLEAR 2006, Southampton, UK, April 6–7, 2006, Revised Selected Papers 1*; Springer: Berlin/Heidelberg, Germany, 2007; pp. 1–44.
35. Gioele, C.; Luque, S.F.; Siham, T.; Troiano, L.; Tagliaferri, R.; Herrera, F. Deep learning in video multi-object tracking: A survey. *Neurocomputing* **2020**, *381*, 61–88.

sensors

MDPI

Article

Application of Adaptive Robust Kalman Filter Base on MCC for SINS/GPS Integrated Navigation

Linfeng Li [1,2], Jian Wang [3], Zhiming Chen [1] and Teng Yu [1,*]

1 School of Integrated Circuits and Electronics, Beijing Institute of Technology, Beijing 100081, China; 3420225072@bit.edu.cn (L.L.); czm@bit.edu.cn (Z.C.)
2 Beijing Institute of Electronic System Engineering, Beijing 100854, China
3 Beijing Institute of Control and Electronic Technology, Beijing 100038, China; wangjianseu@163.com
* Correspondence: 7520230152@bit.edu.cn; Tel.: +86-152-1003-9018

Abstract: In this paper, an adaptive and robust Kalman filter algorithm based on the maximum correntropy criterion (MCC) is proposed to solve the problem of integrated navigation accuracy reduction, which is caused by the non-Gaussian noise and time-varying noise of GPS measurement in complex environment. Firstly, the Grubbs criterion was used to remove outliers, which are contained in the GPS measurement. Then, a fixed-length sliding window was used to estimate the decay factor adaptively. Based on the fixed-length sliding window method, the time-varying noises, which are considered in integrated navigation system, are addressed. Moreover, a MCC method is used to suppress the non-Gaussian noises, which are generated with external corruption. Finally, the method, which is proposed in this paper, is verified by the designed simulation and field tests. The results show that the influence of the non-Gaussian noise and time-varying noise of the GPS measurement is detected and isolated by the proposed algorithm, effectively. The navigation accuracy and stability are improved.

Keywords: adaptive and robust Kalman filter; maximum correntropy criterion (MCC); sliding window method; global positioning system (GPS); strapdown inertial navigation system (SINS)

1. Introduction

The global positioning system (GPS) outputs the velocity and position information of the carrier based on the satellite signals, which contain the characteristics of continuity, real-time, stability, and so on, but its positioning accuracy is easily corrupted by the external environment. Moreover, it cannot output the positioning information without the external satellite signals. Thus, the self-contained characteristic of the GPS is poor [1–4]. The strapdown inertial navigation system (SINS) is an autonomous navigation system based on Newton's law of kinematics [5,6]. The three-dimensional attitude, velocity, and position information of the carrier can be obtained by integrating the specific force and angular velocity with inertial navigation algorithm. Its characteristics, which are strong autonomy, anti-interference, and high sampling rate, are the major advantages of the integrated navigation system. But the errors of the SINS accumulate over time, and its stability is poor [7]. Therefore, the integration of the GPS and SINS can effectively make up for the defects of each and improve the stability and reliability of the integrated navigation system [8].

However, in practical applications, the GPS measurements are easily corrupted by the external environment, resulting in measurement noises and outliers, which are no longer obeying Gaussian distribution. Under the assumption of the noises with the Gaussian distribution, the Kalman filter is the optimal unbiased estimation of information fusion. When the measurement noise is no longer obeying Gaussian distribution, the estimation accuracy of the Kalman filter cannot be maintained [9]. To solve this problem, many researchers put forward several methods to improve the performance of the Kalman filter.

The Kalman filter based on Huber's core function, which is based on the $l1$ and $l2$ norms, is designed to suppress non-Gaussian noise in [10]. However, these methods still assign some weight to invalid measurements. Using a maximum correlation entropy improvement, the Kalman filter can be improved to address non-Gaussian noises [11–13], and the weights of measurement information are updated in real time. The Kalman filter obtains better performance when the non-Gaussian noises are suppressed. Bayesian estimation uses the maximum posterior probability to update the measurement covariance matrix [14], which improves the adaptive characteristics of the Kalman filter. Other methods, which are based on student-t distribution models, are devised in [15], and these methods give the optimal solution with Bayesian estimation, which requires accurate prior information. However, the mixing probability in the time-varying non-Gaussian case does not change fast. At this time, the prior information is no longer reliable.

The above methods deal with one of the non-Gaussian characteristics, and the time-varying characteristics, of the GPS measurement noises, but they fail to deal with both non-Gaussian noise and time-varying noise. Therefore, an adaptive and robust Kalman filter algorithm, which is based on the MCC criterion, is proposed in this paper to suppress non-Gaussian noise and time-varying noise at the same time. Moreover, before giving weight to the different types of measurement information with the MCC criterion, the new outliers in the fixed-length sliding window were detected and eliminated by the Grubbs test method, and the measurement noise covariance matrix was estimated adaptively. In this way, the time variability and non-Gaussian characteristics of measurement noise were processed simultaneously. Moreover, the estimation accuracy of integrated navigation system was improved.

The rest of this paper is organized as follows. Section 2 gives the models of the SINS/GPS integrated navigation system. All of the system equations and the measurement models are shown in detail. In Section 3, the robust Kalman filter, which is based on the MCC criterion, is designed. Moreover, the adaptive filter method, which is based on the adaptive decay factor, is devised with the fixed-length sliding windows method. The simulation and field tests are designed to verify the performance of the proposed method in Section 4. Finally, the conclusions are drawn in Section 5.

2. Models of SINS/GPS Integrated Navigation System

In this section, the model of the SINS/GPS integrated system is derived, and the derivations of the model are also investigated. The inertial coordinate frame, which is a non-rotation frame, is defined as i-frame. The earth coordinate frame, which is the earth-centered earth-fixed (ECEF) frame, is defined as e-frame. The navigation coordinate frame, which is the east–north-up frame, is defined as n-frame. The carrier coordinate frame, which is the right–forward–up frame, is defiend as b-frame. And the coordinate frame, which is obtained by SINS calculation, is n'-frame.

2.1. The System Model of the SINS/GPS Integration

In this paper, the state vector of SINS/GPS integrated navigation is defined as follows:

$$\boldsymbol{x} = \begin{bmatrix} (\boldsymbol{\varphi})^{\top} & (\boldsymbol{v}^{n})^{\top} & (\delta \boldsymbol{p})^{\top} & \left(\boldsymbol{\varepsilon}^{b}\right)^{\top} & \left(\boldsymbol{\nabla}^{b}\right)^{\top} \end{bmatrix}^{\top} \tag{1}$$

where $\boldsymbol{\varphi} = \begin{bmatrix} \varphi_E & \varphi_N & \varphi_U \end{bmatrix}$ denotes the attitude error between the n-frame and calculated n'-frame. $\delta v^n = \begin{bmatrix} \delta v_E & \delta v_N & \delta v_U \end{bmatrix}$ denotes the velocity error. $\delta \boldsymbol{p} = \begin{bmatrix} \delta L & \delta \lambda & \delta h \end{bmatrix}$ is the position error, being, respectively, the latitude error, longitude error, and height error. $\boldsymbol{\varepsilon}^b$ and $\boldsymbol{\nabla}^b$ are the dynamic bias of triaxis gyroscopes and triaxis accelerometers.

Based on the aforementioned analysis, the system model of SINS/GPS integrated navigation is expressed as follows:

$$\dot{\boldsymbol{x}} = \boldsymbol{F}\boldsymbol{x} + \boldsymbol{w} \tag{2}$$

where $\boldsymbol{F}$ denotes the state transition matrix; $\boldsymbol{w}$ denotes the process noise; and $\boldsymbol{F}$ can be calculated by the error equation, which is obtained by the SINS calculation. And the error equation can be expressed as follows:

$$\begin{cases} \dot{\boldsymbol{\varphi}} = -\boldsymbol{\omega}_{in}^{n} \times \boldsymbol{\varphi} + \delta\boldsymbol{\omega}_{in}^{n} + \boldsymbol{C}_{b}^{n}\boldsymbol{\varepsilon}^{b} \\ \delta\dot{\boldsymbol{v}}^{n} = \boldsymbol{C}_{b}^{n}\boldsymbol{f}^{b} \times \boldsymbol{\varphi} - (2\delta\boldsymbol{\omega}_{ie}^{n} + \delta\boldsymbol{\omega}_{en}^{n}) \times \boldsymbol{v}^{n} \\ \quad - (2\boldsymbol{\omega}_{ie}^{n} + \boldsymbol{\omega}_{en}^{n}) \times \delta\boldsymbol{v}^{n} + \boldsymbol{C}_{b}^{n}\boldsymbol{\nabla}^{b} \\ \delta\dot{L} = \dfrac{\delta v_{N}}{R_{M} + h} \\ \delta\dot{\lambda} = \dfrac{\delta v_{E}}{R_{N} + h}\sec L + \dfrac{v_{E}}{R_{N} + h}\sec L \tan L \delta L \\ \delta\dot{h} = \delta v_{U} \\ \dot{\boldsymbol{\varepsilon}}^{b} = 0 \\ \dot{\boldsymbol{\nabla}}^{b} = 0 \end{cases} \tag{3}$$

where $\boldsymbol{\omega}_{in}^{n}$ denotes the angular velocity of the n-frame with respect to the i-frame expressed in the n-frame, and $\delta\boldsymbol{\omega}_{in}^{n}$ is the corresponding error. $\boldsymbol{\omega}_{ie}^{n}$ denotes the angular velocity of the e-frame with respect to the i-frame expressed in the n-frame, and $\delta\boldsymbol{\omega}_{ie}^{n}$ is the corresponding error. $\boldsymbol{C}_{b}^{n}$ denotes the direction cosine matrix (DCM) from the b-frame to the n-frame. $\boldsymbol{f}^{b}$ denotes the specific force expressed in the b-frame. R_{M} is the radius of curvature in meridian, and R_{N} is the radius of curvature in prime vertical.

2.2. The Measurement Model of the SINS/GPS Integration

In this paper, the measurement model is defined as the errors of the velocity and position between the SINS and GPS. Thus, the measurement model of the SINS/GPS integrated navigation system can be given as follows:

$$\boldsymbol{z} = \boldsymbol{H}\boldsymbol{x} + \boldsymbol{v} \tag{4}$$

where $\boldsymbol{z}$ is the observable vector. $\boldsymbol{H}$ denotes observable matrix, and $\boldsymbol{v}$ denotes measurement noise.

The observable vector can be expressed as:

$$\boldsymbol{z} = \begin{bmatrix} \boldsymbol{v}_{SINS}^{n} - \boldsymbol{v}_{GPS}^{n} \\ \boldsymbol{p}_{SINS} - \boldsymbol{p}_{GPS} \end{bmatrix} \tag{5}$$

The observable matrix can be expressed as:

$$\boldsymbol{H} = \begin{bmatrix} \boldsymbol{0}_{3\times3} & \boldsymbol{I}_{3\times3} & \boldsymbol{0}_{3\times3} & \boldsymbol{0}_{3\times3} & \boldsymbol{0}_{3\times3} \\ \boldsymbol{0}_{3\times3} & \boldsymbol{0}_{3\times3} & \boldsymbol{I}_{3\times3} & \boldsymbol{0}_{3\times3} & \boldsymbol{0}_{3\times3} \end{bmatrix} \tag{6}$$

From the above derivation, it can be found that the accuracy of the integrated system is determined by the external velocity and position, which are obtained from the GPS. However, the positioning accuracy of the GPS is determined by the external circumstances. If the vehicle is moving in the opening area, the positioning accuracy will be well. If the vehicle is moving in the occluded area, the positioning accuracy will be poor. To solve this problem, an adaptive and robust Kalman filter algorithm, which is based on the MCC criterion, is proposed in this paper to suppress non-Gaussian noise and time-varying noise, which are caused by the signal block of the GPS. The derivations of the proposed method are shown in the next sections.

3. Adaptive and Robust Kalman Filter Based on MCC

In this section, a robust Kalman filter, which is based on the MCC criterion, is proposed for outliers detection and isolation. Moreover, an adaptive method, which is based on the adaptive decay factor method, is proposed to improve the estimated accuracy of the robust Kalman filter.

3.1. Robust Kalman Filter Based on MCC

Correlation entropy is defined as the similarity between two random variables. Define two random variables X and Y. The correlation entropy between them can be expressed as:

$$V(X,Y) = E[\boldsymbol{\kappa}(X,Y)] = \int \boldsymbol{\kappa}(x,y)dF_{XY}(x,y) \tag{7}$$

where $E[\cdot]$ denotes the expectation of the variables, $\boldsymbol{\kappa}(\cdot,\cdot)$ denotes the kernel function, and $F_{XY}(x,y)$ denotes the joint probability density function between X and Y. In this paper, we select the Gaussian kernel as kernel function:

$$\boldsymbol{\kappa}(x,y) = G_\sigma(x-y) = \exp\left(-\frac{\|x-y\|^2}{2\sigma^2}\right) \tag{8}$$

where $\sigma > 0$ denotes the kernel bandwidth of correntropy. From Equation (8), we can see that the correntropy of random variable X and Y with the Gaussian kernel reaches its maximum if and only if $X = Y$. In practice, the joint probability density is always unknown, so it is common to estimate an approximation of the correlation entropy using the following sampling points:

$$\hat{V}(X,Y) = \frac{1}{N}\sum_{i=1}^{N} G_\sigma(x_i - y_i) \tag{9}$$

Then, the covariance of measurement noise is expressed as:

$$E\left[\boldsymbol{v}_k\boldsymbol{v}_k^\top\right] = \boldsymbol{\lambda}_k\boldsymbol{R}_k = \boldsymbol{B}_k\boldsymbol{B}_k^\top \tag{10}$$

where $\boldsymbol{\lambda}_k$ is the decay factor.

Left multiplying both sides of Equation (4) by $\boldsymbol{B}_k^{-1}$ yields the linear regression model:

$$\boldsymbol{\xi}_k = \boldsymbol{D}_k - g(\boldsymbol{x}_k) \tag{11}$$

where, $\boldsymbol{\xi}_k = \boldsymbol{B}_k^{-1}\boldsymbol{v}_k$, $\boldsymbol{D}_k = \boldsymbol{B}_k^{-1}\boldsymbol{z}_k$, $g(\boldsymbol{x}_k) = \boldsymbol{B}_k^{-1}\boldsymbol{H}_k\boldsymbol{x}_k$.

According to the least squares principle and maximum correlation entropy criterion, the cost function is constructed as follows:

$$J_L(\boldsymbol{x}_k) = \left\|\boldsymbol{x}_k - \hat{\boldsymbol{x}}_{k|k-1}\right\|^2_{(\boldsymbol{P}_{k|k-1})^{-1}} + \sum_{i=1}^{m}\rho_{MCC}(\xi_{k,i}) \tag{12}$$

where $\|\boldsymbol{x}\|_{\boldsymbol{A}}^2 = \boldsymbol{x}^\top\boldsymbol{A}\boldsymbol{x}$ denotes the quadratic form with respect to $\boldsymbol{A}$, $\boldsymbol{x}_k$ is the actual state at time instant k, $\hat{\boldsymbol{x}}_{k|k-1}$ is the prediction of state at time instant k, $\boldsymbol{P}_{k|k-1}$ represents the corresponding error covariance, $\xi_{k,i}$ is the $i-th$ element of $\boldsymbol{\xi}_k$, and m refers to the dimension of the measurement vector. The core function $\rho_{MCC}(\cdot)$ can be given by:

$$\rho_{MCC}(\xi_{k,i}) = \frac{\left(1 - \exp\left(-\xi_{k,i}^2/2\sigma^2\right)\right)}{\sqrt{2\pi}\sigma} \tag{13}$$

The optimal estimate of the state vector is obtained by minimizing the cost function:

$$\hat{\boldsymbol{x}}_k = \arg\min_{\boldsymbol{x}_k}\left(\left\|\boldsymbol{x}_k - \hat{\boldsymbol{x}}_{k|k-1}\right\|^2_{(\boldsymbol{P}_{k|k-1})^{-1}} + \sum_{i=1}^{m}\rho_{MCC}(\xi_{k,i})\right) \tag{14}$$

The optimal solution can thus be found by differencing the cost function with respect to $\boldsymbol{x}_k$ as:

$$\left(\boldsymbol{P}_{k|k-1}\right)^{-1}\left(\boldsymbol{x}_k-\hat{\boldsymbol{x}}_{k|k-1}\right)-\sum_{i=1}^{m}\psi(\xi_{k,i})\frac{\partial\xi_{k,i}}{\partial\boldsymbol{x}_k}=0 \tag{15}$$

where $\psi(e_{k,i})=-G_\sigma(\xi_{k,i})e_{k,i}$; by defining the function and dialog matrix $C_{k,i}=-\frac{\psi(\xi_{k,i})}{\xi_{k,i}}=G_\sigma(\xi_{k,i})$, Equation (14) can be rewritten as:

$$\left(\boldsymbol{P}_{k|k-1}\right)^{-1}\left(\boldsymbol{x}_k-\hat{\boldsymbol{x}}_{k|k-1}\right)-\boldsymbol{H}_k^\top\boldsymbol{B}_k^{-\top}C_k\boldsymbol{B}_k^{-1}(z_k-\boldsymbol{H}_k\boldsymbol{x}_k)=\boldsymbol{0} \tag{16}$$

Let $\overline{\boldsymbol{R}}_k=\boldsymbol{B}_k\boldsymbol{C}_k^{-1}\boldsymbol{B}_k^\top$ and $\hat{\boldsymbol{x}}_{k|k}=\boldsymbol{x}_k$; thus,

$$\boldsymbol{P}_{k|k-1}^{-1}\left(\hat{\boldsymbol{x}}_{k|k}-\hat{\boldsymbol{x}}_{k|k-1}\right)=\boldsymbol{H}_k^\top\boldsymbol{R}_k^{-1}\left(z_k-\boldsymbol{H}_k\hat{\boldsymbol{x}}_{k|k}\right) \tag{17}$$

Referring to [12] for the equation simplification and derivation process, the measurement update can be given by:

$$\begin{aligned}\hat{\boldsymbol{x}}_{k|k}&=\hat{\boldsymbol{x}}_{k|k-1}+\boldsymbol{K}_k\left(z_k-\boldsymbol{H}_k\hat{\boldsymbol{x}}_{k|k-1}\right)\\ \boldsymbol{K}_k&=\boldsymbol{P}_{k|k-1}\boldsymbol{H}_k^\top\left(\boldsymbol{H}_k\boldsymbol{P}_{k|k-1}\boldsymbol{H}_k^\top+\overline{\boldsymbol{R}}_k\right)^{-1}\\ \boldsymbol{P}_{k|k}&=(\boldsymbol{I}-\boldsymbol{K}_k\boldsymbol{H}_k)\boldsymbol{P}_{k|k-1}(\boldsymbol{I}-\boldsymbol{K}_k\boldsymbol{H}_k)^\top+\boldsymbol{K}_k\overline{\boldsymbol{R}}_k\boldsymbol{K}_k^\top\end{aligned} \tag{18}$$

Using the aforementioned method, the outliers, which are contained in the GPS outputs, can be suppressed. Then, the estimated results will be stable. However, the accuracy of the robust Kalman filter will degrade when there are outliers. To address this problem, an adaptive filter method is investigated in the next subsection.

3.2. Adaptive Estimation with the Decay Factor

In order to improve the estimation accuracy of the MCC algorithm when there is time-varying noise, the unknown decay factor $\boldsymbol{\lambda}_k$ is necessary to determine in advance. In this paper, the outliers are eliminated with the residual sliding window by the Grubbs test, and then we use two different calculation methods of the new information covariance matrix to estimate the decay factor adaptively. It is noted that the Grubbs test serves to calculate the adaptive decay factor and does not eliminate the measured outliers, so it does not play a direct role in suppressing time-varying noise.

The Grubbs test, also known as the extreme studentized deviate test, is the most effective method to distinguish outliers in the case of normal distribution. It can be described as:

$$G_i=\frac{x_i-\mu}{\sigma} \tag{19}$$

where x_i represents the residual in a fixed-length sliding window, μ represents the mean value, and σ represents the standard deviation. The confidence level was set as 0.95, and the length of the sliding window M was determined by the minimization variance parameter [16]. When $G_i>G_p(M)$, the data are considered as outliers and removed from the sliding window method to be calculated.

According to the measurement equation, the new information covariance matrix can be expressed as:

$$\hat{\boldsymbol{P}}_{zz,k|k-1}=\boldsymbol{H}_k\boldsymbol{P}_{k|k-1}\boldsymbol{H}_k^\top+\boldsymbol{\lambda}_k\boldsymbol{R}_k \tag{20}$$

where $\boldsymbol{\lambda}_k=\mathrm{diag}\{\lambda_1,\lambda_2,\cdots,\lambda_m\}$.

Under the assumption that the residual is stable, the residual covariance matrix $\hat{P}_{zz,k|k-1}$ can also be obtained by using the residual $\varepsilon_k = z_k - H_k\hat{x}_{k|k-1}$ in the fixed-length sliding window:

$$\hat{P}_{zz,k|k-1} = \frac{1}{M}\sum_{i=1}^{M}\varepsilon_{k-i}\varepsilon_{k-i}^{\top} \tag{21}$$

Let $N_k = \hat{P}_{zz,k|k-1} - H_k P_{k|k-1} H_k^{\top}$, at this time, decay factor can be expressed as:

$$\lambda_i = \frac{\sum_{j=1}^{m} R_k[i,j] N_k[i,j]}{\sum_{j=1}^{m} R_k^2[i,j]}, \quad i = 1, 2, \cdots, m \tag{22}$$

Based on the aforementioned method, an adaptive and robust Kalman filtering method for SINS/GPS is devised. In the next section, the simulation and field tests are designed for verifying the performance of the proposed method.

4. Simulation and Field Tests

In order to verify the performance of the algorithm, which is proposed in this paper, the comparison tests between simulation data and field tests data are presented respectively, and the position errors of KF [6], VBKF [14], MCCKF [11], and MCCRKF are designed for comparison.

4.1. Simulation Test

In the simulation test, a test, which lasts 996 s, is designed for the verification of the SINS/GPS integrated navigation system. It is noted that the errors of the integrated navigation system will converge to a stable value when the filter converges to stable states. Thus, it is not necessary to extend the integrated time. Taking the actual sensor parameters as an example, the simulated parameters of the designed test are shown in Table 1. The initial position of the carrier is set as 108.9° E, 34.2° N, and 380 m height; the initial velocity is set as 0 m/s; and the initial attitude angle is set as 0°. The iterations of the VBKF algorithm are set as 3, and the Gauss kernel bandwidth of the MCCKF algorithm and the MCCRKF algorithm is set as 3. Four groups of comparison experiments were set up. Gaussian noise was added to the measurement in test 1, non-Gaussian noise was added to the measurement in test 2, time-varying noise was added to the measurement in test 3, and time-varying non-Gaussian noise was added to the measurement in test 4.

Table 1. Experimental parameters.

Types	Error Terms	Error Value
Errors of Instruments	bias of gyroscopes	3°/h
	random walk of gyroscopes	$0.03°/\sqrt{h}$
	bias of accelerometers	1000 μg
	random walk of accelerometers	$100\ \mu g\ \sqrt{Hz}$
	velocity errors of GPS	1 m/s
	positioning errors of GPS	2.5 m
Initial Errors of SINS	attitude errors	$[60''\ \ 60''\ \ 60']$
	velocity errors	1 m/s
	positioning errors	2.5 m

In test 1, the measurement noises of the sensors and the GPS errors are shown in Table 1. Using the designed methods, the positioning errors, which are summarized with the different four algorithms, are shown in Figure 1. With the test results, the RMSE errors of the positioning error are shown in Table 2.

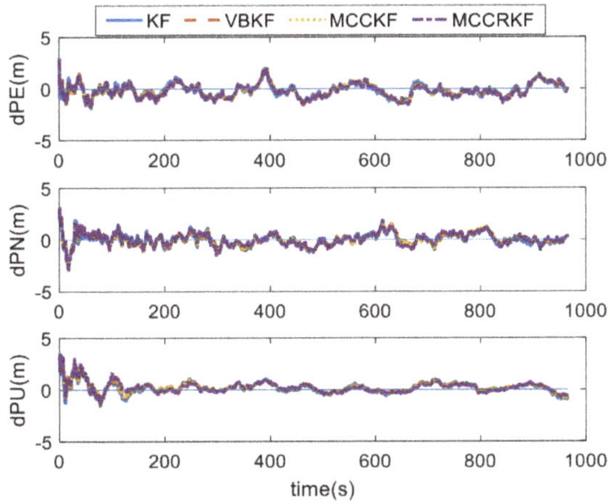

Figure 1. The positioning error of different algorithms under Gaussian noise.

Table 2. The RMSE errors of four methods.

Types	KF	VBKF	MCCKF	MCCRKF
Latitude Error (m)	0.6142	0.6156	0.6146	0.6264
Longitude Error (m)	0.6427	0.6601	0.6599	0.6524
Height Error (m)	0.5853	0.6591	0.5938	0.6251

It can be found that, when the measurement noise conforms to Gaussian distribution, the RMSE errors of the positioning errors of the four algorithms are equivalent.

The measured noise distribution in test 2 is shown in Formula (23), which represents the heavy-tail distribution. Using these type noises, the positioning errors of the four different algorithms are shown in Figure 2, and the RMSE errors of the positioning error statistical are shown in Table 3.

$$v_k \sim 0.99\mathcal{N}(\mathbf{0}, \boldsymbol{R}_k) + 0.01\mathcal{N}(\mathbf{0}, 400\boldsymbol{R}_k) \tag{23}$$

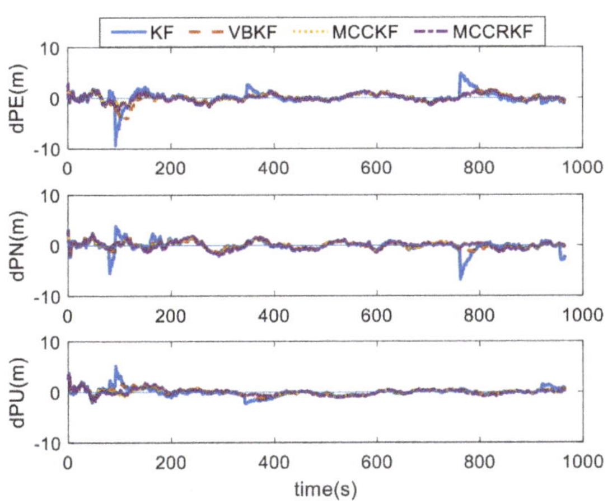

Figure 2. The positioning error of different algorithms under non-Gaussian noise.

Table 3. The RMSE errors of four methods.

Types	KF	VBKF	MCCKF	MCCRKF
Latitude Error (m)	1.7963	0.8570	0.7238	0.7747
Longitude Error (m)	1.1676	1.0291	0.6381	0.6260
Height Error (m)	1.1109	0.6222	0.6153	0.6188

It can be found that, when the measurement noises are non-Gaussian, the estimated errors of the KF algorithm increase by one time; the estimated errors of the VBKF algorithm also increase significantly; while the estimated errors of MCCKF algorithm and MCCRKF algorithm are equivalent, and their estimated accuracy is consistent with the same one when the measurement noises are Gaussian.

In test 3, the measurement noise distribution is shown in Equation (24), where t_k represents the time-varying noise coefficient. The specific value is shown in Figure 3. The positioning errors of the four different algorithms are shown in Figure 4, and the RMSE errors of the positioning error statistics are shown in Table 4.

$$v_k \sim t_k \mathcal{N}(\mathbf{0}, \boldsymbol{R}_k) \tag{24}$$

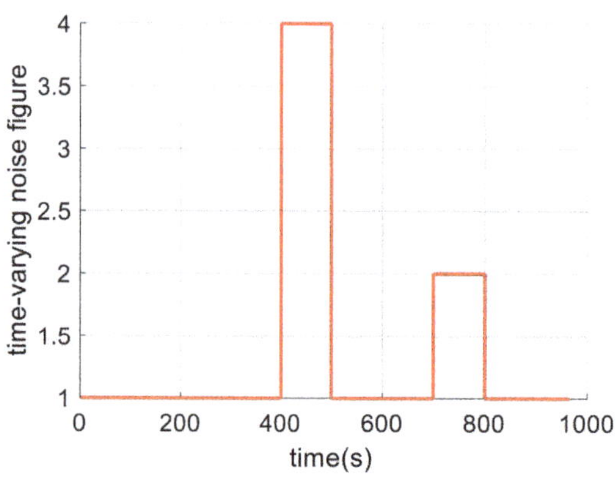

Figure 3. Time-varying parameters.

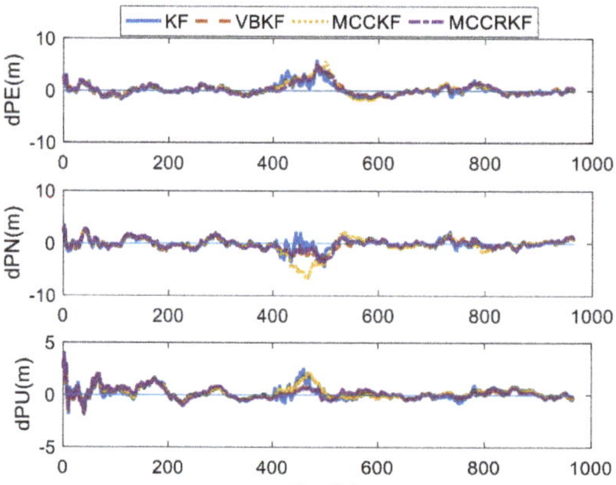

Figure 4. The positioning error of different algorithms under time-varying noise.

Table 4. The RMSE errors of four methods.

Types	KF	VBKF	MCCKF	MCCRKF
Latitude Error (m)	0.9561	0.7685	0.6886	0.7897
Longitude Error (m)	1.1705	0.7881	1.7656	0.7795
Height Error (m)	0.7313	0.6989	0.6340	0.6941

It can be found that, when the measurement noises are time-varying, the estimated errors of the four algorithms increase due to the degradation of GPS accuracy. However, due to the time-varying noise, the MCCKF algorithm is corrupted by these type noises, and the estimated accuracy is even lower than the same results of the Kalman filter, while the estimation errors of the VBKF algorithm and the MCCRKF algorithm have a small increase, which is within the normal range.

The measurement noise distribution, which is used in test 4, is shown in Formula (25). The positioning errors of the four different algorithms are shown in Figure 5, and the RMSE errors of the error statistics are shown in Table 5.

$$v_k \sim t_k(0.99\mathcal{N}(\mathbf{0}, \boldsymbol{R}_k) + 0.01\mathcal{N}(\mathbf{0}, 400\boldsymbol{R}_k)) \tag{25}$$

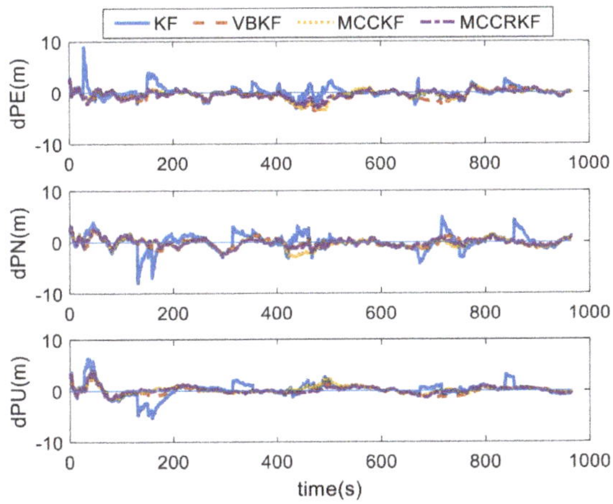

Figure 5. The positioning error of different algorithms under time-varying and non-Gaussian noises.

Table 5. The RMSE errors of four methods.

Types	KF	VBKF	MCCKF	MCCRKF
Latitude Error (m)	1.5112	1.1097	1.2023	0.8520
Longitude Error (m)	1.2113	1.0052	0.8344	0.7199
Height Error (m)	0.9170	0.6098	0.6296	0.5890

It can be found that the estimated errors of the MCCRKF algorithm are minimal when the noises are time-varying and non-Gaussian. And the time-varying noise is suppressed to a certain extent in the case of non-Gaussian noise suppression.

4.2. Field Test

To further validate the performance of the proposed method, the field test, which is carried on a vehicle in Beijing, is designed. During the test, the bias of the gyroscope is under 3°/h. The random walk of the gyroscope is close to $0.03°/\sqrt{\text{h}}$, and the bias of the accelerometer is under 1000 µg. The random walk of the accelerometer is close to $100\ \mu\text{g}/\sqrt{\text{Hz}}$. The output rates of all inertial sensors are 200 Hz. In the test, the GPS receiver, which is a product of the NovAtel Corporation, is adopted. The velocity error of the GPS receiver is around 1 m/s. The positioning error of the GPS receiver is about 2.5 m. The reference system is a high-level navigation system, which is equipped with triaxial fiber-optical gyroscopes and a triaxial quartz accelerometers. The bias of the fiber-optical gyroscopes is under $0.01^{\circ}/\mathrm{h}$. The random walk of the fiber-optical gyroscope is close to $0.001^{\circ}/\sqrt{\mathrm{h}}$, and the bias of the quartz accelerometer is under 100 μg. The random walk of the quartz accelerometer is close to $10\ \mu\mathrm{g}/\sqrt{\mathrm{Hz}}$. The velocity error of the reference navigation is 0.01 m/s, and the position error of the reference system is 0.1 m.

The output rates of the GPS receiver is 1 Hz. The whole field test lasts for 1824 s. In Figures 6 and 7, the measurement errors of the GPS receiver are shown. In Figures 6 and 7, the outliers are contained. Lastly, the in-motion trajectory of the vehicle is shown in Figure 8.

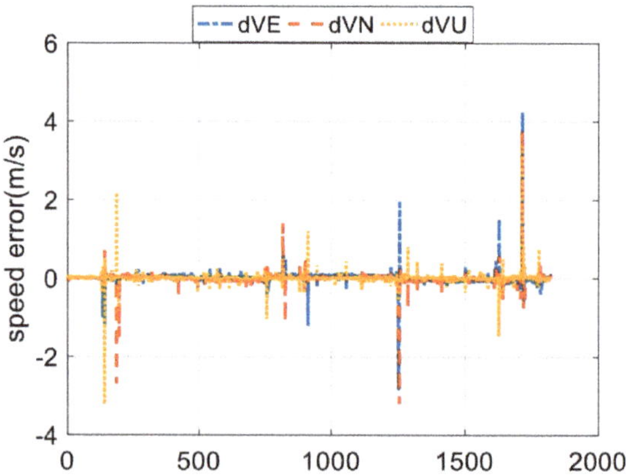

Figure 6. The velocity errors of the GPS outputs.

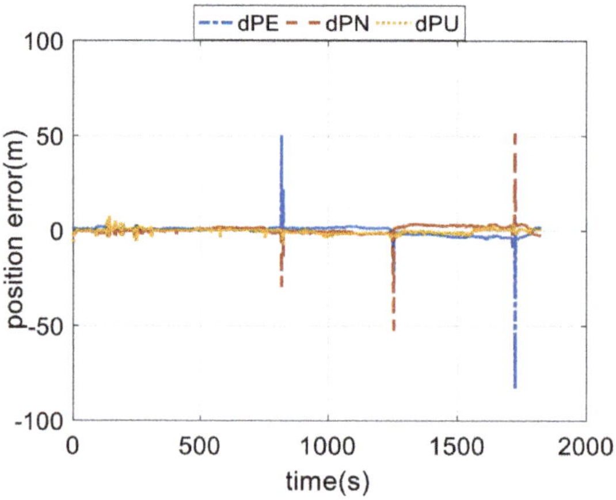

Figure 7. The positioning errors of the GPS outputs.

Figure 8. The moving trajectory of the vehicle.

In Figures 9–11, the attitude, velocity, and positioning errors of the four methods are shown, and the RMSE errors of the four methods are lists in Tables 6 and 7. As can be seen from the attitude errors in Figure 9, the horizontal attitude errors do not diverge; this is because the horizontal attitude is constrained by the accelerometers' measurements. However, the yaw angles diverge when the integrated system lasts for 1250 s; this is because the yaw angle is determined by the GPS measurements. When the outliers are contained in the GPS outputs, the accuracy of the yaw angle degrades.

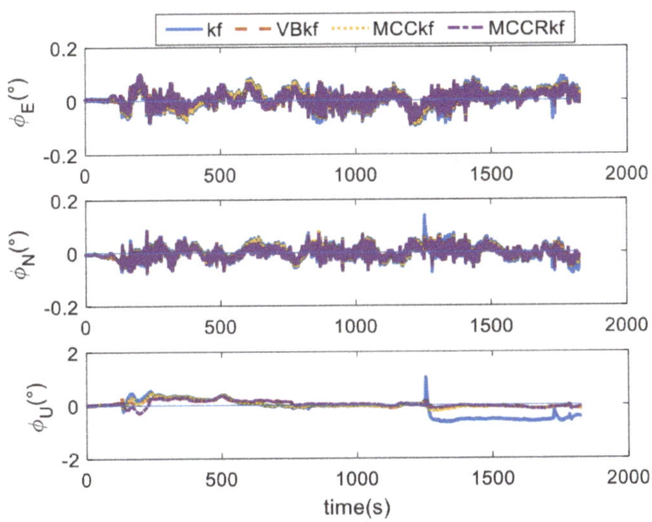

Figure 9. The attitude errors.

In Figure 10, it can be found that the velocity errors converge to a small value when there are no outliers in the GPS outputs. However, when there are outliers, the number of errors of the KF method increase. With the robust filter method, the errors of VBKF, MCCKF, and MCCRKF are not corrupted by the outliers. It is noted that the up velocity is not easily corrupted by the outliers; this is because the height channel is damped in this integrated system.

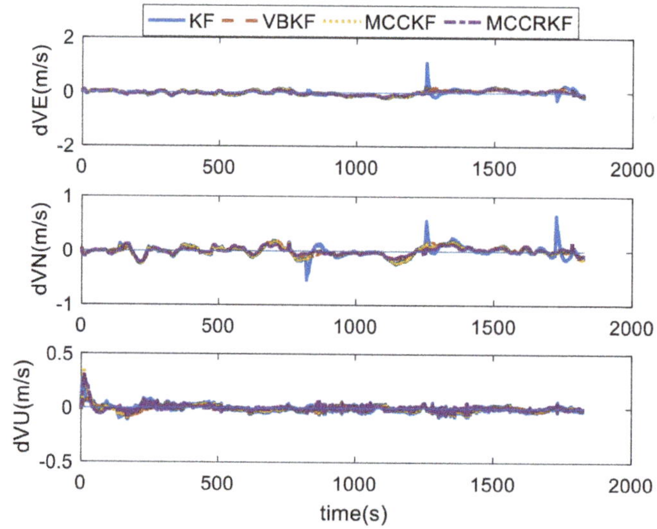

Figure 10. The velocity errors.

Table 6. The RMSE errors of velocity.

Types	KF	VBKF	MCCKF	MCCRKF
East Velocity Error (m/s)	0.1069	0.0782	0.0842	0.0700
North Velocity Error (m/s)	0.1061	0.0762	0.0886	0.0666
Up Velocity Error (m/s)	0.0285	0.0275	0.0332	0.0309

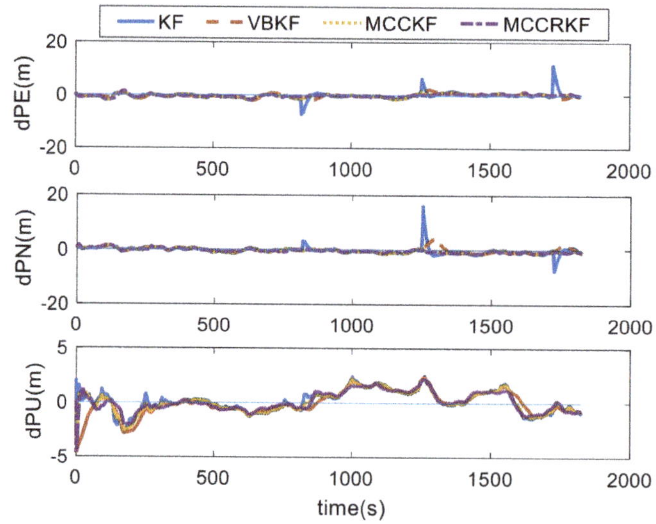

Figure 11. The positioning errors.

Table 7. The RMSE errors of position.

Types	KF	VBKF	MCCKF	MCCRKF
East Velocity Error (m/s)	1.2671	0.7943	0.7580	0.6481
North Velocity Error (m/s)	1.3406	0.8658	0.6964	0.6254
Up Velocity Error (m/s)	0.9871	1.1223	1.0069	0.9295

As can be seen from the position error comparison diagram, the KF algorithm and VBKF algorithm diverge to different degrees when there are outliers in the GPS horizontal position, while the MCCKF algorithm and the MCCRKF algorithm are almost not affected by outliers. There is obvious time-varying noise during 130 s to 230 s of GPS celestial position, which leads to the decrease in GPS accuracy. The estimation errors of the four algorithms all increase, but the MCCRKF algorithm has the smallest increase, indicating that the MCCRKF algorithm has good adaptability and robustness.

5. Conclusions

In this paper, an adaptive and robust Kalman filtering algorithm based on maximum correlation entropy is proposed to solve the problem that the SINS/GPS integrated navigation system estimation accuracy decreases due to the influence of non-Gaussian and time-varying noises in complex environments. Based on the MCCKF algorithm, the decay factor was calculated by a fixed-length sliding window. Then, the $\boldsymbol{R}$ matrix was estimated by using the decay factor to suppress the time-varying noises, which are contained in the GPS outputs. The performance of the proposed algorithm is verified by simulation and field tests. The tests results show that the MCCRKF algorithm can not only keep the estimation accuracy of the SINS/GPS integrated navigation system under the condition of non-Gaussian noise but also avoid the decrease in the estimation accuracy of the SINS/GPS integrated navigation system under the condition of time-varying noise.

Author Contributions: Conceptualization, L.L. and T.Y.; methodology, Z.C.; software, J.W.; validation, L.L., T.Y. and J.W.; and writing, L.L. and T.Y. All authors have read and agreed to the published version of the manuscript.

Funding: This research was funded by National Natural Science Foundation of China under grant number 52271346 and in part by the Aeronautical Science Foundation of China under Grant 20220019059001.

Institutional Review Board Statement: Not applicable.

Informed Consent Statement: Not applicable.

Data Availability Statement: Not applicable.

Conflicts of Interest: The authors declare no conflict of interest.

References

1. Wang, L.; Ma, G.; Jia, R.; Liu, P.; Zhi, Q. Application of High Precision GNSS/INS Smooth Algorithm in the Mobile Mapping Area. *J. Ordnance Equip. Eng.* **2019**, *40*, 112–115.
2. Liu, L.; Zhao, J.; Wang, C.. Application Research on BDs in Spacecraft Space-based TT&C. *Comput. Meas. Control* **2018**, *26*, 107–109+121.
3. Hu, G.; Gao, B.; Zhong, Y.; Gu, C. Unscented kalman filter with process noise covariance estimation for vehicular INS/GPS integration system. *Inf. Fusion* **2020**, *64*, 194–204. [CrossRef]
4. Liu, J.; Ren, W.; Wang, D.. Research on Positioning Accuracy of Pseudo Range Single Point Based on GPS/BDS Satellite Navigation System. *Coal Technol.* **2020**, *317*, 68–70.
5. Qin, Y. *Inertial Navigation*, 2nd ed.; Science Press: Beijing, China, 2014; pp. 154–196.
6. Qin, Y. *Kalman Filtering and Combinatorial Navigation Principle*, 1st ed.; Northwestern Polytechnical University Press: Xi'an, China, 2014; pp. 180–186.
7. Hu, G.; Wang, W.; Zhong, Y.; Gao, B.; Gu, C. A new direct filtering approach to INS/GNSS integration. *Aerosp. Sci. Technol.* **2018**, *77*, 755–764. [CrossRef]
8. Chen, G.; Cheng, J.; Yang, J.; Liu, H.; Zhang, L. Improved Neural Network Enhanced Navigation System of Adaptive Unsented Kalman Filter. *J. Electron. Inf. Technol.* **2019**, *41*, 246–253.

9. Liu, X.; Liu, X.; Yang, Y.; Guo, Y.; Zhang, W. Robust variational Bayesian method-based SINS/GPS integrated system. *Measurement* **2022**, *193*, 1. [CrossRef]
10. Xu, X.; Zhong, L. Robust adaptive multiple model integrated navigation algorithm based on M-estimation. *J. Chin. Inert. Technol.* **2021**, *29*, 482–490.
11. Lu, H.; Hao, S.; Peng, Z.; Huang, G. Application of Robust High-Degree CKF Based on MCC in Integrated Navigation. *Comput. Eng. Appl.* **2020**, *56*, 257–264.
12. Feng, K.; Li, J.; Zhang, D.; Wei, X.; Yin, J. Robust Cubature Kalman Filter for SINS/GPS Integrated Navigation Systems with Unknown Noise Statistics. *IEEE Access* **2021**, *9*, 9101–9116. [CrossRef]
13. Feng, K.; Li, J.; Zhang, D.; Wei, X.; Yin, J. AGMC-Based Robust Cubature Kalman Filter for SINS/GNSS Integrated Navigation System with Unknown Noise Statistics. *IEEE Access* **2021**, *9*, 1693–1706. [CrossRef]
14. Shen, C.; Xu, D.; Shen, F.; Cai, J. Variational bayesian adaptive Kalman filtering for GPS/INS integrated navigation. *J. Harbin Inst. Technol.* **2014**, *46*, 59–65.
15. Jia, G.; Huang, Y.; Bai, M.B.; Zhang, Y. A Novel Robust Kalman Filter With Non-stationary Heavy-tailed Measurement Noise. *IFAC-PapersOnLine* **2020**, *53*, 368–373. [CrossRef]
16. Hu, G.; Xu, L.; Gao, B.; Chang, L.; Zhong, Y. Robust unscented Kalman filter based decentralized multi-sensor information fusion for INS/GNSS/CNS integration in hypersonic vehicle navigation. *IEEE Trans. Instrum. Meas.* **2023**, *72*, 8504011. [CrossRef]

sensors

Article

ConGPS: A Smart Container Positioning System Using Inertial Sensor and Electronic Map with Infrequent GPS

Shan Huang, Zihan Song, Hyung-Rim Choi, Jae-Joong Kim, Do-Myung Park and Byung-Kwon Park *

Smart Logistics Research Center, Dong-A University, Busan 49315, Republic of Korea; 2270923@donga.ac.kr (S.H.); 1973741@donga.ac.kr (Z.S.); hrchoi@dau.ac.kr (H.-R.C.); jjkgb@dau.ac.kr (J.-J.K.); home21cc@gmail.com (D.-M.P.)
* Correspondence: bpark@dau.ac.kr; Tel.: +82-10-3254-9260

Abstract: Real-time global positioning is important for container-based logistics. However, a challenge in real-time global positioning arises from the frequency of both global positioning system (GPS) calls and GPS-denied environments during transportation. This paper proposes a novel system named ConGPS that integrates both inertial sensor and electronic map data. ConGPS estimates the speed and heading direction of a moving container based on the inertial sensor data, the container trajectory, and the speed limit information provided by an electronic map. The directional information from magnetometers, coupled with map-matching algorithms, is employed to compute container trajectories and current positions. ConGPS significantly reduces the frequency of GPS calls required to maintain an accurate current position. To evaluate the accuracy of the system, 280 min of driving data, covering a distance of 360 km, are collected. The results demonstrate that ConGPS can maintain positioning accuracy within a GPS-call interval of 15 min, even if using low-cost inertial sensors in GPS-denied environments.

Keywords: GPS-denied area; multi-sensor technology; inertial navigation; real-time positioning; map-matching algorithm; smart container

1. Introduction

With the increasing complexity and scale of supply chains, the real-time visibility of container position is crucial in today's global logistics industry [1–3]. It enhances supply chain coordination and minimizes operational risks such as theft or damage, as well as enabling efficient problem-solving in emergencies. A smart container is a container that is equipped with global positioning system (GPS) and a telecommunication system to send its position data to a remote server. The major technology currently employed for real-time positioning is GPS [4].

Although the widespread availability of precise positioning services provided by GPS can significantly enhance traceability in container transportation, GPS is subject to two limitations imposed during transportation: the battery life of the containers and the existence of GPS-denied areas [5]. Battery life is important because a battery cannot be replaced at the time of discharge during transportation. A GPS-denied area is where a satellite signal is obstructed or attenuated. For example, areas with dense buildings or long tunnels impede the seamless acquisition of GPS data and necessitate alternative positioning methods or hybrid positioning systems.

To address these limitations, various solutions have been proposed. As shown in Table 1, many researchers have sought to address the issue of positioning loss due to GPS-denied areas by employing inertial sensor data to predict trajectories [6–9]. This approach involves continuously tracking a vehicle's position based on acceleration, direction, and attitude data provided by inertial sensors, assuming an initial position is available. Rogne et al. [10] applied strap-down inertial navigation technology to infer travel trajectories in GPS-denied areas. However, solutions including strap-down inertial navigation require high-precision inertial sensors and continuous calls. For container positioning systems, the cost associated with such a solution is prohibitively high. Furthermore, a high-precision inertial navigation system (INS) requires high-precision inertial measurement unit (IMU) sensors and high sampling frequency, which increase cost and power consumption.

Table 1. Comparison of techniques for container positioning.

Works	Features	Advantage	Limitation
[6–9]	INS, GPS	Precise position	INS can only maintain short-range position guessing
[11–13]	Low-power WIFI, Bluetooth, and ZigBee	Low battery consumption, accurate position	Need to connect to edge nodes
[14,15]	SLAM	No additional infrastructure required	Requires substantial allocation of computational resources

Reducing the frequency of GPS calls to save power (e.g., by increasing the call interval to 1 h) introduces a significant issue of potentially missing important waypoint information. Typically, containers pass through or enter important intermediate stations/waypoints during transportation. If the GPS call frequency is too low, the container may fail to report those important events. A more innovative approach involves utilizing vehicle state data to predict vehicle location during GPS unavailability [9,16]. This approach primarily relies on state prediction through the application of Kalman filtering. Nevertheless, the system is susceptible to rapid error accumulation due to the utilization of only directional data as reference upon entering GPS-denied areas.

On the other hand, considerable research has been conducted to determine container positions through low-power Bluetooth/Wi-Fi/ZigBee connection technologies [11–13]. These technologies are viable alternatives for acquiring location data when GPS signals are unavailable. This approach is relevant within urban context monitoring scenarios. Nevertheless, it is imperative to acknowledge that integrating such alternative network connections engenders a heightened reliance upon existing network infrastructure.

An additional promising avenue pertains to visual simultaneous localization and mapping (SLAM), a technology that offers navigational and positional capabilities in contexts devoid of GPS signals [14,15]. Within GPS-denied areas, visual SLAM achieves navigation and localization through recognizing and tracking visual features such as key points and edges within the environment, coupled with an analysis of inter-frame motion relationships. This methodology operates independent of GPS signals, affording an accurate positioning service within GPS-denied settings such as indoor environments, urban canyons, and forested regions. Nonetheless, it is important to underscore that the processing demands of visual SLAM entail substantial computing resources, potentially constraining its applicability to the scenarios characterized by embedded devices with limited resource availability.

This study aimed to maintain real-time container positioning in GPS-denied areas while minimizing reliance on GPS calls. We propose a novel system named ConGPS that integrates inertial sensor data and electronic map data. ConGPS estimates the speed and heading direction of a moving container based on the inertial sensor data, the container trajectory, and the speed limit information provided by an electronic map. A container velocity estimation method was developed to estimate the current velocity based on the current heading direction, previous average speed, and map speed limit data. Furthermore, to enhance the precision of the position estimation, we introduced a map-matching algorithm. This algorithm aligns the estimated trajectory with the road network depicted on the electronic map to refine the position estimates. Through the results of a series of experiments, we demonstrate the superiority of the proposed algorithm in maintaining real-time position accurately while reducing the system's dependence on GPS. In summary, the contributions of this research can be summarized as follows:

- Integrating the accelerometer and magnetometer data of low-cost inertial sensors with the speed limit data of electronic map for continuously predicting container velocity during inter-GPS calls or in GPS-denied environments.
- Devising a map-matching method utilizing the distance and direction similarity of container trajectory with the roads of electronic maps for adjusting container position.
- Evaluating performance via a field experiment over a 360 km course with 280 min of transportation.

The rest of this paper is organized as follows: proposing a positioning system in Section 2; analyzing accuracy in Section 3; A discussion in Section 4; concluding the paper in Section 5.

2. ConGPS

We propose a novel real-time container positioning system called ConGPS (a smart container positioning system using inertial sensor data and electronic map data with infrequent GPS). ConGPS can maintain positioning with infrequent GPS calls as well as in GPS-denied areas based on acceleration data, orientation data, and electronic map data.

2.1. Flowchart of ConGPS

Figure 1 depicts the flowchart of ConGPS. Firstly, the GPS call interval is set to 15 min. During this period, the container's velocity estimation algorithm and the map-matching algorithm are continuously engaged to update the container's position. After a GPS call and an update of container position and trajectory, the acceleration and orientation data are obtained from inertial sensors. The acceleration data are used to identify the current state of motion (Section 2.2.1). Based on the previous position, the speed limit data from the electronic map are retrieved to estimate the container speed (Section 2.2.2). The current position and trajectory are updated according to the estimated speed and direction. To reduce the estimation error, the current position and trajectory of the container are adjusted using the map-matching algorithm (Section 2.3). The above process is repeated during the GPS call interval. On reaching the interval, we call the GPS and correct the current position and trajectory in accordance with the GPS data.

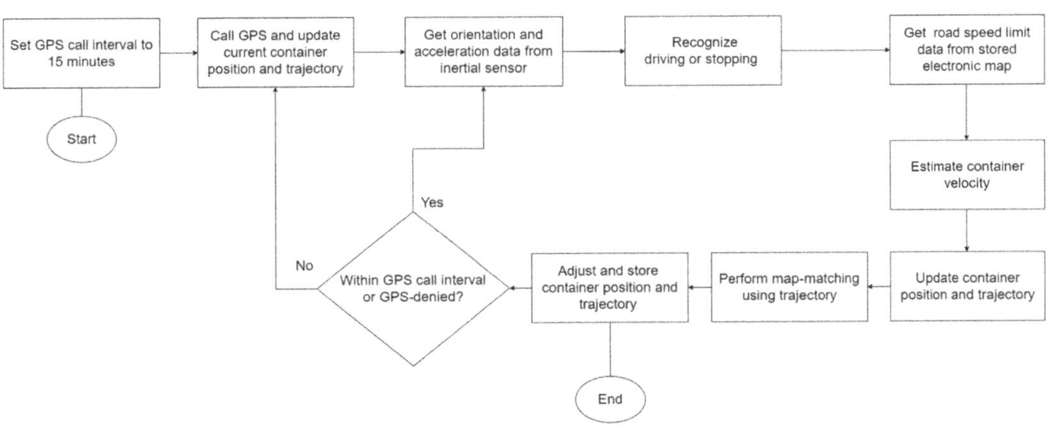

Figure 1. The flowchart of ConGPS.

2.2. Estimation of Container Positions using a Velocity-Estimating Algorithm

2.2.1. Recognizing Container Motion State

Based on the acceleration data from an accelerometer, we can determine the current motion state of the container, i.e., whether it is in motion or at a standstill. As accelerometers are sensitive to various environmental influences and may produce noisy readings, it is essential to address these challenges to ensure the accuracy and reliability of the data. The simple moving average (SMA) method is employed to mitigate the impact of noise on the accelerometer data [17,18]. The SMA is defined as follows:

$$SMA = \frac{\sum_{i=1}^{n} A_i}{n} \tag{1}$$

where n is window size; A_i is the acceleration data in a single window i.

The window size is set to discriminate different movement states through experimentation and validation to achieve optimal performance. Based on the magnitude of the acceleration values, the container's current state is classified as either moving or stationary. This binary classification provides a basis for subsequent modules in the system.

Assuming that the y-axis of the accelerometer points to the direction of container's forward movement, the filtered y-axis acceleration, after eliminating the effect of gravity, can effectively reflect the current motion state of the container. As illustrated in Figure 2, the acceleration threshold for movement state detection was set to 0.05 G (green line). When the consecutive acceleration readings for at least 3 s fall below the threshold, we assume that it is the noise induced by vibration and that the container is in a stationary state. Furthermore, in the ideal condition of uniform linear motion, the acceleration would theoretically be zero, excepting acceleration due to gravity. However, we assume that, in real transportation scenarios, containers experience minor acceleration and deceleration phases. Consequently, the proposed method is sufficiently robust to detect the container's motion status.

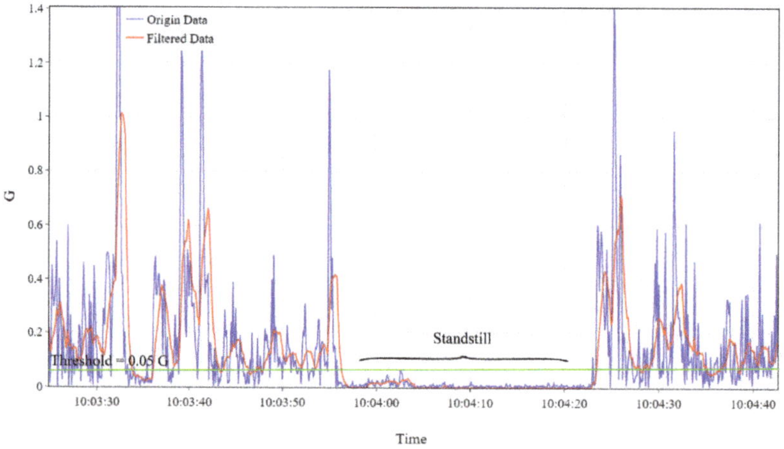

Figure 2. Acceleration data processed using the SMA.

2.2.2. Estimating Velocity and Next Position of Container

Due to error accumulation, the accuracy of inertial navigation systems rapidly deteriorates within a few minutes [19,20]. This leads to the issue that a navigation system based on inertial sensors cannot be used for positioning over a long period. We propose a dynamic velocity estimation algorithm that incorporates electronic map data to address this issue, assuming the following throughout the transportation:

- The container is traveling only on the road.
- The traveling speed of the container remains relatively constant around the speed limit of the road.

We utilize the container's stationary states as reference points. The phase between two consecutive stationary states encompasses the container's acceleration, constant speed, and deceleration stages. We employ the road's speed limit multiplied by a dynamic weight to estimate the container's velocity. The algorithm can be summarized as follows:

(1) Initialization: Perform a GPS call to obtain the current position and retrieve the speed limit of the road segment on which the container is currently traveling.

(2) Retrieve the weight w and estimate the speed $V_{estimate}$ as the road's speed limit times the weight w (in this study, let the initial weight be 0.75).

$$V_{estimate} = SpeedLimit \times w \tag{2}$$

(3) Read the orientation data from the magnetometer, $HeadingAngle$, which is the angle between the container's current direction and true north. When the container faces true north, the heading angle is 0 degrees. For true east, it is 90 degrees; for true south, 180 degrees; and for true west, 270 degrees. Next, calculate $NextPosition$ after Δt, which is the time interval between the current position and the next position. In this study, the frequency of calling the magnetometer was 1 Hz; therefore, Δt was equal to 1 s.

$$NextPosition = (PreviousLatitude + \Delta\varphi, PreviousLongitude + \Delta\lambda) \tag{3}$$

$$\Delta\lambda = \frac{\Delta x}{R \times \cos(PreviousLatitude)} \tag{4}$$

$$\Delta\varphi = \frac{\Delta y}{R} \tag{5}$$

$$\Delta x = V_{estimate} \times \cos(HeadingAngle) \times \Delta t \tag{6}$$

$$\Delta y = V_{estimate} \times \sin(HeadingAngle) \times \Delta t \tag{7}$$

where $\Delta\varphi$ the is amount of change in latitude; $\Delta\lambda$ the is amount of change in longitude; Δx is the displacement in the x-direction (east–west) in meters; Δy is the displacement in the y-direction (north–south) in meters; R is the Earth's radius, approximately 6371 km.

(4) Store $HeadingAngle$ as the latest trajectory point.

(5) Iterate Step 2 to 4 until the next GPS call is made. Note that GPS calls are made in an interval (in this study, 15 min).

(6) After making the next GPS call, estimate the average GPS speed V_{gps} from the distance D_{gps} (estimated by the map-matching algorithm; see Section 2.3) divided by the interval time between the two GPS calls.

(7) Adjust the weight w.

$$w = \frac{v_{gps}}{v_{estimate}} \tag{8}$$

(8) Repeat Steps 2 to 7 until the container comes to a stop.

(9) Upon container stoppage, store w. When the container resumes travel, retrieve w for subsequent calculations.

2.3. Adjustment of Container Positions using Trajectory-Based Map-Matching Algorithm

A trajectory-based map-matching algorithm is a method aimed at associating trajectory data with the geographical road network [21,22]. It facilitates the alignment of continuous spatiotemporal location points, constituting a trajectory, to the corresponding road segments represented in digital maps. Compared to the conventional distance-based approaches [23], a trajectory-based map-matching algorithm determines the precise path traversed by containers by matching trajectory points to the road network topology. Python's OSMnx library [24] can be used to obtain offline map data from OpenStreetMap [25].

Our trajectory-based map-matching algorithm is composed of three subalgorithms. The first involves distance-similarity computing, the second involves direction-similarity computing, and the final involves next-position adjustment. These algorithms can be summarized as follows:

2.3.1. Distance-Similarity Computing

(1) Acquire road data $R = \{r_i, i \in N^*\}$ and trajectory data $P = \{p_i, i \in N^*\}$, where r_i and p_i are composed of the latitude and longitude coordinates, respectively. R refers to all roads within 1 km of the current position. P is a series of location points produced by ConGPS within 1 km of the current position. Figure 3 illustrates an example data structure. It is noteworthy that R is sparser than P because road data typically only consist of endpoints, turning points, and intersections of roads, while trajectory data comprise a series of closely spaced waypoints.

(2) Compose a set of paths T, as shown in Figure 3. A path in T comprises the road data points such that the start point is within 300 m of the start point of P, and the end point is within 300 m of the end point of P. In Figure 3, the start point is r_1, and the end point is r_5 in a relevant path. We can use the breadth-first search algorithm [26] to obtain T.

(3) For each path in T, the path is evenly divided to have the same size as that of P, resulting in a set $K = \{k_i, i \leq |P|\}$. This process aims to transform the sparse road data into a format that allows for comparison with P. As shown in Figure 4, P contains 9 points and the path $\{r_1, r_2, r_3\}$ in T contains 3 points (representing the start, turning, and end of the road). Then, to compare the distance similarity between roads and trajectories, the path needs to be evenly divided into 9 points, as shown in Figure 4.

(4) Calculate the distance between K and P using the Haversine formula [27]. The computation formula is as follows:

$$d = 2 \times arctan(\sqrt{a}, \sqrt{1-a}) \cdot R \tag{9}$$

$$a = sin^2(\Delta\varphi/2) + cos\ \varphi_1 \times cos\ \varphi_2 \times sin^2(\Delta\lambda/2) \tag{10}$$

where d is the great-circle distance between two points; φ_1 and φ_2 are the latitudes of two points; $\Delta\varphi$ is the difference in latitude; $\Delta\lambda$ is the difference in longitude; R is the Earth's radius (mean radius = 6371 km).

(5) Find all d for each K, resulting in the distance similarity denoted as $D = \{d_i, i \leq |T|\}$.

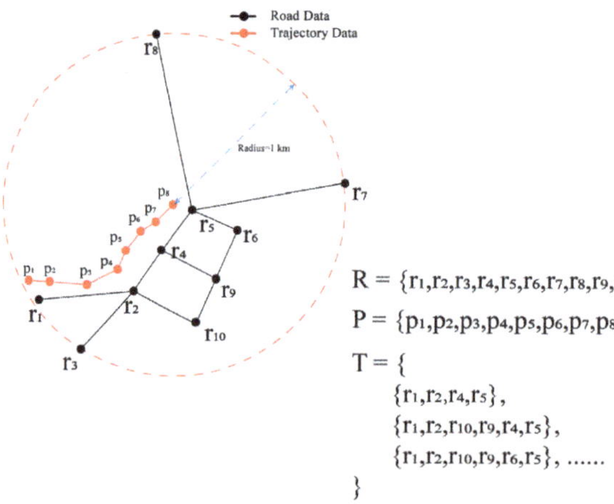

Figure 3. Example of R, P, and T.

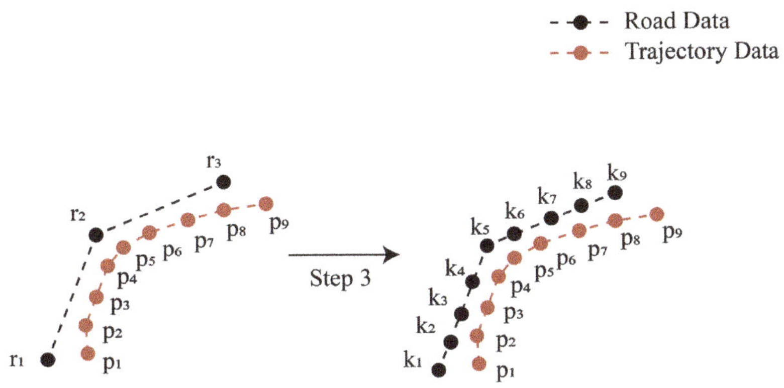

Figure 4. Example of dividing path data based on the number of trajectory points.

2.3.2. Direction-Similarity Computing

(1) to (3) The same as for distance-similarity computing.

(4) The elements in K are sequentially paired and transformed into vectors. As shown in Figure 5, assuming that K contains 7 points, the paired vectors are $\left\{\overrightarrow{k_1k_2}, \overrightarrow{k_2k_3}, \overrightarrow{k_3k_4}, \overrightarrow{k_4k_5}, \overrightarrow{k_5k_6}, \overrightarrow{k_6k_7}\right\}$. The *HeadingAngle* of each vector is then computed, resulting in a set $V = \{\theta_i, i \leq |K| - 1\}$. We define *HeadingAngle* as the angle between the vector and the true north as follows:

$$HeadingAngle = arctan(x, y) \tag{11}$$

$$x = \cos(\varphi_b) \times \sin(\Delta\lambda) \tag{12}$$

$$y = \cos(\varphi_a) \times \sin(\varphi_b) - \sin(\varphi_a) \times \cos(\varphi_b) \times \cos(\Delta\lambda) \tag{13}$$

where a and b are two points of the vector; φ is latitude; λ is longitude.

(5) Apply Step 4 to P in the same way as K, obtaining $Q = \{\theta_i, i \leq |P| - 1\}$.

(6) Compute the Pearson correlation coefficient ρ for V and Q.

$$\rho = \frac{\text{cov}(V, Q)}{\sigma_V \sigma_Q} \tag{14}$$

where σ_V is the standard deviation of V; σ_Q is the standard deviation of Q.

(7) For each K, repeat Steps 4 to 6, resulting in the direction similarity denoted as $W = \{\rho_i, i \leq |T|\}$.

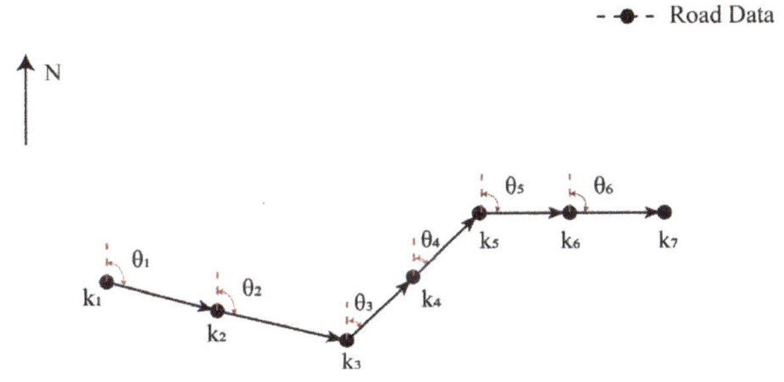

Figure 5. Example of heading angle representations.

2.3.3. Next-Position Adjustment

(1) We compute trajectory similarity s_i for each K as follows:

$$s_i = \rho_i - k \times d_i \quad (15)$$

where d_i is the distance similarity for each K; ρ_i is the direction similarity for each K; k is the weight used to balance the direction similarity and distance similarity (in this study, k was set to 0.5).

(2) We select K having maximum s_i as the matching path with P. After finding the matching path, we adjust the current position to the point in the matching path that has the shortest distance to the estimated position using the method in Section 2.2.

3. Experimental Evaluation

We employed two Android smartphones of an identical model to collect data about vehicular movement. Table 2 presents the relevant parameters of the inertial sensors installed in the smartphones. One smartphone was used to collect GPS data at a frequency of 1 Hz, thereby delineating an accurate representation of the authentic trajectory. The other smartphone was used to capture acceleration data at a rate of 10 Hz and magnetometer data at a rate of 1 Hz. The dataset encompassed a comprehensive temporal span of 360 min (337 min of travel time and 23 min of stop time) over a total vehicular travel distance of 280 km. Preceding each instance of data collection, a calibration procedure of the smartphone's magnetometer was conducted at a locale devoid of pronounced magnetic field interferences.

Table 2. Sensor specifications.

Sensors	Range	Resolution
Accelerometer	±16 g	0.5 mg/LSB
Magnetometer	±2 Gauss	0.0667 mG/LSB
GPS	<2.5 m	<0.1 m/s

Figure 6 illustrates the practical performance of the proposed algorithm along an exemplary trajectory. The container travels from the upper-right corner to the lower-left corner of the map. Figure 6a depicts the trajectory obtained only through the velocity estimation algorithm in Section 2.2. Figure 6b depicts the trajectory adjusted after undergoing the map-matching algorithm in Section 2.3, resulting in the path better aligned with road on which the container is traveling. The green dots denote the positions corrected with GPS calls with a time interval of 15 min. In Figure 6a, although the trajectory is similar to the road in direction, the trajectory travels outside the road due to the limited accuracy of the magnetometer and the algorithm. In contrast, the trajectory in Figure 6b is located inside the road owing to the adjustment produced by the map-matching algorithm.

On the other hand, the errors appearing in the middle and in the end in Figure 6b are caused by the accumulation of acceleration error and magnetometer error. The errors are regularly corrected at each GPS point. Figure 7 illustrates the average distance error of the trajectory generated by the proposed algorithm under varying intervals of GPS call. The average distance error of the trajectory is the average of all the distances between each pair of points (one on the trajectory and the other on the road). The distance is calculated according to the Haversine formula [27]. The average distance error at the 5 min, 10 min, and 15 min intervals increases slowly. However, after the 15 min interval, the error increases rapidly. Specifically, at the 30 min interval, the average distance error increases to 2.85 times greater than that at the 15 min interval.

In addition, we evaluated the performance of the motion status detection and velocity estimation algorithm in Section 2.2. Figure 8a compares V_{gps} and $V_{estimate}$ along an exemplar trajectory. V_{gps} is computed from GPS data sampled at 1 Hz. $V_{estimate}$ is derived through the velocity estimation algorithm and the container motion status recognition algorithm in Section 2.2. The trend of change in $V_{estimate}$ is similar to that in V_{gps}. Figure 8b shows the distribution of error between $V_{estimate}$ and V_{gps} in the dataset. The average discrepancy between $V_{estimate}$ and V_{gps} rests at approximately −1 m per second, showing that $V_{estimate}$ effectively corresponds to V_{gps}.

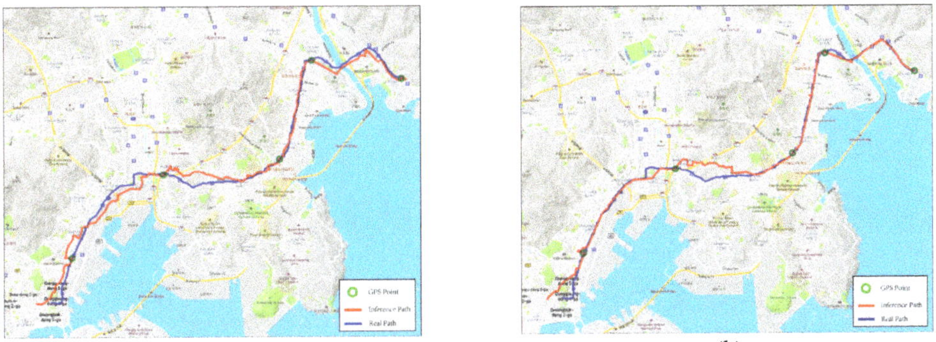

(a) (b)

Figure 6. (**a**) Trajectory using the velocity-estimating algorithm only. (**b**) Trajectory using both the velocity-estimating and map-matching algorithm.

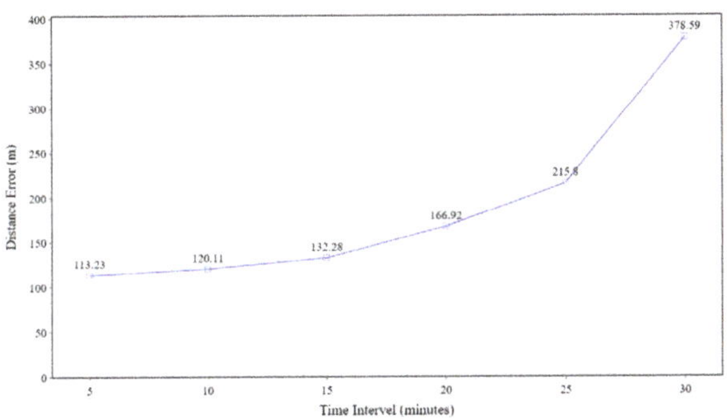

Figure 7. The average distance error between the estimated trajectory and the real trajectory along the GPS call interval.

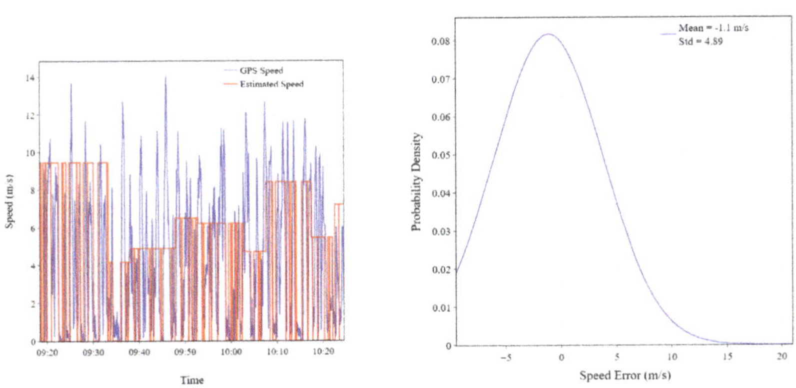

(a) (b)

Figure 8. (**a**) Comparison between GPS speed (V_{gps}) and estimated speed ($V_{estimate}$). (**b**) The probability density distribution of the error between V_{gps} and $V_{estimate}$.

Figure 9 presents the probability density distribution of distance errors between the estimated trajectories and the actual trajectory in 15 min intervals. Curve 1 corresponds to the error between the trajectory obtained through ConGPS and the actual trajectory. Curve 2 corresponds to the error when the trajectory was obtained through ConGPS without the map-matching algorithm. Curve 3 corresponds to the error when the trajectory was obtained through a direct connection of GPS points. Curve 1 exhibits the most favorable performance, with an average distance error of 132.28 m. This implies that, in comparison to the trajectory obtained via directly connecting GPS points, the proposed algorithm enhances the localization accuracy by approximately twofold. On the other hand, the average distance error of curve 2 is 1.12 times that of curve 1, which illustrates the improvement in the map-matching algorithm on positioning accuracy.

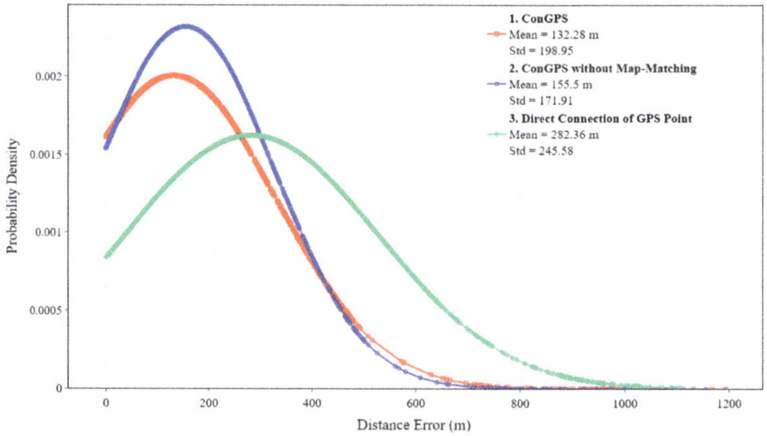

Figure 9. The probability density distribution of distance errors between the estimated trajectories and the actual trajectory.

4. Discussion

In this study, we developed a container positioning system that integrates both inertial sensor data and electronic map data. Our findings illustrate that the incorporation of road data from electronic maps enables an accurate prediction of container trajectory, employing low-precision accelerometers and magnetometers. We observed that the proposed system can maintain positioning accuracy up to a 15 min interval of GPS calls with an average distance error of 132.28 m. Due to the relatively small difference in distance error and to minimize the frequency of GPS calls, we employed a 15 min interval to assess the performance of the proposed system.

The performance partially stems from the dynamic adjustment of the estimated velocity based on the utilization of the road speed limit data obtained from electronic maps, resulting in an average speed error of approximately 1 m per second, as shown in Figure 8b. It is noteworthy that the algorithm circumvents the error accumulation inherent to traditional INS algorithms, thereby averting significant deviation between estimated and actual speed. In addition, the use of accelerometer data for detecting container motion states also contributes to avoiding error accumulation. In particular, the proposed algorithm obviates the necessity for high-precision sensors and the acquisition of supplementary sensor data.

Compared to prior research, this study's novelty resides in utilizing speed limit data and trajectory data to estimate container positions during GPS-denied periods. Sun et al. [16] proposed a navigation solution for GPS-absent scenarios, reporting errors below 100 m within 60 min. However, this solution necessitates employing relatively sophisticated onboard INS equipment and is confined to a singular testing route. Chunhakam et al. [9] introduced a navigation algorithm employing magnetometers and speed data retrieved via truck on-board diagnostics (OBD), yielding an evaluated error of 30 m. The reliance on speed information obtained exclusively from the truck's OBD system limits their method's applicability. In contrast, our work integrates information such as road speed limits and road trajectories with inertial sensor data, enhancing accuracy during GPS-denied periods. Furthermore, it reduces the frequency of GPS calls required while upholding positioning accuracy.

There are some limitations worth mentioning. As the GPS call interval increases, the proposed system's distance error also escalates. This is attributed to the simplicity of the velocity estimation algorithm. Future research endeavors may therefore contemplate refining the velocity estimation algorithm amalgamated with road and traffic information. For instance, incorporating intersections and road congestion levels could elevate the positioning accuracy.

ConGPS can provide a real-time positioning service for moving containers. The algorithm for recognizing container motion state just scans the input data to calculate the SMA and recognize a stationary or moving state, so that its time complexity is $O(n)$, where n is the input data size. The algorithm for estimating the velocity and next position of a container scans the magnetometer data and calculates the velocity and the next position, so that its time complexity is also $O(n)$. The algorithm for distance-similarity computing scans the road data and the trajectory data, making all possible paths from the road data to compute the distance between the path and the trajectory. When there are n points in the road data, maximal $n!$ paths should be considered, so that the time complexity in making all possible paths is $O(n!)$. Therefore, we should control the road data size to prohibit the path-making time from growing exponentially. In this study, we retrieved the road data only from within 1 km of the current position. Furthermore, we considered only those paths where the start point was within 300 m of the start point of the trajectory and where the end point was within 300 m of the end point of the trajectory. The algorithm for direction-similarity computing sequentially scans the path data to compute the heading angles, so that its time complexity is $O(n)$, where n is the number of points in the path. Overall, we can control the time complexity of the proposed algorithm to be $O(n)$ so as to provide a real-time positioning service for a moving container.

5. Conclusions

In this study, we developed a smart container positioning system named ConGPS that combines inertial sensor data and electronic map information. ConGPS innovatively leverages road trajectory data and road speed limit data from electronic maps as well as inertial sensor data to achieve accurate positioning with reduced GPS calls. ConGPS also cost-effectively furnishes a real-time positioning service with low-precision inertial sensors.

The limitation of this study is that the distance error of ConGPS starts to increase rapidly when the GPS request interval exceeds 15 min. This is attributed to the simplicity of the velocity estimation algorithm. Subsequent research endeavors could consider incorporating additional road- or traffic-related data to reduce distance error. For instance, such data as traffic congestion levels, road construction information, and the number of available lanes could be explored to refine the velocity estimation.

Author Contributions: Conceptualization by S.H., Z.S. and B.-K.P.; methodology development by S.H., Z.S. and B.-K.P.; software implementation by S.H. and Z.S.; validation by Z.S. and B.-K.P.; formal analysis by S.H.; investigation, S.H., Z.S., H.-R.C., J.-J.K., D.-M.P. and B.-K.P.; resources, H.-R.C., J.-J.K., D.-M.P. and B.-K.P.; data curation by S.H. and Z.S.; draft by S.H. and Z.S.; review and editing by S.H., Z.S. and B.-K.P.; supervision by H.-R.C., J.-J.K., D.-M.P. and B.-K.P.; project administration by D.-M.P. and B.-K.P.; funding acquisition by H.-R.C., J.-J.K. and D.-M.P. All authors have read and agreed to the published version of the manuscript.

Funding: This research was supported by Korea Institute of Marine Science & Technology Promotion (KIMST) funded by the Ministry of Oceans and Fisheries, Korea (20210154).

Institutional Review Board Statement: Not applicable.

Informed Consent Statement: Not applicable.

Data Availability Statement: The data presented in this study are available on request from the corresponding author. The data are not publicly available due to privacy and confidentiality concerns.

Conflicts of Interest: The authors declare no conflict of interest.

References

1. Salah, K.; Alfalasi, A.; Alfalasi, M.; Alharmoudi, M.; Alzaabi, M.; Alzyeodi, A.; Ahmad, R.W. IoT-Enabled Shipping Container with Environmental Monitoring and Location Tracking. In Proceedings of the 2020 IEEE 17th Annual Consumer Communications & Networking Conference (CCNC), Las Vegas, NV, USA, 10–13 January 2020; Volume 3, pp. 1–6.
2. Mahmood, S.; Hasan, R.; Ullah, A.; Sarker, K.U. SMART Security Alert System for Monitoring and Controlling Container Transportation. In Proceedings of the 2019 4th MEC International Conference on Big Data and Smart City (ICBDSC), Muscat, Oman, 15–16 January 2019; pp. 1–5.
3. Chan, A.S.; Sutapa, I.N. Truck management integrated information system in a shipping line company. In Proceedings of the 2017 International Conference on Soft Computing, Intelligent System and Information Technology (ICSIIT), Bali, Indonesia, 26–29 September 2017; pp. 257–262. [CrossRef]
4. Wang, Y.; Ning, X.; Xu, X. An Improved In-Motion Coarse Alignment Method for SINS/GPS Integration with Initial Velocity Error Suppression. *Sensors* **2023**, *23*, 3662. [CrossRef]
5. Li, X.; Zhang, X.; Chen, K.; Feng, S. Measurement and analysis of energy consumption on Android smartphones. In Proceedings of the 2014 4th IEEE International Conference on Information Science and Technology, Guangdong, China, 26–28 April 2014; pp. 242–245.
6. Ibrahim, A.; Abosekeen, A.; Azouz, A.; Noureldin, A. Enhanced Autonomous Vehicle Positioning Using a Loosely Coupled INS/GNSS-Based Invariant-EKF Integration. *Sensors* **2023**, *23*, 6097. [CrossRef]
7. Cao, Y.; Bai, H.; Jin, K.; Zou, G. An GNSS/INS Integrated Navigation Algorithm Based on PSO-LSTM in Satellite Rejection. *Electronics* **2023**, *12*, 2905. [CrossRef]
8. Davidson, P.; Vazquez, M.A.; Piche, R. Uninterrupted portable car navigation system using GPS, map and inertial sensors data. In Proceedings of the 2009 IEEE 13th International Symposium on Consumer Electronics, Kyoto, Japan, 25–28 May 2009; pp. 836–840.
9. Chunhakam, P.; Pummarin, P.; Jeen-Im, P.; Wardkien, P.; Wisartpong, P.; Lertteerada, K. GPS Positon Predicting System by Kalman Filter with Velocity from OBD and Direction from Magnetometer. In Proceedings of the 2021 9th International Electrical Engineering Congress (iEECON), Pattaya, Thailand, 10–12 March 2021; pp. 444–447. [CrossRef]
10. Rogne, R.H.; Bryne, T.H.; Fossen, T.I.; Johansen, T.A. On the Usage of Low-Cost MEMS Sensors, Strapdown Inertial Navigation, and Nonlinear Estimation Techniques in Dynamic Positioning. *IEEE J. Ocean. Eng.* **2021**, *46*, 24–39. [CrossRef]
11. Ni, P. Intelligent Containers for the Goods Transport. In Proceedings of the 2021 4th International Conference on Information Systems and Computer Aided Education, Dalian, China, 24–26 September 2021; ACM: New York, NY, USA, 2021; pp. 1829–1832.
12. Ungurean, I.; Chi, J.; Wang, K.; Gaitan, N.C.; Yao, H.; Yang, Y. Mobile ZigBee Network in a High RF Interference Environment. In Proceedings of the 2019 International Conference on Sensing and Instrumentation in IoT Era (ISSI), Lisbon, Portugal, 2–30 August 2019; pp. 1–5.
13. Talukder, N.; Ahamed, S.I.; Abid, R.M. Smart Tracker: Light Weight Infrastructure-less Assets Tracking solution for Ubiquitous Computing Environment. In Proceedings of the 2007 Fourth Annual International Conference on Mobile and Ubiquitous Systems: Networking & Services (MobiQuitous), Philadelphia, PA, USA, 6–10 August 2007; pp. 1–8.
14. Luo, F.; Liu, Z.; Zou, F.; Liu, M.; Cheng, Y.; Li, X. Robust Localization of Industrial Park UGV and Prior Map Maintenance. *Sensors* **2023**, *23*, 6987. [CrossRef]
15. Wells, L.A.; Chung, W. Vision-Aided Localization and Mapping in Forested Environments Using Stereo Images. *Sensors* **2023**, *23*, 7043. [CrossRef]
16. Sun, C.; Li, K. A Vehicle-Carried INS Positioning Accuracy Improvement Method by Using Lateral Constraint in GPS-Denied Environment. *IEEE Trans. Veh. Technol.* **2023**, *72*, 205–213. [CrossRef]
17. Moving Average. Available online: https://en.wikipedia.org/wiki/Moving_average (accessed on 2 August 2023).
18. Song, Z.; Park, H.-J.; Thapa, N.; Yang, J.-G.; Harada, K.; Lee, S.; Shimada, H.; Park, H.; Park, B.-K. Carrying Position-Independent Ensemble Machine Learning Step-Counting Algorithm for Smartphones. *Sensors* **2022**, *22*, 3736. [CrossRef]
19. Chen, C.; Lu, X.; Markham, A.; Trigoni, N. IONet: Learning to Cure the Curse of Drift in Inertial Odometry. *Proc. AAAI Conf. Artif. Intell.* **2018**, *32*, 6468–6476. [CrossRef]
20. El-Sheimy, N.; Hou, H.; Niu, X. Analysis and Modeling of Inertial Sensors Using Allan Variance. *IEEE Trans. Instrum. Meas.* **2008**, *57*, 140–149. [CrossRef]
21. Chawathe, S.S. Segment-Based Map Matching. In Proceedings of the 2007 IEEE Intelligent Vehicles Symposium, Istanbul, Turkey, 13–15 June 2007; pp. 1190–1197.
22. Hashemi, M.; Karimi, H.A. A weight-based map-matching algorithm for vehicle navigation in complex urban networks. *J. Intell. Transp. Syst.* **2016**, *20*, 573–590. [CrossRef]

23. Dakai, Y.; Baigen, C.; Yifang, Y. An improved map-matching algorithm used in vehicle navigation system. In Proceedings of the 2003 IEEE International Conference on Intelligent Transportation Systems, Shanghai, China, 12–15 October 2003; Volume 2, pp. 1246–1250.
24. Boeing, G. OSMnx: New methods for acquiring, constructing, analyzing, and visualizing complex street networks. *Comput. Environ. Urban Syst.* **2017**, *65*, 126–139. [CrossRef]
25. OpenStreetMap Planet Dump. Available online: https://www.openstreetmap.org/ (accessed on 3 September 2023).
26. Breadth-First Search. Available online: https://en.wikipedia.org/wiki/Breadth-first_search (accessed on 16 October 2023).
27. Haversine Formula. Available online: https://en.wikipedia.org/wiki/Haversine_formula (accessed on 13 August 2023).

sensors

MDPI

Article

Constrained Cubature Particle Filter for Vehicle Navigation

Li Xue [1], Yongmin Zhong [2,*] and Yulan Han [1]

[1] School of Electronic and Electrical Engineering, Ningxia University, Yinchuan 750021, China; xueli20221@nxu.edu.cn (L.X.); han_yl@nxu.edu.cn (Y.H.)
[2] School of Engineering, RMIT University, Bundoora, VIC 3082, Australia
* Correspondence: yongmin.zhong@rmit.edu.au

Abstract: In vehicle navigation, it is quite common that the dynamic system is subject to various constraints, which increases the difficulty in nonlinear filtering. To address this issue, this paper presents a new constrained cubature particle filter (CCPF) for vehicle navigation. Firstly, state constraints are incorporated in the importance sampling process of the traditional cubature particle filter to enhance the accuracy of the importance density function. Subsequently, the Euclidean distance is employed to optimize the resampling process by adjusting particle weights to avoid particle degradation. Further, the convergence of the proposed CCPF is also rigorously proved, showing that the posterior probability function is converged when the particle number $N \to \infty$. Our experimental results and the results of a comparative analysis regarding GNSS/DR (Global Navigation Satellite System/Dead Reckoning)-integrated vehicle navigation demonstrate that the proposed CCPF can effectively estimate system state under constrained conditions, leading to higher estimation accuracy than the traditional particle filter and cubature particle filter.

Keywords: vehicle navigation; constraints; nonlinear filtering; particle filter; constrained cubature particle filter

1. Introduction

As the "eye" of a vehicle, the navigation system provides navigation information for vehicle maneuvers to reach their target. Since navigation systems commonly involve nonlinear structural characteristics and dynamics, nonlinear filtering is an important means for navigation computations [1–3]. In practice, vehicle navigation systems are subject to constraints and non-Gaussian uncertainties, making nonlinear filter even more challenging [4–6].

Currently, the extended Kalman filter (EKF), polynomial filter, unscented Kalman filter (UKF), and cubature Kalman filter (CKF) are commonly used for nonlinear state estimation. The extended Kalman filter involves a system linearization error and requires the calculation of the Jacobian matrix [7,8]. The polynomial filter is sensitive to outliers and requires a large amount of computing for large-scale signal processing [9]. The UKF and CKF select typical sampling points according to prior state estimation to apply the unscented transformation [10,11] and cubature rules [12,13] to obtain sample points, respectively. Subsequently, the obtained sampling points are weighted to acquire system state estimations. Compared to the UKF, the CKF is superior in both estimation accuracy and stability for high-dimensional systems, since the Gaussian-weighted integrals are calculated through third-degree spherical-radial cubature [12–14]. In general, these improved nonlinear Kalman filters are subject to the condition that system noises are Gaussian. However, in practical applications (such as multi-sensor integrated systems for vehicle navigation), non-Gaussian noises are commonly involved in nonlinear systems, and thus the direct use of these improved Kalman filters will lead to divergent solutions.

The particle filter (PF) is a typical method used to handle nonlinear systems with non-Gaussian noises [15–17]. Instead of integral calculation, PF uses sample mean to implement the Bayesian estimation under nonlinear dynamics. When the particle number is sufficiently large, the posterior probability density function of particles will be sufficiently accurate to guarantee the accuracy of the state mean and variance. However, the particle filter suffers from particle degradation, and there can be difficulty in selecting an appropriate importance density function for importance sampling [18–20]. Research efforts have been dedicated to improving the PF. Pitt et al. used auxiliary samples to adjust the resampling order to improve the PF's performance [21]. D. Liu et al. studied an adaptive partial resampling method to prevent particle degradation [22]. T.H. Liu et al. proposed an adaptive processing scheme to lower particle weights based on a genetic algorithm [23]. However, these improvements cannot guarantee an optimal resultant particle distribution.

The cubature rules provide an effective way to improve sampling accuracy, and thus have been used in the PF to obtain the importance density function, leading to the cubature particle filter (CPF). Xia et al. studied a CPF in the context of system state estimation, where the second-order resistor–capacitor equivalent circuit model was used to verify the filter accuracy [24]. Shi et al. proposed a robust CPF by introducing Huber's M-estimation theory to handle poor observation data [25]. Feng et al. studied a CPF based on the artificial bee colony for target tracking via underwater wireless sensor networks [26]. Zhang et al. combined a CPF with truncation adaptation to generate recommendation distributions for strong nonlinear systems [27]. Liu et al. investigated a hybrid CPF for stable-state estimation in harsh charging or discharging schedules [28]. Xing et al. constructed an adaptive CPF to estimate navigation parameters for high-dimensional nonlinear vehicle systems [29]. In general, the existing CPFs still suffer from the problem of particle degradation in the PF.

Further, the existing CPFs do not consider constraints. In practical applications (such as multi-sensor integrated systems for vehicle navigation), it is quite common that a nonlinear system is subject to various conditions. Considering these conditions as state constraints in the CPF filter process can effectively improve the estimation accuracy. However, since the existence of state constraints changes the probability structure of a dynamic system, which increases the difficulty of filtering estimation, the related research is still limited. Gao et al. proposed a constrained unscented particle filter for SINS/GNSS/ADS-integrated airship navigation in the presence of wind field disturbances [30]. Seifzadeh et al. studied a constrained particle filter based on soft data to improve filtering performance for target tracking [31]. In complex and highly dynamic environments, Xu et al. presented a particle filter based on spatial–temporal constraints for cooperative target tracking [32]. Zhang et al. presented a constrained multiple model based on a PF for target tracking [33]. However, the above studies are not based on the advanced cubature rules. They also focus on particular applications with specific constraints rather than on general constraint problems.

This paper presents a novel constrained cubature particle filter (CCPF) to improve the CPF's performance for vehicle navigation. This method incorporates constraints in the cubature transformation to improve the CPF's importance sampling, resulting in an improved importance density distribution. It also improves the CPF resampling process by using the Euclidean distance of the measurement residual to adjust particle weights to ensure the diversity of particles, thus avoiding particle degradation. The convergence of the proposed CCPF has been rigorously proved. Experiments and a comparative analysis with the traditional PF and CPF were conducted to evaluate the CCPF's efficiency for GNSS/DR-integrated vehicle navigation.

2. Cubature Particle Filter

Consider the nonlinear dynamic system

$$\begin{aligned} x_k &= f(x_{k-1}) + v_k \\ y_k &= h(x_k) + n_k \end{aligned} \tag{1}$$

where $x_k \in R^n$ is the n-dimensional system state vector at time point k, $y_k \in R^n$ is the m-dimensional system measurement vector, $v_k \in R^n$ is the process noise with covariance

$\boldsymbol{R}$, $n_k \in R^m$ is the measurement noise with covariance $\boldsymbol{Q}$, and $f(\cdot)$ and $h(\cdot)$ represent the nonlinear system and measurement functions. Obviously, the system described by (1) does not involve the assumption of Gaussian noises.

Like the unscented transform, CPF uses the third-order spherical radial cubature rules to approximate the nonlinear system functions. According to the prior mean and covariance of the system state, CPF selects the cubature points via the cubature rules and further approximates the state mean and covariance by the weighting sum of the cubature points [34–36]. Consider the dynamic system described by (1), the CPF procedure is given as below [37,38]:

Step 1: Initialization

Set the initial state estimate together with its covariance and weights as follows:

$$\begin{aligned} &\hat{\boldsymbol{x}}_0^j = E[\boldsymbol{x}_0^j] \\ &\boldsymbol{P}_0^j = E[(\boldsymbol{x}_0^j - \hat{\boldsymbol{x}}_0^j)(\boldsymbol{x}_0^j - \hat{\boldsymbol{x}}_0^j)]^T \quad (j = 1, 2, \cdots, N) \\ &\boldsymbol{w}_0^j = \tfrac{1}{N} \end{aligned} \tag{2}$$

where N is the number of particles, $\hat{\boldsymbol{x}}_0^j \sim p(\boldsymbol{x}_0)$ denotes the initial state estimate of the jth particle, and $p(\boldsymbol{x}_0)$ denotes the initial state distribution.

For $k = 1, 2, \cdots, M$, execute Steps 2–3.

Step 2: Importance sampling

For $j = 1, 2, \cdots, N$, execute Steps (a)–(c).

(a) Calculate the cubature points as follows:

$$\boldsymbol{x}_{i,k-1}^j = \boldsymbol{S}_{k-1}^j \varsigma_i + \hat{\boldsymbol{x}}_{k-1}^j \tag{3}$$

$$\boldsymbol{P}_{k-1}^j = \boldsymbol{S}_{k-1}^j \left(\boldsymbol{S}_{k-1}^j\right)^T \tag{4}$$

where $i = 1, 2, \cdots, 2n$, $\boldsymbol{P}_{k-1}^j$ is the state covariance at $k-1$, $\boldsymbol{S}_{k-1}^j$ is the lower triangular matrix obtained via the Cholesky decomposition of $\boldsymbol{P}_{k-1}^j$, $\boldsymbol{x}_{i,k-1}^j$ is the ith cubature point of the jth particle at $k-1$, $\hat{\boldsymbol{x}}_{k-1}^j$ is the state estimation, and ς_i is defined as follows:

$$\varsigma_i = \begin{cases} \sqrt{n} I_i & i = 1, 2, \cdots, n \\ -\sqrt{n} I_{i-n} & i = n+1, n+2, \cdots, 2n \end{cases} \tag{5}$$

where I_i denotes the ith column vector of the $n \times n$ identity matrix.

(b) Time update

Calculate the state prediction covariance and one-step measurement prediction via the cubature points

$$\boldsymbol{x}_{i,k|k-1}^j = \boldsymbol{f}(\boldsymbol{x}_{i,k-1}^j) \tag{6}$$

$$\hat{\boldsymbol{x}}_{i,k|k-1}^j = \frac{1}{2n} \sum_{i=1}^{2n} \boldsymbol{x}_{i,k|k-1}^j \tag{7}$$

$$\boldsymbol{P}_{k|k-1}^j = \frac{1}{2n} \sum_{i=1}^{2n} \left[\left(\boldsymbol{x}_{i,k|k-1}^j - \hat{\boldsymbol{x}}_{i,k|k-1}^j \right) \left(\boldsymbol{x}_{i,k|k-1}^j - \hat{\boldsymbol{x}}_{i,k|k-1}^j \right)^T \right] + \boldsymbol{R} \tag{8}$$

$$\boldsymbol{y}_{k|k-1}^j = \boldsymbol{h}(\hat{\boldsymbol{x}}_{k|k-1}^j) \tag{9}$$

$$\hat{\boldsymbol{y}}_{k|k-1}^j = \frac{1}{2n} \sum_{i=1}^{2n} \boldsymbol{y}_{k|k-1}^j \tag{10}$$

where $\hat{x}^j_{k|k-1}$, $\hat{y}^j_{k|k-1}$ and $P^j_{k|k-1}$, denote the predicted state mean, measurement mean, and measurement covariance of the jth cubature point, respectively.

(c) Measurement update

Calculate the cubature point auto-correlation covariance and cross-correlation covariance

$$\begin{aligned} P^j_{y_k y_k} &= \frac{1}{2n}\sum_{i=1}^{2n}\left[\left(y^j_{i,k|k-1}-\hat{y}^j_{k|k-1}\right)\left(y^j_{i,k|k-1}-\hat{y}^j_{k|k-1}\right)^T\right]+Q \\ P^j_{x_k y_k} &= \frac{1}{2n}\sum_{i=1}^{2n}\left[\left(x^j_{i,k|k-1}-\hat{x}^j_{k|k-1}\right)\left(y^j_{i,k|k-1}-\hat{y}^j_{k|k-1}\right)^T\right] \end{aligned} \tag{11}$$

Calculate the cubature point estimates

$$\begin{aligned} \hat{x}^j_k &= \hat{x}^i_{k|k-1}+K^j_k(y_k-\hat{y}^j_{k|k-1}) \\ \hat{P}^j_k &= P^j_{k|k-1}-K^j_k P^j_{y_k y_k}(K^j_k)^T \\ K^j_k &= P^j_{x_k y_k}(P^j_{y_k y_k})^{-1} \end{aligned} \tag{12}$$

where K^j_k denotes the filter gain at k of the jth particle, and $\hat{x}^j_k$ and $\hat{P}^j_k$ denote the estimate and its covariance of the jth cubature point after cubature transformation.

The particle importance density function is calculated using $(\hat{x}^j_k, \hat{P}^j_k)$, and the particles are sampled by $\overline{x}^j_k \sim q(x^j_k \middle| x^j_{0:k-1}, y_{1:k}) = N(\hat{x}^j_k, \hat{P}^j_k)$, $\overline{x}^j_{0:k} \triangleq (x^j_{0:k-1}, \overline{x}^j_k)$, and $\hat{P}^j_{0:k} \triangleq (P^j_{0:k-1}, \hat{P}^j_k)$.

Calculate the particle weights and normalize them as follows:

$$w^j_k = w^j_{k-1}\frac{p(y_k \middle| \overline{x}^j_k)p(\overline{x}^j_k \middle| x^j_{k-1})}{q(\overline{x}^j_k \middle| x^j_{k-1}, y_k)} \tag{13}$$

$$\widetilde{w}^i_k = w^i_k \Big/ \sum_{i=1}^{n} w^i_k \tag{14}$$

Step 3: Resampling

According to the approximately distributed $p\overline{x}^j_{0:k} \middle| y_{1:k}$ $(j = 1, 2, \cdots N)$, generate N new particles by duplicating the particles of high weights and further assigning the same weight $\frac{1}{N}$ to them.

Calculate the state estimate and its covariance

$$\begin{aligned} \hat{x}_k &= \sum_{j=1}^{N}\widetilde{w}^j_k\overline{x}^j_k \\ P_k &= \sum_{j=1}^{N}\widetilde{w}^j_k[(\overline{x}^j_k-x_k)(\overline{x}^j_k-x_k)^T] \end{aligned} \tag{15}$$

As mentioned previously, the existence of state constraints changes the probability structure of a dynamic system. In the importance sampling step, if the system state is subject to constraints, (7) will be biased. The bias in (7) will be propagated through (8)–(14), and eventually, the state estimate from (15) will be biased. Therefore, it is necessary to incorporate the state changes caused by constraints in the importance sampling step. In addition, in the resampling process, the duplication of the particles with high weights from (13) will lose the particle diversity. Since the useful particle samples are reduced, the filtering solution will deteriorate. A straightforward solution is to adjust the weights of useful particles to increase their contributions to the state estimation. This paper addresses the above two issues in CPF, leading to a new CCPF for state estimation under constraints.

3. Constrained Cubature Particle Filter

The proposed CCPF combines Euclidean distance-based resampling and constraints to improve the CPF's performance. In the importance sampling process, the proposed CCPF improves the CPF importance sampling density function with state constraints. In the resampling process, the proposed CCPF optimizes the CPF weights by adopting the Euclidean distance to adjust the particle weights to ensure the diversity of particles, preventing particle degradation while enabling the easy acquisition of an importance density function which is close to the true density function.

3.1. Importance Sampling

In the importance sampling process of CCPF, we perform constrained projection calculations on the cubature point estimates obtained by (12). Suppose the system described by (1) is subject to the following constraint:

$$\boldsymbol{D}x_k \leq \boldsymbol{d} \tag{16}$$

where $\boldsymbol{D}$ represents the constrained matrix, and $\boldsymbol{d}$ represents the constrained vector [39,40].

The estimation problem of the constrained state can be transformed into the following optimization problem:

$$\min J(\tilde{x}_k^j) = (\tilde{x}_k^j - \hat{x}_k^j)^T \boldsymbol{W}(\tilde{x}_k^j - \hat{x}_k^j) \tag{17}$$

$$\boldsymbol{D}\tilde{x}_k^j = \boldsymbol{d}_k^j \tag{18}$$

where $\tilde{x}_k^j$ represents the state estimation under the constraint, $\boldsymbol{W}$ represents an arbitrary symmetric positive definite matrix for constructing the Lagrange function, and $\hat{x}_k^j$ represents the state estimation without the constraint.

Using the Lagrange multiplier to solve (17) and (18), we can obtain the following:

$$J(\tilde{x}_k^j, \lambda) = (\tilde{x}_k^j - \hat{x}_k^j)^T \boldsymbol{W}(\tilde{x}_k^j - \hat{x}_k^j) + 2\lambda^T(\boldsymbol{D}\tilde{x}_k^j - \boldsymbol{d}_k^j) \tag{19}$$

where λ represents a vector for constructing the Lagrange function.

Finding the partial derivative of (19), we obtain

$$\tilde{x}_k^j = \hat{x}_k^j - \boldsymbol{W}^{-1}\boldsymbol{D}^T(\boldsymbol{D}\boldsymbol{W}^{-1}\boldsymbol{D}^T)^{-1}(\boldsymbol{D}\hat{x}_k^j - \boldsymbol{d}_k^j) \tag{20}$$

Using (20), the cubature point estimate described by (12) becomes

$$\begin{aligned} \hat{x}_k^j &= \hat{x}_{k|k-1}^j + \boldsymbol{K}_k^j(y_k - \hat{y}_{k|k-1}^j) \\ \hat{\boldsymbol{P}}_k^j &= \boldsymbol{P}_{k|k-1}^j - \boldsymbol{K}_k^j \boldsymbol{P}_{y_k y_k}^j (\boldsymbol{K}_k^j)^T \\ \boldsymbol{K}_k^j &= \boldsymbol{P}_{x_k y_k}^j (\boldsymbol{P}_{x_k y_k}^j)^{-1} \\ \tilde{x}_k^j &= \hat{x}_k^j - \boldsymbol{W}^{-1}\boldsymbol{D}^T(\boldsymbol{D}\boldsymbol{W}^{-1}\boldsymbol{D}^T)^{-1}(\boldsymbol{D}\hat{x}_k^j - \boldsymbol{d}_k^j) \end{aligned} \tag{21}$$

From (21), we can obtain the importance sampling density function under constraints, i.e., $\bar{x}_k^j \sim N(\tilde{x}_k^j, \hat{\boldsymbol{P}}_k^{*j})$, where $\hat{\boldsymbol{P}}_k^{*j}$ denotes the covariance under constraints.

3.2. Resampling

Calculate the Euclidean distance of the measurement residual. Using (13), record the maximum weights $w_k^{j_{\max}}$ and minimum weights $w_k^{j_{\min}}$

$$L_{\max} = (r_k^{j_{\min}} - r_k^{j_{\max}})^T \cdot (r_k^{j_{\min}} - r_k^{j_{\max}}) \tag{22}$$

$$L^j = (r_k^j - r_k^{j_{\max}})^T \cdot (r_k^j - r_k^{j_{\max}}) \tag{23}$$

where $L_{\max}$ and L^j denote the Euclidean distances with the subscripts "$j_{\max}$" and "$j_{\min}$", representing the index numbers of the particles with the maximum and minimum weights, and r denotes the measurement residual, i.e.,

$$\begin{aligned} r_k^{j_{\max}} &= y_k - h(\overline{x}_k^{j_{\max}}) \\ r_k^{j_{\min}} &= y_k - h(\overline{x}_k^{j_{\min}}) \end{aligned} \tag{24}$$

Accordingly, the weights are calculated as

$$w_k^{j^*} = w_k^j + (\frac{w_k^{j_{\max}}}{N}) \cdot \sin(\frac{L^j}{L_{\max}} \cdot \frac{\pi}{2}) \cdot \beta \tag{25}$$

where $\beta \geq 0$ is a coefficient related to measurement characteristics [41]. The larger β is, the larger the adjustment to the weight will be. When $\beta = 0$, there will be no adjustment to the weight. The CCPF procedure is shown in Figure 1.

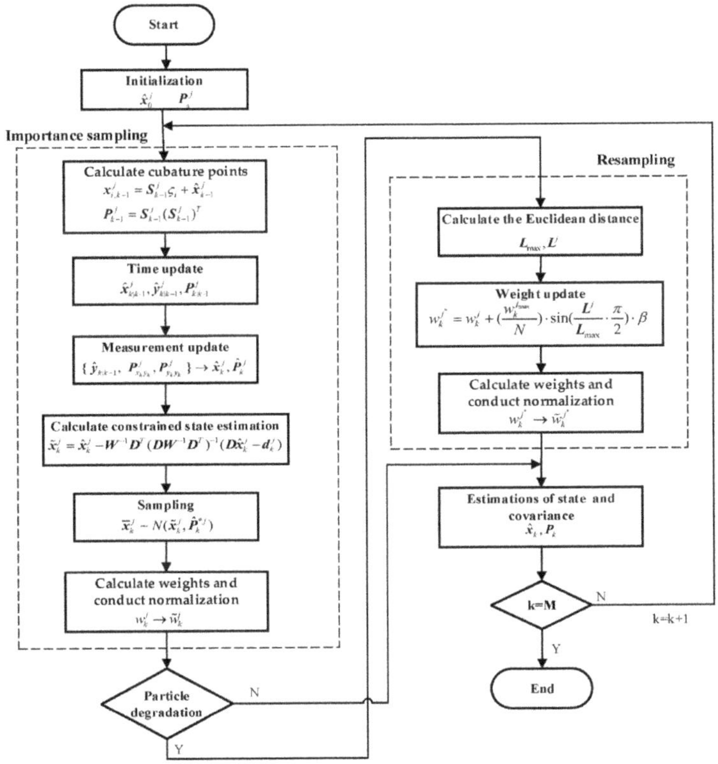

Figure 1. The CCPF procedure.

3.3. Convergence Analysis

In this section, we will study the convergence of the proposed CCPF when the sample size N is sufficiently large. Since the posterior density of the system state corresponds to the empirical measure of particle samples, we conduct a convergence analysis based on an empirical measure which is defined by a probability measure.

Similar to the PF, assuming that the transfer kernel function ϕ satisfies the Feller process and the state likelihood function g is a continuous bounded positive definite function (i.e., $\|g\| < \infty$), for $k \geq 0$, when $N \to \infty$, we can have [42]

$$\lim_{N\to\infty} \pi_{k|k}^N = \lim_{N\to\infty} \phi_{1:k}^N(\eta^N) = \phi_{1:k}(\eta) = \pi_{k|k} \tag{26}$$

where $\pi_{k|k}$ denotes the ideal distribution of the system state, η represents the initial distribution, and $\pi_{k|k}^N$ is the posterior probability density, which is expressed as

$$\pi_{k|k}^N(dx_k) = \frac{1}{N}\sum_{i=1}^{N} \delta_{x_k^i}(dx_k) \tag{27}$$

where δ denotes the Dirichlet function.

The time and measurement updates of CCPF can be represented as follows [43]:

$$\pi_{k|k-1} = \int_{\mathbb{R}^{n_x}} \pi_{k-1|k-1}(\boldsymbol{dx}_{k-1})\phi(\boldsymbol{dx}_k \big| \boldsymbol{x}_{k-1}) \tag{28}$$

$$\pi_{k|k}(\boldsymbol{dx}_k) = \frac{g(\boldsymbol{y}_k \big| \boldsymbol{x}_k)\pi_{k|k-1}(\boldsymbol{dx}_k)}{\int_{\mathbb{R}^{n_x}} g(\boldsymbol{y}_k \big| \boldsymbol{x}_k)\pi_{k|k-1}(\boldsymbol{dx}_k)} \tag{29}$$

Equations (28) and (29) can be rewritten as

$$(\pi_{k|k-1}, \varphi) = (\pi_{k-1|k-1}, \phi\varphi) \tag{30}$$

$$(\pi_{k|k}, \varphi) = (\pi_{k|k-1}, g)^{-1}(\pi_{k|k-1}, \varphi g) \tag{31}$$

where $\varphi \subset B(\mathbb{R}^{n_x})$ can be any bounded function, and $(\pi_{k|k-1}, g) > 0$.

As shown in Section 3.1, the CCPF involves three key steps: the initialization, importance sampling, and resampling. Therefore, the convergence analysis of CCPF will be applied to each step. Denote the posterior probability density in the importance sampling process using $\pi_{k|k}^N$ and that in the resampling process using $\widetilde{\pi}_{k|k}^N$.

Theorem 1. *For any bounded function $\varphi \subset B(\mathbb{R}^{n_x})$; if $\|g\|$ is bounded, then*

$$E[((\pi_{k-1|k-1}^N, \varphi) - (\pi_{k-1|k-1}, \varphi))^2] \leq c_{k-1|k-1}^N \frac{\|\varphi\|^2}{N} \tag{32}$$

Proof. Using (31), we obtain

$$\begin{aligned}
(\pi_{k-1|k-1}^N, \varphi) - (\pi_{k-1|k-1}, \varphi) &= \frac{(\pi_{k-1|k-2}^N, g\varphi)}{(\widetilde{\pi}_{k-1|k-2}^N, g)} - \frac{(\pi_{k-1|k-2}, g\varphi)}{(\pi_{k-1|k-2}, g)} \\
&= \frac{(\pi_{k-1|k-1}^N, g\varphi)}{(\pi_{k-1|k-1}^N, g)} - \frac{(\pi_{k-1|k-2}, g\varphi)}{(\pi_{k-1|k-2}^N, g)} + \frac{(\pi_{k-1|k-2}, g\varphi)}{(\pi_{k-1|k-2}^N, g)} - - \frac{(\pi_{k-1|k-2}, g\varphi)}{(\pi_{k-1|k-2}^N, g)} \\
&= \frac{(\pi_{k-1|k-2}^N, g\varphi) - (\pi_{k-1|k-2}, g\varphi)}{(\pi_{k-1|k-2}^N, g)} \\
&+ \frac{(\pi_{k-1|k-2}, g\varphi)((\pi_{k-1|k-2}, g\varphi) - (\pi_{k-1|k-2}^N, g))}{(\pi_{k-1|k-2}^N, g)(\pi_{k-1|k-2}, g)}
\end{aligned} \tag{33}$$

According to the Minkowski inequality, we have

$$
\begin{aligned}
&\mathrm{E}\left[\left(\frac{(\pi^N_{k-1|k-2},g\varphi)-(\pi_{k-1|k-2},g\varphi)}{(\pi^N_{k-1|k-2},g)}+\frac{(\pi_{k-1|k-2},g\varphi)((\pi_{k-1|k-2},g\varphi)-(\pi^N_{k-1|k-2},g))}{(\pi^N_{k-1|k-2},g)(\pi_{k-1|k-2},g)}\right)^2\right]^{\frac{1}{2}} \\
&\leq \mathrm{E}[(\frac{(\pi^N_{k-1|k-2},g\varphi)-(\pi_{k-1|k-2},g\varphi)}{(\pi^N_{k-1|k-2},g)})^2]^{\frac{1}{2}}+\mathrm{E}[(\frac{(\pi_{k-1|k-2},g\varphi)((\pi_{k-1|k-2},g\varphi)-(\pi^N_{k-1|k-2},g))}{(\pi^N_{k-1|k-2},g)(\pi_{k-1|k-2},g)})^2]^{\frac{1}{2}} \\
&\leq \frac{\mathrm{E}[((\pi^N_{k-1|k-2},g\varphi)-(\pi_{k-1|k-2},g\varphi))^2]^{\frac{1}{2}}}{(\pi^N_{k-1|k-2},g)}+\frac{(\pi_{k-1|k-2},g\varphi)\mathrm{E}[((\pi_{k-1|k-2},g\varphi)-(\pi^N_{k-1|k-2},g\varphi))^2]^{\frac{1}{2}}}{(\pi^N_{k-1|k-2},g)(\pi_{k-1|k-2},g)} \\
&\leq \frac{\sqrt{c_{k-1|k-2}}\|g\varphi\|}{(\tilde{\pi}^N_{k-1|k-2},g)\sqrt{N}}+\frac{\mathrm{E}[((\pi^N_{k-1|k-2},g\varphi)-(\pi_{k-1|k-2},g\varphi))^2]^{\frac{1}{2}}}{(\pi^N_{k-1|k-2},g)}
\end{aligned}
\tag{34}
$$

Since $\|g\| < \infty$, we have

$$
\begin{aligned}
&\frac{\sqrt{c_{k-1|k-2}}\|g\varphi\|}{(\tilde{\pi}^N_{k-1|k-2},g)\sqrt{N}}+\frac{\mathrm{E}[((\pi^N_{k-1|k-2},g\varphi)-(\pi_{k-1|k-2},g\varphi))^2]^{\frac{1}{2}}}{(\pi^N_{k-1|k-2},g)} \\
&\leq \frac{\sqrt{c_{k-1|k-2}}\|g\varphi\|}{(\pi^N_{k-1|k-2},g)\sqrt{N}}+\frac{\sqrt{c_{k-1|k-2}}\|g\varphi\|}{(\pi^N_{k-1|k-2},g)\sqrt{N}} \\
&\leq \frac{2\sqrt{c_{k-1|k-2}}\|g\varphi\|}{(\pi^N_{k-1|k-2},g)\sqrt{N}} = \sqrt{c^N_{k-1|k-1}}\frac{\|\varphi\|}{\sqrt{N}}
\end{aligned}
\tag{35}
$$

where $\sqrt{c^N_{k-1|k-1}} = \frac{2\sqrt{c_{k-1|k-2}}\|g\|}{(\pi^N_{k-1|k-2},g)}$.

Equation (32) follows by substituting (35) into (34) and further applying the square operation. Thus, the proof of (32) is completed.

Theorem 1 shows that the initialization process of the CCPF is converged when $N \to \infty$. □

Theorem 2. *If $\|\phi\varphi\| \leq \|\varphi\|$ holds, the importance sampling process of the CCPF yields*

$$
E[((\pi^N_{k|k-1},\varphi)-(\pi_{k|k-1},\varphi))^2] \leq c^N_{k|k-1}\frac{\|\varphi\|^2}{N}
\tag{36}
$$

Proof. Using (30), we can obtain

$$
\begin{aligned}
&E[((\pi^N_{k|k-1},\varphi)-(\pi_{k|k-1},\varphi))^2]^{\frac{1}{2}} \\
&= E[((\pi^N_{k|k-1},\varphi)-(\pi^N_{k-1|k-1},\phi\varphi)+(\pi^N_{k-1|k-1},\phi\varphi)-(\pi_{k|k-1},\varphi))^2]^{\frac{1}{2}} \\
&\leq E[((\pi^N_{k|k-1},\varphi)-(\pi^N_{k-1|k-1},\phi\varphi))^2]^{\frac{1}{2}}+E[(\pi^N_{k-1|k-1},\phi\varphi)-(\pi_{k-1|k-1},\phi\varphi))^2]^{\frac{1}{2}} \\
&= E[((\pi^N_{k-1|k-1},\phi\varphi^2)-(\pi^N_{k-1|k-1},\phi\varphi))^2]^{\frac{1}{2}}+E[(\pi^N_{k-1|k-1},\phi\varphi)-(\pi_{k-1|k-1},\phi\varphi))^2]^{\frac{1}{2}}
\end{aligned}
\tag{37}
$$

According to Theorem 1 and letting $(1+\sqrt{c_{k-1|k-1}})^2 = c^N_{k|k-1}$, and when $\|\phi\varphi\| \leq \|\varphi\|$ holds, we can have

$$
\begin{aligned}
&E[((\pi^N_{k-1|k-1},\phi\varphi^2)-(\pi^N_{k-1|k-1},\phi\varphi))^2]^{\frac{1}{2}}+E[(\pi^N_{k-1|k-1},\phi\varphi)-(\pi_{k-1|k-1},\phi\varphi))^2]^{\frac{1}{2}} \\
&\leq \frac{}{\sqrt{N}}+\sqrt{c_{k-1|k-1}}\frac{\|\varphi\|}{\sqrt{N}} \\
&= (1+\sqrt{c_{k-1|k-1}})\frac{\|\varphi\|}{\sqrt{N}}
\end{aligned}
\tag{38}
$$

Equation (36) follows by substituting (38) into (37) and further applying the square operation. Thus, the proof of Theorem 2 is completed.

Theorem 2 shows that the importance sampling process of the CCPF is converged when $N \to \infty$. □

The resampling process of the CCPF involves the adjustment of the particle weights. In theory, the prior distribution should be close to the importance sampling distribution.

Then, using (13), we have $w_k^j \propto g$. From (25) and $\left|(\frac{w_k^{j\max}}{N}) \cdot \sin(\frac{L^j}{L_{\max}} \cdot \frac{\pi}{2})\right| < 1$, it is readily known that the particle weights are bounded. By normalization, we can readily have the particle weights within [0, 1]. Without the loss of generality and for the concise description purpose, we write $g \to Fg$, where F is a linear transformation, and Fg is bounded. Thus, we can derive the following theorem.

Theorem 3. *For the resampling process of the CCPF, we have*

$$E[((\tilde{\pi}_{k|k}^N, \varphi) - (\pi_{k|k}, \varphi))^2] \leq c_{k|k}^N \frac{\|\varphi\|^2}{N} \tag{39}$$

Proof. Using (31) and $g \to Fg$, we can obtain the following:

$$\begin{aligned}
&E[((\tilde{\pi}_{k|k}^N, \varphi) - (\pi_{k|k}, \varphi))^2]^{\frac{1}{2}} \\
&= E[((\tilde{\pi}_{k|k}^N, \varphi) - (\tilde{\pi}_{k|k}^N, \varphi) + (\tilde{\pi}_{k|k}^N, \varphi) - (\pi_{k|k}, \varphi))^2]^{\frac{1}{2}} \\
&= E[(\frac{(\tilde{\pi}_{k|k-1}^N, (Fg)\varphi)}{(\tilde{\pi}_{k|k-1}^N, Fg)} - \frac{(\pi_{k|k-1}, (Fg)\varphi)}{(\tilde{\pi}_{k|k-1}^N, Fg)} + \frac{(\pi_{k|k-1}, (Fg)\varphi)}{(\tilde{\pi}_{k|k-1}^N, Fg)} - \frac{(\pi_{k|k-1}, (Fg)\varphi)}{(\pi_{k|k-1}, Fg)})^2]^{\frac{1}{2}} \\
&= E[((\frac{(\tilde{\pi}_{k|k-1}^N, (Fg)\varphi)}{(\tilde{\pi}_{k|k-1}^N, Fg)} - \frac{(\pi_{k|k-1}, (Fg)\varphi)}{(\tilde{\pi}_{k|k-1}^N, Fg)}) + (\frac{(\pi_{k|k-1}, (Fg)\varphi)}{(\tilde{\pi}_{k|k-1}^N, Fg)} - \frac{(\pi_{k|k-1}, (Fg)\varphi)}{(\pi_{k|k-1}, Fg)}))^2]^{\frac{1}{2}} \\
&\leq E[(\frac{(\tilde{\pi}_{k|k-1}^N, (Fg)\varphi)}{(\tilde{\pi}_{k|k-1}^N, Fg)} - \frac{(\pi_{k|k-1}, (Fg)\varphi)}{(\tilde{\pi}_{k|k-1}^N, Fg)})^2]^{\frac{1}{2}} + E[(\frac{(\pi_{k|k-1}, (Fg)\varphi)}{(\tilde{\pi}_{k|k-1}^N, Fg)} - \frac{(\pi_{k|k-1}, (Fg)\varphi)}{(\pi_{k|k-1}, Fg)})^2]^{\frac{1}{2}}
\end{aligned} \tag{40}$$

According to Theorem 2 and the letting constant $c_{k|k}^N = \frac{2\sqrt{c_{k|k-1}}\|Fg\|}{(\tilde{\pi}_{k|k-1}^N, Fg)}$, we have

$$\begin{aligned}
&E[(\frac{(\tilde{\pi}_{k|k-1}^N, (Fg)\varphi)}{(\tilde{\pi}_{k|k-1}^N, Fg)} - \frac{(\pi_{k|k-1}, (Fg)\varphi)}{(\tilde{\pi}_{k|k-1}^N, Fg)})^2]^{\frac{1}{2}} + E[(\frac{(\pi_{k|k-1}, (Fg)\varphi)}{(\tilde{\pi}_{k|k-1}^N, Fg)} - \frac{(\pi_{k|k-1}, (Fg)\varphi)}{(\pi_{k|k-1}, Fg)})^2]^{\frac{1}{2}} \\
&= E[(\frac{(\tilde{\pi}_{k|k-1}^N, (Fg)\varphi) - (\pi_{k|k-1}, (Fg)\varphi)}{(\tilde{\pi}_{k|k-1}^N, Fg)})^2]^{\frac{1}{2}} + E[(\frac{(\pi_{k|k-1}, (Fg)\varphi)((\pi_{k|k-1}, Fg) - (\tilde{\pi}_{k|k-1}^N, Fg))}{(\tilde{\pi}_{k|k-1}^N, Fg)(\pi_{k|k-1}, Fg)})^2]^{\frac{1}{2}} \\
&\leq \frac{\sqrt{c_{k|k-1}}\|Fg\| \cdot \|\varphi\|}{(\tilde{\pi}_{k|k-1}^N, g)\sqrt{N}} + \frac{\|\varphi\| \cdot E[((\pi_{k|k-1}, Fg) - (\tilde{\pi}_{k|k-1}^N, Fg))^2]^{\frac{1}{2}}}{(\tilde{\pi}_{k|k-1}^N, g)} \\
&\leq \frac{2\sqrt{c_{k|k-1}}\|Fg\| \cdot \|\varphi\|}{(\tilde{\pi}_{k|k-1}^N, Fg)\sqrt{N}} = \sqrt{c_{k|k}^N} \frac{\|\varphi\|}{\sqrt{N}}
\end{aligned} \tag{41}$$

Since $\|Fg\| < \infty$, (39) follows by substituting (41) into (40) and further applying the square operation. Thus, the proof of (39) is completed.

Theorem 3 shows that the resampling process of the CCPF is converged when $N \to \infty$. □

From Theorems 1–3, it can be inferred that the posterior probability function of CCPF will be converged when $N \to \infty$, which indicates the convergence of the proposed CCPF.

4. Experimental Results

4.1. GNSS/DR Vehicle Navigation System

Experiments on a GNSS/DR-integrated navigation system of a sports car were conducted to evaluate the performance of the proposed CCPF. The state vector of the GNSS/DR-integrated navigation system is defined as follows:

$$\boldsymbol{x}(t) = \begin{bmatrix} p_E & v_E & a_E & p_N & v_N & a_N & \varepsilon & \psi \end{bmatrix}^T \tag{42}$$

where p_E, v_E, and a_E are the position, velocity, and acceleration in east; p_N, v_N, and a_N are the position, velocity, and acceleration in the north; ε is the gyro drift error; and ψ is the DR calibration coefficient.

The system state equation is as follows [44,45]:

$$\dot{x} = \begin{bmatrix} 0 & 1 & 0 & 0 & 0 & 0 & 0 & 0 \\ 0 & 0 & 1 & 0 & 0 & 0 & 0 & 0 \\ 0 & 0 & -\frac{1}{\tau_{a_E}} & 0 & 0 & 0 & 0 & 0 \\ 0 & 0 & 0 & 0 & 1 & 0 & 0 & 0 \\ 0 & 0 & 0 & 0 & 0 & 1 & 0 & 0 \\ 0 & 0 & 0 & 0 & 0 & -\frac{1}{\tau_{a_N}} & 0 & 0 \\ 0 & 0 & 0 & 0 & 0 & 0 & \frac{1}{\tau_\varepsilon} & 0 \\ 0 & 0 & 0 & 0 & 0 & 0 & 0 & 0 \end{bmatrix} x(t) + u + v \tag{43}$$

where v is the process noise with covariance R; τ_{a_E} and τ_{a_N} are the correlation time constants for the variation rates of the accelerations in the east and north; τ_ε is the correlation time constant for the first-order Markov process in the gyro drift; and u is the control input, defined as

$$u = \begin{bmatrix} 0 & 0 & \frac{\overline{a}_E}{\tau_{a_E}} & 0 & 0 & \frac{\overline{a}_N}{\tau_{a_N}} & 0 & 0 \end{bmatrix}^T \tag{44}$$

where $\overline{a}_E$ and $\overline{a}_N$ are the means of the accelerations in the east and north at the current time.

The measurement vector y is defined as

$$y = \begin{bmatrix} p_E & p_N & \omega & s \end{bmatrix}^T \tag{45}$$

where p_E and p_N represent the positions in the east and north from the GNSS receiver, ω is the angular rate from the gyroscope, and s is the output distance from the DR.

The system measurement equation is described as follows:

$$y = h(x) + n \tag{46}$$

where $h(\cdot)$ is the nonlinear measurement function, and n is the measurement noise with covariance Q.

$$h(x) = \begin{cases} p_E \\ p_N \\ \frac{v_N a_E - v_E a_N}{v_E^2 + v_N^2} + \varepsilon \\ \psi T\sqrt{v_E^2 + v_N^2} \end{cases} \tag{47}$$

where ε is the first-order Markov process component of the gyro drift error, and T is the sampling period.

According to the road conditions restricting the system state, the car travelling direction was constrained on the road to the east by the angle θ. The constraint equation is expressed as follows:

$$Dx = d = 0$$
$$D = \begin{bmatrix} -tg\theta & 0 & 0 & 1 & 0 & 0 & 0 & 0 \\ 0 & -tg\theta & 0 & 0 & 1 & 0 & 0 & 0 \end{bmatrix} \tag{48}$$

4.2. Experimental Setup

The experimental setup is shown in Figure 2. The high-precision differential GNSS receiver UT-206 with the positioning accuracy less than 10 cm was used to obtain the actual vehicle position as the reference to calculate the estimation error.

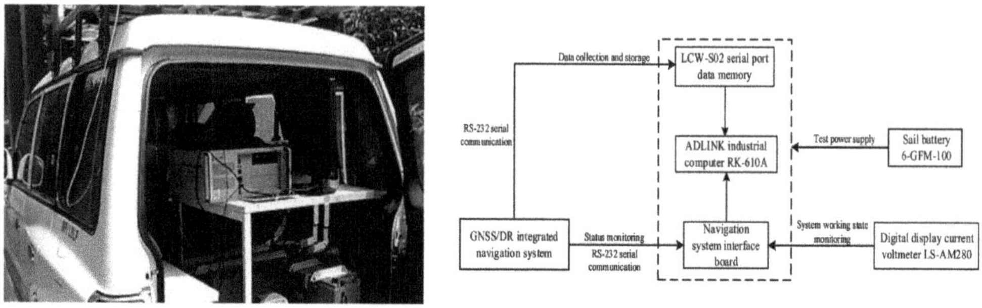

(**a**) Vehicle navigation experiment sports car (**b**) Navigation system equipment

Figure 2. Experimental setup.

The GNSS data update rate was 1 Hz. The GNSS single-point L1/L2 accuracy was 1.5 m in the vertical direction and 1 m in the horizontal direction. The GNSS receiver accuracy was 0.5 m in the vertical direction and 0.3 m in the horizontal direction. The GNSS velocity accuracy was 0.05 m/s. During the test, at least seven navigation satellite signals were available for the GNSS measurement. The experimental data were collected from the car, which was travelling within a continuous time of 1000 s. After initialization for 1 min, the car moved stably at an average velocity of 40 km/h. The car travelling direction was constrained to the east by $\theta = 65^{\circ}$. The calibration coefficient of DR was $\psi = 0.5$. The drift error of the gyroscope was $0.1^{\circ}/\text{h}$, and $\tau_E = \tau_N = 300$ s. The car's initial position, velocity, and acceleration were (0 m, 0 m), (5 m/s, 5 m/s), and (0 m/s^2, 0 m/s^2). The total mileage was 15 km. $\boldsymbol{R} = diag[(10\,\text{m})^2 \quad (1\,\text{m/s})^2 \quad (0.1\,\text{m/s}^2)^2 \quad (10\,\text{m})^2 \quad (1\,\text{m/s})^2 \quad (0.1\,\text{m/s}^2)^2 \quad (0.1^{\circ}/\text{h})^2 \quad (1\,\text{m})^2]$, and $Q = diag[(20\,\text{m})^2\,(20\,\text{m})^2\,(4\,\text{m/s})^2\,(4\,\text{m/s})^2]$.

For comparison purposes, experiments were conducted using the PF, CPF, and CCPF to estimate the car's position and velocity errors. The root mean squared error (RMSE) was used as the metric for accuracy evaluation. The RMSE is defined as follows:

$$\text{RMSE} = \sqrt{\frac{1}{F}\sum_{i=1}^{F}(\hat{x}_i - x_{ref})} \tag{49}$$

where F is the number of Monte Carlo runs, and x_{ref} is the reference value. For the experimental analysis, 1000 Monte Carlo runs were conducted for the RMSE calculation. The overall RMSE is defined as follows:

$$\text{RMSE}_{overall} = \sqrt{\text{RMSE}_E^2 + \text{RMSE}_N^2} \tag{50}$$

where RMSE_E and RMSE_N denote the RMSEs in the east and north.

Figures 3–5 show the position errors obtained by the PF, CPF, and CCPF. It can be seen that the position error curves of all the three methods involve large fluctuations within the initial 100 s due to the system initialization. After the initial 100 s, the PF curves of position error still fluctuate greatly, leading to a position error of (−13.1 m, 14.0 m) in the north and a position error of (−14.5 m, 16.1 m) in the east. The CPF improves the PF with the cubature rules for particle sampling, leading to the position errors within (−13.0 m, 11.1 m) the north and (−13.0 m, 12.1 m) east. However, since the CPF still suffers from particle degradation and does not involve constraints in the state estimation, its improvement is limited. In contrast, since the CCPF optimizes the CPF sampling procedures, its position error is (−8.1 m, 9.6 m) in the north and (−7.0 m, 7.1 m) in the east. Table 1 lists the position RMSEs of the PF, CPF, and CCPF. The position RMSEs of the PF, CPF, and CCPF are 4.1719 m, 3.7231 m, and 2.7061 m in the north and 4.7618 m, 3.7504 m, and 2.5168 m in the east. The overall position RMSEs of the PF, CPF, and CCPF are 6.3308 m, 5.2846 m, and 3.6956 m, showing that the accuracy of the CCPF is 41.6% higher than that of the PF and 30.1% higher than that of the CPF.

Figures 6–8 illustrate the velocity errors obtained by the PF, CPF, and CCPF. The velocity errors of the three methods have a similar trend to their position errors shown in Figures 3–5. The velocity errors of the PF, which are (−0.24 m/s, 0.22 m/s) in the north and (−0.25 m/s, 0.21 m/s) in the east, is the largest. The CPF slightly improves the PF, leading to velocity errors of (−0.11 m/s, 0.16 m/s) in the north and (−0.18 m/s, 0.19 m/s) in the east. In contrast, the velocity errors of the CCPF, which are (−0.14 m/s, 0.15 m/s) (north) and (−0.10 m/s, 0.15 m/s) (east), are the smallest. Table 2 summarizes the velocity RMSEs of the three methods. The velocity RMSEs of the PF, CPF, and CCPF are 0.0912 m/s, 0.0892 m/s, and 0.0769 m/s in the north and 0.1023 m/s, 0.0887 m/s, and 0.0674 m/s in the east. The overall velocity RMSEs of the PF, CPF, and CCPF are 0.1371 m/s, 0.1258 m/s, and 0.1022 m/s, showing that the accuracy of the CCPF is 25.5% higher than that of the PF and 18.8% higher than that of the CPF.

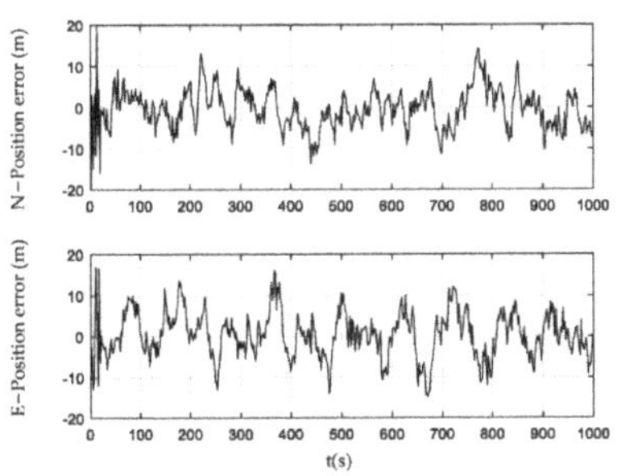

Figure 3. Position error obtained using the PF.

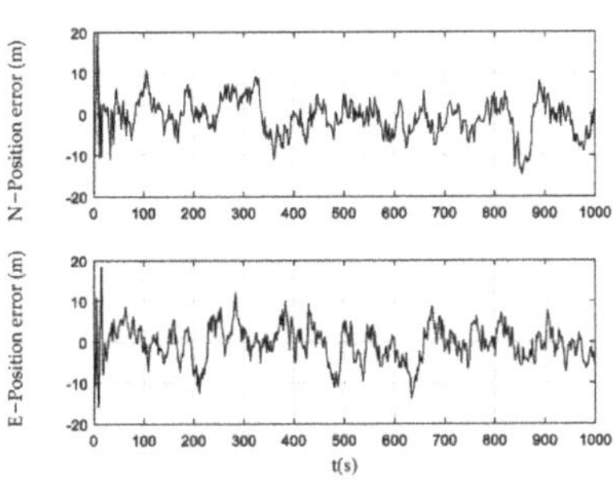

Figure 4. Position error obtained using the CPF.

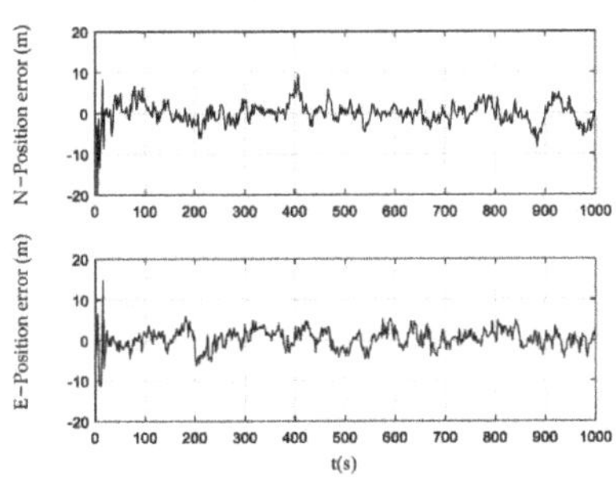

Figure 5. Position error obtained using the CCPF.

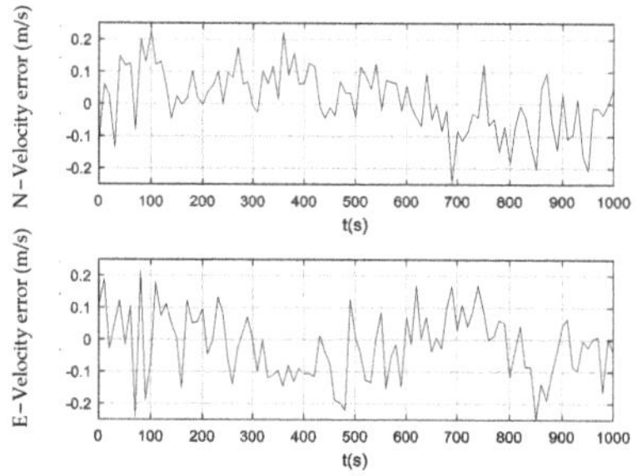

Figure 6. Velocity error obtained using the PF.

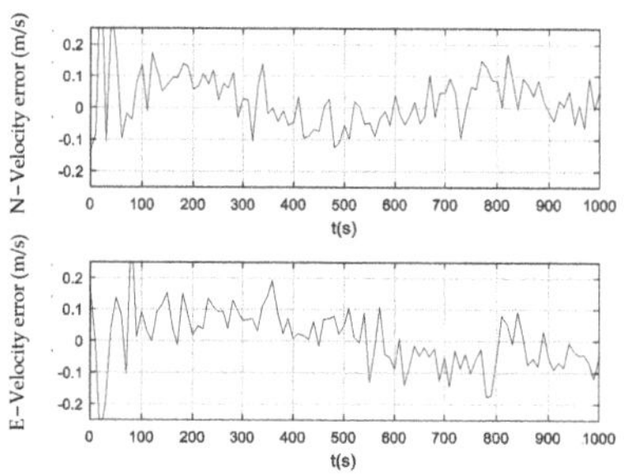

Figure 7. Velocity error obtained using the CPF.

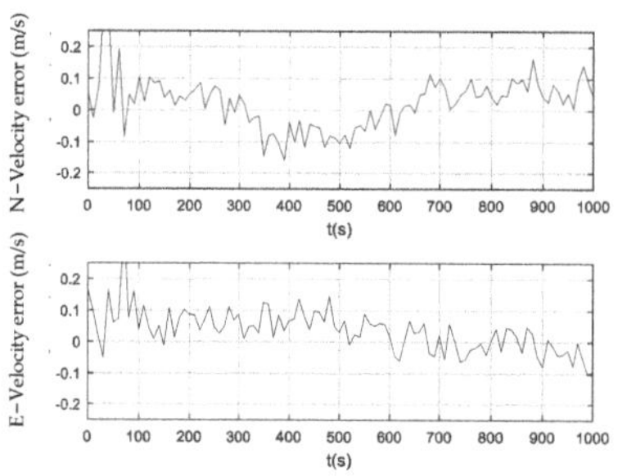

Figure 8. Velocity error obtained using the CCPF.

Table 1. Position RMSEs of the PF, CPF, and CCPF.

Filter	RMSE in East (m)	RMSE in North (m)	Error Range (m)	Overall RMSE (m)
PF	4.1719	4.7618	(−13.1, 14.0), (−14.5, 16.1)	6.3308
CPF	3.7231	3.7504	(−13.0, 11.1), (−13.0, 12.1)	5.2846
CCPF	2.7061	2.5168	(−8.1, 9.6), (−7.0, 7.1)	3.6956

Table 2. Velocity RMSEs of the PF, CPF, and CCPF.

Filter	RMSE in East (m/s)	RMSE in North (m/s)	Error Range (m/s)	Overall RMSE (m/s)
PF	0.0912	0.1023	(−0.24, 0.22), (−0.25, 0.21)	0.1371
CPF	0.0892	0.0887	(−0.11, 0.16), (−0.18, 0.19)	0.1258
CCPF	0.0769	0.0674	(−0.14, 0.15), (−0.10, 0.15)	0.1022

From the above results, it can be concluded that the CCPF can achieve more accurate state estimations under constrained conditions than the CPF and PF for GNSS/DR-integrated navigation. It is also shown that the filtering solution of CCPF is bounded, which further verifies the theoretical convergence analysis of the CCPF (described in Section 3.3).

The above RMSEs, together with the theoretical convergence analysis described in Section 3.3, demonstrate the consistent performance of the proposed CCPF. To further evaluate the filter's consistency, Figure 9 illustrates quantile–quantile plots of the position residuals derived from the use of the PF, CPF, and CCPF. As shown in Figure 9a, it is obvious that the quantile of the position residuals of the PF does not coincide with its normal quantile. Although the residual quantiles of both the CPF and CCPF are in good agreement with their normal quantiles, as shown in Figure 9b,c, the residual quantile of the CCPF is much closer to its normal quantile compared to that of the CPF. Therefore, it is clear that the proposed CCPF possesses the consistency for state estimation.

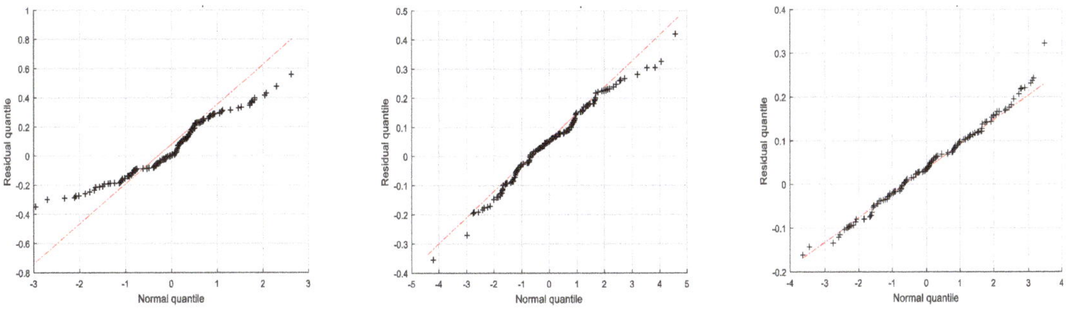

(a) (b) (c)

Figure 9. Quantiles of the position residuals derived from the use of (**a**) the PF, (**b**) the CPF, and (**c**) the CCPF. The quantiles of the position residuals are indicated by the black dots and "+", and the normal quantiles are indicated by the dashed red line.

5. Conclusions

Nonlinear systems with constraints and non-Gaussian uncertainties are commonly encountered in vehicle navigation. To address this issue, this paper proposes a new CCPF to estimate system state parameters for vehicle navigation. It enhances the CPF importance sampling process using constraints to improve the importance density distribution. Subsequently, it improves the CPF resampling process by using the Euclidean distance of the measurement residual to adjust particle weights to ensure the diversity of particles, thus decreasing particle degradation. Theories were also established to prove the convergence of the proposed CCPF. The experimental results and the results derived from our comparative analysis of GNSS/DR-integrated vehicle navigation demonstrate that the proposed CCPF can effectively estimate system state parameters in constrained environments, resulting in it having enhanced accuracy compared to the PF and CPF.

Our future research work will focus on the improvement of the proposed CCPF. The proposed CCPF will be combined with advanced artificial intelligence technologies and genetic algorithms to achieve nonlinear state estimation in the environments of more complex constraints.

Author Contributions: Conceptualization, L.X. and Y.Z.; formal analysis, L.X.; investigation, Y.Z.; methodology, Y.H.; project administration, L.X.; resources, L.X.; software, L.X.; supervision, Y.H.; writing—original draft, L.X. and Y.Z. All authors have read and agreed to the published version of the manuscript.

Funding: This research was funded by Ningxia Natural Science Foundation, China (Grant number 2022AAC03118).

Institutional Review Board Statement: Not applicable.

Informed Consent Statement: Not applicable.

Data Availability Statement: The data used to support the findings of this study are available from the corresponding author upon request.

Conflicts of Interest: The authors declare that they have no conflicts of interest.

References

1. Wang, W.; Xing, C.Y.; Feng, S. State of the art and perspectives of autonomous navigation technology. *Acta Aeronaut. Astronaut. Sin.* **2021**, *42*, 525049. [CrossRef]
2. Zhao, X.; Min, H.; Xu, Z.; Wang, W. An ISVD and SFFSD-based vehicle ego-positioning method and its application on indoor parking guidance. *Transp. Res. Part C Emerg. Technol.* **2019**, *108*, 29–48. [CrossRef]
3. Yuan, Y.; Li, F.; Chen, J.; Wang, Y.; Liu, K. An improved Kalman filter algorithm for tightly GNSS/INS integrated navigation system. *Math. Biosci. Eng.* **2024**, *21*, 963–983. [CrossRef] [PubMed]
4. Brossard, M.; Barrau, A.; Bonnabel, S. AI-IMU dead reckoning. *IEEE Trans. Intell. Veh.* **2020**, *5*, 585–595. [CrossRef]
5. Hu, G.; Ni, L.; Gao, B.; Zhu, X.; Wang, W.; Zhong, Y. Model predictive based unscented Kalman filter for hypersonic vehicle navigation with INS/GNSS integration. *IEEE Access* **2020**, *8*, 4814–4823. [CrossRef]
6. Min, H.; Wu, X.; Cheng, C.; Zhao, X. Kinematic and dynamic vehicle model-assisted global positioning method for autonomous vehicles with low-cost GPS/camera/in-vehicle sensors. *Sensors* **2019**, *19*, 5430. [CrossRef] [PubMed]
7. Wang, W.; Min, H.; Wu, X.; Hou, X.; Li, Y.; Zhao, X. High accuracy and low complexity LiDAR place recognition using unitary invariant frobenius norm. *IEEE Sens. J.* **2022**, *23*, 11205–11207. [CrossRef]
8. Gustafsson, F.; Hendeby, G. Some relations between extended and unscented Kalman filters. *IEEE Trans. Signal Process.* **2012**, *60*, 545–555. [CrossRef]
9. Stepanov, O.A.; Litvinenko, Y.A.; Vasiliev, V.A.; Toropov, A.B.; Basin, M.V. Polynomial filtering algorithm applied to navigation data processing under quadratic nonlinearities in system and measurement equations. Part 1. description and comparison with Kalman type algorithms. *Gyroscopy Navig.* **2021**, *12*, 205–223. [CrossRef]
10. Gao, B.; Gao, S.; Hu, G.; Zhong, Y.; Gu, C. Maximum likelihood principle and moving horizon estimation based adaptive unscented Kalman filter. *Aerosp. Sci. Technol.* **2018**, *73*, 184–196. [CrossRef]
11. Hu, G.; Gao, B.; Zhong, Y.; Gu, C. Unscented Kalman filter with process noise covariance estimation for vehicular INS/GPS integration system. *Inf. Fusion* **2020**, *64*, 194–204. [CrossRef]
12. Gao, B.; Hu, G.; Zhang, L.; Zhong, Y.; Zhu, X. Cubature Kalman filter with closed-loop covariance feedback control for integrated INS/GNSS navigation. *Chin. J. Aeronaut.* **2023**, *36*, 363–376. [CrossRef]
13. Gao, B.; Li, W.; Hu, G.; Zhong, Y.; Zhu, X. Mahalanobis distance-based fading cubature Kalman filter with augmented mechanism for hypersonic vehicle INS/CNS autonomous integration. *Chin. J. Aeronaut.* **2022**, *35*, 114–128. [CrossRef]
14. Kulikov, G.Y.; Kulikova, M.V. Stability analysis of extended, cubature and unscented Kalman filters for estimating stiff continuous-discrete stochastic systems. *Automatica* **2018**, *90*, 91–97. [CrossRef]
15. Wei, W.; Gao, S.; Zhong, Y.; Gu, C.; Hu, G. Adaptive square-root unscented particle filtering algorithm for dynamic navigation. *Sensors* **2018**, *18*, 2337. [CrossRef] [PubMed]
16. Speekenbrink, M. A tutorial on particle filters. *J. Math. Psychol.* **2016**, *73*, 140–152. [CrossRef]
17. Closas, P.; Guillamon, A. Sequential estimation of intrinsic activity and synaptic input in single neurons by particle filtering with optimal importance density. *EURASIP J. Adv. Signal Process.* **2017**, *2017*, 65. [CrossRef]
18. Gordon, N.J.; Salmond, D.J.; Smith, A.F.M. Novel approach to nonlinear/non-Gaussian Bayesian state estimation. *IEE Proc. F Radar Signal Process.* **1993**, *140*, 107–113. [CrossRef]
19. Del Moral, P.; Doucet, A.; Jasra, A. On adaptive resampling strategies for sequential Monte Carlo methods. *Bernoulli* **2012**, *18*, 252–278. [CrossRef]
20. Jia, K.; Pei, Y.; Gao, Z.; Zhong, Y.; Gao, S.; Wei, W.; Hu, G. A quaternion-based robust adaptive spherical simplex unscented particle filter for MINS/VNS/GNS integrated navigation system. *Math. Probl. Eng.* **2019**, *2019*, 8532601. [CrossRef]
21. Pitt, M.; Shephard, N. Filtering via simulation: Auxiliary particle filters. *J. Am. Stat. Assoc.* **1999**, *94*, 590–599. [CrossRef]
22. Liu, D.; Duan, J.; Shi, H. A strong tracking square root central difference FastSLAM for unmanned intelligent vehicle with adaptive partial systematic resampling. *IEEE Trans. Intell. Transp. Syst.* **2016**, *17*, 3110–3120. [CrossRef]
23. Liu, H.T.; Lin, Y.M.; Chen, Y.H.; Zhou, E.M.; Peng, B. A study on resampling strategy of intelligent particle filter based on genetic algorithm. *J. Electron. Inf. Technol.* **2021**, *43*, 3459–3466. [CrossRef]
24. Xia, B.; Sun, Z.; Zhang, R.; Lao, Z. A cubature particle filter algorithm to estimate the state of the charge of lithium-ion batteries based on a second-order equivalent circuit model. *Energies* **2017**, *10*, 457. [CrossRef]
25. Shi, Q.; Liu, M.; Hang, L. A novel distribution system state estimator based on robust cubature particle filter used for non-Gaussian noise and bad data scenarios. *IET Gener. Transm. Distrib.* **2022**, *16*, 1385–1399. [CrossRef]
26. Feng, H.; Cai, Z. Target tracking based on improved cubature particle filter in UWSNs. *IET Radar Sonar Navig.* **2019**, *13*, 638–645. [CrossRef]

27. Zhang, Y.G.; Cheng, R.; Huang, Y.L.; Li, N. Truncated adaptive cubature particle filter. *Syst. Eng. Electron.* **2016**, *38*, 382–391. [CrossRef]
28. Liu, M.; He, M.; Qiao, S.; Liu, B.; Cao, Z.; Wang, R. A high-order state-of-charge estimation model by cubature particle filter. *Measurement* **2019**, *146*, 35–42. [CrossRef]
29. Xing, D.; Wei, M.; Zhao, W.; Wang, Y.; Wu, S. Vehicle state estimation based on adaptive cubature particle filtering. *J. Nanjing Univ. Aeronaut. Astronaut.* **2020**, *52*, 445–453. [CrossRef]
30. Gao, Z.; Mu, D.; Zhong, Y.; Gu, C. Constrained Unscented Particle Filter for SINS/GNSS/ADS Integrated Airship Navigation in the Presence of Wind Field Disturbance. *Sensors* **2019**, *19*, 471. [CrossRef]
31. Seifzadeh, S.; Khaleghi, B.; Karray, F. Distributed soft-data-constrained multi-model particle filter. *IEEE Trans. Cybern.* **2015**, *45*, 384–394. [CrossRef]
32. Xu, C.H.; Wang, X.; Duan, S.; Wan, J. Spatial-temporal constrained particle filter for cooperative target tracking. *J. Netw. Comput. Appl.* **2021**, *176*, 102913. [CrossRef]
33. Zhang, H.; Li, L.; Xie, W. Constrained multiple model particle filtering for bearings-only maneuvering target tracking. *IEEE Access* **2018**, *6*, 51721–51734. [CrossRef]
34. Xu, A.; Zou, X.; Qing, Y.; Fang, Q.; Sui, X. The improved cubature Kalman filter in GNSS/INS tightly coupled mode. *Sci. Surv. Mapp.* **2022**, *47*, 22–28. [CrossRef]
35. Arasaratnam, I.; Haykin, S.; Hurd, T.R. Cubature Kalman filtering for continuous discrete systems: Theory and simulations. *IEEE Trans. Signal Process.* **2010**, *58*, 4977–4993. [CrossRef]
36. Arasaratnam, I.; Haykin, S. Cubature Kalman filters. *IEEE Trans. Autom. Control* **2009**, *54*, 1254–1269. [CrossRef]
37. Gao, B.; Hu, G.; Zhu, X.; Zhong, Y. Robust cubature Kalman filter with abnormal observations identification using Mahalanobis distance criterion for vehicular INS/GNSS integration. *Sensors* **2019**, *19*, 5149. [CrossRef]
38. Andrieu, C.; Doucet, A.; Holenstein, R. Particle markov chain Monte Carlo methods. *J. R. Stat. Soc. Ser. B* **2010**, *72*, 269–342. [CrossRef]
39. Chen, J.G.; Li, J.; Gao, X.B. Particle filtering with equality state constraints. *J. Univ. Electron. Sci. Technol. China* **2011**, *40*, 596–601. [CrossRef]
40. Simon, D.; Tien, L.C. Kalman filtering with state equality constraints. *IEEE Trans. Aerosp. Electron. Syst.* **2002**, *38*, 128–136. [CrossRef]
41. Cui, P.Y.; Zheng, L.F.; Pei, F.J. Research on integrated navigation system based on self-adjust particle filter. *Comput. Eng.* **2008**, *34*, 185–187. [CrossRef]
42. Crisan, D.; Doucet, A. A survey of convergence results on particle filtering methods for practitioners. *IEEE Trans. Signal Process.* **2002**, *50*, 736–746. [CrossRef]
43. Hu, X.; Schön, T.B.; Ljung, L. A basic convergence result for particle filtering. *IEEE Trans. Signal Process.* **2008**, *56*, 1337–1348. [CrossRef]
44. Fang, J.C.; Li, X.E.; Shen, G.X.; Fang, J.G.; Wang, Q.; Wan, D. Study of GPS/DR integrated navigation system for urban vehicle. *China J. Highw. Transp.* **1999**, *12*, 84–89.
45. Zhao, Y.; Wang, N.; Ye, J.K. Constraints unscented particle filter and its application in vehicle navigation. *J. Chang. Univ. Nat. Sci. Ed.* **2020**, *40*, 109–116.

sensors

MDPI

Article

Improving Optical Flow Sensor Using a Gimbal for Quadrotor Navigation in GPS-Denied Environment

Jonathan Flores †, Ivan Gonzalez-Hernandez †, Sergio Salazar, Rogelio Lozano * and Christian Reyes

Department of Research and Multidisciplinary Studies, Program of Aerial and Submarine Autonomous Navigation Systems, Center for Research and Advanced Studies of the National Polytechnic Institute, Mexico City 07360, Mexico; jonathan.flores@cinvestav.mx (J.F.); ivan.gonzalez@cinvestav.mx (I.G.-H.); sesalazar@cinvestav.mx (S.S.); christian.reyes@cinvestav.mx (C.R.)

* Correspondence: rogelio.lozano@cinvestav.mx

† These authors contributed equally to this work.

Abstract: This paper proposes a new sensor using optical flow to stabilize a quadrotor when a GPS signal is not available. Normally, optical flow varies with the attitude of the aerial vehicle. This produces positive feedback on the attitude control that destabilizes the orientation of the vehicle. To avoid this, we propose a novel sensor using an optical flow camera with a 6DoF IMU (Inertial Measurement Unit) mounted on a two-axis anti-shake stabilizer mobile aerial gimbal. We also propose a robust algorithm based on Sliding Mode Control for stabilizing the optical flow sensor downwards independently of the aerial vehicle attitude. This method improves the estimation of the position and velocity of the quadrotor. We present experimental results to show the performance of the proposed sensor and algorithms.

Keywords: UAV; navigation; robust control; GPS-denied

1. Introduction

Research around Unmanned Aerial Vehicles (UAVs) has increased in recent decades due to their capacity for experimentation in different disciplines, low cost, and ease of repair compared to classical helicopter configurations. Sensor improvements and robust control algorithm developments allow for accuracy tasks in vehicles of different configurations such as fixed-wing aircrafts, rotorcrafts, and hybrids. Sensor-based navigation approaches are developed in specific environments and conditions for civil, military, or scientific applications.

In the early stages of development, optical flow-based navigation was limited by the range and field of view of the sensors. The frame-fixed optical flow module uses a downward-facing camera and a distance sensor for velocity estimation. Optical flow detection is affected due to angle changes caused by UAV maneuvers. In hover flights the maneuvers are reduced, and the velocity estimation is reliable.

Motion capture can be used for high-precision indoor flight. In [1], the position of the multirotor aircraft is estimated by a set of cameras in a room; the tracking of the movement of the UAV was controlled offline to follow a ground vehicle, but the cost of the motion capture system is highlighted. The works [2,3] present open-air flights using GPS in precise long-range missions, analyze the sensitivity of GPS to interference, and propose a return-to-home flight mode using the heading angle to home. In [4], vision-based navigation is used for indoor and outdoor flights with high-cost computational requirements and lighting changes sensitivity. Moreover, navigation using optical flow and distance sensors requires low-cost computational and provides good accuracy in local environments.

In [5], the authors present an eight-rotor UAV using optical flow, and this rotorcraft uses the main four rotors for altitude and attitude displacement and another four rotors for horizontal maneuvers. This prototype uses the optical flow for estimating the horizontal position avoiding angular tilt. In [6] in the authors proposed improved motion compensation using feature block selection, look-ahead rotation, fault case detection and filtering using PX4FLOW hardware by tunning the camera resolution, the interval of adaptive boxes and advanced search algorithm. In [7] presents a UAV position estimation using an optical flow approach using a Gated Recurrent Unit (GRU) network-based pointing angles and Magnetic, Angular Rate, and Gravity (MARG) sensors improving robustness and performance in real-time experimentation.

Optical flow sensors used for position control in quadrotor aircraft are facing downwards and fixed to the airframe. In a flight based on the optical flow sensor, the roll and pitch maneuvers are restricted to $\pm 15^{\circ}$, while maneuvers in flights without optical flow sensors are up to $\pm 35^{\circ}$. The bounded angles reduce the horizontal displacement velocity of the vehicle. Quadrotor rotorcraft in an ideal hover flight would maintain the Euler angles of pitch and roll aligned to the horizontal and with a linear displacement velocity near zero. However, in practice (even in hover flight), a quadrotor rotorcraft vehicle is sensitive to disturbances and the responses of the attitude controller can be considered as optical noise. Furthermore, it is well known that tilting an optical flow module produces positive feedback on vertical position estimates and linear velocity estimates.

The rotorcraft in Figure 1a shows an optical flow module mounted directly on the quadrotor's frame. When the quadrotor tilts it causes a distortion in the estimation of the altitude and horizontal velocity. Figure 1b shows an optical flow module stabilized by a gimbal. The gimbal feedback control stabilizes pitch and roll angles on the horizontal plane, providing reliable altitude and horizontal velocity estimations with respect to the ground and avoiding optical distortions.

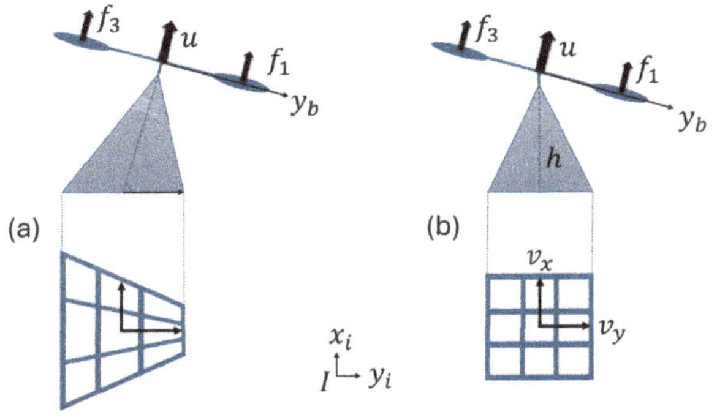

Figure 1. (**a**) Optical flow without stabilization and (**b**) stabilized optical flow.

The main contribution of the present paper is improving the optical flow sensing to obtain a reliable set of image data using a gimbal that in general is used to stabilize camera videos for recording or First-Person View (FPV) flights. The optical flow sensor is improved for trajectory control in GPS-Denied environments. This approach allows for flight maneuvers with large changes in the attitude angles; in other cases this produces positive feedback that destabilizes the orientation of the vehicle. Furthermore, we use a robust control technique to enable hover flights and trajectory following with tracking velocity without affecting the sensed images.

The paper is organized as follows: Section 2 describes the system mathematical models. Section 3 is devoted to the control algorithms. Materials and Methods are given in Section 4. The experimental results are shown in Section 5. A brief discussion about our approach is presented in section 6. Final remarks are given in the conclusions.

2. Mathematical Models

Mathematical models are approximations of the real systems. The subsystems include the dynamics, estimations, control inputs, and degrees of freedom. In the quadrotor aircraft, the forces and angular moments are induced by four actuators to move in the inertial reference frame (x, y, z) and rotate in their Euler angles (ϕ, θ, ψ). In the gimbal, the angular moments are induced by its three actuators that allow it to tilt around its Euler angles $(\phi_g, \theta_g, \psi_g)$.

The optical flow module estimates the lateral and longitudinal displacement velocity and altitude of the complete system $(\dot{x}, \dot{y}, z)$. The following subsections present the mathematical models of the quadrotor aircraft, gimbal, and optical flow module to describe the complete system.

2.1. Quadrotor Aircraft Model

Figure 2 shows the representation of the quadrotor aircraft. Let us define the total thrust u as follows:

$$u = (f_1 + f_2 + f_3 + f_4) \tag{1}$$

where $f_i = k\omega_i^2$ is the force produced by i-rotor where $i = 1, 2, 3, 4$. k represents a set of aerodynamic constants, and ω_i is the angular velocity. We obtain the Euler angles rates as

$$\begin{bmatrix} \dot{\phi} \\ \dot{\theta} \\ \dot{\psi} \end{bmatrix} = T \begin{bmatrix} p \\ q \\ r \end{bmatrix}$$

where

$$T = \begin{bmatrix} 1 & \sin\phi \tan\theta & \cos\phi \tan\theta \\ 0 & \cos\phi & -\sin\phi \\ 0 & \sin\phi \sec\theta & \cos\phi \sec\theta \end{bmatrix}$$

The rotor torque $\tau_{(\phi,\theta,\psi)}$ is defined as

$$\begin{array}{rl} \tau_\phi = & (f_1 - f_3)l \\ \tau_\theta = & (f_2 - f_4)l \\ \tau_\psi = & \sum_{i=1}^4 \tau_i \end{array} \tag{2}$$

where l is the distance from the rotors to the center of the gravity and τ_i is the torque produced by the i-rotor. Therefore, according to [8] the quadrotor aircraft model can be defined as

$$\begin{array}{l} m\ddot{x} = -u \sin\theta \\ m\ddot{y} = u \cos\theta \sin\phi \\ m\ddot{z} = u \cos\theta \cos\phi - mg \\ I\ddot{\eta} + C(\eta, \dot{\eta})\dot{\eta} = \tilde{\tau} \end{array} \tag{3}$$

where m is the mass, g is the acceleration due to the gravity, I represents the inertia matrix, C is the Coriolis matrix, $\eta = (\phi, \theta, \psi)^T$ is the attitude vector, and $\tilde{\tau}$ is the control input vector.

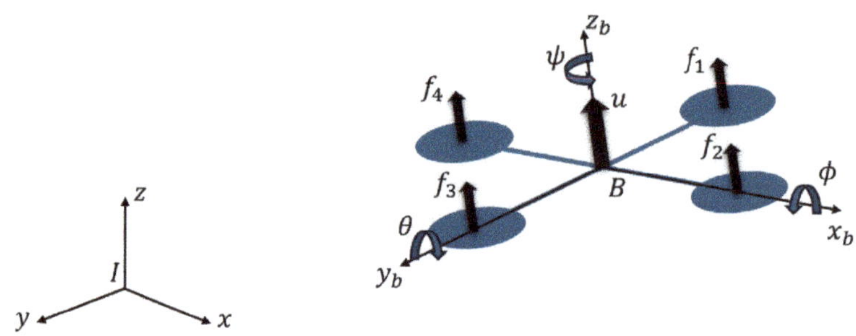

Figure 2. Quadrotor aircraft diagram.

2.2. Gimbal Model

Figure 3 shows the gimbal diagram, where the $G = (x_g, y_g, z_g)$ represents the frame fixed to the optical flow sensor. In [9,10], the authors propose the gimbal mathematical model as follows:

$$\begin{array}{l} J_x\ddot{\phi}_g + \xi_1\dot{\phi}_g = v_\phi + d_\phi \\ J_y\ddot{\theta}_g + \xi_2\dot{\theta}_g = v_\theta + d_\theta \\ J_z\ddot{\psi}_g + \xi_3\dot{\psi}_g = v_\psi + d_\psi \end{array} \quad (4)$$

where $J = D(J_x, J_y, J_z)$ is the gimbal rotational inertia matrix, and $\eta_g = (\phi_g, \theta_g, \psi_g)$ represents the torques in the pitch, roll, and yaw gimbal angles. $\xi = (\xi_1, \xi_2, \xi_3)$ are the friction coefficients for each angle. Moreover, $v = (v_\phi, v_\theta, v_\psi)$ is the control input, and $d = (d_\phi, d_\theta, d_\psi)$ is the nonlinear dynamic effect of the couplings that are considered as a disturbance.

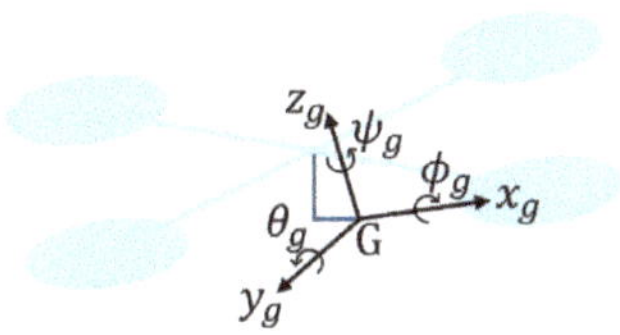

Figure 3. Gimbal diagram.

2.3. Optical Flow Model

Optical flow is a monocular vision method that uses images of a moving object to estimate the displacement velocity.

Figure 4 shows the optical flow diagram, where the $O = (x_o, y_o, z_o)$ represents the frame fixed of the optical flow sensor. P is a direct point on the ground from the optical axis, f is the focal length from the centre of projection O to the image plane, and h is the altitude from the image plane to the ground. θ_o is the optical axis angle and the z_i axis. The authors of [11] present the approximate pixel displacements as follows:

$$\begin{array}{l} V_x = \frac{T_z x - T_x f}{h} - \omega_y f + \omega_z y + \frac{\omega_x xy - \omega_y x^2}{f} \\ V_y = \frac{T_z y - T_y f}{h} + \omega_x f - \omega_z x + \frac{\omega_x y^2 - \omega_y xy}{f} \end{array} \quad (5)$$

where (V_x, V_y) are the estimate optical flow translation velocities, $(\omega_x, \omega_y, \omega_z)$ are the angular velocities, and (T_x, T_y, T_z) are the linear velocities.

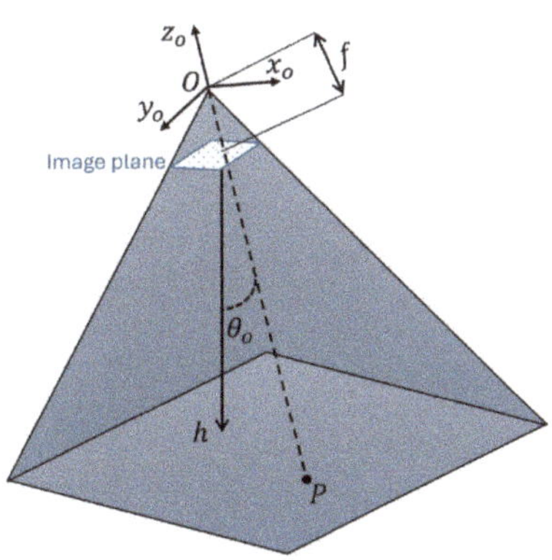

Figure 4. Optical flow diagram.

3. Control Strategies

This section presents a feedback control algorithm implemented to stabilize the quadrotor aircraft altitude and the robust control algorithms that guarantee finite-time convergence for the quadrotor aircraft pitch and roll angles and for the gimbal pitch and roll angles. The main goal of the algorithms based on Sliding Mode Control is to stabilize the quadrotor aircraft attitude and the optical flow module attitude independently. The quadrotor aircraft altitude and attitude are controlled by the aircraft's four rotors angular velocity, and the gimbal is controlled by its two rotor torques that stabilize the optical flow module on the horizontal line.

3.1. *Quadrotor Aircraft Control*

The quadrotor aircraft robust control approach enables stable flight by damping external disturbances or attitude tilts for horizontal vehicle displacement. This behavior is very important for the entire aerial vehicle system.

3.1.1. Vertical Displacement and Yaw Control (z, ψ)

A feedback control algorithm has been selected to stabilize the vertical position, and it is obtained using the following input:

$$u = \left(\frac{c_1 + mg}{\cos\phi \cos\theta} \right) \tag{6}$$

where c_1 is defined as

$$c_1 = -k_p(z - z_d) - k_d\dot{z} \tag{7}$$

where $k_p, k_d > 0$ and z_d is the desired altitude. Introducing (6) into (3)

$$\ddot{z} = -k_p(z - z_d) - k_d\dot{z} \tag{8}$$

When the quadrotor aircraft reaches the desired altitude $z \to z_d$ then $c_1 \to 0$.
The yaw angle ψ is controlled by

$$\tau_\psi = -k_a(\psi - \psi_d) - k_b\dot{\psi} \tag{9}$$

where $k_a, k_b > 0$ and ψ_d means the desired yaw angle.

3.1.2. Lateral Displacement and Roll Control (y, ϕ)

By applying (6) into Equation (3), after a convergence time, the lateral dynamics become

$$\ddot{y} \approx mg \tan\phi \tag{10}$$

In order to guarantee the robust tracking in y-axis, differentiating (10) twice gives

$$y^{(4)} = mg(\ddot{\phi} + 2\dot{\phi}^2 \tan\phi) \sec^2\phi \tag{11}$$

Introducing (3) into (11) and applying a variable change $\tilde{\tau} = \tau + C(\eta, \dot{\eta})\dot{\eta}$ and inertia are considered the identity matrix

$$\begin{aligned} y^{(4)} &= mg(\tau_\phi + 2\dot{\phi}^2 \tan\phi) \sec^2\phi \\ &= \tau_\phi mg \sec^2\phi + 2mg\dot{\phi}^2 \tan\phi \sec^2\phi \end{aligned} \tag{12}$$

Let us define the following f_a and f_b as

$$f_a(\phi) = (mg \sec^2\phi), \quad f_b(\phi, \dot{\phi}) = 2mg\dot{\phi}^2 \tan\phi \sec^2\phi \tag{13}$$

where $f_a(\cdot), f_b(\cdot)$ represents terms obtained after a second derivative. We have considered

$$f_a(\phi) > 0 \tag{14}$$

Then, the y dynamics are

$$y^{(4)} = \tau_\phi f_a(\phi) + f_b(\phi, \dot{\phi}) \tag{15}$$

In order to reach the desired trajectory, we propose the robust control based on the Sliding Mode Control algorithm. Let us define the following variables:

$$\begin{aligned} \dot{y}_1 &= \dot{y} \\ \dot{y}_2 &= \ddot{y} \\ \dot{y}_3 &= y^{(3)} \\ \dot{y}_4 &= y^{(4)} = \tau_\phi f_a(\phi) + f_b(\phi, \dot{\phi}) \end{aligned} \tag{16}$$

and the following errors as

$$\begin{aligned} e_1 &= y - y_d(t) \\ e_2 &= \dot{y} - \dot{y}_d(t) \\ e_3 &= \ddot{y} - \ddot{y}_d(t) \\ e_4 &= y^{(3)} - y_d^{(3)}(t) \end{aligned} \tag{17}$$

Differentiating

$$\begin{aligned} \dot{e}_1 &= e_2 \\ \dot{e}_2 &= e_3 \\ \dot{e}_3 &= e_4 \\ \dot{e}_4 &= y^{(4)} - y_d^{(4)}(t) \end{aligned} \tag{18}$$

The sliding surface is defined as a function of error

$$s = (k_1 e_1 + k_2 e_2 + k_3 e_3) + e_4 \tag{19}$$

where $k_1, k_2, k_3 > 0$. The objective is to reach the surface $s = 0$. We propose the following positive function:

$$V = \frac{1}{2} s^2$$

we have

$$\dot{V} = s(k_1 e_2 + k_2 e_3 + k_3 e_4 + \tau_\phi f_a(\phi) + f_b(\phi, \dot{\phi}))$$

By choosing τ_ϕ like

$$\tau_\phi = -\frac{1}{f_a(\phi)} \left[f_b(\phi, \dot{\phi}) + (k_1 e_2 + k_2 e_3 + k_3 e_4) + \nu_\phi \right] \tag{20}$$

where ν_ϕ is defined as

$$\nu_\phi = \rho_\phi \, sign(s) \tag{21}$$

where $\rho_\phi > 0$ and ν_ϕ denote the Sliding Mode Controller, then $\dot{V} = -\rho_\phi s \; sign(s)$. Then, $s \to 0$ in a finite time. Then, from (19)

$$e_4 = -(k_1 e_1 + k_2 e_2 + k_3 e_3)$$

this is an error linear system, where the Hurwitz dynamics characteristic polynomial is

$$p^3 + k_3 p^2 + k_2 p + k_1 = 0$$

then from (18) $e_1, e_2, e_3, e_4 \to 0$.

3.1.3. Longitudinal Displacement and Pitch Control (x, θ)

After a transient time, the longitudinal dynamics become

$$\ddot{x} \approx -mg\frac{\tan\theta}{\cos\phi} \tag{22}$$

Consider it diagonal and compensate the torque. Considering a small angle ϕ, (22) becomes approximately $\ddot{x} \approx -mg\tan\theta$. Indeed, since the quadrotor aircraft configuration is a symmetrical frame, it is also possible to apply the Sliding Mode Controller to the pitch angle (θ) and to the longitudinal displacement in the x-axis.

3.2. Gimbal Control

The gimbal robust control approach ensures that the attitude of the optical flow module is stable and independent of the quadrotor aircraft attitude. This performance is very important for optical flow sensing.

3.2.1. Gimbal Pitch Control (θ_g)

An algorithm based on Nonsingular Terminal Sliding Mode Control (NTSMC) is selected to stabilize the gimbal pitch angle (θ_g) due to the robustness to disturbances.

Since the tracking error converges to zero, we propose the following sliding surface:

$$s_g = \theta_g + \alpha\, sign^{\gamma_1}(\theta_g) + \beta\, sign^{\gamma_2}(\dot{\theta}_g) \tag{23}$$

where $sign^{\gamma}(\theta_g) = sign\theta_g|\theta_g|^{\gamma}$, θ_g is the gimbal pitch angle, and the variables satisfying $\alpha, \beta > 0, 1 > \gamma_1 > 2, \gamma_1 > \gamma_2$ and $sign(\cdot)$ are the signum function. Then, s_g guarantees the convergence to zero in finite time. The algorithm control law v_θ is defined as

$$v_\theta = -J_y\left[\frac{1}{\beta\gamma_2}sign^{2-\gamma_2}\dot{\theta}_g\left(1+\alpha\gamma_1|\theta_g|^{\gamma-1}\right) + \beta\dot{\theta}_g + K_1 sign(s) + K_2 s\right] \tag{24}$$

where $K_1, K_2 > 0$. The control law v_θ for the pitch (θ_g) angle guarantees the finite-time convergence of the tracking error to zero. The stability proof can be found in [12].

3.2.2. Gimbal Roll Control (ϕ_g)

Indeed, since the gimbal configuration is symmetrical and orthogonal in xy-axes, we applied the same control strategy to the angular displacement roll control (ϕ).

The algorithm control law v_θ is defined as

$$v_\phi = -J_x\left[\frac{1}{\beta\gamma_2}sign^{2-\gamma_2}\dot{\phi}_g\left(1+\alpha\gamma_1|\phi_g|^{\gamma-1}\right) + \beta\dot{\phi}_g + K_1 sign(s) + K_2 s\right] \tag{25}$$

The control law v_θ for the roll (ϕ_g) angle guarantees the finite-time convergence of the tracking error to zero.

3.3. Chattering Avoidance

It is well known that the implementation of the control approach based on sliding modes brings with it the disadvantage of the vibration effect present in the real-time experimentation of the sign function. In implemented control systems, it is important to ensure the performance of the system in the face of high working frequency and that the electronic controller can command in real time considering the electronic and mechanical wear of its actuators. However, in [8] the authors propose replacing the sign function with a high slope saturation function $sign(\cdot) \approx sat(\cdot)$ to solve this issue.

In our approach, in order to avoid the chattering effect caused by the signum function in robust control algorithms used in the quadotor attitude and gimbal attitude, the approximation function $sign(\cdot) \approx \tanh(\cdot)$ is used to smooth the signal. The signum function and hyperbolic tangent function are shown in Figure 5.

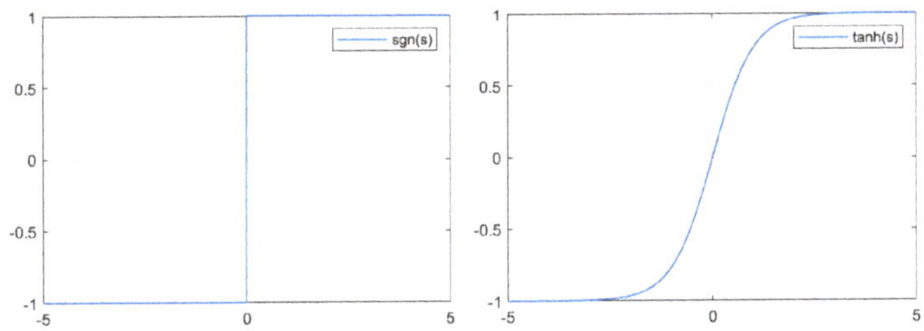

Figure 5. Signum function and hyperbolic tangent function.

4. Materials and Methods

In this section, we will describe the different hardware parts of the embedded control system, namely, sensors, actuators, pilot communication, and the flight controller board that links them together. The total weight of the Unmanned Aerial Vehicle is 2.5 kg; the weight of the gimbal is clearly considered as the payload. The setup presented can be scaled to a smaller quadrotor aircraft by using a smaller gimbal or using servomotors instead of brushless motors for stabilization purposes. However, as in the case of Unmanned Aerial Vehicles, the gains must be adjusted in flight tests.

In order to guarantee the stability of the aerial vehicle for the development of the experiments through the optical flow module, an advanced autopilot called: Pixhawk 6X is used, which provides the corresponding measurements (attitude, position) for the high-performance control of our quadrotor aircraft.

Finally, the optical flow sensor presented in this work uses a stabilized gimbal to avoid the positive feedback produced when the vehicle tilts. The vehicle platform was built with the components shown in Table 1.

Table 1. UAV components.

Carbon fiber airframe 550 mm	Pixhawk 6X Autopilot
Carbon fiber Propellers 1355 mm	Radio Control Receiver DSM
Outrunner Brushless Motors 335 KV	Telemetry Radio Sik
Brushless Electronic Speed Controllers 30 A	Rangefinder Lidar-Lite
LiPo Batery 5500 mAh 6S	Optical Flow PMW3901 Based Sensor

Figure 6 shows the system diagram. A quadrotor aircraft and optical flow module are stabilized by gimbal, where $I = (x_i, y_i, z_i)$ represents the inertial frame, $B = (x_b, y_b, z_b)$ represents the body frame fixed to the center of mass of the quadrotor aircraft, and $O = (x_o, y_o, z_o)$ is the frame fixed to the optical flow module. Finally, h is the altitude and (V_x, V_y) are the horizontal velocity estimations.

The flight controller and custom PX4 firmware are used to control the quadrotor aircraft and stabilized gimbal. The modular flight controller has redundant embedded sensors (IMU's), communication modules, and peripherals sensors. Figure 7 shows the experimental platform using an optical flow module stabilized by a gimbal. In this configuration, the experimental setup does not include GPS or MoCam.

The gimbal system configuration consists of two tilting joints around the x_o and y_o axis to control θ_o and ϕ_o close to zero. The gimbal mathematical model and robust control algorithm are important to ensure the performance through the torque stability of the motors. The gimbal robust control algorithm is implemented on a PX4 compatible controller board. In order to avoid the chattering effect caused by the signum function in robust control algorithms, we use the approximation function $sign(\cdot) \approx \tanh(\cdot)$.

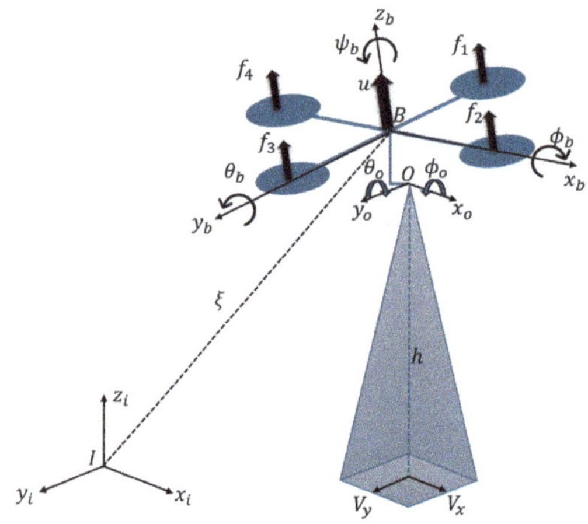

Figure 6. Navigation system: A quadrotor aircraft and a stabilized optical flow module.

The improved optical flow module placed on the gimbal ensures that the estimation of the vertical position and the horizontal displacement velocity are reliably estimated regardless of changes in pitch-roll angles caused by the attitude disturbances or by the quadrotor aircraft displacement. Therefore, the optical flow angular velocities are close to zero $\omega_x, \omega_y \approx 0$; Equation (5) can be rewritten as

$$\begin{aligned} V_x &= \frac{T_z x - T_x f}{h} \\ V_y &= \frac{T_z y - T_y f}{h} \end{aligned} \tag{26}$$

Therefore, the distance and optical flow sensor estimates can be used to control the position of the quadrotor aircraft considering the altitude $h \approx z$ and the horizontal displacement as

$$\begin{aligned} V_x &\approx \dot{x} \\ V_y &\approx \dot{y} \end{aligned} \tag{27}$$

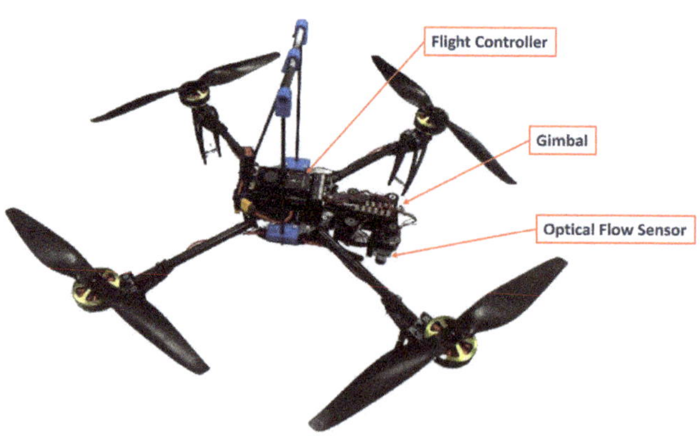

Figure 7. Experimental UAV platform.

5. Experimental Results

To validate the proposed control strategies for robust attitude control implemented in the quadrotor aircraft and the stabilized gimbal, we will summarize the various real-time experimental tracking tests at different stages of development, see Figure 8.

Figure 8. Stabilized gimbal in hover flight test with external perturbations.

The tracking performance can be observed as it reaches the desired position. The controller parameters were adjusted by trial and error until the best performance was obtained. The parameters used are shown in Table 2.

Table 2. System parameters.

Parameter	Value	Parameter	Value	Parameter	Value	Parameter	Value
k_p	0.36	δ_1	1.45	k_b	0.86	γ_1	1.23
k_d	0.044	δ_2	1.05	α	0.71	γ_2	0.95
z_d	2.0 m	k_1	0.54	β	1.61	K_1	0.68
k_a	0.084	k_2	0.33	ρ	1.23	K_2	1.33

Figure 9 shows an indoor flight tracking control test, and Figure 10 shows the trajectory in three dimensions of the quadrotor aircraft, which demonstrates the correct execution of the indoor flight mission using the optical flow sensor to track the programmed trajectory in the quadrotor aircraft at an altitude of 2 m and with a distance of 7 m.

Figure 9. Indoor flight tracking control test with stabilized optical flow.

In the following link, we share a video to show the disturbance rejection tests of the unmanned aerial system to demonstrate the quadrotor aircraft performance and the stabilized gimbal performance. Additionally, hover stability and a good performance in optical flow-based trajectory tracking control in a GPS-Denied environment are ensured.

→ https://youtu.be/SX1zxz5x9ko (accessed on 20 March 2024)

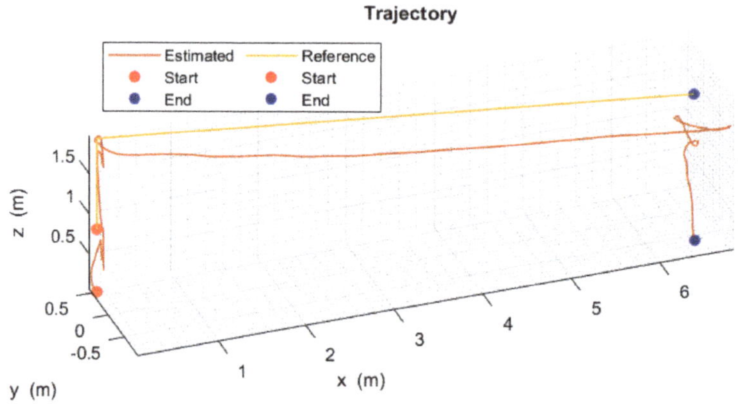

Figure 10. Three-dimensional trajectory tracking of the quadrotor aircraft (x, y, z).

5.1. Quadrotor Aircraft Pitch and Roll Attitude (θ, ϕ)

The attitude angles performance of the quadrotor aircraft during real time flight are presented below. Figure 11 shows the robust control law algorithm (20) behavior applied on the pitch angle stability of the quadrotor aircraft with disturbances added over the whole trajectory.

Figure 12 shows the robust control law algorithm (20) behavior applied on the roll angle stability of the quadrotor aircraft, with disturbances added over the whole trajectory.

The quadrotor aircraft attitude angles (θ, ϕ) back to the reference in a short period of time, achieving normal operation after disturbances, with some oscillatory behavior around the horizontal reference values. Therefore, the results obtained are satisfactory.

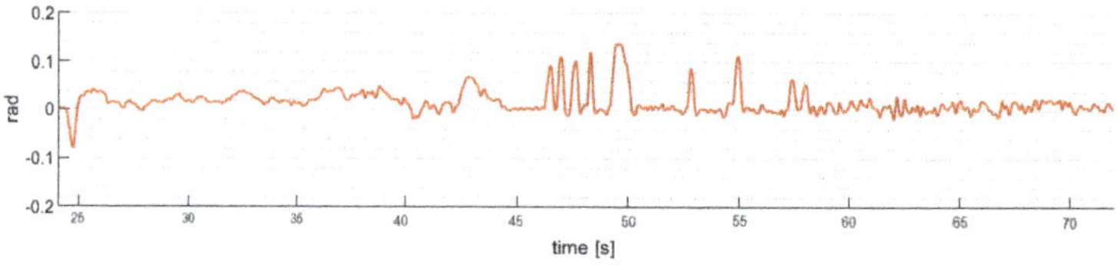

Figure 11. Quadrotor aircraft pitch angle (θ).

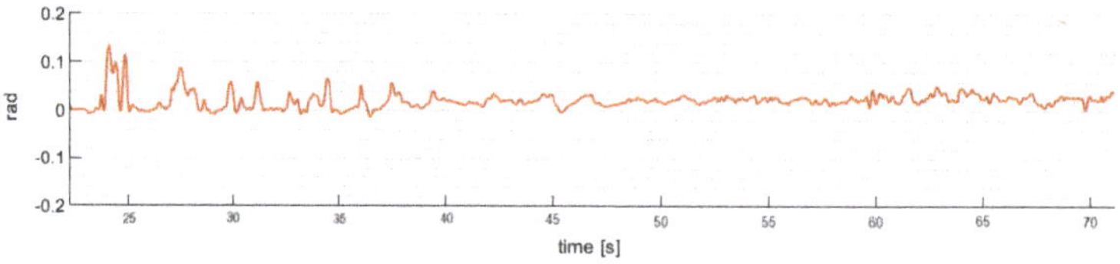

Figure 12. Quadrotor aircraft roll angle (ϕ).

5.2. Gimbal Pitch and Roll Attitude (θ_g, ϕ_g)

The attitude angles performance of the stabilized gimbal during real time flight are presented below.

Figure 13 shows the robust control law algorithm (24) behavior applied on the pitch angle stability of the gimbal, with disturbances added over the fight.

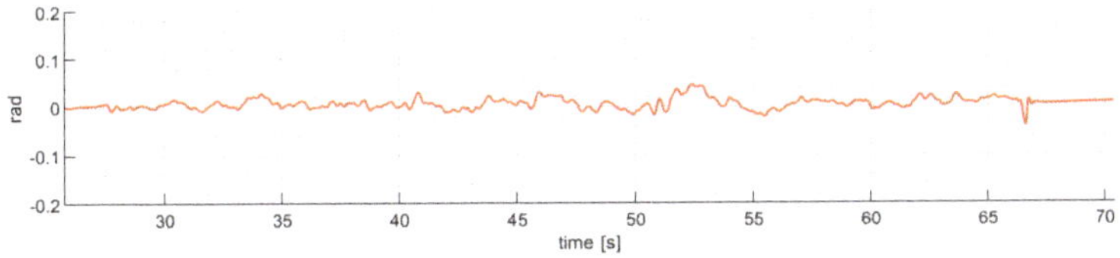

Figure 13. Gimbal pitch angle (θ_g).

Figure 14 shows the robust control law algorithm (25) behavior applied on the roll angle stability of the gimbal, with disturbances added over the fight.

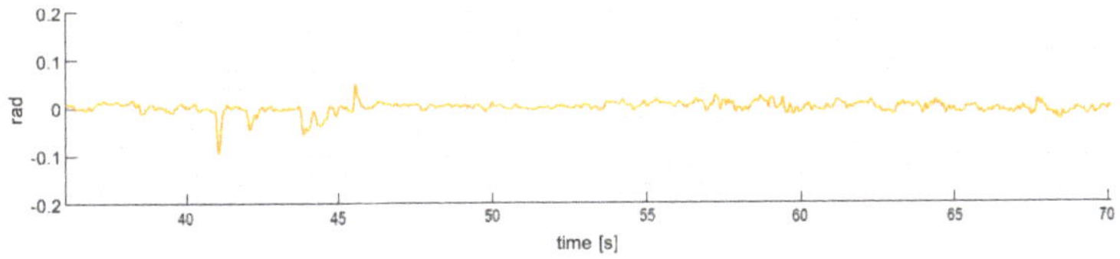

Figure 14. Gimbal roll angle (ϕ_g).

The gimbal attitude angles (θ_g, ϕ_g) back to the reference in a short period of time to normal operation after rejecting the angular disturbances induced by the quadrotor aircraft, with some oscillatory behavior around the horizontal reference values. Therefore, the results obtained are satisfactory.

5.3. Optical Flow Sensing (z, V_x, V_y)

The altitude and displacement velocity sensing in real time flight are presented below.

Figure 15 shows the altitude estimate by the distance sensor to reach the desired altitude ($z_d = 2$ m).

Figure 16 shows the optical flow estimation of lateral velocity during indoor flight. An increase in speed of almost 0.04 m/s is observed during its lateral path of 7 m.

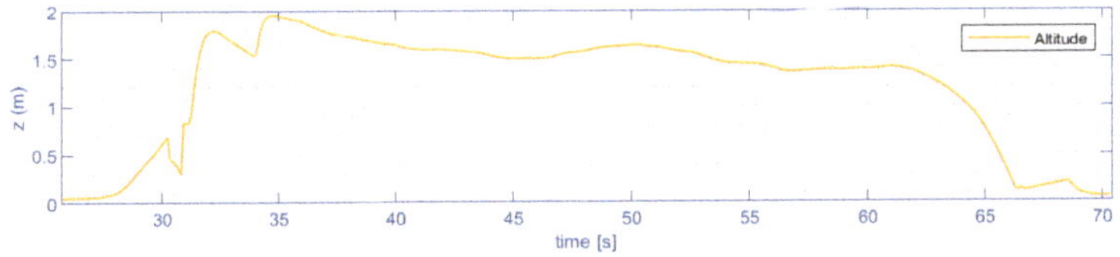

Figure 15. Vertical displacement (z).

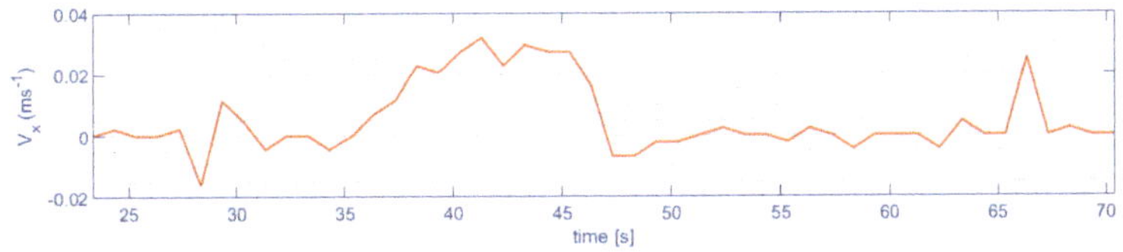

Figure 16. Lateral velocity (V_x).

Figure 17 shows the optical flow estimates of the lateral and longitudinal velocities, respectively, during the indoor flight. A lateral velocity of 0.01 m/s during a longitudinal path of 1 m is observed.

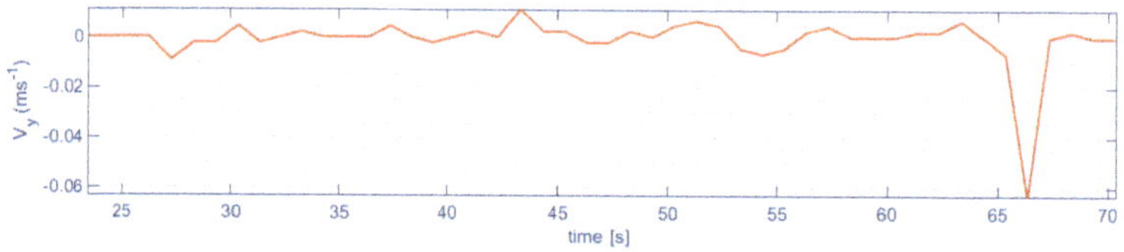

Figure 17. Longitudinal velocity (V_y).

It is possible to observe a peak in the horizontal displacement velocity at the start and at the end of the flight. This occurs when the quadrotor aircraft does not take off or land with its full landing gear (due to disturbances, like the ground effect, or because the terrain is not flat and horizontal); the optical flow can detect a small horizontal displacement that increases with short focal length.

6. Discussion

There are many approaches around GPS-Denied navigation; some are mentioned in Section 1 of this work. The stabilized optical flow avoids the positive feedback of the velocity and position estimation. Regardless of the inclination rate and attitude change of the quadrotor, the stabilized optical flow will always maintain the downward orientation; this is the main goal of this work. However, a handicap of using a stabilized optical flow is the extra weight of the gimbal itself. Normally, the use of conventional optical flow sensors considers limitations in the translational velocity and the altitude of the sensor with respect to the ground as this altitude increases the estimation of the velocity, which is more sensitive to changes in sensor orientation. In our proposal, we are avoiding these attitude changes due to the stabilization of the optical flow. To carry out safe flight tests, it is required to know the operating limits of the sensors to produce reliable local estimates of altitude and horizontal velocity. For example, the maximum distance sensed by the optical flow module is 2 m, so we use a distance sensor with a longer operating range (40 m).

7. Conclusions

This paper has proposed a novel technique for navigating in a GPS-Denied environment using an optical flow module stabilized by a gimbal system. The proposed approach solves the problem of having a positive feedback in the pitch angle when the optical module is attached to the body of the aircraft. The technique consists of a control algorithm that keeps the optical module facing downwards so that the optical flow is independent of the pitch angle of the aircraft. We have successfully tested the proposed approach in an experimental platform in outdoor and indoor environments. The control algorithms of both the aircraft and gimbal were designed robustly to reduce the effect of external perturbations. The experiments showed that in spite of disturbances, the aircraft Euler angles remained close to the origin.

Author Contributions: Investigation, J.F.; project administration, I.G.-H.; supervision, S.S.; data curation, R.L.; and conceptualization, C.R. All authors have read and agreed to the published version of the manuscript.

Funding: This project was funded by Department of Research and Multidisciplinary Studies of Research and Advanced Studies Center of the National Polytechnic Institute (CINVESTAV-IPN).

Institutional Review Board Statement: Not applicable.

Informed Consent Statement: Not applicable.

Data Availability Statement: Dataset available on request from the authors.

Acknowledgments: The authors are grateful to the National Council of Science, Humanities, and Technology (CONAHCYT) for its support.

Conflicts of Interest: The authors declare no conflicts of interest.

Abbreviations

The following abbreviations are used in this manuscript:

UAV	Unmanned Aerial Vehicle
VTOL	Vertical Take-Off and Landing
IMU	Inertial Measurement Unit
GPS	Global Position System
GRU	Gated Recurrent Unit
MARG	Magnetic, Angular Rate, and Gravity
FPV	First Person View
SMC	Sliding Mode Control
NTSMC	Nonsingular Terminal Sliding Mode Control
NTS	Nonsingular Terminal Sliding Surface

References

1. Mashood, A.; Dirir, A.; Hussein, M.; Noura, H.; Awwad, F. Quadrotor object tracking using real-time motion sensing. In Proceedings of the 2016 5th International Conference on Electronic Devices, Systems and Applications (ICEDSA), Ras Al Khaimah, United Arab Emirates, 6–8 December 2016; pp. 1–4.
2. Okulski, M.; Ławryńczuk, M. A Small UAV Optimized for Efficient Long-Range and VTOL Missions: An Experimental Tandem-Wing Quadplane Drone. *Appl. Sci.* **2022**, *12*, 7059. [CrossRef]
3. Van den Bergh, B.; Pollin, S. Keeping UAVs under control during GPS jamming. *IEEE Syst. J.* **2018**, *13*, 2010–2021. [CrossRef]
4. Schmid, K.; Tomic, T.; Ruess, F.; Hirschmüller, H.; Suppa, M. Stereo vision based indoor/outdoor navigation for flying robots. In Proceedings of the 2013 IEEE/RSJ International Conference on Intelligent Robots and Systems, Tokyo, Japan, 3–7 November 2013; pp. 3955–3962.
5. Romero, H.; Salazar, S.; Lozano, R. Real-time stabilization of an eight-rotor UAV using optical flow. *IEEE Trans. Robot.* **2009**, *25*, 809–817. [CrossRef]
6. Sun, K.; Yu, Y.; Zhou, W.; Zhou, G.; Wang, T.; Li, Z. A low-cost and robust optical flow CMOS camera for velocity estimation. In Proceedings of the 2013 IEEE International Conference on Robotics and Biomimetics (ROBIO), Shenzhen, China, 12–14 December 2013; pp. 1181–1186.
7. Liu, X.; Li, X.; Shi, Q.; Xu, C.; Tang, Y. UAV attitude estimation based on MARG and optical flow sensors using gated recurrent unit. *Int. J. Distrib. Sens. Netw.* **2021**, *17*, 15501477211009814. [CrossRef]
8. González, I.; Salazar, S.; Lozano, R. Chattering-free sliding mode altitude control for a quad-rotor aircraft: Real-time application. *J. Intell. Robot. Syst.* **2014**, *73*, 137–155. [CrossRef]
9. Altan, A.; Hacıoğlu, R. Model predictive control of three-axis gimbal system mounted on UAV for real-time target tracking under external disturbances. *Mech. Syst. Signal Process.* **2020**, *138*, 106548. [CrossRef]
10. Mousavi, Y.; Zarei, A.; Jahromi, Z.S. Robust adaptive fractional-order nonsingular terminal sliding mode stabilization of three-axis gimbal platforms. *ISA Trans.* **2022**, *123*, 98–109. [CrossRef] [PubMed]
11. Honegger, D.; Meier, L.; Tanskanen, P.; Pollefeys, M. An open source and open hardware embedded metric optical flow cmos camera for indoor and outdoor applications. In Proceedings of the 2013 IEEE International Conference on Robotics and Automation, Karlsruhe, Germany, 6–10 May 2013; pp. 1736–1741.
12. Yang, L.; Yang, J. Nonsingular fast terminal sliding-mode control for nonlinear dynamical systems. *Int. J. Robust Nonlinear Control* **2011**, *21*, 1865–1879. [CrossRef]

sensors

MDPI

Article

VA-LOAM: Visual Assist LiDAR Odometry and Mapping for Accurate Autonomous Navigation

Tae-Ki Jung and Gyu-In Jee *

Department of Electronic Engineering, Konkuk University, 120 Neungdong-ro, Gwangjin-gu, Seoul 05029, Republic of Korea; taeki222@konkuk.ac.kr
* Correspondence: gijee@konkuk.ac.kr

Abstract: In this study, we enhanced odometry performance by integrating vision sensors with LiDAR sensors, which exhibit contrasting characteristics. Vision sensors provide extensive environmental information but are limited in precise distance measurement, whereas LiDAR offers high accuracy in distance metrics but lacks detailed environmental data. By utilizing data from vision sensors, this research compensates for the inadequate descriptors of LiDAR sensors, thereby improving LiDAR feature matching performance. Traditional fusion methods, which rely on extracting depth from image features, depend heavily on vision sensors and are vulnerable under challenging conditions such as rain, darkness, or light reflection. Utilizing vision sensors as primary sensors under such conditions can lead to significant mapping errors and, in the worst cases, system divergence. Conversely, our approach uses LiDAR as the primary sensor, mitigating the shortcomings of previous methods and enabling vision sensors to support LiDAR-based mapping. This maintains LiDAR Odometry performance even in environments where vision sensors are compromised, thus enhancing performance with the support of vision sensors. We adopted five prominent algorithms from the latest LiDAR SLAM open-source projects and conducted experiments on the KITTI odometry dataset. This research proposes a novel approach by integrating a vision support module into the top three LiDAR SLAM methods, thereby improving performance. By making the source code of VA-LOAM publicly available, this work enhances the accessibility of the technology, fostering reproducibility and transparency within the research community.

Keywords: Simultaneous Localization and Mapping (SLAM); sensor fusion

1. Introduction

Simultaneous Localization and Mapping (SLAM) is a process in which autonomous systems like robots or cars determine their location and simultaneously create a map of an unknown environment. This technology is a critical component in navigation systems such as unmanned aerial vehicles (UAVs) and autonomous driving vehicles. Visual Odometry (VO) and LiDAR (Light Detection and Ranging) Odometry are extensively used in these systems for tracking location and constructing maps. Understanding the unique strengths of each technology is vital for designing effective SLAM systems.

Vision sensors, including monocular cameras, stereo cameras, and RGB-D cameras, collect high-frequency visual data (i.e., 30~60 Hz) and provide detailed spatial analysis. Monocular cameras alone face challenges in estimating depth, thus fusion with additional sensors is used to achieve more accurate Visual Odometry. For instance, studies [1–4] have enhanced the depth estimation accuracy of monocular cameras by integrating accelerometers and gyroscopes. Research [5,6] proposed new methods for performing Visual SLAM using stereo cameras. Stereo cameras can measure depth by aligning images from both lenses, allowing for more precise environmental mapping. To compensate for the limited depth range of stereo cameras, methods have expanded it using RGB-D cameras, as proposed in [7,8]. RGB-D cameras offer a more accurate and wider depth measurement range than stereo cameras but can face difficulties in outdoor environments due to their sensitivity to light.

LiDAR measures distances by detecting light reflected from objects, creating highly precise three-dimensional point clouds. Various studies [9–14] have proposed methods for Odometry and Mapping using LiDAR. However, as LiDAR operates at a low frequency (i.e., 10 Hz) and only provides 3D points and intensity data, spatial analysis can be challenging. While accurate mapping with LiDAR enables precise pose estimation, significant mapping errors may occur leading to serious pose estimation inaccuracies if sufficient descriptors are not provided.

Many studies propose methods that combine and complement the strengths of vision sensors and LiDAR sensors, which have contrasting characteristics. The approaches suggested in [15–18] involve extracting visual features from the vision sensor and measuring depth with the LiDAR sensor. Although these methods leverage the advantages of both sensors, the point cloud generated with the LiDAR sensor is less dense compared to the vision sensor, resulting in 3D–2D depth association errors. Particularly, these depth association errors become more pronounced with objects that are further away. Such errors can degrade the precision of LiDAR Odometry's pose estimation. Moreover, vision sensors are highly dependent on environmental conditions such as weather, changes in lighting, shadows, and light reflections. Methods that use the vision sensor as the primary sensor in Visual–LiDAR Fusion are significantly affected by environmental changes, which can lead to substantial errors. In [19], deep learning is used to fuse the two sensors, while in [20], a method is employed to adjust the weights of each sensor's measurements based on environmental conditions.

This paper proposes a new method that utilizes visual information from vision sensors to enhance the accuracy of LiDAR Odometry. We suggest a technique to reduce 3D–2D depth association errors and enable more precise pose estimation in LiDAR Odometry. By using only LiDAR features and assigning image descriptors to them, we enhance the uniqueness of the LiDAR points. Employing LiDAR as the primary sensor allows the system to maintain performance in LiDAR Odometry and Mapping even when vision sensors fail or environmental conditions change. This approach offers the advantage of maintaining the high precision of LiDAR sensors while minimizing the environmental limitations faced by vision sensors. To achieve this, we analyzed the performance of various open-source LiDAR Odometry methods using the KITTI dataset [21] and developed the Visual Assist LiDAR Odometry and Mapping (VA-LOAM) method, which integrates visual information into the top three methods with the lowest root mean square error (RMSE) on location.

To summarize, the main contributions of this work are fourfold:

(1) Visual Information Integration: This study proposes a new method that utilizes visual information collected from vision sensors to enhance the precision of LiDAR Odometry. This approach reduces 3D–2D depth association errors and enables accurate pose estimation in LiDAR Odometry. By integrating vision sensor data with LiDAR data, this method achieves better performance compared to traditional LiDAR Odometry. The rich environmental information provided via vision sensors complements the limitations of LiDAR, maintaining high accuracy even in complex environments;

(2) Enhanced LiDAR Odometry through Vision Sensor Support: We have clarified that this contribution focuses on using LiDAR as the primary sensor while utilizing vision sensors as a supplementary aid. Traditional methods that fuse LiDAR and vision sensors often rely on the vision sensor as the main sensor, which can fail in environments where vision sensors are weak (e.g., dark conditions and reflective surfaces). Our method ensures that in typical environments vision sensors assist in matching LiDAR feature points, improving accuracy. However, in challenging conditions for vision sensors, the system can operate using only LiDAR, maintaining the performance of traditional LiDAR-based Odometry. This approach ensures stable and consistent performance across various environments by leveraging the strengths of LiDAR while mitigating the weaknesses of vision sensors;

(3) Validation and Performance Improvement of VA-LOAM: This paper develops and validates the Visual Assist LiDAR Odometry and Mapping (VA-LOAM) method,

which integrates visual information into existing LiDAR Odometry techniques. This method was tested using the publicly available KITTI dataset, demonstrating improved performance over existing LiDAR Odometry methods;

(4) Open-Source Contribution: By making the source code of VA-LOAM publicly available, this work ensures the reproducibility and transparency of the research across the community, enhancing the accessibility of the technology. This fosters collaboration and innovation in research and development.

2. Related Work

Vision sensors and LiDAR sensors are widely used for estimating the 6 degrees of freedom (6DOF) position and orientation. They are essential for accurately determining the position and orientation in UAV autopilot systems, robot SLAM, and autonomous vehicle navigation systems. There are three primary methods of SLAM that utilize these sensors: Visual SLAM, which uses only vision sensors; LiDAR SLAM, which uses only LiDAR sensors; and Visual–LiDAR SLAM, which integrates both vision and LiDAR sensors.

(1) Visual SLAM: Refs. [1–8] pertain to this method. LSD-SLAM [1] and SVO [2] match continuous images using photogrammetric consistency without preprocessing the sensor data. This approach is particularly useful in environments lacking distinct image features. Methods [3–8] extract local image feature points (edges, corner points, and lines) and track them to estimate their positional changes. By analyzing the camera's motion and the feature points' positional shifts, the distance to these points and the camera's pose can be calculated. Additionally, image features are used to perform loop detection. ORB-SLAM [3] performs feature matching using fast features and brief descriptors. The pose is calculated based on matched feature points using the Perspective-n-Point (PnP) algorithm. VINS-Mono [4] is a tightly coupled sensor fusion method that uses visual sensors and IMUs. It performs visual–inertial odometry using tracked features from a monocular camera and pre-integrated IMU measurements. While accurate feature matching can be achieved through image descriptors, the process of estimating feature point depth involves significant errors. Research has been conducted using stereo cameras and RGB-D cameras to reduce these errors in depth estimation. TOMONO [5] and ENGEL [6] proposed methods using stereo cameras. TOMONO [5] introduced an edge point-based SLAM method using stereo cameras, which is particularly effective in non-textured environments where it detects edges and performs edge-based SLAM. ENGEL [6] improved SLAM accuracy by estimating pixel depth using a fixed-baseline stereo camera and motion from a multiview stereo. KERL [7] and SCHOPS [8] proposed RGB-D SLAM using entropy-based keyframe selection and loop closure detection;

(2) LiDAR SLAM: Refs. [9–14] apply to this method. This technique uses point clouds containing three-dimensional points and intensities, employing feature extraction and matching to estimate position. LiDAR provides accurate 3D points, enabling precise pose estimation. However, a lack of sufficient descriptors can lead to matching errors. Methods [10–14] have evolved from LOAM [9]. F-LOAM [14] offers faster processing speeds and lower memory usage, enabling real-time SLAM on lower-performance devices. A-LOAM [11] enhances accuracy by incorporating loop closure functionality and reducing mapping errors caused by obstacles. LeGO-LOAM [10] proposes a lightweight, terrain-optimized method for ground vehicles, classifying and processing the terrain accordingly. ISC-LOAM [13] addresses the issue of insufficient descriptors in LiDAR by proposing an intensity-based scan context, which improves performance in loop closure detection. LIO-SAM [12] tightly integrates LiDAR sensors with Inertial Measurement Units (IMUs), using pre-integrated IMU-derived estimated motion as initial values for point cloud corrections and LiDAR Odometry optimization;

(3) Visual-LiDAR SLAM: This technology is broadly classified into loosely coupled and tightly coupled approaches. The loosely coupled approach includes LiDAR-assisted Visual SLAM and Visual Assist LiDAR SLAM. LiDAR-assisted Visual SLAM [15–18]

addresses one of the main issues in Visual SLAM—depth inaccuracies of image feature points—by correcting them with LiDAR data. While this greatly enhances the accuracy of feature points, the relatively low resolution of LiDAR data can lead to errors in 3D–2D mapping. To mitigate these issues, more precise data integration techniques are required. Visual Assist LiDAR SLAM utilizes LiDAR Odometry to estimate position and uses image data from the corresponding location to recognize the environment. It employs image feature points for loop closure detection, minimizing errors that may occur over time. The tightly coupled approach combines the advantages of LiDAR Odometry and Visual Odometry. In this method, rapid pose estimation from Visual-assisted Visual Odometry is refined through LiDAR Odometry. Although tightly coupled systems are robust, they require substantial computational resources;

Loosely Coupled Systems: These systems are relatively simple to implement and offer high accuracy. However, they can be vulnerable to sensor errors and changes in dynamic environments;

Tightly Coupled Systems: These systems are strong against uncertainty in sensor data and environmental changes, enabling precise position estimation. Nevertheless, they necessitate high-performance processing capabilities and sophisticated data integration techniques.

3. Preliminaries

3.1. Coordinate Systems

As seen in Figure 1, we utilize three coordinate systems: $(\cdot)^w$ denotes the world coordinate system, $(\cdot)^c$ denotes the camera coordinate system, and $(\cdot)^l$ denotes the LiDAR coordinate system. The world coordinate system provides a fixed reference frame and serves as the reference point for all other coordinate systems. The LiDAR and camera coordinate systems are defined based on the sensor's position and orientation and can be transformed from the world coordinate system through rotations and translations. The origin of the world coordinate system coincides with the initial measurement position of the LiDAR coordinate system (odometry's initial position). At the same time, t_i, the relationship between the camera coordinate system and the LiDAR coordinate system, can be expressed with the transformation matrix $T_{l_i}^{c_i} \in SE(3)$, which belongs to the special Euclidean group. T_l^c represents the extrinsic parameters between the two sensors, consisting of a rotation matrix and a translation matrix. These extrinsic parameters, which indicate the relative position and direction between the two sensors, can be obtained through calibration [22].

$$T_l^c = \begin{bmatrix} R_l^c & t_l^c \\ 0 & 1 \end{bmatrix} \tag{1}$$

$$R \in SO(3), t \in \mathbb{R}^3, \tag{2}$$

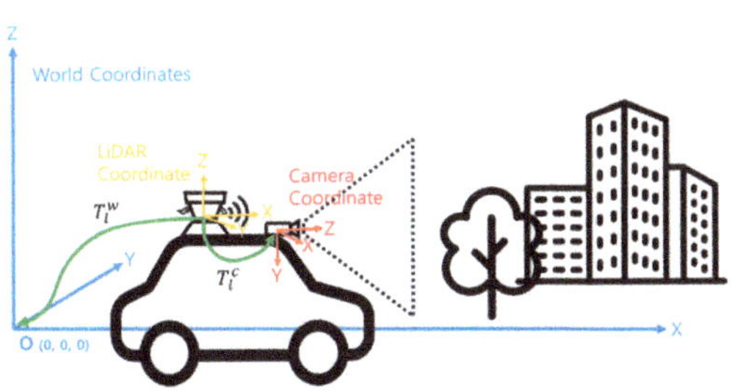

Figure 1. This illustration visually represents the process by which each sensor's coordinate system is transformed into the world coordinate system. Data collected in each sensor coordinate system can be expressed in another sensor's coordinate system or integrated into the world coordinate system through transformation matrices.

3.2. Camera Projection Model

Data collected via LiDAR are measured as 3D points in the LiDAR coordinate system. Each coordinate, X^l, from the LiDAR point cloud is transformed into the camera coordinate system using a transformation matrix. Subsequently, these 3D coordinates are projected onto a 2D plane using the pinhole camera model matrix K. During this process, the depth information along the Z-axis is removed, resulting in 2D coordinates that correspond to the camera's image plane. In Equation (5), u and v represent the coordinates in the image plane, f denotes the focal length of the camera, and c refers to the principal point of the camera.

$$X^l = (x, y, z)^T \in \mathbb{R}^3 \tag{3}$$

$$X^c = T_l^c X^l \tag{4}$$

$$x^c = (u, v)^T = \begin{bmatrix} f & 0 & c_x \\ 0 & f & c_y \\ 0 & 0 & 1 \end{bmatrix} T_l^c X^l \in \mathbb{R}^2 \tag{5}$$

4. Proposed Methodology

4.1. System Overview

This study aims to enhance localization performance by integrating LiDAR Odometry and Mapping (LOAM) methods with a visual module, as shown in Figure 2. The overall system configuration is as follows. First, LiDAR data are used to detect LiDAR features (surf and edge). Next, the visual assist module uses images from the camera sensor to generate image descriptors for the LiDAR features. Then, to compute the motion displacement, the features detected in the previous frame are matched with those detected in the current frame. Edge features are matched using image descriptors, while surf features are matched using the conventional LiDAR Odometry method. If no matching data for the image descriptors are available, indicating a sensor failure or an environment where the camera sensor cannot operate, the edge features are also matched using the conventional LiDAR Odometry method, similar to the surf features. Since vision sensors are sensitive to environmental changes, LiDAR is utilized as the main sensor, with the vision sensor serving a supportive role. This configuration is implemented in a loosely coupled manner, allowing the visual module to enhance the performance of LiDAR Odometry under normal conditions, while maintaining its effectiveness even in environments where vision sensors are vulnerable.

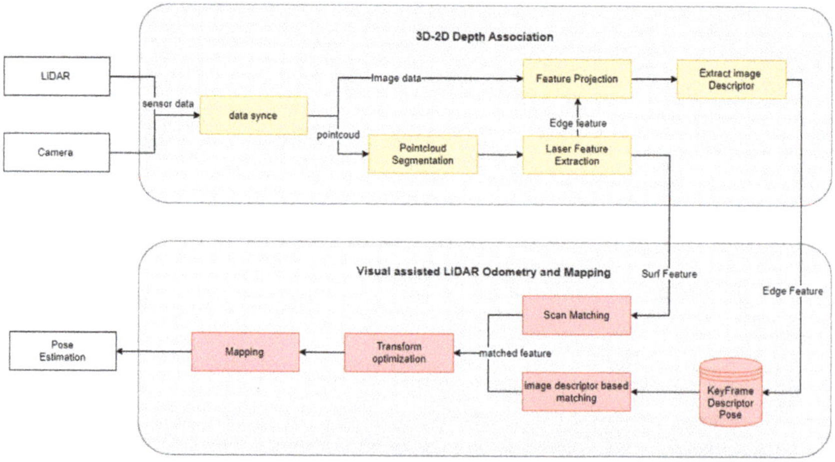

Figure 2. System flowchart for Visual Assist LiDAR Odometry and Mapping.

4.2. Point Cloud Preprocessing

During the point cloud preprocessing stage, the detection of LiDAR features is a crucial task. The primary LiDAR features are categorized as edges and planars, based on the curvature calculations among adjacent point clouds. If the curvature exceeds a predefined threshold, the point is classified as an edge feature, indicative of significant local changes. Conversely, if the curvature does not surpass the threshold, it is classified as a planar feature, representing a relatively flat surface. This curvature-based extraction method plays a key role in environmental mapping and obstacle detection within LiDAR Odometry. This equation calculates the curvature c around a specific point $X^l_{(k,i)}$. The curvature is determined by considering the differences in distances between the point and its neighboring points. Higher curvature values indicate a more abrupt change in the geometric structure around the point.

$$c = \frac{1}{|S|\left|X^l_{(k,i)}\right|}||\sum_{j \in S, j \neq i}(X^l_{(k,i)} - X^l_{(k,j)})|| \tag{6}$$

4.3. Visual Assist LiDAR Feature

Previous studies [15–18] have utilized LiDAR-assisted visual odometry based on image feature points. However, as illustrated in Figure 3, due to the density differences between vision sensors and LiDAR sensors, not all image feature points match with LiDAR 3D points.

Figure 3. This figure demonstrates the depth association issues arising from the density differences between vision sensors and LiDAR sensors. The LiDAR point cloud is shown in green, image features in red, and corresponding features in blue.

As illustrated in Figure 4, depth association errors frequently occur at image features such as edges and corners. Image features are detected where there is a significant contrast from surrounding areas, particularly at edges or corners. LiDAR is not as dense as vision sensors, which makes it challenging to generate point clouds at precise locations like edges and corners. This discrepancy leads to incorrect matches between image feature points and LiDAR 3D points, resulting in errors in pose estimation.

In this study, we propose a visual assist LiDAR feature matching method to resolve the depth association errors between image feature points and LiDAR 3D points. During the point cloud preprocessing phase, both edge and planar features are detected, and image descriptors corresponding to the edge features are extracted. These edge features, equipped with image descriptors, are defined as visual assist LiDAR features (VALFs). While planar features generally represent non-textured objects and do not provide significant differences in image feature vectors, edge features are suitable for image descriptor extraction due to their texture presence and curvature changes. Unlike LiDAR features, VALFs are detected based on three-dimensional terrain characteristics and can be matched through image descriptors.

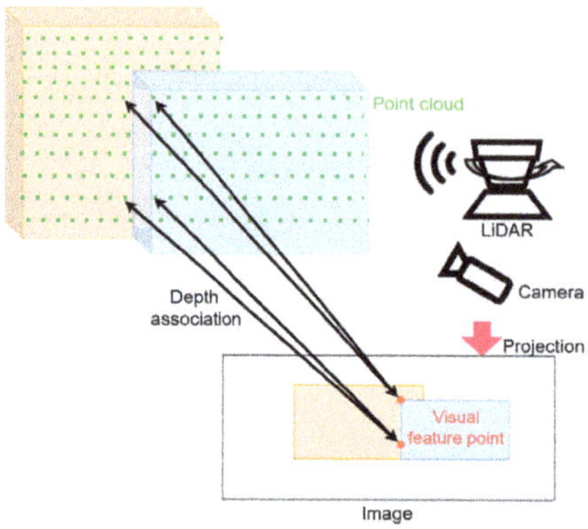

Figure 4. Illustration of depth association challenges in LiDAR-assisted Visual SLAM. This figure highlights the potential for multiple LiDAR point clouds to be projected onto a single image feature point.

Table 1 summarizes the advantages and disadvantages of various image descriptors. In this study, considering the presence of 200–300 LiDAR feature points projected onto urban area camera images, ORB and BRISK descriptors were selected for their balance of matching accuracy and processing time. Using Equation (6), edge features are extracted and then projected onto the camera image through Equation (5). If these features are present within the image region, visual descriptors are extracted to generate VALFs. Unlike traditional LiDAR edge/planar features, VALFs utilizes these descriptors to perform matching across consecutive frames, thereby enhancing the continuity and accuracy of the feature tracking process. We conducted experiments on 3D–2D depth association methods. Specifically, we tested the conventional method of estimating the depth of image features and the proposed method of applying image descriptors to LiDAR features. As seen in Figure 5, VALF generates a diverse and large number of 3D features compared to the conventional method of estimating the 3D depth of image features. The diversity and number of features are crucial factors in pose estimation.

Table 1. Advantages and disadvantages of image descriptors.

Descriptor	Advantages	Disadvantages
SIFT	Scale invariance, rotation invariance, robustness to illumination changes	Slow computation speed
SURF	Performance similar to SIFT Faster speed	Vulnerable to specific lighting conditions
ORB	Fast speed–Real-time processing Low computational cost	Slightly lower accuracy
BRISK	Shares characteristics with ORB Superior performance in texture-rich images	Vulnerable to specific deformations
AKAZE	Performance similar to SIFT Fast speed	Vulnerable to specific textures
Deep Learning-based Descriptors	High accuracy	Requires large amounts of training data and computation

Figure 6 displays the results of matching the current frame and previous frame features obtained through the two 3D–2D depth association methods. As illustrated in Figure 6, it is evident that the number of matched features using VALFs is significantly higher.

(a)

(b)

Figure 5. (**a**) Shows the 3D image features obtained by estimating the depth of image features using LiDAR point clouds, which is the conventional method. The LiDAR point cloud is shown in green, image features in red, and corresponding features in blue. (**b**) Illustrates the proposed VALF method presented in this paper.

(a)

(b)

Figure 6. Shows the results of 3D feature matching between the current and previous frames. (**a**) Shows the matching of 3D image features, while (**b**) shows the matching of VALF features.

4.4. Pose Estimation

Pose estimation utilizes VALFs (F_v^l), edge features $\left(F_e^l\right)$, planar features (F_p^l), and a Global feature map (M^g). The Global feature map comprises an edge feature map (M_e^g), a planar feature map (M_p^g), and a VALF map (M_v^g). The edge and planar feature maps are stored in a 3D kd-tree structure, while the VALF map is stored as a vector containing descriptors and 3D points.

To estimate the optimal pose between the current frame and the Global map, the distance between matched Local and Global features is minimized. Local features consist of edge feature points $p_e^l \epsilon F_e^g$, planar feature points $p_p^l \epsilon F_p^g$, and VALF points $p_v^l \epsilon F_v^g$; Global features are composed of Global lines $p_e^g \epsilon M_e^g$, Global planars $p_p^g \epsilon M_p^g$, and Global VALFs $p_v^g \epsilon M_v^g$. The Gauss–Newton method, a nonlinear optimization technique, is employed to minimize the distances between these matched points in order to accurately estimate the optimal pose. This process involves point-to-line, point-to-planar, and point-to-point matching. The mathematical expressions for each matching type are as follows:

$$\text{Residual for edge features}: \ f_e(\text{p}_e) = \frac{\left\| (T\text{p}_e^l - \text{p}_{e_u}^g) \times (T\text{p}_e^l - \text{p}_{e_v}^g) \right\|}{\left\| \text{p}_{e_u}^g - \text{p}_{e_v}^g \right\|} \tag{7}$$

$$\text{Residual for planar features}: \ f_p(\text{p}_p) = (T\text{p}_p^l - \text{p}_p^g) \cdot n_p^g \tag{8}$$

$$\text{Residual for VALFs}: \ f_v(\text{P}_v) = \left\| T\text{p}_v^l - \text{p}_v^g \right\| \tag{9}$$

Global features are estimated by collecting points adjacent to the edge/planar feature points. Edge/planar features determine the position and direction of lines and planars by calculating their covariance matrices. The local smoothness of each feature point is used to compute weights that enhance accuracy by reflecting the importance of consistently extracted features during matching. VALFs, with their descriptors, facilitate finding matching local and global features. In environments where visual sensors are challenging to operate, optimization is performed using edge and planar features when VALFs do not match, and using VALFs and planar features when VALFs match. The optimization equations are as follows:

$$\text{When VALFs do not match}: \ \min_{T_k} \sum f_e(\text{p}_e) + \sum f_e(\text{p}_p), \ \text{if non} - \text{matched p}_v \tag{10}$$

$$\text{When VALFs match}: \ \min_{T_k} \sum f_v(\text{p}_v) + \sum f_e(\text{p}_p), \ \text{if matched p}_v \tag{11}$$

Optimization does not simultaneously use all three types—VALFs and edge and planar features—due to the significant difference in the number of VALFs and edge features. LiDAR features cover a full 360-degree range, but cameras only cover the front, leading to a narrower field of view. Many edge features projected from the LiDAR onto the image fall outside the camera's Field of View (FOV), resulting in a discrepancy in the number of points between VALFs and edge features.

5. Experimental Setup and Validation

We employed the KITTI dataset [21] to evaluate the performance of Visual Assist Lidar Odometry and Mapping (VA-LOAM). The odometry benchmark from this dataset was used as the test set, initially experimenting with algorithms such as F-LOAM [14], A-LOAM [11], LeGO-LOAM [10], ISC-LOAM [13], and LIO-SAM [12]. From these, the top three performing algorithms were selected, and further experiments were conducted by integrating our proposed Visual Assist module. The accuracy of each experiment was quantitatively analyzed using the root mean square error (RMSE) as a measure of positional error.

5.1. Evaluation on Public Datasets

We conducted a performance evaluation of both traditional LiDAR-based Odometry and Mapping methods and the enhanced Visual Assist Lidar Odometry and Mapping (VA-LOAM) using the renowned KITTI dataset [21], which is notable for its road driving scenarios. This dataset comprises sensor data collected from vehicles equipped with stereo and mono cameras, Velodyne HDL-64 LiDAR, and high-precision GPS/INS, capturing LiDAR point clouds and camera images at a frequency of 10Hz. As shown in Table 2, the tests used the odometry benchmark from the KITTI training set, which contains ground truth position data. This evaluation aimed to compare the discrepancies between the positions estimated via each method and their ground truth positions. The dataset features a variety of environments, ranging from urban areas densely packed with buildings to rural areas lush with vegetation, and highways devoid of nearby objects. It also includes scenarios both with and without loop closures.

Table 2. This table presents the various environments and driving distances that comprise the KITTI dataset.

Dataset	Seq No.	Data Name	Environment	Path Len. (m)
KITTI Odometry Set	1	2011_09_30_drive_0016	Country	397
	2	2011_09_30_drive_0018	Urban	2223
	3	2011_09_30_drive_0020	Urban	1239
	4	2011_09_30_drive_0027	Urban	695
	5	2011_09_30_drive_0033	Urban + Country	1717
	6	2011_09_30_drive_0034	Urban + Country	919
	7	2011_10_03_drive_0027	Urban	3714
	8	2011_10_03_drive_0034	Urban + Country	5075
	9	2011_10_03_drive_0042	Highway	4268
KITTI Raw Data Set	1	2011_09_26_drive_0001	Urban	107
	2	2011_09_26_drive_0005	Urban	69
	3	2011_09_26_drive_0009	Urban	332
	4	2011_09_26_drive_0011	Urban	114

5.2. Performance of LiDAR Odometry and Mapping

We conducted a comparative analysis of existing LiDAR Odometry algorithms on the KITTI dataset [21], including F-LOAM [14], A-LOAM [11], LEGO-LOAM [10], ISC-LOAM [13], and LIO-SAM [12]. We compared their performances both with and without the integration of loop closure detection. In Tables 3 and 4, the numbers highlighted in bold represent the algorithms that achieved the highest performance in each dataset sequence. Using only LiDAR for odometry, F-LOAM [14] and ISC-LOAM [13] showed equivalent performance. A-LOAM [11] did not exhibit the best performance in any of the datasets, while ISC-LOAM [13] performed well in 8 out of 13 datasets, LEGO-LOAM [10] in 2, and LIO-SAM [12] in 3. The RMSE of LiDAR Odometry was evaluated in the absence and presence of loop detection in the KITTI odometry dataset.

Table 3. The root mean square error (RMSE) of LiDAR Odometry was evaluated without loop detection in the KITTI odometry dataset.

Dataset	Seq No.	F-LOAM [14]	A-LOAM [11]	LEGO-LOAM [10]	ISC-LOAM [13]	LIO-SAM [12]
KITTI Odometry Set	1	**2.67**	5.00	3.65	**2.67**	5.78
	2	10.89	113.08	**9.53**	10.90	10.25
	3	9.25	21.28	**4.91**	9.25	21.34
	4	1.85	5.94	1.63	1.85	**1.55**
	5	**17.79**	124.16	22.94	**17.79**	20.08
	6	**8.67**	61.02	17.85	**8.67**	12.05
	7	**19.76**	182.32	171.39	**19.76**	fail
	8	**31.30**	183.91	294.73	**31.30**	41.76
	9	**145.49**	261.86	373.20	**145.49**	167.86
KITTI Raw Data Set	1	1.70	5.70	5.04	1.70	**1.22**
	2	1.53	1.77	1.76	1.53	**1.51**
	3	**5.86**	6.15	13.4	**5.86**	7.05
	4	**1.27**	3.66	2.2	**1.27**	1.71

Note: The numbers highlighted in bold indicate the algorithm that achieved the highest performance on each dataset sequence.

5.3. Performance of Visual Assist LiDAR Odometry and Mapping

In our performance comparison experiment using the KITTI dataset [21], we enhanced the top-performing methods—LEGO-LOAM [10], ISC-LOAM [13], and LIO-SAM [12]—by integrating a visual module, resulting in the development of VA-LEGO-LOAM, VA-ISC-LOAM, and VA-LIO-SAM. Each modified method maintains its original point cloud processing capabilities while incorporating an additional visual assist Lidar feature (VALF), which utilizes image descriptors to further refine the estimation of position and orientation.

Tables 5 and 6 detail the comparative performance of these enhanced methods against their original counterparts, showcasing the benefits of integrating the visual module.

Table 4. The root mean square error (RMSE) of LiDAR Odometry was evaluated with loop detection in the KITTI odometry dataset.

Dataset	Seq No.	F-LOAM [14]	A-LOAM [11]	LEGO-LOAM [10]	ISC-LOAM [13]	LIO-SAM [12]
KITTI Odometry Set	1	**2.67**	2.70	3.74	**2.67**	5.77
	2	10.89	10.43	**7.61**	10.89	8.40
	3	9.25	8.46	**4.92**	9.25	19.02
	4	1.85	1.62	1.76	1.85	**1.50**
	5	**17.79**	18.71	22.93	**17.79**	20.15
	6	**8.67**	9.82	18.53	**8.67**	11.35
	7	**19.76**	20.77	158.65	**19.76**	fail
	8	**31.30**	101.64	275.88	**31.30**	39.79
	9	**145.49**	166.58	376.63	**145.49**	166.19
KITTI Raw Data Set	1	1.70	5.70	5.04	1.70	**1.22**
	2	1.53	1.77	1.76	1.53	**1.51**
	3	**5.86**	6.15	13.4	**5.86**	7.05
	4	**1.27**	3.66	2.2	**1.27**	1.71

Note: The numbers highlighted in bold indicate the algorithm that achieved the highest performance on each dataset sequence.

Table 5. Evaluation of RMSE for LiDAR Odometry with the proposed visual aid module (without loop detection). The visual aid module was incorporated into the three LiDAR odometers that exhibited the most favorable performance in the Kitti test set.

Dataset	Seq No.	LEGO-LOAM [10]	VA-LEGO-LOAM	ISC-LOAM [13]	VA-ISC-LOAM	LIO-SAM [12]	VA-LIO-SAM
KITTI Odometry Set	1	3.65	3.22	**2.67**	2.89	5.78	5.74
	2	9.53	8.74	10.90	**8.53**	10.25	9.49
	3	4.91	**4.31**	9.25	9.20	21.34	21.17
	4	1.63	1.58	1.85	2.22	1.55	**1.43**
	5	22.94	20.03	17.79	**14.18**	20.08	18.59
	6	17.85	17.92	**8.67**	9.35	12.05	12.29
	7	171.39	165.12	19.76	**13.81**	fail	fail
	8	294.73	230.55	31.30	**24.13**	41.76	34.35
	9	373.20	361.16	145.49	**129.41**	167.86	163.62
KITTI Raw Data Set	1	5.04	4.91	1.70	**0.36**	1.22	1.22
	2	1.76	1.68	1.53	**1.50**	1.51	1.50
	3	13.4	13.24	5.86	**5.83**	7.05	6.98
	4	2.2	2.05	1.27	**1.15**	1.71	1.71

Note: The numbers highlighted in bold indicate the algorithm that achieved the highest performance on each dataset sequence.

Table 6. Evaluation of RMSE for LiDAR Odometry with the proposed visual aid module (with loop detection). The visual aid module was incorporated into the three LiDAR odometers that exhibited the most favorable performance in the Kitti test set.

Dataset	Seq No.	LEGO-LOAM [10]	VA-LEGO-LOAM	ISC-LOAM [13]	VA-ISC-LOAM	LIO-SAM [12]	VA-LIO-SAM
KITTI Odometry Set	1	3.74	3.74	**2.67**	2.89	5.77	5.74
	2	7.61	**7.54**	10.89	8.39	8.40	7.97
	3	4.92	**4.31**	9.25	9.20	19.02	19.00
	4	1.76	1.66	1.85	2.22	1.50	**1.42**
	5	22.93	19.87	17.79	**14.18**	20.15	18.68
	6	18.53	18.69	**8.67**	9.35	11.35	12.17
	7	158.65	152.04	19.76	**13.05**	fail	fail
	8	275.88	214.67	31.30	**23.82**	39.79	33.15
	9	376.63	362.46	145.49	**129.04**	166.19	164.48
KITTI Raw Data Set	1	5.04	4.91	1.70	**0.36**	1.22	1.22
	2	1.76	1.68	1.53	**1.50**	1.51	1.50
	3	13.4	13.24	5.86	**5.83**	7.05	6.98
	4	2.2	2.05	1.27	**1.15**	1.71	1.71

Note: The numbers highlighted in bold indicate the algorithm that achieved the highest performance on each dataset sequence.

Tables 7 and 8 show the improvement rates in position accuracy for each method, demonstrating a reduction in RMSE. These figures indicate that position estimation accuracy has improved and highlight the effectiveness of the visual module in reducing position errors under various conditions. Specifically, VA-LEGO-LOAM showed an average reduction of 7.02%, VA-ISC-LOAM showed 12.43%, and VA-LIO-SAM showed 3.67%. Examining each data sequence, in most cases, the visual assist module provided missing descriptors in the existing LiDAR feature matching process, thereby enhancing LiDAR feature matching performance and consequently improving position estimation accuracy. However, in some instances, performance degradation occurred. This was due to incorrect matching during the VALF process. In the KITTI dataset, as shown in Figure 7a, significant changes between the previous frame image and the current frame image occur due to the synchronization of the vision sensor and the LiDAR sensor, resulting in VALF matching errors. For example, if the vision sensor's output frequency is 30Hz, like most vision sensors, and the LiDAR sensor's output frequency is 10Hz, as illustrated in Figure 7b, there can be two additional images between LiDAR data frames. This allows for improved VALF matching through image-based VALF tracking. Future research will aim to address this issue by applying this method.

Table 7. The efficacy of the proposed visual aid module evaluated by comparing it with the traditional LiDAR Odometry method to assess the percentage improvement in accuracy without loop detection.

Dataset	Seq No.	Algorithm	Accuracy Improvement (%)	Algorithm	Accuracy Improvement (%)	Algorithm	Accuracy Improvement (%)
Dataset KITTI Odometry Set	1	VA-LEGO-LOAM	11.79	VA-ISC-LOAM	−8.24	VA-LIO-SAM	0.65
	2		8.24		21.73		7.41
	3		12.22		0.63		0.78
	4		2.91		−19.73		7.75
	5		12.68		20.28		7.46
	6		−0.41		−7.86		−1.98
	7		3.66		30.11		-
	8		21.78		22.90		17.75
	9		3.23		11.05		2.52
KITTI Raw Data Set	1	VA-LEGO-LOAM	2.58	VA-ISC-LOAM	78.82	VA-LIO-SAM	0.00
	2		4.55		1.96		0.66
	3		1.19		0.51		0.99
	4		6.82		9.45		0.00
Avg			7.02		12.43		3.67

Table 8. The efficacy of the proposed visual aid module evaluated by comparing it with the traditional LiDAR Odometry method to assess the percentage improvement in accuracy with loop detection.

Dataset	Seq No.	Algorithm	Accuracy Improvement (%)	Algorithm	Accuracy Improvement (%)	Algorithm	Accuracy Improvement (%)
Dataset KITTI Odometry Set	1	VA-LEGO-LOAM	0.09	VA-ISC-LOAM	−8.24	VA-LIO-SAM	0.55
	2		0.88		23.01		5.11
	3		12.38		0.63		0.08
	4		5.60		−19.73		5.42
	5		13.33		20.28		7.31
	6		−0.84		−7.86		−7.17
	7		4.17		33.98		-
	8		22.19		23.90		16.69
	9		3.76		11.31		1.03
KITTI Raw Data Set	1	VA-LEGO-LOAM	2.58	VA-ISC-LOAM	78.82	VA-LIO-SAM	0.00
	2		4.55		1.96		0.66
	3		1.19		0.51		0.99
	4		6.82		9.45		0.00
Avg			5.90		5.95		2.42

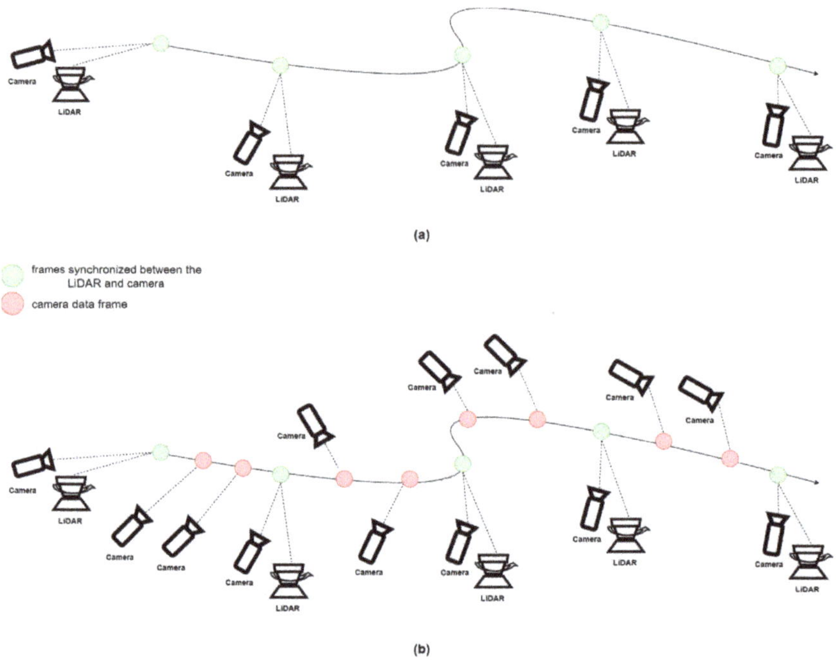

Figure 7. (**a**) shows the KITTI dataset consisting of synchronized sensor data, while (**b**) represents a typical asynchronous data system.

Table 9 presents the results of experiments conducted under varying environmental conditions. The test environments include daytime, night-time, and camera sensor failure. Night-time and camera sensor failures are challenging conditions for the visual assist module. The night-time data were generated by reducing the brightness of the original daytime image data and applying image processing, as illustrated in Figure 8. We tested VA-LEGO LOAM, VA-ISC LOAM, and VA-LIO-SAM, which integrate the sensor fault visual assist module. While the performance in night-time conditions was lower compared to daytime, the error did not increase significantly. This is attributed to the feature matching performance; although there is a difference in feature matching performance between images with a large time gap, the impact on feature matching between images with a small-time gap is minimal. Additionally, in cases where the camera sensor failed, the LiDAR Odometry performance remained stable. This is because the visual assist module cannot operate without image-based descriptors, but LiDAR-based Odometry can still be performed without image descriptors, maintaining the original LiDAR Odometry performance. This result demonstrates the advantages of a loosely coupled approach in sensor fusion, showing that using LiDAR as the main sensor, compared to the conventional method of obtaining depth using visual features, provides robustness against camera sensor failures and environmental changes.

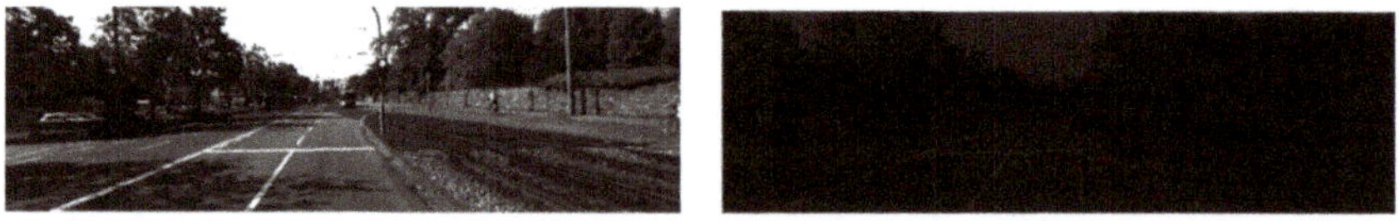

(a) (b)

Figure 8. (**a**) is the daytime image, and (**b**) is the night-time image.

Table 9. This demonstrates the performance of Visual Assist LiDAR Odometry under sensor faults and environmental changes.

Dataset	Seq No.	Environment	VA-LEGO-LOAM	VA-ISC-LOAM	VA-LIO-SAM
KITTI Raw Data set	1	Only LiDAR (Traditional)	5.042	1.698	**1.221**
		Daytime/Urban	4.913	**0.362**	1.215
		Night-time/Urban	4.922	**0.365**	1.218
		Image Sensor Fault	5.042	1.698	**1.215**
	2	Only LiDAR (Traditional)	1.76	1.529	**1.507**
		Daytime/Urban	1.684	**1.496**	1.503
		Night-time/Urban	1.686	**1.497**	1.504
		Image Sensor Fault	1.759	1.529	**1.509**
	3	Only LiDAR (Traditional)	13.375	**5.859**	7.047
		Daytime/Urban	13.241	**5.832**	6.98
		Night-time/Urban	13.297	**5.868**	7.379
		Image Sensor Fault	13.375	**5.859**	6.99
	4	Only LiDAR (Traditional)	2.203	**1.266**	1.71
		Daytime/Urban	2.053	**1.149**	1.705
		Night-time/Urban	2.052	**1.159**	1.706
		Image Sensor Fault	2.203	**1.267**	1.707

Note: The numbers highlighted in bold indicate the algorithm that achieved the highest performance on each dataset sequence.

6. Conclusions

This paper provides several significant contributions. Firstly, it proposes a novel method to reduce 3D–2D depth association errors and enable accurate pose estimation in LiDAR Odometry by utilizing only LiDAR features instead of image key points and enhancing their uniqueness through image descriptors. Furthermore, it demonstrates the capability to maintain performance even in the event of vision sensor failure or environmental changes by leveraging LiDAR as the primary sensor. Through evaluations on the KITTI dataset [21], the top three methods with the lowest position RMSE are selected to develop Visual Assist LiDAR Odometry and Mapping (VA-LOAM), which evolves into subversions such as VA-LeGo-LOAM, VA-ISC-LOAM, and VA-LIO-SAM. The results of RMSE reduction in each version clearly indicate the potential of visual assistance modules to enhance LiDAR Odometry performance. This research lays a crucial foundation for the advancement of precise mapping and localization techniques using LiDAR and visual sensor data and provides broader research and application possibilities by making these methods publicly available.

Author Contributions: Project administration, T.-K.J.; supervision, G.-I.J. All authors have read and agreed to the published version of the manuscript.

Funding: This research received no external funding.

Data Availability Statement: Restrictions apply to the availability of these data. Data were obtained from KITTI and are available https://www.cvlibs.net/datasets/kitti/ (accessed on 3 March 2023) with the permission of KITTI.

Conflicts of Interest: The authors declare no conflict of interest.

References

1. Engel, J.; Schöps, T.; Cremers, D. LSD-SLAM: Large-Scale Direct Monocular SLAM. In *European Conference on Computer Vision*; Springer International Publishing: Cham, Switzerland, 2014; pp. 834–849. [CrossRef]
2. Forster, C.; Pizzoli, M.; Scaramuzza, D. SVO: Fast Semi-Direct Monocular Visual Odometry. In Proceedings of the 2014 IEEE International Conference on Robotics and Automation (ICRA), Hong Kong, China, 31 May–7 June 2014; pp. 15–22. [CrossRef]
3. Mur-Artal, R.; Montiel, J.M.M.; Tardós, J.D. ORB-SLAM: A Versatile and Accurate Monocular SLAM System. *IEEE Trans. Robot.* **2015**, *31*, 1147–1163. [CrossRef]

4. Qin, T.; Li, P.; Shen, S. VINS-Mono: A Robust and Versatile Monocular Visual-Inertial State Estimator. *IEEE Trans. Robot.* **2018**, *34*, 1004–1020. [CrossRef]
5. Tomono, M. Robust 3D SLAM with a Stereo Camera Based on an Edge-Point ICP Algorithm. In Proceedings of the 2009 IEEE International Conference on Robotics and Automation, Kobe, Japan, 12–17 May 2009; pp. 4306–4311. [CrossRef]
6. Engel, J.; Stückler, J.; Cremers, D. Large-Scale Direct SLAM with Stereo Cameras. In Proceedings of the 2015 IEEE/RSJ International Conference on Intelligent Robots and Systems (IROS), Hamburg, Germany, 28 September–2 October 2015; pp. 1935–1942. [CrossRef]
7. Kerl, C.; Sturm, J.; Cremers, D. Dense Visual SLAM for RGB-D Cameras. In Proceedings of the 2013 IEEE/RSJ International Conference on Intelligent Robots and Systems, Tokyo, Japan, 3–7 November 2013; pp. 2100–2106. [CrossRef]
8. Schöps, T.; Sattler, T.; Pollefeys, M. BAD SLAM: Bundle Adjusted Direct RGB-D SLAM. In Proceedings of the IEEE/CVF Conference on Computer Vision and Pattern Recognition, Long Beach, CA, USA, 15–20 June 2019; pp. 134–144. [CrossRef]
9. Zhang, J.; Singh, S. LOAM: Lidar Odometry and Mapping in Real-Time. In Proceedings of the Robotics: Science and Systems, Berkeley, CA, USA, 12–16 July 2014; pp. 1–9. [CrossRef]
10. Shan, T.; Englot, B. LEGO-LOAM: Lightweight and Ground-Optimized Lidar Odometry and Mapping on Variable Terrain. In Proceedings of the 2018 IEEE/RSJ International Conference on Intelligent Robots and Systems (IROS), Madrid, Spain, 1–5 October 2018; pp. 4758–4765. [CrossRef]
11. Shen, T.; Li, X.; Wang, G.; Wei, B.; Hu, H. A-LOAM: Tightly Coupled Lidar-Odometry and Mapping for Autonomous Vehicles. In Proceedings of the IEEE/RSJ International Conference on Intelligent Robots and Systems (IROS), Las Vegas, NV, USA, 24 October 2020–24 January 2021; pp. 8747–8754.
12. Shan, T.; Englot, B.; Meyers, D.; Wang, W.; Ratti, C.; Rus, D. LIO-SAM: Tightly-Coupled Lidar Inertial Odometry via Smoothing and Mapping. In Proceedings of the 2020 IEEE/RSJ International Conference on Intelligent Robots and Systems (IROS), Las Vegas, NV, USA, 24 October 2020–24 January 2021; pp. 5135–5142. [CrossRef]
13. Wang, H.; Wang, C.; Xie, L. Intensity Scan Context: Coding Intensity and Geometry Relations for Loop Closure Detection. In Proceedings of the 2020 IEEE International Conference on Robotics and Automation (ICRA), Paris, France, 31 May–31 August 2020; pp. 2095–2101.
14. Wang, H.; Wang, C.; Chen, C.L.; Xie, L. F-LOAM: Fast Lidar Odometry and Mapping. In Proceedings of the 2021 IEEE/RSJ International Conference on Intelligent Robots and Systems (IROS), Prague, Czech Republic, 27 September–1 October 2021; pp. 4390–4396. [CrossRef]
15. Zhang, J.; Singh, S. Visual-Lidar Odometry and Mapping: Low-Drift, Robust, and Fast. In Proceedings of the 2015 IEEE International Conference on Robotics and Automation (ICRA), Seattle, WA, USA, 26–30 May 2015; pp. 2174–2181. [CrossRef]
16. Zhang, J.; Kaess, M.; Singh, S. A Real-Time Method for Depth Enhanced Visual Odometry. *Auton. Robot.* **2017**, *41*, 31–43. [CrossRef]
17. Graeter, J.; Wilczynski, A.; Lauer, M. LIMO: Lidar-Monocular Visual Odometry. In Proceedings of the 2018 IEEE/RSJ International Conference on Intelligent Robots and Systems (IROS), Madrid, Spain, 1–5 October 2018; pp. 7872–7879. [CrossRef]
18. Huang, S.S.; Ma, Z.Y.; Mu, T.J.; Fu, H.; Hu, S.M. Lidar-Monocular Visual Odometry Using Point and Line Features. In Proceedings of the 2020 IEEE International Conference on Robotics and Automation (ICRA), Paris, France, 31 May–31 August 2020; pp. 1091–1097.
19. Wong, C.-C.; Feng, H.-M.; Kuo, K.-L. Multi-Sensor Fusion Simultaneous Localization Mapping Based on Deep Reinforcement Learning and Multi-Model Adaptive Estimation. *Sensors* **2024**, *24*, 48. [CrossRef] [PubMed]
20. Cai, Y.; Qian, W.; Dong, J.; Zhao, J.; Wang, K.; Shen, T. A LiDAR–Inertial SLAM Method Based on Virtual Inertial Navigation System. *Electronics* **2023**, *12*, 2639. [CrossRef]
21. Geiger, A.; Lenz, P.; Urtasun, R. Are We Ready for Autonomous Driving? The KITTI Vision Benchmark Suite. In Proceedings of the 2012 IEEE Conference on Computer Vision and Pattern Recognition, Providence, RI, USA, 16–21 June 2012; pp. 3354–3361. [CrossRef]
22. De Aguiar, A.S.; de Oliveira, M.A.; Pedrosa, E.F.; dos Santos, F.B. A Camera to LiDAR Calibration Approach through the Optimization of Atomic Transformations. *Expert Syst. Appl.* **2021**, *176*, 114894. [CrossRef]

sensors

MDPI

Article

Multiscale Bayes Adaptive Threshold Wavelet Transform Geomagnetic Basemap Denoising Taking Residual Constraints into Account

Pan Xiong, Gang Bian *, Qiang Liu, Shaohua Jin and Xiaodong Yin

Department of Military Oceanography and Hydrography and Cartography, Dalian Naval Academy, Dalian 116018, China; zeros0411@163.com (P.X.); liuqiang3931@163.com (Q.L.); jsh_1978@163.com (S.J.); triest@163.com (X.Y.)

* Correspondence: trighosts@163.com; Tel.: +86-17323540073

Abstract: To achieve high-precision geomagnetic matching navigation, a reliable geomagnetic anomaly basemap is essential. However, the accuracy of the geomagnetic anomaly basemap is often compromised by noise data that are inherent in the process of data acquisition and integration of multiple data sources. In order to address this challenge, a denoising approach utilizing an improved multiscale wavelet transform is proposed. The denoising process involves the iterative multiscale wavelet transform, which leverages the structural characteristics of the geomagnetic anomaly basemap to extract statistical information on model residuals. This information serves as the a priori knowledge for determining the Bayes estimation threshold necessary for obtaining an optimal wavelet threshold. Additionally, the entropy method is employed to integrate three commonly used evaluation indexes—the signal-to-noise ratio, root mean square (RMS), and smoothing degree. A fusion model of soft and hard threshold functions is devised to mitigate the inherent drawbacks of a single threshold function. During denoising, the Elastic Net regular term is introduced to enhance the accuracy and stability of the denoising results. To validate the proposed method, denoising experiments are conducted using simulation data from a sphere magnetic anomaly model and measured data from a Pacific Ocean sea area. The denoising performance of the proposed method is compared with Gaussian filter, mean filter, and soft and hard threshold wavelet transform algorithms. The experimental results, both for the simulated and measured data, demonstrate that the proposed method excels in denoising effectiveness; maintaining high accuracy; preserving image details while effectively removing noise; and optimizing the signal-to-noise ratio, structural similarity, root mean square error, and smoothing degree of the denoised image.

Keywords: multiscale wavelet transform; geomagnetic reference map; adaptive thresholding; Bayesian estimation; residuals

1. Introduction

Due to the increasingly complex electromagnetic confrontation environment in modern warfare, satellite navigation is vulnerable to interference and damage during wartime [1]. As a result, there is a growing focus on geomagnetic matching navigation as a means to enhance the anti-jamming capabilities of weapon platform navigation. Geomagnetic matching navigation is gaining prominence because of the stability of the geomagnetic field, its immunity to the geographic location and weather conditions, and its classification as a passive detection signal with significant concealment properties, making it a valuable auxiliary navigation tool in wartime scenarios. This approach involves the utilization of real-time magnetic anomaly data and an established geomagnetic anomaly reference map for precise positioning, necessitating the highest level of accuracy in the geomagnetic anomaly reference map [2].

To create the geomagnetic anomaly reference map, geomagnetic anomaly data from the region are gathered and compiled into a map using shipborne or aeromagnetic measurements. Throughout this process, several corrections are essential, including adjustments for daily variation, ship magnetic effects, position normalization, and instrument errors. However, even after applying these corrections, residual noise resulting from these errors tends to persist, making complete eradication challenging. Given that the compilation of the geomagnetic anomaly reference map involves amalgamating magnetic data from various time periods, instruments, and voyages, the introduction of noise is inevitable [3]. This integration of data from different sources necessitates the filtering of noise in the geomagnetic anomaly reference map to enhance the accuracy of geomagnetic matching navigation.

At present, commonly used denoising methods in measurement data processing include Gaussian filtering, mean value filtering, and soft and hard threshold wavelet denoising methods. Gaussian filtering is a commonly used image denoising technique with the advantages of a good smoothing effect and simple calculation, but its filtering process is also a fuzzy process of image processing, which will cause the loss of the original image edge information [4]. Mean value filtering has the advantages of being simple and easy to implement and better preservation of the overall color tone and brightness of the image, but it is unable to effectively remove large outliers that cause artifacts in the image; because mean value filtering is essentially a linear smoothing filter, the edges of the feature region and texture information will be lost when filtering [5]. Wavelet transform was firstly proposed by Mallat et al. in 1989, and after decades of development, it has become one of the most important methods for denoising all kinds of data; wavelet transform has the advantages of strong tunability, multiscale processing, time-frequency localization, etc. [6]. However, wavelet denoising involves many positive transforms and inverse transforms, which makes the computational quantity larger, and the size of the threshold, the setting of the threshold function, and the selection of wavelet bases has a greater impact on the experimental results. Therefore, the appropriate threshold, threshold function, and wavelet base are the key to the denoising effect [7]. The design of soft and hard threshold functions has also become a hot research topic; the soft threshold function focuses on smoothing the image and retaining the image details, while the hard threshold function can remove the noise more thoroughly, but it will lose some detail information, resulting in a loss of accuracy [8]. Researchers have spent a lot of effort to integrate the respective advantages of the soft threshold function and the hard threshold function [9] to obtain a moderate threshold function. And whether it is a soft threshold function or a hard threshold function, their denoising effect is dependent on setting the appropriate size of the threshold, and the selection of the threshold is very critical for the denoising effect [10]. The threshold can be determined by the minimum mean square error criterion to minimize the overall distortion of the noise and the signal and also by analyzing the statistical properties of the noise-containing data to distinguish the wavelet coefficients from the noise to determine the appropriate threshold. The Bayesian estimation threshold principle is a signal processing method based on Bayesian theory, which is used to determine the appropriate signal threshold in the process of wavelet transform denoising, and its basic idea is to determine the optimal threshold by maximizing the a posteriori probability or minimizing the expected loss [11]. However, the Bayesian determination of the threshold requires estimation of the probability distributions of the signal and noise, but in general, this part of the a priori information is more difficult to obtain.

In this paper, we propose to use the residual information as the a priori information of the Bayesian estimation threshold for the iterative derivation of the threshold and put forward a fusion model of soft and hard threshold functions to take into account the advantages of each of the soft and hard threshold functions. Considering that the geomagnetic reference map has higher requirements for detail information retention, a regularization term is introduced in the selection of wavelet coefficients to improve the accuracy of the filtering, and the superiority of the algorithm in this paper compared with the traditional algorithms is verified by the simulation data and the measured data in this paper.

2. Model Building

2.1. Bayesian Threshold Modeling with Residuals in Mind

The key to the denoising effect of multiscale wavelet transform lies in the appropriate selection of the wavelet bases, the calculation of the optimal threshold, and the determination of the optimal decomposition scale, while the a priori distribution information of the parameters needs to be obtained in the Bayesian estimation [12]. Therefore, in this paper, in view of the noise characteristics of the magnetic reference map, the a priori information of the noise distribution is determined by designing the cyclic residual model, and the optimal information in the wavelet decomposition is further projected.

Firstly, the original Bayesian threshold wavelet transform is used to obtain the residual information, and the obtained residuals are used to calculate their statistical distribution information, which is used as the prior distribution of the Bayesian estimation and to estimate the thresholds, and the obtained new thresholds are used as the update parameters to carry out the multiscale wavelet transform again. The peak signal-to-noise ratio can represent the quality of the denoised image, and when the peak signal-to-noise ratio of the resulting image reaches the inflection point, it is recognized as the optimal threshold under the current wavelet basis and decomposition scales, while the wavelet basis and decomposition scales are alternately optimized by constructing the binary tree model and using the depth-first search until the convergence condition is reached. On the basis of the optimal parameters computed by the model, a regularization term is introduced to limit the influence of the denoising process on the magnetic image and to improve the denoising accuracy. The model can calculate the optimal wavelet basis, decomposition scale, and threshold value adapted to the noise of the target geomagnetic reference map and ensure the reliability and accuracy of the data quality on the basis of denoising. The specific formulas which take into account the residual Bayesian wavelet threshold are as follows:

Let the mean of the residuals be μ, the variance is σ_S^2, and the formula is as follows:

$$p(S) = (1/\sqrt{2\pi}\sigma_S)\exp(-S^2/2\sigma_S^2) \tag{1}$$

where $p(S)$ denotes the information about the statistical distribution of the residual S. Let X be the original image with distribution $GGD_{(\alpha_X,\beta)}$, σ_X is the standardized variance of X, and β is the shape parameter, which here takes the value 2.

By calculating the resulting prior information, find the threshold that minimizes the Bayesian risk T, let $\hat{X} = \eta_t(y)$, $y/X \sim N(X,{\sigma_S}^2)$, and the Bayesian risk is expressed as follows:

$$r(T) = E(\hat{X} - X)^2 = E_x E_{y|x}(\hat{X} - X)^2 \tag{2}$$

$$\begin{aligned} E_X E_{Y|X}(\hat{X} - X) &= \int_{-\infty}^{+\infty}\int_{-\infty}^{+\infty} (\eta_t(\mathrm{y}) - x)^2 p(y|x)p(x)d_y d_x \\ &= {\sigma_S}^2\omega(\sigma_X^2/{\sigma_S}^2, T/\sigma_S) \end{aligned} \tag{3}$$

$$\omega(\sigma_X^2, T) = \sigma_X^2 + 2(T^2+1-\sigma_X^2)\overline{\phi}(T/\sqrt{1+\sigma_X^2}) - 2T(1+\sigma_X^2)\phi(T,1+\sigma_X^2) \tag{4}$$

The probability density function is:

$$\phi(x,{\sigma_S}^2) = (1/\sqrt{2\pi{\sigma_S}^2})\exp(-(x^2/2{\sigma_S}^2)) \tag{5}$$

$$\overline{\phi}(x) = \int_x^{\infty}\phi(t,1)dt \tag{6}$$

In the above equations, $\phi(x, \sigma_S{}^2)$ is the probability density function, $\hat{X}$ is the estimate of X, Y is the observed image, $r(T)$ denotes the Bayesian risk, $E(\hat{X} - X)^2$ is the error risk function, and the image denoising is performed to obtain an estimate Y of X using image $\hat{X}$. To minimize the error risk function, it can be obtained:

$$T^* = \arg\min_T r(T) = \frac{\sigma_S^2}{\sigma_X} \tag{7}$$

Equation (7) minimized the Bayesian risk function introduced by Equations (2)–(6).

$$\sigma_Y^2 = \frac{1}{n^2} \sum_{s,q=1}^{m} Y_{sq}^2 \tag{8}$$

where Y_{sq} is the subband coefficient of the wavelet transform.

$$\sigma_X = \sqrt{\max(\sigma_Y^2 - \sigma_S, 0)} \tag{9}$$

The threshold for each iteration is calculated as:

$$T = \sigma_S{}^2 / \sigma_X \tag{10}$$

The optimal threshold is calculated by bringing Equations (8) and (9) into Equation (10).

In order to improve the accuracy and stability of the denoising results, Elastic Net regularization is used to optimize the wavelet coefficients during the wavelet transform, and the regularization formula is as follows:

$$\min(X - L \times C_i)^2 + \lambda_1 \times |C_i| + \lambda_1 \times C_i{}^2 \tag{11}$$

where L is the wavelet transform matrix, C_i denotes the wavelet coefficients of the ith rewavelet transform, and λ_1 and λ_2 are the sparse regularization and ridge regression parameters, respectively.

The flowchart of the improved multiscale variational wavelet transform algorithm is shown below (see Figure 1).

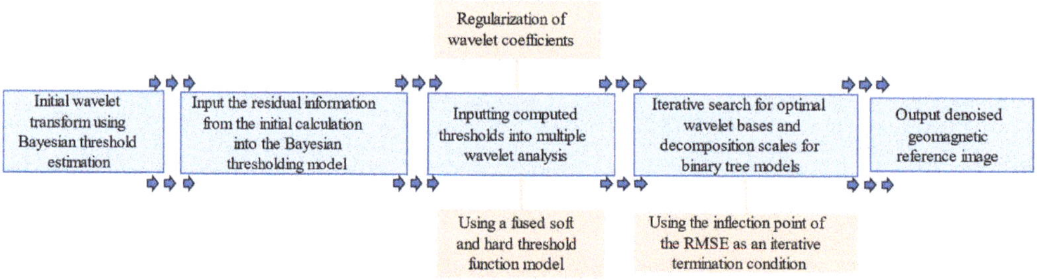

Figure 1. Flowchart of the improved multiscale wavelet transform algorithm.

2.2. Soft and Hard Threshold Fusion Function Models

Soft threshold function denoising will make the image tend to be smooth, does not produce ringing, have a pseudo-Gibbs effect, and at the same time can better retain the details of the signal and can provide a higher signal-to-noise ratio. But with the increase in the number of layers of decomposition, it will make the image's boundaries blurred, with a distortion phenomenon, and the computational complexity is higher. Hard threshold function denoising can retain the sharp changes in the signal and can effectively filter out low-amplitude noise, and the calculation speed is faster, but in the case of a low signal-to-noise ratio, it is prone to signal loss and will produce artifacts. Therefore, in order to synthesize the advantages of the two proposed soft and hard threshold function fusion models, the signal-to-noise ratio, root mean square, and smoothness of the denoised image, respectively, are calculated using the entropy method for the fusion of the indicators [13]. The weighted fusion of the soft and hard threshold function calculations are determined using the calculated fusion metrics [14]. The specific expression is:

$$F(m) = \omega_{vrm} C_{vrm}(m) + \omega_{vsnr} C_{vsnr}(m) + \omega_{vs} C_{vs}(m) \tag{12}$$

where $F(m)$ is the calculated integration indicator; ω_{vrm}, ω_{vsnr}, ω_{vs} are the weights of the root mean square error, the signal-to-noise ratio, and the change in smoothness obtained from the entropy calculation, respectively; m denotes the wavelet decomposition scale; C_{vrm}, C_{snr}, C_{vs} denote the normalization result of each quantity, respectively, and the expression is:

$$C_{vrm}(m) = [vrm(m) - vrm_{\min}]/(vrm_{\max} - vrm_{\min}) \tag{13}$$

$$C_{vsnr}(m) = [vsnr(m) - vsnr_{\min}]/(vsnr_{\max} - vsnr_{\min}) \tag{14}$$

$$C_{vs}(m) = [vs(m) - vs_{\min}]/(vs_{\max} - vs_{\min}) \tag{15}$$

The amount of change in the signal-to-noise ratio error is given as an example of how the weights of the indicators are calculated as:

$$\omega_{C_{vsnr}} = \frac{1 - H_{C_{vsnr}}}{(1 - H_{C_{vsnr}}) + (1 - H_{C_{vrm}}) + (1 - H_{C_{vs}})} \tag{16}$$

$$H_{C_{vsnr}} = -(\frac{1}{\ln(N)}) \sum_{i=1}^{N} P_i^{C_{vsnr}} \ln P_i^{C_{vsnr}} \tag{17}$$

$$P_i^{C_{vsnr}} = C_{vsnr}(i) / \sum_{i=1}^{N} C_{vsnr}(i) \tag{18}$$

And the limit difference threshold is set. When the difference between the fusion indexes of the denoised images of different function models is less than the threshold, the model is combined by calculating the fusion weights of each index, and when the difference is greater than the threshold, the fusion is not carried out and the optimal model is used directly. The realization process is as follows:

$$W_{hard} = \frac{F(m)_{hard}}{F(m)_{hard} + F(m)_{soft}} \tag{19}$$

$$W_{soft} = \frac{F(m)_{soft}}{F(m)_{hard} + F(m)_{soft}} \tag{20}$$

where W_{hard}, W_{soft} are the fusion model weights of the soft and hard threshold functions, and $F(m)_{soft}$, $F(m)_{hard}$ are the fusion metrics computed by the soft and hard threshold functions, respectively.

$$\frac{\left|F(m)^{hard} - F(m)^{soft}\right|}{\max(F(m)^{hard} - F(m)^{soft})} > \theta \tag{21}$$

In order to improve the generalization ability of the model, by setting a threshold value θ, when the fusion index calculated by the soft and hard threshold functions satisfies the above equation, the model fusion is not carried out, and the model with the large fusion index of the two is directly selected.

3. Test and Analysis

3.1. Assessment Method

In order to verify the effectiveness of the proposed multimodal Bayes adaptive threshold wavelet geomagnetic basemap denoising algorithm considering residuals, the proposed algorithm is compared and analyzed by using mean filtering, Gaussian filtering, soft threshold wavelet filtering, and hard threshold wavelet filtering; in order to evaluate the performance of different denoising methods, four evaluation metrics, namely, the peak signal-to-noise ratio (PSNR), structural similarity index (SSIM), root mean square error (RMSE), and smoothness, are used to analyze and discuss the experiments.

The PSNR can be used to evaluate the degree of distortion of the denoised image; the larger the PSNR, the better the quality of the denoised image [15]. The PSNR is calculated using the following formula:

$$PSNR = 10 \times \log_{10}\left(\frac{N \times MAX_I^2}{\sum_{i=1}^{N}(I_{orig(i)} - I_{comp(i)^2})}\right) \tag{22}$$

where MAX_I^2 represents the maximum value of the basemap, $I_{orig(i)}$ is the value of the original base image, and I_{comp} represents the value of the processed image.

The structural similarity index is used to evaluate the denoising quality of an image by comparing the similarity, brightness, and structure between images, which can be used to compare the difference between geomagnetic datum maps. The larger value of the SSIM represents the higher quality of the image denoising. The formula for calculating the SSIM can be expressed as:

$$SSIM(x,y) = l(x,y)^{\alpha}c(x,y)^{\beta}s(x,y)^{\gamma} \tag{23}$$

$$SSIM(x,y) = \frac{(2\mu_x\mu_y + C_1)(2\sigma_x\sigma_y + C_2)}{(\mu_x^2 + \mu_y^2 + C_1)(\sigma_x^2 + \sigma_y^2 + C_2)} \tag{24}$$

where $l(x,y)$ denotes the brightness, $c(x,y)$ denotes the contrast, $s(x,y)$ denotes the structure, x denotes the original image, and y denotes the noisy image.

The accuracy of different denoising algorithms can be reflected by calculating the root mean square error between the denoised image and the original image and examining the effect of denoising algorithms on the accuracy of geomagnetic reference maps [16].

$$RMSE = \sqrt{\frac{\sum_{i=1}^{n}(X_{obs,i} - X_{\text{model},i})^2}{n}} \tag{25}$$

where $X_{obs,i}$ represents the value of the original basemap and $X_{\text{model},i}$ represents the value of the processed basemap.

The smoothness can evaluate the image texture and the degree of change in the magnetic reference map after denoising, and in this paper, the gradient calculation method is used to calculate the smoothness.

$$[G_x, G_y]^T = [\frac{\delta f}{\delta x}, \frac{\delta f}{\delta y}]^T \tag{26}$$

$$g(x,y) = \sqrt{\frac{\partial^2 f}{\partial x^2} + \frac{\partial^2 f}{\partial y^2}} \tag{27}$$

where $G(x)$, $G(y)$ is the gradient of the image $f(x,y)$ in the x direction and the y direction at the point (x,y), and $g(x,y)$ is the smoothness of the point.

3.2. Results and Discussion

3.2.1. Simulation Experiment

The magnetic anomaly basemap generated by the sphere magnetic anomaly model is used for the simulation experiments, and the parameters of the sphere are set as follows: a magnetization rate of 0.15, magnetic field strength of 50 nT, radius of the sphere of 250 m, and burial depth of 100 m. In order to visualize the denoising effect, the Gaussian noise with a variance of 0.03, 0.05, 0.07, and 0.09 is added to the magnetic anomaly reference map generated by the sphere magnetic anomaly model, and the mean filter, Gaussian filter, soft threshold wavelet filter, and hard threshold wavelet filter are used to compare the algorithms with the proposed algorithms, and the denoising results are analyzed to judge the five filtering algorithms on the processing of the noise of the magnetic reference map. The ability of the five filtering algorithms to deal with the noise of the magnetic reference map is analyzed. The experimental environment of this paper is configured as follows: the CPU is an AMD Ryzen 7 4800 H (Advanced Micro Devices, Santa Clara, CA, America), the GPU is an NVIDIA GeForce RTX 2060 (NVIDIA, Santa Clara, CA, America), and the development language is 2023b MATLAB.

In Figures 1–5, The purple area in the graph above is a positive magnetic anomaly, which slowly turns to a negative anomaly as it moves closer to the blue color. In Figure 2(g1,g2), Figure 3(g1,g2), Figure 4(g1,g2), and Figure 5(g1,g2), it can be seen that the simulated magnetic anomaly map benchmark map processed by the algorithm proposed in this paper has clear boundaries and high smoothness, and the denoising effect is better than the remaining four algorithms.

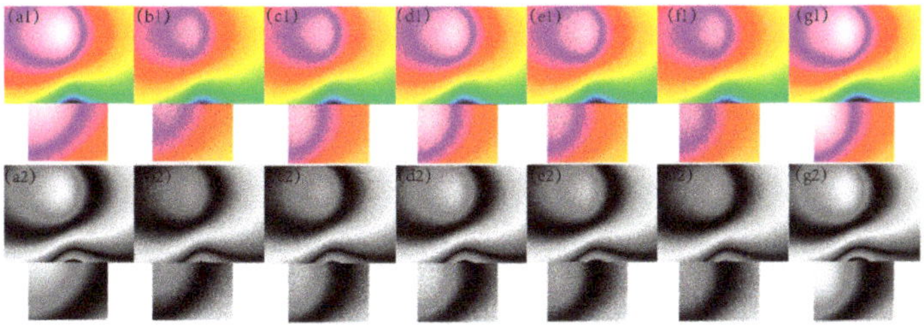

Figure 2. The filtering effect of the simulated data of each algorithm when the variance of the Gaussian noise is 0.003, (**a1**) represents the original magnetic anomaly benchmark image, (**b1**) the added noise magnetic anomaly benchmark image, (**c1**) the Gaussian filter denoising effect image, (**d1**) the mean filtering effect image, (**e1**) the soft-thresholding wavelet filtering effect image, (**f1**) the hard-thresholding wavelet filtering effect image, (**g1**) the filtering effect image of this paper's algorithm, respectively, and (**a2**), (**b2**), (**c2**), (**d2**), (**e2**), (**f2**) and (**g2**) are the corresponding grayscale images.

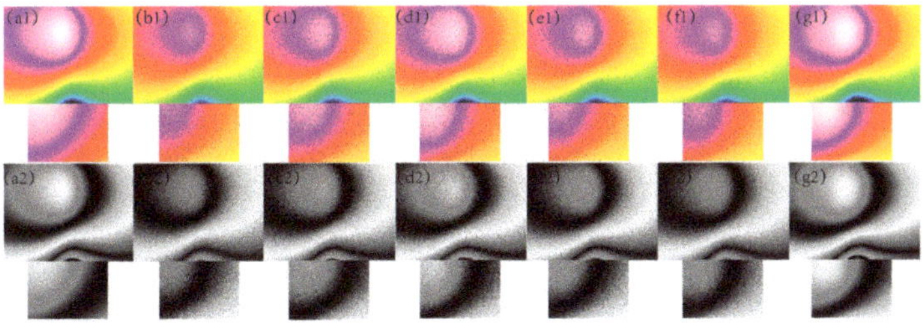

Figure 3. The filtering effect of the simulated data of each algorithm when the variance of the Gaussian noise is 0.005, (**a1**) represents the original magnetic anomaly benchmark image, (**b1**) the added noise magnetic anomaly benchmark image, (**c1**) the Gaussian filter denoising effect image, (**d1**) the mean filtering effect image, (**e1**) the soft-thresholding wavelet filtering effect image, (**f1**) the hard-thresholding wavelet filtering effect image, (**g1**) the filtering effect image of this paper's algorithm, respectively, and (**a2**), (**b2**), (**c2**), (**d2**), (**e2**), (**f2**) and (**g2**) are the corresponding grayscale images.

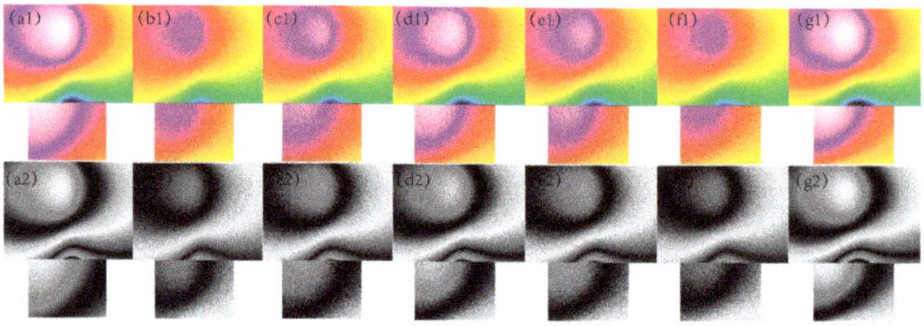

Figure 4. The filtering effect of the simulated data of each algorithm when the variance of the Gaussian noise is 0.007, (**a1**) represents the original magnetic anomaly benchmark image, (**b1**) the added noise magnetic anomaly benchmark image, (**c1**) the Gaussian filter denoising effect image, (**d1**) the mean filtering effect image, (**e1**) the soft-thresholding wavelet filtering effect image, (**f1**) the hard-thresholding wavelet filtering effect image, (**g1**) the filtering effect image of this paper's algorithm, respectively, and (**a2**), (**b2**), (**c2**), (**d2**), (**e2**), (**f2**) and (**g2**) are the corresponding grayscale images.

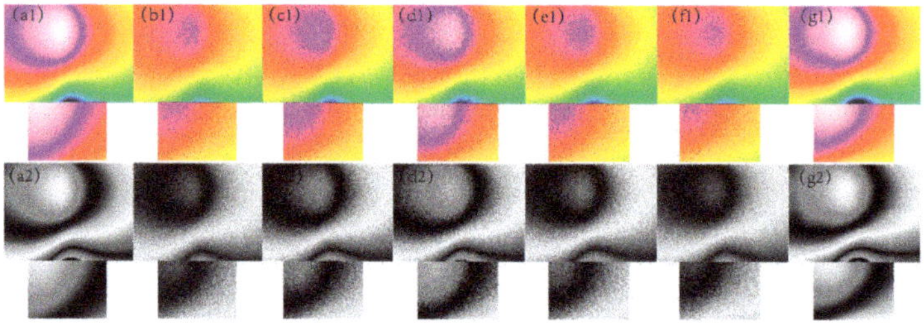

Figure 5. The filtering effect of the simulated data of each algorithm when the variance of the Gaussian noise is 0.009, (**a1**) represents the original magnetic anomaly benchmark image, (**b1**) the added noise magnetic anomaly benchmark image, (**c1**) the Gaussian filter denoising effect image, (**d1**) the mean filtering effect image, (**e1**) the soft-thresholding wavelet filtering effect image, (**f1**) the hard-thresholding wavelet filtering effect image, (**g1**) the filtering effect image of this paper's algorithm, respectively, and (**a2**), (**b2**), (**c2**), (**d2**), (**e2**), (**f2**), and (**g2**) are the corresponding grayscale images.

In order to better observe the denoising effect of the different algorithms in the magnetic anomaly simulation datum map under different noise levels, the changes in four commonly used denoising judgments, the PSNR, SSIM, RMSE, and smoothness, are used to compare the denoising ability of each algorithm [17]. As seen in Figures 6–9, the algorithm proposed in this paper is better than the remaining four algorithms in all the indicators, especially in the indicator of smoothness where it is far better than the rest of the denoising algorithms. In Figure 6, the curve of this paper's model decreases the curvature of the PSNR most slowly with the increase in noise, and it can be seen that it is less affected by noise. In Figure 7, the RMSE of this paper's model increases most slowly with the increase in noise, and it can also be concluded that this paper's model is less affected by noise and is more robust. In Figure 9, the similarity of this paper's model decreases by only 0.06% with the increase in noise, which is the lowest degree of decrease compared to the rest of the algorithms.

With the magnetic anomaly reference map in the geomagnetic matching navigation for the data of the higher accuracy requirements and the root mean square error that can be a better response to the different denoising algorithms for the accuracy of the original data to maintain the level of Figure 6, it can be seen that after using the algorithm of this paper that denoising the calculation of the root mean square error is the lowest, and the denoised data can be used to calculate the accuracy of the data of the original data. The root mean square error is the lowest and the accuracy of the denoised image is the highest.

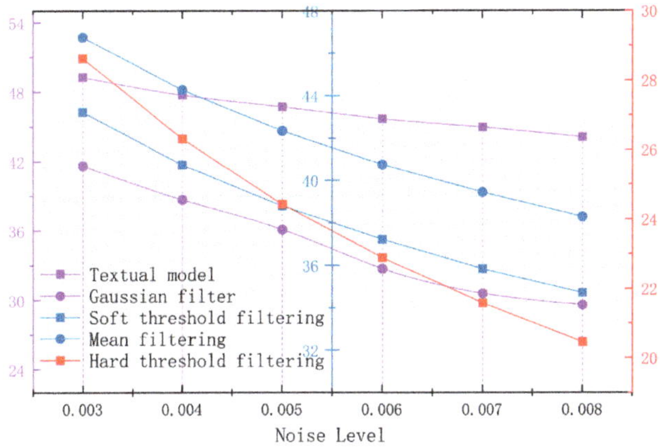

Figure 6. Variation in image PSNR with noise after denoising with different algorithms.

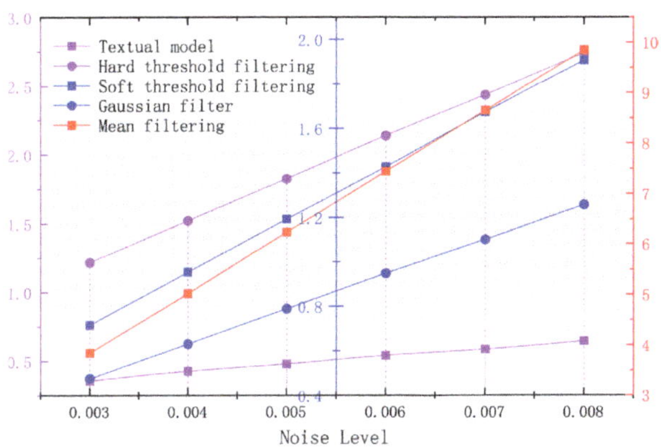

Figure 7. Variation in image RMSE with noise after denoising by different algorithms.

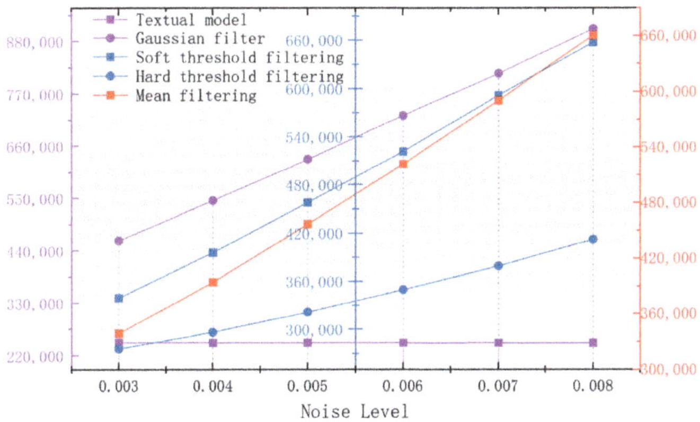

Figure 8. Variation in image smoothness with noise after denoising by different algorithms.

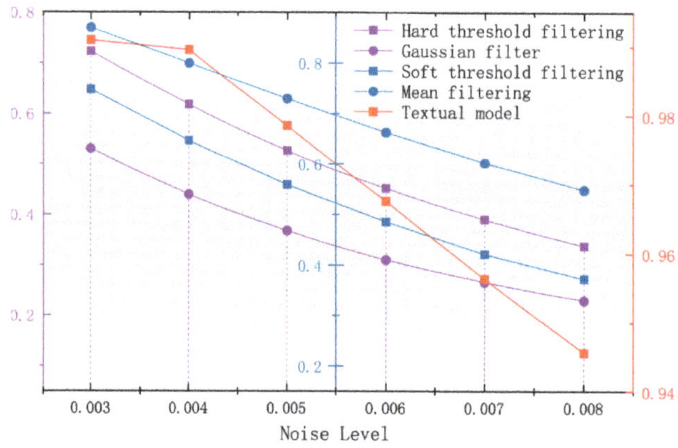

Figure 9. Variation in image SSIM with noise after denoising by different algorithms.

3.2.2. Measured Data Analysis and Validation

In order to further test the denoising effect of this paper's algorithm in the actual situation, the measured geomagnetic anomaly benchmark map of a sea area in the Pacific Ocean is selected for the denoising comparison experiments; the size of the region is 110 km $\times$ 110 km, the spacing of the measurement lines is 1 km, and the original magnetic anomaly map is shown in Figure 5(a1). And in order to more intuitively illustrate the denoising effect, this paper uses the artificial inclusion of the variance of 0.05, 0.08, 0.11, and 0.14 Gaussian noise in the original data; the same mean filtering, Gaussian filtering, soft threshold wavelet filtering, and hard threshold wavelet filtering; and the proposed algorithm for the comparison experiments, analyzing the denoising results to judge the ability of the five filtering algorithms to deal with the noise of the magnetic reference map.

As can be seen from Figures 10–13, the denoising effect of this paper's algorithm is better than the rest of the algorithms in the measured magnetic anomaly benchmark map under different noise levels. After the denoising of this paper's algorithm, the edge information of the benchmark map is better retained, and the edge information of the magnetic anomaly features is clearer, which can be observed intuitively in the advantages of this paper's algorithm, both in the true-color and grayscale images. In Figure 10(g1,g2), Figure 11(g1,g2), Figure 12(g1,g2) and Figure 13(g1,g2), it can be seen that this paper's algorithm basically does not lose the details of the reference image while removing the noise, and it has the highest retention, while the other methods lose a small amount of detail and appear distorted.

In Figures 10–13, The purple area in the graph above is a positive magnetic anomaly, which slowly turns to a negative anomaly as it moves closer to the blue color. In order to objectively analyze the denoising effect, in addition to the visual effect being better than the remaining four algorithms, the denoising effect is also evaluated by calculating the PSNR, SSIM, RMSE, and smoothness of the denoised image, and the experimental results are shown in Figures 14–17. As seen in Figures 10 and 11, this paper's algorithm has a lower level of smoothness on the image and maintains the highest level of accuracy, indicating that this paper's algorithm loses less detail information on the geomagnetic datum map, and the rest of the comparative algorithms denoise the geomagnetic datum map excessively, removing the detail information as noisy data, resulting in a lower level of accuracy and a higher level of smoothness.

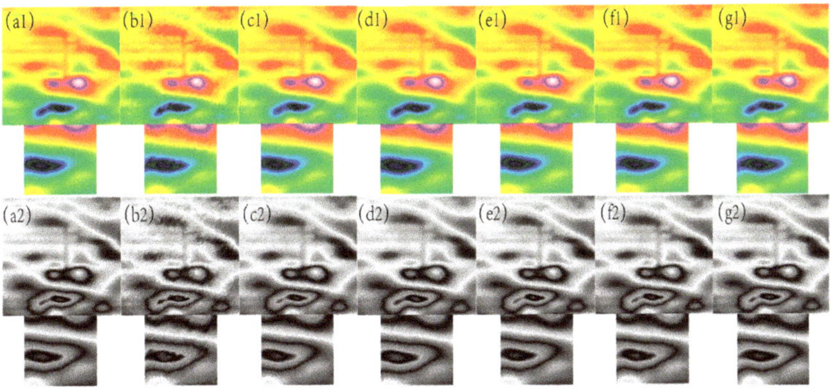

Figure 10. The filtering effect of the measured data of each algorithm when the variance of the Gaussian noise is 0.05, (**a1**) represents the original magnetic anomaly benchmark image, (**b1**) the added noise magnetic anomaly benchmark image, (**c1**) the Gaussian filter denoising effect image, (**d1**) the mean filtering effect image, (**e1**) the soft-thresholding wavelet filtering effect image, (**f1**) the hard-thresholding wavelet filtering effect image, (**g1**) the filtering effect image of this paper's algorithm, respectively, and (**a2**), (**b2**), (**c2**), (**d2**), (**e2**), (**f2**) and (**g2**) are the corresponding grayscale images.

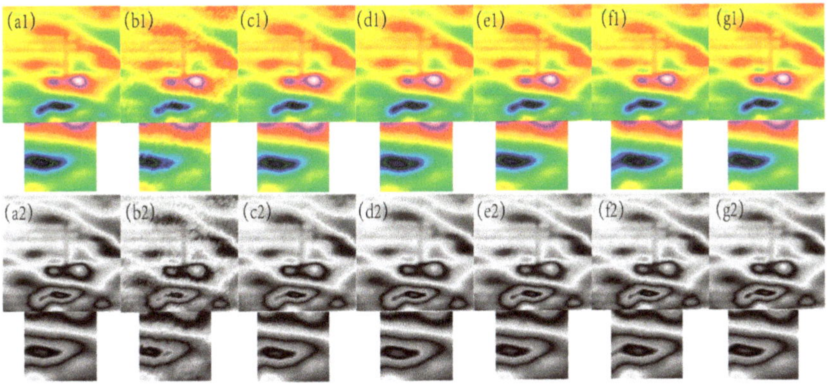

Figure 11. The filtering effect of the measured data of each algorithm when the variance of the Gaussian noise is 0.08, (**a1**) represents the original magnetic anomaly benchmark image, (**b1**) the added noise magnetic anomaly benchmark image, (**c1**) the Gaussian filter denoising effect image, (**d1**) the mean filtering effect image, (**e1**) the soft-thresholding wavelet filtering effect image, (**f1**) the hard-thresholding wavelet filtering effect image, (**g1**) the filtering effect image of this paper's algorithm, respectively, and (**a2**), (**b2**), (**c2**), (**d2**), (**e2**), (**f2**) and (**g2**) are the corresponding grayscale images.

Similarly, in order to evaluate the accuracy level of each algorithm in maintaining the original measured data, the size of the root mean square error of the denoising results of different algorithms can be seen in Figure 15. The algorithm of this paper has the smallest root mean square error, and with the increase in the noise, the root mean square error rate of change is smaller, and its performance is better for the robustness of the noise. The SSIM reflects the structural similarity between the denoised image and the original image, and the higher the SSIM, the better the visual effect of the denoised image is. The SSIM can reflect the structural similarity between the denoised image and the original image; the higher the SSIM, the better the visual effect after denoising the benchmark image, which can also reflect the effect of the denoising algorithm in retaining the details of the benchmark image. In Figure 17, it can be seen that the algorithm in this paper, in the case of the increasing noise level, results in the SSIM decreasing more slowly than the rest of the algorithms, and the denoising effect is less affected by the noise level.

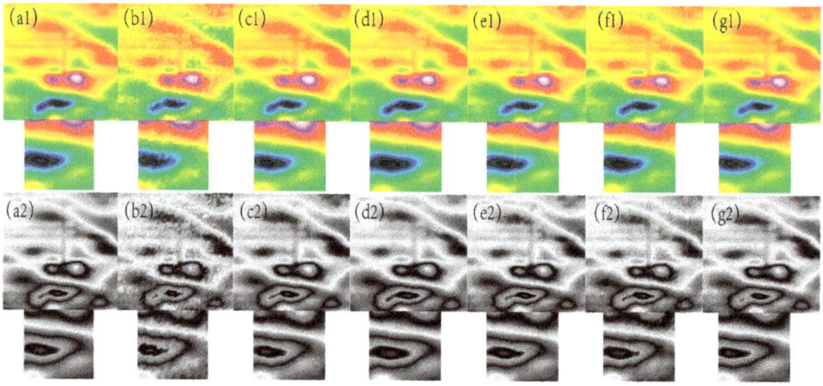

Figure 12. The filtering effect of the measured data of each algorithm when the variance of the Gaussian noise is 0.11, (**a1**) represents the original magnetic anomaly benchmark image, (**b1**) the added noise magnetic anomaly benchmark image, (**c1**) the Gaussian filter denoising effect image, (**d1**) the mean filtering effect image, (**e1**) the soft-thresholding wavelet filtering effect image, (**f1**) the hard-thresholding wavelet filtering effect image, (**g1**) the filtering effect image of this paper's algorithm, respectively, and (**a2**), (**b2**), (**c2**), (**d2**), (**e2**), (**f2**), and (**g2**) are the corresponding grayscale images.

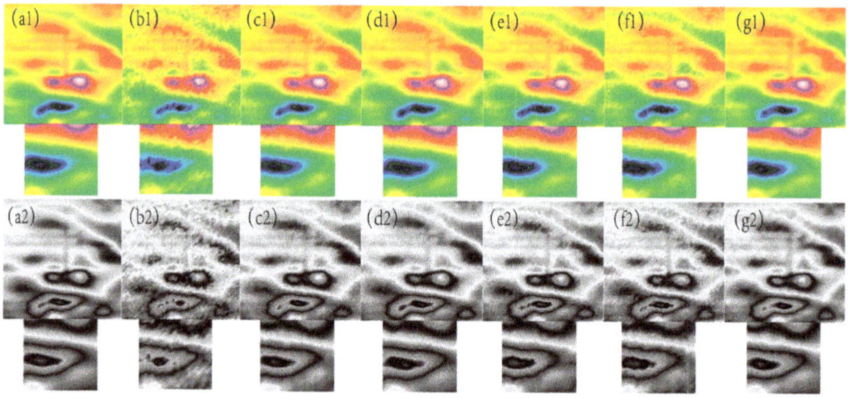

Figure 13. The filtering effect of the measured data of each algorithm when the variance of the Gaussian noise is 0.14, (**a1**) represents the original magnetic anomaly benchmark image, (**b1**) the added noise magnetic anomaly benchmark image, (**c1**) the Gaussian filter denoising effect image, (**d1**) the mean filtering effect image, (**e1**) the soft-thresholding wavelet filtering effect image, (**f1**) the hard-thresholding wavelet filtering effect image, (**g1**) the filtering effect image of this paper's algorithm, respectively, and (**a2**), (**b2**), (**c2**), (**d2**), (**e2**), (**f2**), and (**g2**) are the corresponding grayscale images.

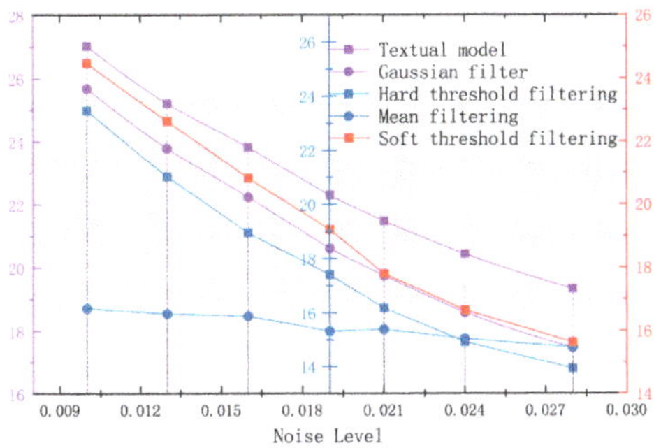

Figure 14. Variation in image PSNR with noise after denoising with different algorithms.

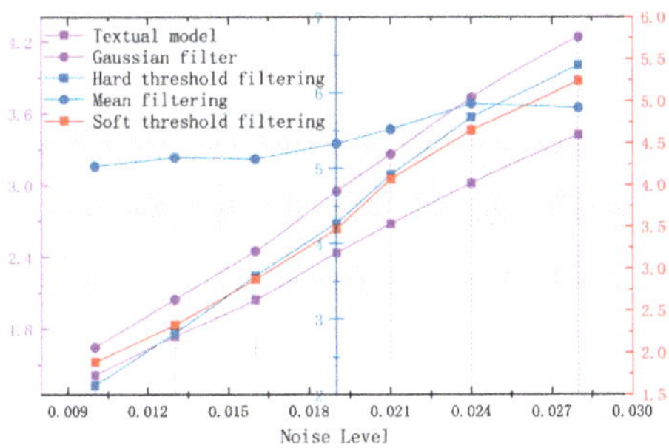

Figure 15. Variation in image RMSE with noise after denoising by different algorithms.

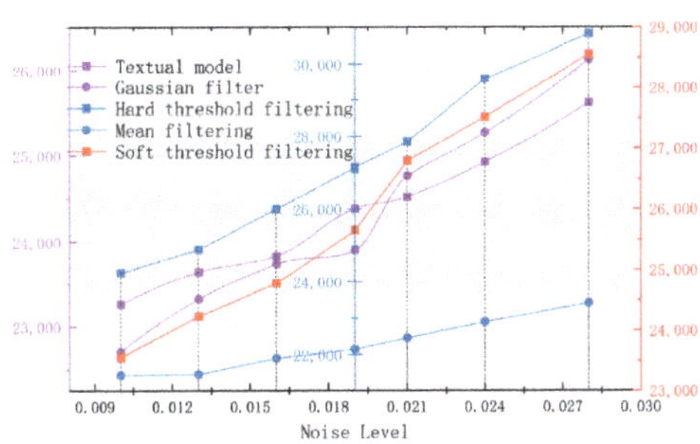

Figure 16. Variation in image smoothness with noise after denoising by different algorithms.

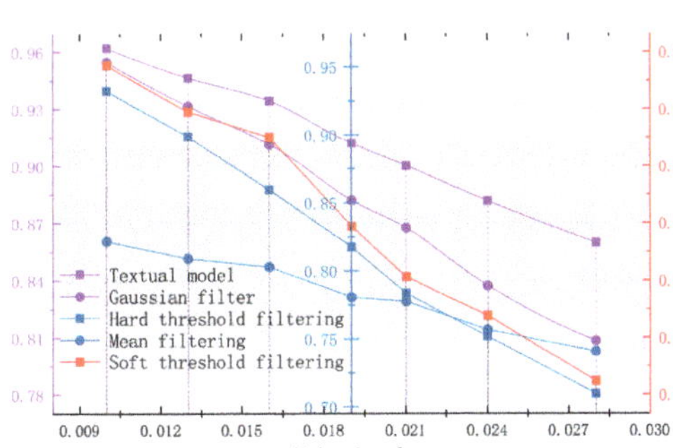

Figure 17. Variation in image SSIM with noise after denoising by different algorithms.

4. Conclusions

This paper focuses on the denoising method of a geomagnetic anomaly basemap and proposes a multimodal Bayes adaptive threshold wavelet geomagnetic basemap denoising method taking residuals into account. It is shown by the simulation experiments and measured data that the denoising effect of this method is better than that of mean value filtering, Gaussian filtering, soft-thresholding wavelet filtering, and hard-thresholding wavelet filtering under the same conditions. In addition, the method limits the modification of the original measurements by the filtering algorithm by setting the inclusion of the regularization term, which effectively reduces the root mean square error, which is conducive to the improvement in the accuracy of the reference map in geomagnetic matching navigation. Through simulation experiments and real data verification, it can be seen that this paper's algorithm, in the case of the increasing noise level, results in the denoising effect decreasing more slowly, and the robustness to noise is better. This method can play an important role in practical application scenarios, such as geomagnetic data denoising, the preparation of geomagnetic reference maps, and geomagnetic matching navigation. In the next study, we will continue to optimize the computational efficiency of the algorithm, deeply explore the denoising effect of different wavelet bases in the context of a geomagnetic field, and use the optimization algorithm to further improve the denoising efficiency of the algorithm.

Author Contributions: Data curation, Q.L.; Formal analysis, G.B.; Investigation, G.B.; Methodology, P.X.; Project administration, X.Y.; Resources, S.J.; Software, P.X.; Visualization, Q.L.; Writing—original draft, P.X. All authors have read and agreed to the published version of the manuscript.

Funding: This research received no external funding.

Institutional Review Board Statement: Not applicable.

Informed Consent Statement: Not applicable.

Data Availability Statement: Data are contained within the article.

Conflicts of Interest: The authors declare no conflicts of interest.

References

1. Zhang, K.; Zeng, F.; Zhao, Y.; Zeng, R. Study on Index System of GPS Interference Effect Evaluation. In *AIP Conference Proceedings*; AIP Publishing: Melville, NY, USA, 2018; Volume 1967. [CrossRef]
2. Wang, Q.; Zheng, C.; Wu, P.; Wang, X. Geomagnetic/Inertial Navigation Integrated Matching Navigation Method. *Heliyon* **2022**, *8*, e11249. [CrossRef] [PubMed]
3. Yu, H.; Li, Z.; Yang, W.; Shen, T.; Liang, D.; He, Q. Underwater Geomagnetic Localization Based on Adaptive Fission Particle-Matching Technology. *J. Mar. Sci. Eng.* **2023**, *11*, 1739. [CrossRef]
4. Shan, Z.; Zhou, J.; Chen, J. Background Noise Suppression of Magnetic Anomaly Signal Based on Wavelet Transform. In Proceedings of the 2017 IEEE 17th International Conference on Communication Technology (ICCT), Chengdu, China, 27–30 October 2017; pp. 333–337.
5. Ito, K.; Xiong, K. Gaussian Filters for Nonlinear Filtering Problems. *IEEE Trans. Automat. Contr.* **2000**, *45*, 910–927. [CrossRef]
6. Cao, N.; Liu, Y. High-Noise Grayscale Image Denoising Using an Improved Median Filter for the Adaptive Selection of a Threshold. *Appl. Sci.* **2024**, *14*, 635. [CrossRef]
7. Chalghoumi, S.; Smiti, A. Median Filter for Denoising MRI: Literature Review. In Proceedings of the 2022 International Conference on Decision Aid Sciences and Applications (DASA), Chiangrai, Thailand, 23 March 2022; pp. 1603–1606.
8. Ďuriš, V.; Chumarov, S.G.; Semenov, V.I. Increasing the Speed of Multiscale Signal Analysis in the Frequency Domain. *Electronics* **2023**, *12*, 745. [CrossRef]
9. Halidou, A.; Mohamadou, Y.; Ari, A.A.A.; Zacko, E.J.G. Review of Wavelet Denoising Algorithms. *Multimed Tools Appl.* **2023**, *82*, 41539–41569. [CrossRef]
10. Gao, W.Z.; Chen, Z.Y.; Zeng, Q.M. Improved algorithm of wavelet thresholding for image denoising. *J. East China Norm. Univ. (Nat. Sci.)* **2013**, *2013*, 83–92.
11. Rivoirard, V. Thresholding Procedure with Priors Based on Pareto Distributions. *Test* **2004**, *13*, 213–246. [CrossRef]
12. Liu, G.; Yang, S.; Wang, G.; Li, F.; You, D. A Decision-Making Method for Machinery Abnormalities Based on Neural Network Prediction and Bayesian Hypothesis Testing. *Electronics* **2021**, *10*, 1610. [CrossRef]

13. Ismael, A.A.; Baykara, M. Image Denoising Based on Implementing Threshold Techniques in Multi-Resolution Wavelet Domain and Spatial Domain Filters. *Trait. Signal* **2022**, *39*, 1119–1131. [CrossRef]
14. Tao, K.; Zhu, J.J. A Hybrid Indicator for Determining the Best Decomposition Scale of Wavelet De-noising. *Acta Geod. Cartogr. Sin.* **2012**, *41*, 749–755.
15. Wang, C.; Zhang, Y.S.; Wang, X.; Ji, S. Stripe noise removal of remote image based on wavelet variational method. *Acta Geod. Cartogr. Sin.* **2019**, *48*, 1025–1037. [CrossRef]
16. Guo, Y.; Zhou, X.; Li, J.; Ba, R.; Xu, Z.; Tu, S.; Chai, L. A Novel and Optimized Sine–Cosine Transform Wavelet Threshold Denoising Method Based on the Sym4 Basis Function and Adaptive Threshold Related to Noise Intensity. *Appl. Sci.* **2023**, *13*, 10789. [CrossRef]
17. Ye, Q.; Pan, H.; Liu, C. A Framework for Final Drive Simultaneous Failure Diagnosis Based on Fuzzy Entropy and Sparse Bayesian Extreme Learning Machine. *Comput. Intell. Neurosci.* **2015**, *2015*, 427965. [CrossRef] [PubMed]

sensors

MDPI

Article

Limited Memory-Based Random-Weighted Kalman Filter

Zhaohui Gao [1,*], Hua Zong [2], Yongmin Zhong [3,*] and Guangle Gao [4]

1 School of Electronic Engineering, Xi'an Shiyou University, Xi'an 710065, China
2 National Key Laboratory of Science and Technology on Aerospace Intelligent Control, Beijing Aerospace Automatic Control Institute, Beijing 100854, China; zonghua3@sina.cn
3 School of Engineering, RMIT University, Bundoora, Melbourne 3083, Australia
4 School of Automatics, Northwestern Polytechnical University, Xi'an 710072, China; gaole1211@163.com
* Correspondence: gzhgzhstation@hotmail.com (Z.G.); yongmin.zhong@rmit.edu.au (Y.Z.)

Abstract: The Kalman filter is an important technique for system state estimation. It requires the exact knowledge of system noise statistics to achieve optimal state estimation. However, in practice, this knowledge is often unknown or inaccurate due to uncertainties and disturbances involved in the dynamic environment, leading to degraded or even divergent filtering solutions. To address this issue, this paper presents a new method by combining the random weighting concept with the limited memory technique to accurately estimate system noise statistics. To avoid the influence of excessive historical information on state estimation, random weighting theories are established based on the limited memory technique to estimate both process noise and measurement noise statistics within a limited memory. Subsequently, the estimated system noise statistics are fed back into the Kalman filtering process for system state estimation. The proposed method improves the Kalman filtering accuracy by adaptively adjusting the weights of system noise statistics within a limited memory to suppress the interference of system noise on system state estimation. Simulations and experiments as well as comparison analysis were conducted, demonstrating that the proposed method can overcome the disadvantage of the traditional limited memory filter, leading to im-proved accuracy for system state estimation.

Keywords: Kalman filter; limited memory; random weighting estimation; noise statistics; unbiased estimation

1. Introduction

The Kalman filter (KF) is widely used in many fields such as guidance and navigation, biological tissue characterization, and power systems [1–3]. Under the condition that system noise statistics are pre-defined accurately, KF can produce accurate state estimation results [4–6]. However, it needs to know the accurate statistical characteristics of system noise. In engineering practice, such as global positioning systems and inertial navigation integrated systems, due to the uncertainties and interferences involved in the dynamic environment, the statistical characteristics of system noise are usually unknown or inaccurate, which leads to biased or even divergent KF solutions [7–9]. Thus, a robust filtering technique is needed to overcome the influence of unknown or inaccurate noise statistics on the system.

In this paper, by combining the concepts of random weighting and limited memory technology into KF, a random-weighted Kalman filter (LM-RWKF) based on limited memory is developed to achieve the accuracy and reliability of system state estimation under unknown or inaccurate noise statistics. This method estimates system noise statistics by adaptive adjusting the weights of noise statistics within a limited memory to restrain the interferences of system noise and excessive historical information on system state estimation. Based on the limited memory technique, random weighting theories are established to estimate and adjust both process noise and measurement noise statistics within a limited memory via random weights. The estimated system noise statistics are further fed back into the Kalman filtering process for system state estimation. Simulations and experiments, as well as comparison analysis with the proposed LM–RWKF and the traditional limited memory KF (LM–KF), were conducted to comprehensively evaluate the LM-RWKF performance.

2. Related Work

To achieve accuracy and reliability of system state estimation under unknown or inaccurate noise statistics, various techniques of noise statistics estimation have been reported for KF. The correlation method estimates system noises statistics based on the innovation sequence. However, it is not optimal in terms of mean square error [10,11]. The Sage–Husa adaptive filtering improves the filtering accuracy by adaptively adjusting system noise covariances [12–15]. However, since system noise covariances are estimated by arithmetic mean, it may not guarantee that the final filtering solution is optimal. As an approximation to Bayesian estimation, multiple-model based robust KF tracks time-varying noise statistics by simultaneously operating a set of KFs. However, this method suffers from substantial computational complexity [16–19]. Fuzzy logic estimates system noise statistics to improve the KF adaptability and robustness [20]. However, since fuzzy rules are developed based on empiricism and heuristic information, the resultant estimation has limited performance. Fault detection is also used to handle inaccurate system noise statistics [21,22]. However, this technique is only able to identify and isolate the occurrence of inaccurate system noise statistics, while being incapable of suppressing their influence on state estimation. The maximum likelihood theory estimates noise statistics through the maximization of their posteriori probability density [23,24]. However, due to the involvement of too many historical residuals, the resultant estimations cannot reflect the current characteristics of system noise.

Various strategies have been reported to limit the memory of historical information in the filtering process to reduce their contributions while increasing the contributions of the latest information to system state estimation. The fading memory technique uses fading factors to increase the use of latest residuals and reduce the effect of historical information on system state estimation, leading to improved filtering accuracy. It is less expensive in computation and has good numerical stability. B. Kwon studied a fading memory Kalman filter [25]. However, because this method uses thc empirical method to determine the fading factor, only suboptimal filtering results can be obtained [26]. As an improvement of the fading memory technique, the limited memory technique can handle the KF divergence caused by model inaccuracy or system noise changes. It uses a limited number of residuals near to the current time to avoid the influence of excessive historical information on system state estimation [27]. Based on the limited memory of residuals, system noise statistics were estimated to improve the KF accuracy [28,29]. However, this method involves a complex update process of state and measurement information, leading to an increased computational load. Wishnef et al. also adopted the limited memory technique to improve KF accuracy, leading to the limited memory-based KF (LM-KF) [30]. Although the limited memory technique reduces the contribution of historical residuals to system state estimation [29,30], since residuals at different epochs within a limited memory are applied with the same weight, it is difficult to distinguish the contributions of residuals at different epochs on system state estimation, leading to a limited improvement of the KF accuracy. Deng et al. also combined the limited memory technique with KF [31]. However, this method determines the initial filtering values based on empiricism. If the initial values are not determined properly, the filtering accuracy will seriously decrease [32]. Yang proposed an adaptive fitting method of system error based on limited memory theory. In this method, the size of system noise statistics is not distinguished, and the system noise and its residual are estimated by the arithmetic mean method (using the same weight) in the selected moving window, which leads to low filtering accuracy [33].

Random-weighted estimation is an advanced statistical calculation method. This method applies random weights to different samples to estimate target parameters. It can achieve unbiased estimation, is simple in computation, and can handle large samples without requiring the accurate distribution of target parameters. It has been used to solve many problems, such as dynamic navigation and positioning [34,35], multi-sensor data fusion [36], M-test in linear models [37], analysis of asymptotic properties of function distribution [38], and estimation of system model error [9,39,40]. Since the application of random weights to different samples provides a solution to address the disadvantage of the limited memory technique due to the use of the same weight to residuals, the combination of the random weighting with limited memory techniques provides a promising solution to estimate unknown system noise statistics for improving KF accuracy.

In this paper, a new random-weighted Kalman filtering method based on limited memory is proposed, and a random-weighted estimation method of system noise statistics is designed. In the moving window, this method adaptively adjusts the weight of noise statistics in limited memory, which suppresses the interference of system noise and excessive historical information on system state estimation, improves the accuracy of filtering calculation, and proves the unbiased nature of random-weighted estimation of system noise statistics. Finally, through simulation and practical experiments, it is proven that the filtering accuracy of the proposed LM-RWKF is much higher than that of KF and LM-KF.

The main contributions of this paper include the following: (i) the random-weighted theories are established based on the concept of limited memory for estimation of system noise statistics; and (ii) a new KF filter is developed by combining the limited memory-based random weighting theories into the KF framework for system state estimation with increased accuracy.

3. Computation Scheme

3.1. The Concept of Random Weight Estimation

Assume that $x_1, x_2, \cdots, x_n$ is a sequence of random variables of independent and identical distribution (iid) with the common distribution function $F(x)$, and its empirical distribution function is defined as follows:

$$F_n(x) = \frac{1}{n}\sum_{i=1}^{N} I_{(X_i \leq x)} \tag{1}$$

where $I_{(X_{i \leq} x)}$ is the indicator function.

The random weight estimate of $F_n(x)$ can be defined as follows:

$$T_n(x) = \sum_{i=1}^{N} \lambda_i I_{(X_{i \leq} x)} \tag{2}$$

where the random-weighted vector $(\lambda_1, \lambda_2, \cdots, \lambda_n)$ obeys Dirichlet distribution $\mathrm{D}(1, 1, \cdots, 1)$, that is, $\sum_{i=1}^{N} \lambda_i = 1$ and the joint density function of $(\lambda_1, \lambda_2, \cdots, \lambda_{n-1})$ is $f(\lambda_1, \lambda_2, \cdots, \lambda_{n-1}) = \Gamma(n)$, in which $\Gamma(\cdot)$ denotes the function, $(\lambda_1, \lambda_2, \cdots, \lambda_{n-1}) \in D_{n-1}$ and $D_{n-1} = \left\{(\lambda_1, \lambda_2, \cdots, \lambda_{n-1}) : \lambda_i \geq 0, i = 1, 2, \cdots n-1, \sum_{i=1}^{N} \lambda_i = 1\right\}$.

3.2. Limited Memory-Based Random-Weighted Estimations of System Noise Statistics

Consider the following dynamic discrete system:

$$\begin{cases} \boldsymbol{x}_{k+1} = \boldsymbol{\Phi}_k \boldsymbol{x}_k + \boldsymbol{w}_k \\ \boldsymbol{y}_k = \boldsymbol{H}_k \boldsymbol{x}_k + \boldsymbol{v}_k \end{cases} \tag{3}$$

where $\boldsymbol{x}_k \in \boldsymbol{R}^n$ is the n-dimensional system state vector at time k, $\boldsymbol{y}_k \in \boldsymbol{R}^m$ the m-dimensional measurement vector, $\boldsymbol{\Phi}_k$ the system state transition matrix, $\boldsymbol{H}_k$ the system measurement matrix, $\boldsymbol{w}_k$ the system process noise, and $\boldsymbol{v}_k$ the measurement noise.

Suppose the statistics of system process noise $\boldsymbol{w}_k$ are unknown, i.e.,

$$\begin{cases} E(\boldsymbol{w}_k) = \boldsymbol{a}_k \\ cov(\boldsymbol{w}_k, \boldsymbol{w}_j) = E\left(\boldsymbol{w}_k \boldsymbol{w}_j^{\mathrm{T}}\right) = \boldsymbol{A}_k \delta_{kj} \end{cases} \tag{4}$$

where $\boldsymbol{a}_k$ and $\boldsymbol{A}_k \geq 0$ are the unknown mean and covariance of the process noise, and δ_{kj} is the Kronecker$-\delta$ function.

Suppose the statistical properties of measurement noise $\boldsymbol{v}_k$ are unknown, i.e.,

$$\begin{cases} E(\boldsymbol{v}_k) = \boldsymbol{b}_k \\ cov(\boldsymbol{v}_k, \boldsymbol{v}_j) = E\left(\boldsymbol{v}_k \boldsymbol{v}_j^{\mathrm{T}}\right) = \boldsymbol{B}_k \delta_{kj} \end{cases} \tag{5}$$

where $\boldsymbol{b}_k$ and $\boldsymbol{B}_k \geq 0$ are the unknown mean and covariance of the measurement noise.

The means of the process and measurement noises are called the first-order noise statistics, while their covariances are called the second-order noise statistics.

In a limited memory period of length $\mathrm{N}(j = 1, 2, \cdots, N)$, the arithmetic mean estimation of the measurement noise mean can be expressed as follows:

$$\hat{\boldsymbol{b}}_k = \frac{1}{N} \sum_{j=1}^{N} \boldsymbol{b}_{k-j} \tag{6}$$

Applying the random-weighted concept to Equation (6), the random-weighted estimation of the measurement noise mean is as follows:

$$\hat{\boldsymbol{b}}_k^* = \sum_{j=1}^{N} \lambda_j \boldsymbol{b}_{k-j} \tag{7}$$

In the limited memory period, the arithmetic mean estimation of the measurement noise covariance can be expressed as follows:

$$\hat{\boldsymbol{B}}_k = \frac{1}{N} \sum_{j=1}^{N} \left[\left(\boldsymbol{b}_{k-j} - \hat{\boldsymbol{b}}_k\right) \left(\boldsymbol{b}_{k-j} - \hat{\boldsymbol{b}}_k\right)^{\mathrm{T}} \right] \tag{8}$$

Applying the random-weighted concept to Equation (8), the random weighting estimation of the measurement noise covariance can be obtained as follows:

$$\hat{\boldsymbol{B}}_k^* = \sum_{j=1}^{N} \lambda_j \left(\boldsymbol{b}_{k-j} - \hat{\boldsymbol{b}}_k\right) \left(\boldsymbol{b}_{k-j} - \hat{\boldsymbol{b}}_k\right)^{\mathrm{T}} \tag{9}$$

Define the measurement residual as follows:

$$\boldsymbol{\beta}_k = \boldsymbol{y}_k - \boldsymbol{H}_k \hat{\boldsymbol{x}}_k \tag{10}$$

According to the first formula in Equation (3), the measurement residual $\boldsymbol{\beta}_k$ can be written as follows:

$$\begin{aligned} \boldsymbol{\beta}_k &= \boldsymbol{y}_k - \boldsymbol{H}_k \hat{\boldsymbol{x}}_k \\ &= \boldsymbol{H}_k \boldsymbol{x}_k + \boldsymbol{v}_k - \boldsymbol{H}_k \hat{\boldsymbol{x}}_k \\ &= \boldsymbol{H}_k (\boldsymbol{x}_k - \hat{\boldsymbol{x}}_k) + \boldsymbol{v}_k \\ &= \boldsymbol{H}_k \widetilde{\boldsymbol{x}}_k + \boldsymbol{v}_k \end{aligned} \tag{11}$$

where $\widetilde{\boldsymbol{x}}_k = \boldsymbol{x}_k - \hat{\boldsymbol{x}}_k$ denotes the state estimation error.

Similarly, in the limited memory period of length $N(j = 1, 2, \cdots, N)$, the arithmetic mean estimation of the process noise mean can be expressed as follows:

$$\hat{a}_k = \frac{1}{N}\sum_{j=1}^{N} a_{k-j} \tag{12}$$

The arithmetic mean estimation of the process noise covariance can be expressed as follows:

$$\hat{A}_k = \frac{1}{N}\sum_{j=1}^{N}\left[\left(a_{k-j} - \hat{a}_k\right)\left(a_{k-j} - \hat{a}_k\right)^{\mathrm{T}}\right] \tag{13}$$

Applying the random-weighted concept to Equations (12) and (13), the random-weighted estimation of a_k and A_k can be written as follows:

$$\hat{a}_k^{*} = \sum_{j=1}^{N} \lambda_j a_{k-j} \tag{14}$$

$$\hat{A}_k^{*} = \sum_{j=1}^{N} \lambda_j \left[\left(a_{k-j} - \hat{a}_k\right)\left(a_{k-j} - \hat{a}_k\right)^{\mathrm{T}}\right] \tag{15}$$

Define the process residual as follows:

$$\delta_k = x_{k+1} - \Phi_k \hat{x}_k \tag{16}$$

According to the second formula in Equation (3), the process residual δ_k can be rewritten as follows:

$$\begin{aligned} \delta_k &= x_{k+1} - \Phi_k \hat{x}_k \\ &= (\Phi_k x_k + w_k) - \Phi_k \hat{x}_k \\ &= \Phi_k (x_k - \hat{x}_k) + w_k \\ &= \Phi_k \tilde{x} + w_k \end{aligned} \tag{17}$$

According to the KF principle, we have the following:

$$D_k = P_k - P_{k|k-1} = P_k - \Phi_k P_{k-1} \Phi_k - A_k \tag{18}$$

where D_k is the process residual covariance, and $P_{k|k-1}$ is the one-step state estimation error variance, which is expressed as follows:

$$P_{k|k-1} = \Phi_k P_{k-1} \Phi_k + A_{k-1} \tag{19}$$

P_k is the state error covariance, which is expressed as follows:

$$P_k = (I - K_k H_k) P_{k|k-1} (I - K_k H_k)^{\mathrm{T}} + \hat{B}_k \tag{20}$$

where K_k is filter gain matrix, which is expressed as follows:

$$K_k = P_{k|k-1} H_k^{\mathrm{T}} \left(H_k P_{k|k-1} H_k^{\mathrm{T}} + \hat{B}_k\right)^{-1} \tag{21}$$

Equations (7), (9), (14), and (15) provide the random-weighted estimations of the process noise statistics and measurement noise statistics, which allow us to adaptively adjust the random weights to suppress the interferences of the process and measurement noises on the state estimation for improving the KF accuracy.

3.3. Unbiasedness of Random-Weighted Estimations of System Noise Statistics

Theorem 1. *The random-weighted estimations* $\hat{\boldsymbol{b}}_k^*$ *and* $\hat{\boldsymbol{a}}_k^*$ *of* $\boldsymbol{b}_k$ *and* $\boldsymbol{a}_k$*, which are given by (7) and (14), are suboptimally unbiased.*

Proof of Theorem 1. From Equation (7), we have the following:

$$\begin{aligned} E\left(\hat{\boldsymbol{b}}_k^*\right) &= E\left(\sum_{j=1}^{N} \lambda_j \boldsymbol{b}_{k-j}\right) \\ &= \sum_{j=1}^{N} \lambda_j E\left(\boldsymbol{b}_{k-j}\right) \\ &\neq \hat{\boldsymbol{b}}_k \end{aligned} \tag{22}$$

where $\sum_{j=1}^{N} \lambda_j = 1$ is used in the last step derivation.

It can be seen from Equation (22) that $\hat{\boldsymbol{b}}_k^*$ is not the optimal unbiased estimation of $\boldsymbol{b}_k$.

If the measurement noise is constant or involves small variations in the limited memory, i.e., $\boldsymbol{b}_k = \boldsymbol{b}_{k-j}$, Equation (22) can be further written as follows:

$$\begin{aligned} E\left(\hat{\boldsymbol{b}}_k^*\right) &= \sum_{j=1}^{N} \lambda_j E\left(\boldsymbol{b}_{k-j}\right) \\ &= \hat{\boldsymbol{b}}_k \end{aligned} \tag{23}$$

It is known from Equations (22) and (23) that the random weighting estimation $\hat{\boldsymbol{b}}_k^*$ of $\boldsymbol{b}_k$ is suboptimally unbiased.

Similarly, from (14), we have the following:

$$\begin{aligned} E\left(\hat{\boldsymbol{a}}_k^*\right) &= E\left(\sum_{j=1}^{N} \lambda_j \boldsymbol{a}_{k-j}\right) \\ &= \sum_{j=1}^{N} \lambda_j E\left(\boldsymbol{a}_{k-j}\right) \\ &\neq \boldsymbol{a}_k \end{aligned} \tag{24}$$

It can be seen from Equation (24) that $\hat{\boldsymbol{a}}_k^*$ is not the optimal unbiased estimation of $\boldsymbol{a}_k$.

If the process noise is constant or involves small variations in the limited memory, i.e., $\boldsymbol{a}_k = \boldsymbol{a}_{k-j}$, Equation (24) can be further written as follows:

$$\begin{aligned} E\left(\hat{\boldsymbol{a}}_k^*\right) &= \sum_{j=1}^{N} \lambda_j E\left(\boldsymbol{a}_{k-j}\right) \\ &= \boldsymbol{a}_k \end{aligned} \tag{25}$$

It is known from Equations (24) and (25) that the random-weighted estimation $\hat{\boldsymbol{a}}_k^*$ of $\boldsymbol{a}_k$ is suboptimally unbiased.

The proof of Theorem 1 is completed. □

Theorem 2. *The random-weighted estimations* $\hat{\boldsymbol{B}}_k^*$ *and* $\hat{\boldsymbol{A}}_k^*$ *of* $\boldsymbol{B}_k$ *and* $\boldsymbol{A}_k$*, which are given by (9) and (15), are suboptimally unbiased.*

Proof of Theorem 2. Since the state estimate $\hat{\boldsymbol{x}}_k$ from KF is unbiased, we have the following:

$$\begin{aligned} E(\tilde{\boldsymbol{x}}_k) &= E(\boldsymbol{x}_k - \hat{\boldsymbol{x}}_k) \\ &= E(\boldsymbol{x}_k) - E(\hat{\boldsymbol{x}}_k) \\ &= 0 \end{aligned} \tag{26}$$

By Equation (11), we have the following:

$$\begin{aligned} E(\boldsymbol{\beta}_k) &= E(\boldsymbol{H}_k\widetilde{\boldsymbol{x}}_k + \boldsymbol{v}_k) \\ &= E(\boldsymbol{H}_k\widetilde{\boldsymbol{x}}_k) + E(\boldsymbol{v}_k) \\ &= \boldsymbol{b}_k \end{aligned} \tag{27}$$

Calculate the measurement residual covariance:

$$\begin{aligned} & E\left\{[\boldsymbol{\beta}_k - E(\boldsymbol{\beta}_k)][\boldsymbol{\beta}_k - E(\boldsymbol{\beta}_k)\,]^{\mathrm{T}}\right\} \\ &= E\left[(\boldsymbol{\beta}_k - \boldsymbol{b}_k)(\boldsymbol{\beta}_k - \boldsymbol{b}_k)^{\mathrm{T}}\,\right] \\ &= E(\boldsymbol{\beta}_k\boldsymbol{\beta}_k^{\mathrm{T}} - \boldsymbol{\beta}_k\boldsymbol{b}_k^{\mathrm{T}} - \boldsymbol{b}_k\,\boldsymbol{\beta}_k^{\mathrm{T}} + \boldsymbol{b}_k\boldsymbol{b}_k^{\mathrm{T}}) \\ &= E\left[(\boldsymbol{H}_k\widetilde{\boldsymbol{x}}_k + \boldsymbol{v}_k)(\boldsymbol{H}_k\widetilde{\boldsymbol{x}}_k + \boldsymbol{v}_k)^{\mathrm{T}}\right] \\ &\quad - E\left[(\boldsymbol{H}_k\widetilde{\boldsymbol{x}}_k + \boldsymbol{v}_k)\boldsymbol{b}_k^{\mathrm{T}}\right] - E\left[\boldsymbol{b}_k(\boldsymbol{H}_k\widetilde{\boldsymbol{x}}_k + \boldsymbol{v}_k)^{\mathrm{T}}\right] \\ &\quad + E(\boldsymbol{b}_k\boldsymbol{b}_k^{\mathrm{T}}) \\ &= E(\boldsymbol{H}_k\widetilde{\boldsymbol{x}}_k\widetilde{\boldsymbol{x}}_k^{\mathrm{T}}\boldsymbol{H}_k^{\mathrm{T}}) + E\left[(\boldsymbol{v}_k - \boldsymbol{b}_k)(\boldsymbol{v}_k - \boldsymbol{b}_k)^{\mathrm{T}}\right] \\ &\quad + E(\boldsymbol{H}_k\widetilde{\boldsymbol{x}}_k\ \boldsymbol{b}_k^{\mathrm{T}}) + E\left(\boldsymbol{v}_k\widetilde{\boldsymbol{x}}_k^{\mathrm{T}}\boldsymbol{H}_k^{\mathrm{T}}\right) - E\left(\boldsymbol{H}_k\widetilde{\boldsymbol{x}}_k\ \boldsymbol{b}_k^{\mathrm{T}}\right) \\ &\quad - E\left(\boldsymbol{b}_k\widetilde{\boldsymbol{x}}_k^{\mathrm{T}}\boldsymbol{H}_k^{\mathrm{T}}\right) \\ &= E\left(\boldsymbol{H}_k\widetilde{\boldsymbol{x}}_k\widetilde{\boldsymbol{x}}_k^{\mathrm{T}}\boldsymbol{H}_k^{\mathrm{T}}\right) + \boldsymbol{B}_k + E(\boldsymbol{H}_k\widetilde{\boldsymbol{x}}_k\ \boldsymbol{v}_k^{\mathrm{T}}) \\ &\quad + E\left(\boldsymbol{v}_k\widetilde{\boldsymbol{x}}_k^{\mathrm{T}}\boldsymbol{H}_k^{\mathrm{T}}\right) - E\left(\boldsymbol{B}_k\widetilde{\boldsymbol{x}}_k\ \boldsymbol{b}_k^{\mathrm{T}}\right) - E\left(\boldsymbol{b}_k\widetilde{\boldsymbol{x}}_k^{\mathrm{T}}\boldsymbol{H}_k^{\mathrm{T}}\right) \end{aligned} \tag{28}$$

where $\boldsymbol{B}_k = E\left[(\boldsymbol{v}_k - \boldsymbol{b}_k)(\boldsymbol{v}_k - \boldsymbol{b}_k)^{\mathrm{T}}\right]$.

Since $\widetilde{\boldsymbol{x}}_k$ and $\boldsymbol{v}_k$ are independent each other, by Equation (26), we have the following:

$$E\left(\boldsymbol{H}_k\widetilde{\boldsymbol{x}}_k\boldsymbol{v}_k^{\mathrm{T}}\right) = E\left(\boldsymbol{v}_k\widetilde{\boldsymbol{x}}_k^{\mathrm{T}}\boldsymbol{H}_k^{\mathrm{T}}\right) = 0 \tag{29}$$

and the following:

$$E\left(\boldsymbol{H}_k\widetilde{\boldsymbol{x}}_k\boldsymbol{b}_k^{\mathrm{T}}\right) = E\left(\boldsymbol{b}_k\widetilde{\boldsymbol{x}}_k^{\mathrm{T}}\boldsymbol{H}_k^{\mathrm{T}}\right) = 0 \tag{30}$$

Substituting Equations (29) and (30) into (28) yields the following:

$$\begin{aligned} E\left[(\boldsymbol{\beta}_k - \boldsymbol{b}_k)(\boldsymbol{\beta}_k - \boldsymbol{b}_k)^{\mathrm{T}}\,\right] &= E(\boldsymbol{H}_k\widetilde{\boldsymbol{x}}_k\widetilde{\boldsymbol{x}}_k^{\mathrm{T}}\boldsymbol{H}_k^{\mathrm{T}}) + \boldsymbol{B}_k \\ &= \boldsymbol{H}_k\boldsymbol{P}_k\boldsymbol{H}_k^{\mathrm{T}} + \boldsymbol{B}_k \end{aligned} \tag{31}$$

where $\boldsymbol{P}_k = E\left[\widetilde{\boldsymbol{x}}_k\,\widetilde{\boldsymbol{x}}_k^{\mathrm{T}}\right]$ is the state error covariance at time k.

From Equation (31), in the limited memory, the arithmetic mean estimation of $\boldsymbol{B}_k$ can be calculated as follows:

$$\begin{aligned} \hat{\boldsymbol{B}}_k &= \frac{1}{N}\sum_{j=1}^{N}\boldsymbol{B}_{k-j} \\ &= \frac{1}{N}\sum_{j=1}^{N}\left[(\boldsymbol{\beta}_{k-j} - \hat{\boldsymbol{b}}_{k-j})(\boldsymbol{\beta}_{k-j} - \hat{\boldsymbol{b}}_{k-j})^{\mathrm{T}} - \boldsymbol{H}_{k-j}\boldsymbol{P}_{k-j}\boldsymbol{H}_{k-j}^{\mathrm{T}}\right] \end{aligned} \tag{32}$$

The random-weighted estimation of $\boldsymbol{B}_k$ can be written as follows:

$$\hat{\boldsymbol{B}}_k^* = \sum_{j=1}^{N}\lambda_j\left[(\boldsymbol{\beta}_{k-j} - \hat{\boldsymbol{b}}_{k-j}^*)(\boldsymbol{\beta}_{k-j} - \hat{\boldsymbol{b}}_{k-j}^*)^{\mathrm{T}} - \boldsymbol{H}_{k-j}\boldsymbol{P}_{k-j}\boldsymbol{H}_{k-j}^{\mathrm{T}}\right] \tag{33}$$

Taking the mathematical expectation on both sides of Equation (33) generates the following:

$$\begin{aligned} E\left(\hat{\boldsymbol{B}}_k^*\right) &= \sum_{j=1}^{N} \lambda_j E\left[(\boldsymbol{\beta}_{k-j} - \hat{\boldsymbol{b}}_{k-j}^*)(\boldsymbol{\beta}_{k-j} - \hat{\boldsymbol{b}}_{k-j}^*)^{\mathrm{T}} - \boldsymbol{H}_{k-j}\boldsymbol{P}_{k-j}\boldsymbol{H}_{k-j}^{\mathrm{T}}\right] \\ &= \sum_{j=1}^{N} \lambda_j \boldsymbol{H}_{k-j}\boldsymbol{P}_{k-j}\boldsymbol{H}_{k-j}^{\mathrm{T}} + \sum_{j=1}^{M} \lambda_j \boldsymbol{B}_{k-j} - \sum_{j=1}^{M} \lambda_j \boldsymbol{H}_{k-j}\boldsymbol{P}_{k-j}\boldsymbol{H}_{k-j}^{\mathrm{T}} \\ &= \sum_{j=1}^{N} \lambda_j \boldsymbol{B}_{k-j} \\ &\neq \boldsymbol{B}_k \end{aligned} \quad (34)$$

It can be seen from Equation (34) that $\hat{\boldsymbol{B}}_k^*$ is not the optimal unbiased estimation of $\boldsymbol{B}_k$.

If the measurement noise is constant or involves small variations in the limited memory period, i.e., $\boldsymbol{B}_k = \boldsymbol{B}_{k-j}$, Equation (34) can be further written as follows:

$$\begin{aligned} E(\hat{\boldsymbol{B}}_k^*) &= \sum_{j=1}^{N} \lambda_j \boldsymbol{B}_{k-j} \\ &= \sum_{j=1}^{N} \lambda_j \boldsymbol{B}_k \\ &= \boldsymbol{B}_k \end{aligned} \quad (35)$$

It is known from (34) and (35) that the random-weighted estimation $\hat{\boldsymbol{B}}_k^*$ of $\boldsymbol{B}_k$ is suboptimally unbiased.

Now let us study the unbiasedness for the random-weighted estimation $\hat{\boldsymbol{A}}_k^*$ of $\boldsymbol{A}_k$.

According to Equation (17), we have the following:

$$\begin{aligned} E(\boldsymbol{\delta}_k) &= E(\boldsymbol{\Phi}_k\widetilde{\boldsymbol{x}}_k + \boldsymbol{w}_k) \\ &= E(\boldsymbol{\Phi}_k\widetilde{\boldsymbol{x}}_k) + E(\boldsymbol{w}_k) \\ &= \boldsymbol{a}_k \end{aligned} \quad (36)$$

Calculate the process residual covariance:

$$\begin{aligned} & E\left\{[\boldsymbol{\delta}_k - E(\boldsymbol{\delta}_k)][\boldsymbol{\delta}_k - E(\boldsymbol{\delta}_k)]^{\mathrm{T}}\right\} \\ &= E\left[(\boldsymbol{\delta}_k - \boldsymbol{a}_k)(\boldsymbol{\delta}_k - \boldsymbol{a}_k)^{\mathrm{T}}\right] \\ &= E(\boldsymbol{\delta}_k\boldsymbol{\delta}_k^{\mathrm{T}} - \boldsymbol{\delta}_k\boldsymbol{a}_k^{\mathrm{T}} - \boldsymbol{a}_k\boldsymbol{\delta}_k^{\mathrm{T}} + \boldsymbol{a}_k\boldsymbol{a}_k^{\mathrm{T}}) \\ &= E\left[(\boldsymbol{\Phi}_k\widetilde{\boldsymbol{x}}_k + \boldsymbol{w}_k)(\boldsymbol{\Phi}_k\widetilde{\boldsymbol{x}}_k + \boldsymbol{w}_k)^{\mathrm{T}}\right] - E\left[(\boldsymbol{\Phi}_k\widetilde{\boldsymbol{x}}_k + \boldsymbol{w}_k)\boldsymbol{a}_k^{\mathrm{T}}\right] \\ &\quad -E\left[\boldsymbol{a}_k(\boldsymbol{\Phi}_k\widetilde{\boldsymbol{x}}_k + \boldsymbol{w}_k)^{\mathrm{T}}\right] + E(\boldsymbol{a}_k\boldsymbol{a}_k^{\mathrm{T}}) \\ &= E(\boldsymbol{\Phi}_k\widetilde{\boldsymbol{x}}_k\widetilde{\boldsymbol{x}}_k^{\mathrm{T}}\boldsymbol{\Phi}_k^{\mathrm{T}}) + E\left[(\boldsymbol{w}_k - \boldsymbol{a}_k)(\boldsymbol{w}_k - \boldsymbol{a}_k)^{\mathrm{T}}\right] \\ &\quad +E(\boldsymbol{\Phi}_k\widetilde{\boldsymbol{x}}_k\,\boldsymbol{w}_k^{\mathrm{T}}) + E(\boldsymbol{w}_k\widetilde{\boldsymbol{x}}_k^{\mathrm{T}}\boldsymbol{\Phi}_k^{\mathrm{T}}) - E(\boldsymbol{\Phi}_k\widetilde{\boldsymbol{x}}_k\,\boldsymbol{a}_k^{\mathrm{T}}) \\ &\quad -E(\boldsymbol{a}_k\widetilde{\boldsymbol{x}}_k^{\mathrm{T}}\boldsymbol{\Phi}_k^{\mathrm{T}}) \\ &= E(\boldsymbol{\Phi}_k\widetilde{\boldsymbol{x}}_k\widetilde{\boldsymbol{x}}_k^{\mathrm{T}}\boldsymbol{\Phi}_k^{\mathrm{T}}) + \boldsymbol{A}_k + E(\boldsymbol{\Phi}_k\widetilde{\boldsymbol{x}}_k\,\boldsymbol{w}_k^{\mathrm{T}}) \\ &\quad +E(\boldsymbol{w}_k\widetilde{\boldsymbol{x}}_k^{\mathrm{T}}\boldsymbol{\Phi}_k^{\mathrm{T}}) - E(\boldsymbol{\Phi}_k\widetilde{\boldsymbol{x}}_k\,\boldsymbol{a}_k^{\mathrm{T}}) - E(\boldsymbol{a}_k\widetilde{\boldsymbol{x}}_k^{\mathrm{T}}\boldsymbol{\Phi}_k^{\mathrm{T}}) \end{aligned} \quad (37)$$

where $\boldsymbol{A}_k = E\left[(\boldsymbol{w}_k - \boldsymbol{a}_k)(\boldsymbol{w}_k - \boldsymbol{a}_k)^{\mathrm{T}}\right]$.

Since $\widetilde{\boldsymbol{x}}_k$ and $\boldsymbol{w}_k$ are independent each other, according to Equation (26), we have the following:

$$E\left(\boldsymbol{\Phi}_k\widetilde{\boldsymbol{x}}_k\boldsymbol{w}_k^{\mathrm{T}}\right) = E\left(\boldsymbol{w}_k\widetilde{\boldsymbol{x}}_k^{\mathrm{T}}\boldsymbol{\Phi}_k^{\mathrm{T}}\right) = 0 \quad (38)$$

and the following:

$$E\left(\boldsymbol{\Phi}_k\widetilde{\boldsymbol{x}}_k\boldsymbol{a}_k^{\mathrm{T}}\right) = E\left(\boldsymbol{a}_k\widetilde{\boldsymbol{x}}_k^{\mathrm{T}}\boldsymbol{\Phi}_k^{\mathrm{T}}\right) = 0 \tag{39}$$

Substituting Equations (38) and (39) into Equation (37) yields the following:

$$\begin{aligned} E\left[(\boldsymbol{\delta}_k - \boldsymbol{a}_k)(\boldsymbol{\delta}_k - \boldsymbol{a}_k)^{\mathrm{T}}\right] &= E(\boldsymbol{\Phi}_k\widetilde{\boldsymbol{x}}_k\widetilde{\boldsymbol{x}}_k^{\mathrm{T}}\boldsymbol{\Phi}_k^{\mathrm{T}}) + \boldsymbol{A}_k \\ &= \boldsymbol{\Phi}_k\boldsymbol{P}_k\boldsymbol{\Phi}_k^{\mathrm{T}} + \boldsymbol{A}_k \end{aligned} \tag{40}$$

where $\boldsymbol{P}_k = E\left[\widetilde{\boldsymbol{x}}_k\widetilde{\boldsymbol{x}}_k^{\mathrm{T}}\right]$ is the state error covariance at time k.

The arithmetic mean estimation of $\boldsymbol{A}_k$ can be calculated as

$$\begin{aligned} \hat{\boldsymbol{A}}_k &= \frac{1}{N}\sum_{j=1}^{N}\boldsymbol{A}_{k-j} \\ &= \frac{1}{N}\sum_{j=1}^{N}\left[(\boldsymbol{\delta}_{k-j} - \hat{\boldsymbol{a}}_{k-j})(\boldsymbol{\delta}_{k-j} - \hat{\boldsymbol{a}}_{k-j})^{\mathrm{T}} - \boldsymbol{\Phi}_{k-j}\boldsymbol{P}_{k-j}\boldsymbol{\Phi}_{k-j}^{\mathrm{T}}\right] \end{aligned} \tag{41}$$

By Equation (41), the random-weighted estimation of $\boldsymbol{A}_k$ can be written as follows:

$$\hat{\boldsymbol{A}}_k^* = \sum_{j=1}^{N}\lambda_i\left[(\boldsymbol{\delta}_{k-j} - \hat{\boldsymbol{a}}_{k-j}^*)(\boldsymbol{\delta}_{k-j} - \hat{\boldsymbol{a}}_{k-j}^*)^{\mathrm{T}} - \boldsymbol{\Phi}_{k-j}\boldsymbol{P}_{k-j}\boldsymbol{\Phi}_{k-j}^{\mathrm{T}}\right] \tag{42}$$

Taking the mathematical expectation on both sides of Equation (42) generates the following:

$$\begin{aligned} E\left(\hat{\boldsymbol{A}}_k^*\right) &= \sum_{j=1}^{N}\lambda_i E\left[(\boldsymbol{\delta}_{k-j} - \hat{\boldsymbol{a}}_{k-j}^*)(\boldsymbol{\delta}_{k-j} - \hat{\boldsymbol{a}}_{k-j}^*)^{\mathrm{T}} - \boldsymbol{\Phi}_{k-j}\boldsymbol{P}_{k-j}\boldsymbol{\Phi}_{k-j}^{\mathrm{T}}\right] \\ &= \sum_{j=1}^{N}\lambda_j\boldsymbol{\Phi}_{k-j}\boldsymbol{P}_{k-j}\boldsymbol{\Phi}_{k-j}^{\mathrm{T}} + \sum_{j=1}^{M}\lambda_j\boldsymbol{A}_{k-j} - \sum_{j=1}^{M}\lambda_j\boldsymbol{\Phi}_{k-j}\boldsymbol{P}_{k-j}\boldsymbol{\Phi}_{k-j}^{\mathrm{T}} \\ &= \sum_{j=1}^{N}\lambda_j\boldsymbol{A}_{k-j} \\ &\neq \boldsymbol{A}_k \end{aligned} \tag{43}$$

It can be seen from Equation (43) that $\hat{\boldsymbol{A}}_k^*$ is not the optimal unbiased estimation of $\boldsymbol{A}_k$.

If the process noise is constant or involves small variations in the limited memory, i.e., $\boldsymbol{A}_k = \boldsymbol{A}_{k-j}$, Equation (43) can be further written as follows:

$$\begin{aligned} E(\hat{\boldsymbol{A}}_k^*) &= \sum_{j=1}^{N}\lambda_j\boldsymbol{A}_{k-j} \\ &= \sum_{j=1}^{N}\lambda_j\boldsymbol{A}_k \\ &= \boldsymbol{A}_k \end{aligned} \tag{44}$$

It is known from Equations (43) and (44) that $\hat{\boldsymbol{A}}_k^*$ are the random-weighted suboptimal unbiased estimation of $\boldsymbol{A}_k$.

The proof of Theorem 2 is completed. □

Based on above, the overview diagram of LM-RWKF is as Figure 1, and the procedure of the proposed LM-RWKF is as follows:

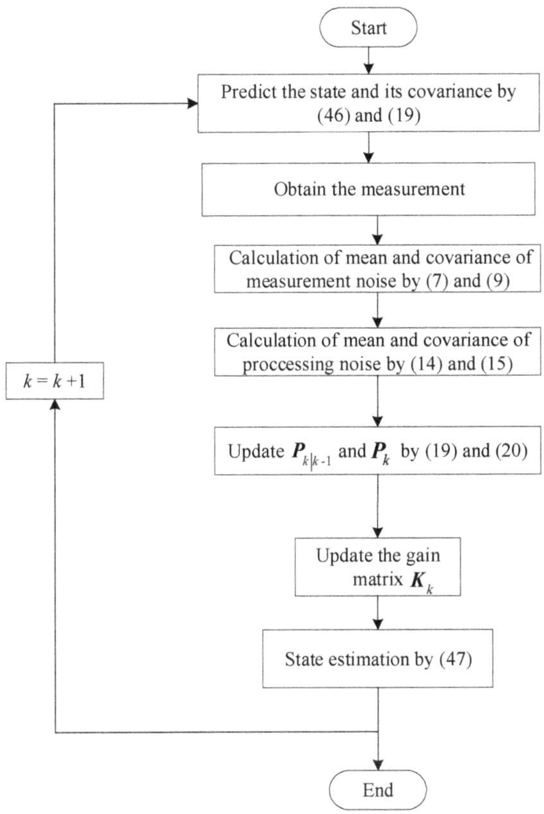

Figure 1. Overview diagram of LM-RWKF.

The procedure of the proposed LM-RWKF is as follows:

(i) Initialize the estimated state and its associated error covariance:

$$\begin{aligned} \hat{x}_0 &= E[x_0] \\ P_0 &= \mathrm{cov}(x_0, x_0^{\mathrm{T}}) = E[(x_0 - \hat{x}_0)(x_0 - \hat{x}_0)^{\mathrm{T}}] \end{aligned} \quad (45)$$

(ii) Calculate predicted state vector:

$$x_{k|k-1} = \Phi_{k|k-1}\hat{x}_{k-1} \quad (46)$$

(iii) Calculate the one-step prediction covariance by (19).
(iv) Estimate the mean and covariance of the process and measurement noise statistics by (7), (9), (14), and (15).
(v) The process and measurement noise statistics are fed back to (19)–(21) to obtain a new filter gain matrix.
(vi) Calculate the new state estimation vector.

$$\hat{x}_k^* = x_{k|k-1}^* + K_k\left(y_k - H_k x_k^* - \hat{b}_k^*\right) \quad (47)$$

(vii) Let k = k + 1 return to (ii) until all iterations are complete.

4. Performance Evaluation and Discussion

Simulations and experiments were conducted to comprehensively evaluate the performance of the proposed LM-RWKF for dynamic vehicle navigation. The comparison analysis of the proposed LM-RWKF with KF and LM-KF [27] was also conducted to demonstrate the improved performance.

4.1. Simulations and Analysis

Computational simulations were conducted to verify the proposed LM-RWKF for tracking the motion of a moving object. The object moves along the three coordinate axes X, Y and Z as per the following equation:

$$\begin{cases} P_x = 10t + \frac{1}{2}a_x t^2 \\ P_y = 10t + \frac{1}{2}a_y t^2 \\ P_z = 10t + \frac{1}{2}a_z t^2 \end{cases} \tag{48}$$

where t is time, (P_x, P_y, P_z) represents the vehicle position, and (a_x, a_y, a_z) denotes the vehicle acceleration.

The state vector is as follows:

$$\boldsymbol{X} = (\boldsymbol{P}_x,\ \boldsymbol{P}_y,\ \boldsymbol{P}_z) \tag{49}$$

and the processing noise is expressed as follows:

$$w = (a_x, a_y, a_z) \tag{50}$$

The moving object is observed in the three axes:

$$\begin{cases} L_x = P_x + v_x \\ L_y = P_y + v_y \\ L_z = P_z + v_z \end{cases} \tag{51}$$

where (L_x, L_y, L_z) represents the measurement vector, and (v_x, v_y, v_z) denotes the measurement noise.

The simulation time was 1200 s and the sampling cycle was 1 s. The limited memory length was set to N = 20. The trajectory of the moving object is shown in Figure 2. The initial parameters are given in Table 1.

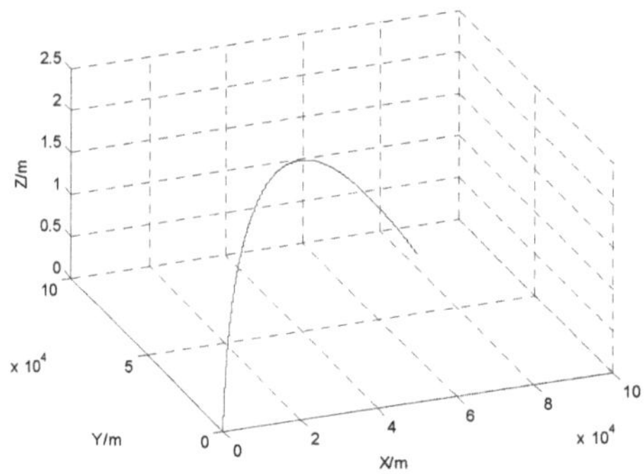

Figure 2. Trajectory of the moving object.

Table 1. Initial parameters.

Parameters		Value
Initial position	East longitude	108.736°
	North latitude	34.246°
	Altitude	3500 m
Initial velocity	East	0 m/s
	North	100 m/s
	Up	0 m/s
Initial attitude	Pitch	0°
	Roll	0°
	Yaw	0°
Initial position error	East longitude	10 m
	North latitude	10 m
	Altitude	10 m
Initial velocity error	East	0.5 m/s
	North	0.5 m/s
	Up	0.5 m/s
Initial attitude error	Pitch	1′
	Roll	1′
	Yaw	1.5′
Initial variances	Position	0.2 m
	Velocity	$9.0 \times 10^{-6} \mathrm{m}^2\mathrm{s}^{-2}$

4.1.1. Estimation of System Noise Statistics

Simulation trials were conducted to examine the accuracy of LM–RWKF for estimation of the measurement noise statistics. The initial state and its estimate were set as follows:

$$x_0 = \begin{bmatrix} 0.1 & 0.1 & 0.1 \end{bmatrix},\ \hat{x}_0 = \begin{bmatrix} 0.1 & 0.1 & 0.1 \end{bmatrix} \tag{52}$$

The initial estimation error covariance was chosen as follows:

$$\hat{\boldsymbol{P}}_0 = \begin{bmatrix} 1 & 0 & 0 \\ 0 & 1 & 0 \\ 0 & 0 & 1 \end{bmatrix} \tag{53}$$

The true values and initial estimates of the measurement noise mean and covariance were as follows:

$$\boldsymbol{b}_k = [1,\ 1,\ 1],\ \boldsymbol{B}_k = \begin{bmatrix} 15 & 0 & 0 \\ 0 & 15 & 0 \\ 0 & 0 & 15 \end{bmatrix} \tag{54}$$

$$\hat{\boldsymbol{b}}_0 = \begin{bmatrix} 0.2 & 0.2 & 0.2 \end{bmatrix},\ \hat{\boldsymbol{B}}_0 = \begin{bmatrix} 0.2 & 0 & 0 \\ 0 & 0.2 & 0 \\ 0 & 0 & 0.2 \end{bmatrix} \tag{55}$$

The true values and initial estimates of the process noise mean and covariance were as follows:

$$\boldsymbol{a}_k = [0, 0, 0],\ \boldsymbol{A}_k = \begin{bmatrix} 3 & 0 & 0 \\ 0 & 3 & 0 \\ 0 & 0 & 3 \end{bmatrix} \tag{56}$$

$$\hat{\boldsymbol{a}}_0 = [0, 0, 0],\ \hat{\boldsymbol{A}}_0 = \begin{bmatrix} 1 & 0 & 0 \\ 0 & 1 & 0 \\ 0 & 0 & 1 \end{bmatrix} \tag{57}$$

Figures 3 and 4 illustrate the measurement noise means and covariances estimated by both LM–KF and LM–RWKF, and their mean absolute errors (MAEs) are listed in Table 2. As shown in Figure 3, the LM–KF estimation curve of the measurement noise mean involves large-magnitude oscillations, leading to the 0.564 m MAE. By contrary, the oscillations involved in the LM–RWKF estimation curve are much smaller in magnitude than those in the LM–KF estimation curve, leading to the 0.161 m MAE. As shown in Figure 4, both LM–KF and LM–RWKF estimations of the measurement noise covariance exhibit a similar trend in the case of the measurement noise mean estimation. The MAE of the measurement noise covariance estimated by LM–KF is 8.137 m^2, while the MAE by LM–RWKF is 1.061 m^2, which is very close to the true value.

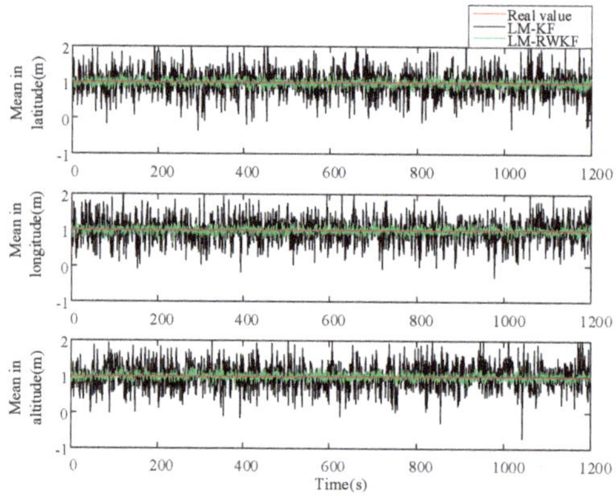

Figure 3. Estimations of the measurement noise mean by LM–KF and LM–RWKF.

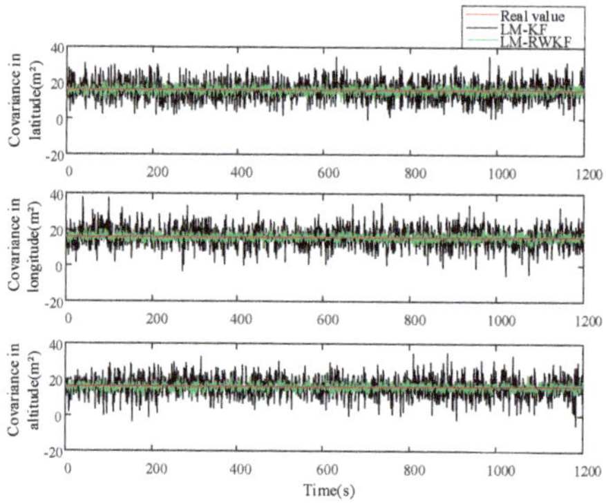

Figure 4. Estimations of the measurement noise covariance by LM–KF and LM–RWKF.

Table 2. MAEs of the system noise statistics estimated by LM–KF and LM–RWKF.

Filtering Methods	Measurement Noise Mean	Measurement Noise Covariance
LM–KF	0.564 m	8.137 m^2
LM–RWKF	0.161 m	1.061 m^2

Figures 5 and 6 show the process noise means and covariances estimated by both LM–KF and LM-RWKF, while their MAEs are listed in Table 3. As shown in Figure 5, the LM–KF estimation curve of the process noise mean involves large-magnitude oscillations, leading to the MAE of 0.122 m. By contrast, the oscillations involved in the LM–RWKF estimation curve are much smaller in magnitude than those in the LM–KF estimation curve, leading to the MAE of 0.057 m. As shown in Figure 6, the LM–KF estimation of the process noise covariance involves large-magnitude fluctuations, leading to the MAE of 1.114 m^2. Nevertheless, the MAE of the process noise covariance estimated by LM–RWKF is 0.369 m^2, much smaller than that of LM-KF.

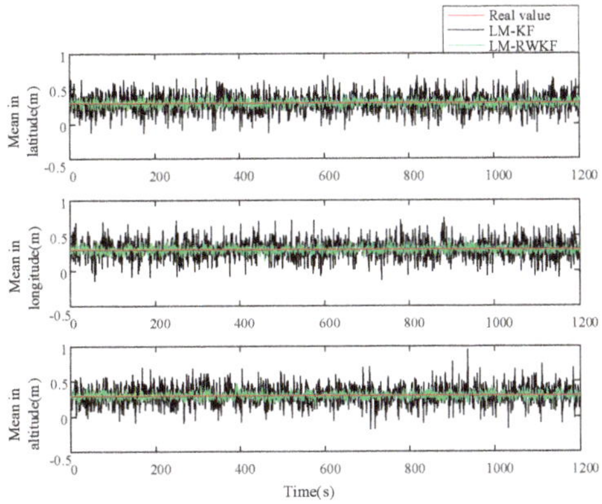

Figure 5. Estimations of the process noise mean by LM–KF and LM–RWKF.

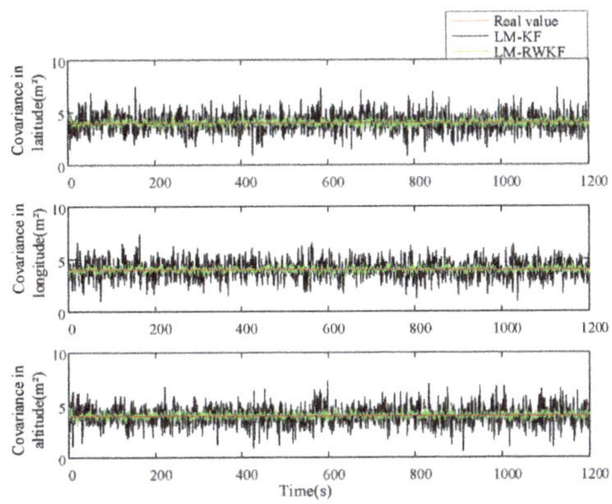

Figure 6. Estimations of the process noise covariance by LM–KF and LM–RWKF.

Table 3. MAEs of the process noise statistics estimated by LM–KF and LM–RWKF.

Filtering Methods	Measurement Noise Mean	Measurement Noise Covariance
LM–KF	0.122 m	1.114 m^2
LM–RWKF	0.057 m	0.369 m^2

The above results demonstrate that LM–RWKF can effectively estimate system noise statistics, leading to higher estimation accuracy than LM–KF.

4.1.2. Vehicle Position Estimation

The position error of the moving vehicle is also estimated under the same conditions by both KF, LM–KF, and LM–RWKF. Figure 7 shows the position error curves of KF, LM–KF, and LM–RWKF. Table 4 also provides the position MAEs of these three methods. As shown in Figure 7, the KF estimation curve involves large-magnitude oscillations, leading to the 8.515 m MAE. The oscillation magnitude is decreased by LM–KF due to its capability of noise statistics estimation. However, since it applies the same weight to the noise statistics within a limited memory, the LM–KF improvement is still limited, leading to the 4.953 m MAE. By contrast, since LM–RWKF adopts the RW estimation of system noise statistics, its mean absolute error in position is 1.235 m, which is much smaller than those of KF and LM–KF.

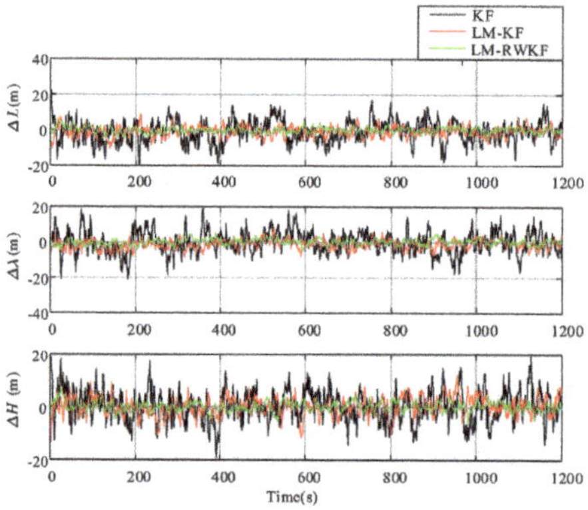

Figure 7. Moving object's position errors estimated by KF, LM–KF and LM–RWKF.

Table 4. Moving object's position MAEs by KF, LM–KF and LM–RWKF.

Filtering Methods	MAE
KF	8.515 m
LM–KF	4.953 m
LM–RWKF	1.235 m

4.1.3. Computational Load Analysis

The computational performances of KF, LM–KF, and LM–RWKF were studied based on the above simulations, which were implemented on a PC (Intel® Core™ i5 12100F CPU, Intel, Santa Clara, CA, USA). As shown in Figure 8, KF has the smallest computational time, leading to a mean of 0198 ms per iteration. Due to the involvement of the noise estimation, the mean computational time of LM–KF is 0.235 ms per iteration, which is larger than that of KF. Since the random weight principle is further involved, the mean computational time of LM–RWKF is 0.276 ms per iteration. However, similar to KF, the computational time of LM–RWKF is much above the threshold of 0.07 s per iteration for real-time performance.

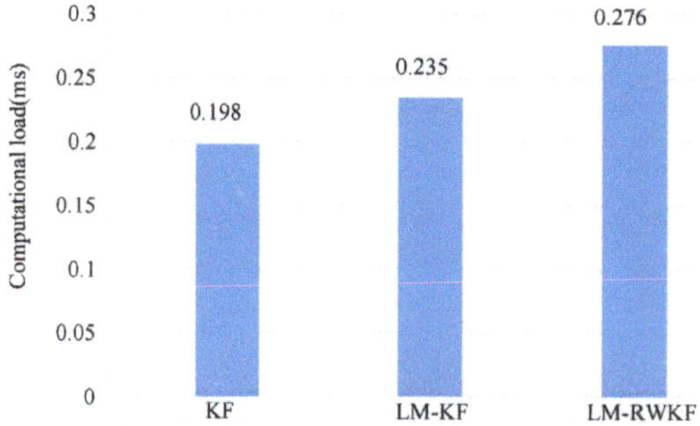

Figure 8. Comparison of computational load.

In practical engineering, the size of limited memory window is closely related to the filtering accuracy. The longer the limited memory window is, the higher the filtering precision and the larger the computational load will be. Therefore, the practical application should select an appropriate length of the limited memory window to balance the filtering accuracy and the calculation amount.

4.2. Experiments and Analysis

Experiments were also conducted to evaluate the performance of LM-RWKF in comparison with LM-KF and KF for navigation of a ground vehicle using a BDS/MEMS IMU (Bei Dou Satellite Navigation System/Micro-Electro-Mechanical System Inertial Measurement Unit) integrated navigation system.

4.2.1. BDS/SINS Integrated Navigation System Mathematical Model

The state vector of the BDS/MEMS IMU integrated navigation system is defined as follows:

$$\boldsymbol{X}(t) = \left[\delta v_E, \delta v_N, \delta v_U, \delta L, \delta h, \delta\lambda,\ \varepsilon_x,\ \varepsilon_y,\ \varepsilon_z, \nabla_x, \nabla_y, \nabla_z\right]^{\mathrm{T}} \tag{58}$$

where $(\delta v_E, \delta v_N, \delta v_U)$ are the velocity errors of the aircraft in East, North, and Up, $(\delta L,\ \delta\lambda,\ \delta h)$ are the errors in latitude, longitude, and altitude, $(\varepsilon_x,\ \varepsilon_y,\ \varepsilon_z)$ is the constant drift of the gyro, and $(\nabla_x,\ \nabla_y,\ \nabla_z)$ is the zero bias of the accelerometer.

The system state equation of the BDS/ MEMS IMU integrated navigation is described by the following:

$$\dot{\boldsymbol{X}}(t) = F(t)\boldsymbol{X}(t) + \boldsymbol{G}(t)\boldsymbol{w}(t) \tag{59}$$

where F(t) is the system function, and $\boldsymbol{w}(t)$ is the system process noise consisting of the gyro's Gaussian white noise $(\boldsymbol{w}_{gx},\ \boldsymbol{w}_{gy},\ \boldsymbol{w}_{gz})$ and accelerometer's Gaussian white noise $(\boldsymbol{w}_{ax},\ \boldsymbol{w}_{ay},\ \boldsymbol{w}_{az})$, i.e.,

$$\boldsymbol{w}(t) = \left[w_{gx}, w_{gy}, w_{gz}, w_{ax}, w_{ay}, w_{az}\right]^{\mathrm{T}}_{6\times1} \tag{60}$$

$\boldsymbol{G}(t)$ is the coefficient matrix of the system noise and is expressed as follows:

$$\boldsymbol{G}(t) = \begin{bmatrix} C_b^n & 0_{3\times3} \\ 0_{3\times3} & C_b^n \\ 0_{9\times3} & 0_{9\times3} \end{bmatrix}_{15\times6} \tag{61}$$

where $\boldsymbol{C}_b^n$ is the conversion matrix from the body coordinate system to navigation coordinate system.

The measurement equation of the BDS/MEMS IMU integrated navigation system is established using the velocity error and position error as measurement information.

The measurement equation of position error is described by the following:

$$\boldsymbol{Z}_p(t) = \boldsymbol{H}_p(t)\boldsymbol{X}(t) + \boldsymbol{V}_p(t) = \begin{bmatrix} R\cos L\delta\lambda + n_E \\ R\delta L + n_N \\ \delta h + n_U \end{bmatrix} \tag{62}$$

where $\boldsymbol{H}_p(t)$ is the position measurement matrix, which is expressed as follows:

$$\boldsymbol{H}_p = \left[0_{3\times3} \vdots\, diag[R\ R\cos L\ 1] \vdots\, 0_{3\times9}\right]_{9\times15} \tag{63}$$

$\boldsymbol{V}_p(t)$ is the position measurement noise, which is expressed as follows:

$$\boldsymbol{V}_p(t) = \begin{bmatrix} n_E & n_N & n_U \end{bmatrix}^{\mathrm{T}} \tag{64}$$

where n_E, n_N, and n_U are the position errors of the BDS receiver in the three axes, respectively.

The velocity error measurement equation can be written as follows:

$$\boldsymbol{Z}_v(t) = \boldsymbol{H}_v(t)\boldsymbol{X}(t) + \boldsymbol{V}_v(t) = \begin{bmatrix} \delta v_E + n_{vE} \\ \delta v_N + n_{vN} \\ \delta v_U + n_{vU} \end{bmatrix} \tag{65}$$

where $\boldsymbol{H}_v(t)$ is the velocity measurement matrix, which is expressed as follows:

$$\boldsymbol{H}_v(t) = \left[diag[1\ 1\ 1] \vdots\ 0_{3\times 12} \right]_{9\times 15} \tag{66}$$

$\boldsymbol{V}_v(t)$ is the velocity measurement noise, which is expressed as follows:

$$\boldsymbol{V}_v(t) = \begin{bmatrix} n_{v_E} & n_{v_N} & n_{v_U} \end{bmatrix}^{\mathrm{T}} \tag{67}$$

where n_{v_E}, n_{v_N}, and n_{v_U} are the velocity measurement errors of the BDS receiver in the three axes, respectively.

According to Equations (62) and (65), the measurement equation of the BDS/MEMS IMU integrated navigation system can be obtained as follows:

$$\boldsymbol{Z}(t) = \begin{bmatrix} \boldsymbol{H}_p(t) \\ \boldsymbol{H}_v(t) \end{bmatrix} \boldsymbol{X}(t) + \begin{bmatrix} \boldsymbol{V}_p(t) \\ \boldsymbol{V}_v(t) \end{bmatrix} = \boldsymbol{H}(t)\boldsymbol{X}(t) + \boldsymbol{V}(t) \tag{68}$$

4.2.2. Experimental Setup

Practical experiments were also conducted to evaluate the performance of the proposed improved LM-RWKF algorithm for vehicle navigation. The vehicle used s BDS/ MEMS IMU integrated system for navigation. The BDS/MEMS IMU integrated navigation system adopted the East-North-Up geography frame as the navigation frame. The test vehicle is a gray Mazda SUV. The BDS/MEMS IMU integrated navigation system is installed on the Mazda off-road vehicle. The BDS/MEMS IMU integrated navigation system includes a set of IMU and Bei Dou receivers with NV-IMU300 model, and the BeiDou receiver outputs C/A code measurement at a data update rate of 1 Hz. The test vehicle also carries related auxiliary equipment, including a DC power supply, small computer, data processor, and voltmeter. Figure 9 shows the structure of the test system.

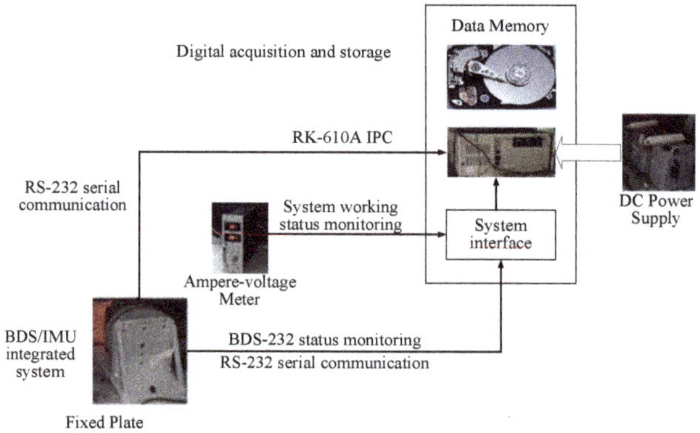

Figure 9. The framework of the experimental system.

The parameters of the BDS/MEMS IMU integrated navigation system are provided in Table 5. The initial flight parameters of the UAV are shown in Table 6. The test time for the filtering calculation was 1200 s. The filtering time step was 0.1 s.

Table 5. The parameters of the BDS / MEMS IMU integrated navigation system.

Sensors	Error Sources	Values (1σ)
IMU	Gyro constant drift	0.1°/h
	Gyro white noise	0.05°/h
	Accelerometer zero bias	1×10^{-3} g
	Accelerometer white noise	1×10^{-4} g
BDS receiver	Data update rate	1 Hz
	Positioning accuracy	15 m
	Velocity accuracy	0.05 m/s

Table 6. Initial position parameters.

Parameters		Value
Initial position	East longitude	108°46′05.89″
	North latitude	34°01′41.24″
	Altitude	2 m
Initial velocity	East	0 m/s
	North	120 m/s
	Up	0 m/s
Initial position error	East longitude	10 m
	North latitude	10 m
	Altitude	12 m
Initial velocity error	East	0.5 m/s
	North	0.5 m/s
	Up	0.5 m/s

The vehicle was traveling along South Qinling North Road in Xi'an, with the initial starting position being 34°01′41.24″ North latitude and 108°46′05.89″ east longitude. After arriving at Qinling Ring Island on Huanshan Road, it turned at 34°03′10.28″ N, 108°49′04.61″ E and returned to the initial position. The driving route and location of the experimental vehicle are shown in Figures 10 and 11. The driving distance is 15.38 km, the driving time is 20 min, and the average speed of the vehicle is 35 km/h. When the vehicle is moving, the Beidou receiver receives more than seven satellite signals and uses the information obtained by the differential Beidou system as reference data for positioning errors.

Figure 10. Vehicle traveling trajectory.

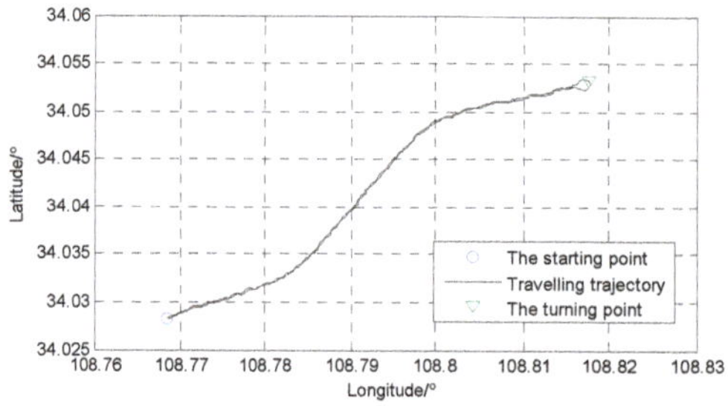

Figure 11. The position coordinates of the vehicle travelling trajectory.

4.2.3. Experimental Results and Analysis

For the purpose of comparison analysis, trials were conducted by using KF, LM–KF, and LM–RWKF, respectively. Figure 12 shows the position errors of these three methods, and Table 5 provides their MAEs.

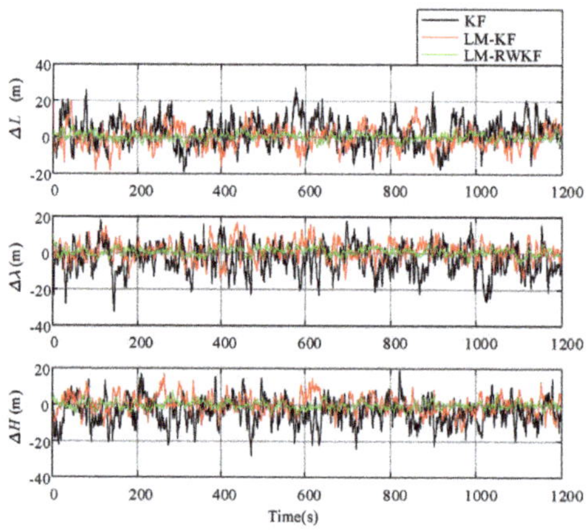

Figure 12. Position errors of KF, LM–KF, and LM–RWKF.

As shown in Table 7 and Figure 10, KF has large-magnitude oscillations in the filtering curve, and its position MAE is 13.612 m. Although LM–KF improves KF, leading to the decreased oscillation magnitude, its improvement is still limited, resulting in the 8.587 m position MAE. In contrast, LM–RWKF significantly decreases the oscillation magnitude, and its position MAE is 2.421 m, which is much smaller than those of the other two methods.

Table 7. MAEs by KF, LM–KF, and LM–RWK for the UAV position.

Filtering Methods	MAE
KF	13.612 m
LM–KF	8.587 m
LM–RWKF	2.421 m

5. Conclusions

This paper proposes a new LM-RWKF for system state estimation in the presence of unknown or inaccurate system noise statistics by combining the random weighting concept with limited memory technique. Random-weighted theories are established based on the limited memory technique to estimate system noise statistics within a limited memory, which are further fed back to KF for system state estimation. The proposed method cannot only adaptively adjust the weights to suppress the interference of noise statistics on system state estimation, but it can also suppress the influence of too much historical information on system state estimation, since the random weighting estimations of system noise statistics are established within a limited memory and embedded with random weights. Simulations, experiments, and comparison analysis demonstrate that the proposed LM-RWKF can effectively estimate system noise statistics, leading to higher accuracy than LM-KF in the presence of unknown or biased noise statistics.

In the existing methods, the length of the moving window is selected according to experience. Future work needs to study how to accurately select the window length, as well as the relationship between the length of the moving window and the filtering calculation accuracy, in order to find a balance between the length of the moving window, the filtering accuracy, and the computational workload. In addition, we will investigate how to combine the proposed improved LM-RWKF algorithm with the concept of artificial intelligence to develop an intelligent filtering algorithm that can accurately estimate sensor errors and system noise statistics.

Author Contributions: Supervision by Z.G., H.Z., Y.Z., Z.G. and H.Z. achieved the conceptualization of this paper; G.G. performed the simulations and analyzed the data; Z.G. and Y.Z. wrote. All authors have read and agreed to the published version of the manuscript.

Funding: This research was funded by the Natural Science Foundation of Shaanxi Province (Project number: 2024 JC-YBMS-574).

Institutional Review Board Statement: Not applicable.

Informed Consent Statement: Not applicable.

Data Availability Statement: Data are contained within the article.

Conflicts of Interest: The authors declare that they have no known competing financial interests or personal relationships that could have appeared to influence the work reported in this paper.

References

1. Schmidt, S.F. The Kalman filter—Its recognition and development for aerospace applications. *J. Guid. Control* **1981**, *4*, 4–7. [CrossRef]
2. Xie, H.; Song, J.; Zhong, Y.; Gu, C. Kalman filter finite element method for real-time soft tissue modeling. *IEEE Access* **2020**, *8*, 53471–53483. [CrossRef]
3. Liu, H.; Hu, F.; Su, J.; Wei, X.; Qin, R. Comparisons on kalman-filter-based dynamic state estimation algorithms of power systems. *IEEE Access* **2020**, *8*, 51035–51043. [CrossRef]
4. Michalski, J.; Kozierski, P.; Zietkiewicz, J. Double hybrid Kalman filtering for state estimation of dynamical systems. *Computer Appl. Electr. Eng.* **2019**, *28*, 01051. [CrossRef]
5. Gao, Z.; Mu, D.; Gao, S.; Zhong, Y.; Gu, C. Robust adaptive filter allowing systematic model errors for transfer alignment. *Aerosp. Sci. Technol.* **2016**, *59*, 32–40. [CrossRef]
6. Yang, Y.; Xu, T. An adaptive kalman filter based on sage windowing weights and variance components. *J. Navig.* **2003**, *56*, 231–240. [CrossRef]
7. Zhang, X.; Yan, Z.; Chen, Y. High-degree cubature Kalman filter for nonlinear state estimation with missing measurements. *Asian J. Control* **2021**, *24*, 1261–1272. [CrossRef]
8. Yang, Y.; He, H.; Xu, G. Adaptively robust filtering for kinematic geodetic positioning. *J. Geod.* **2001**, *75*, 109–116. [CrossRef]
9. Gao, S.; Wei, W.; Zhong, Y.; Subic, A. Sage windowing and random weighting adaptive filtering method for kinematic model error. *IEEE Trans. Aerosp. Electron. Syst.* **2015**, *51*, 1488–1500. [CrossRef]
10. Åkesson, B.M.; Jørgensen, J.B.; Poulsen, N.K.; Jørgensen, S.B. A generalized autocovariance least-squares method for Kalman filter tuning. *J. Process. Control* **2008**, *18*, 769–779. [CrossRef]
11. Abdel-Hafez, M.F. The Autocovariance least-squares technique for gps measurement noise estimation. *IEEE Trans. Veh. Technol.* **2009**, *59*, 574–588. [CrossRef]
12. Zhu, J.; Liu, B.; Wang, H.; Li, Z.; Zhang, Z. State estimation based on improved cubature Kalman filter algorithm. *IET Sci. Meas. Technol.* **2020**, *14*, 536–542. [CrossRef]
13. Narasimhappa, M.; Mahindrakar, A.D.; Guizilini, V.C.; Terra, M.H.; Sabat, S.L. MEMS-based IMU drift minimization: Sage husa adaptive robust kalman filtering. *IEEE Sens. J.* **2019**, *20*, 250–260. [CrossRef]
14. Jiang, H.; Li, B.; Sheng, Z. Design of adaptive Kalman filter based on FPGA implementation. *Infrared Laser Eng.* **2005**, *34*, 89–92.
15. Mohamed, A.H.; Schwarz, K.P. Adaptive kalman filtering for INS/GPS. *J. Geod.* **1999**, *73*, 193–203. [CrossRef]

16. Li, W.; Jia, Y. H-infinity filtering for a class of nonlinear discrete-time systems based on unscented transform. *Signal Process.* **2010**, *90*, 3301–3307. [CrossRef]
17. Jain, A.K.; Katiyar, A. Robust adaptive control design of nonlinear systems with input delay. *Int. J. Adapt. Control Signal Process.* **2023**, *37*, 1193–1202. [CrossRef]
18. Gao, G.; Gao, B.; Gao, S.; Hu, G.; Zhong, Y. A hypothesis test-constrained robust kalman filter for INS/GNSS integration with abnormal measurement. *IEEE Trans. Veh. Technol.* **2022**, *72*, 1662–1673. [CrossRef]
19. Tseng, C.-H.; Lin, S.-F.; Jwo, D.-J. Robust huber-based cubature kalman filter for GPS navigation processing. *J. Navig.* **2016**, *70*, 527–546. [CrossRef]
20. Wang, S.; Zhang, W.; Yin, C.; Feng, Y. Huber-based Unscented Kalman Filters with the q-gradient. *IET Sci. Meas. Technol.* **2017**, *11*, 380–387. [CrossRef]
21. Madeti, S.R.; Singh, S. Online fault detection and the economic analysis of grid-connected photovoltaic systems. *Energy* **2017**, *134*, 121–135. [CrossRef]
22. Zhang, T.; Fu, H.; Cheng, Y. Minimum mixture error entropy-based robust cubature kalman filter for outlier-contaminated measurements. *IEEE Sens. Lett.* **2022**, *6*, 1–4. [CrossRef]
23. Gao, B.; Gao, S.; Hu, G.; Zhong, Y.; Gu, C. Maximum likelihood principle and moving horizon estimation based adaptive un-scented Kalman filter. *Aerosp. Sci. Technol.* **2018**, *73*, 184–196. [CrossRef]
24. Gao, B.; Hu, G.; Li, W.; Zhao, Y.; Zhong, Y. Maximum likelihood-based measurement noise covariance estimation using sequential quadratic programming for cubature kalman filter applied in INS/BDS integration. *Math. Probl. Eng.* **2021**, *2021*, 1–13. [CrossRef]
25. Kwon, B. Adaptive fading-memory receding-horizon filters and smoother for linear discrete time-varying systems. *Appl. Sci.* **2022**, *12*, 6692. [CrossRef]
26. Liu, Y.; Liu, B. Adaptive fading-memory unscented Kalman filter algorithm for passive target tracking. *Sens. Lett.* **2013**, *11*, 2110–2113. [CrossRef]
27. Duan, Z.; Song, X.; Qin, M. Limited memory optimal filter for discrete-time systems with measurement delay. *Aerosp. Sci. Technol.* **2017**, *68*, 422–430. [CrossRef]
28. Gao, Y.; Qian, W.; Guo, J. A limited memory Kalman filter with control term. *Mod. Radar* **2004**, *26*, 44–47.
29. Jazwinski, A.H. Limited memory optimal filtering. *IEEE Trans. Autom. Control* **1968**, *13*, 558–563. [CrossRef]
30. Wishner, R.P.; Larson, R.E.; Athans, M. Status of Radar Tracking Algorithms. *Proc. Symp. Nonlinear Estima-Tion Theory Its Appl.* **1970**, *9*, 32–54.
31. Deng, M.; Cheng, Z.; He, Z. Limited-memory receive filter design for massive MIMO radar in signal-dependenti nterference. *IEEE Signal Process. Lett.* **2022**, *29*, 1536–1540. [CrossRef]
32. Li, R.; Lei, X.; Ma, L.; Yan, Z.; Jia, X. Analysis of etalon filter in quantum memory. *Microw. Opt. Technol. Lett.* **2022**, *65*, 1463–1467. [CrossRef]
33. Yang, Y.; Zhang, S. Adaptive fitting of systematic errors in navigation. *J. Geod.* **2005**, *79*, 43–49. [CrossRef]
34. Gao, Z.; Mu, D.; Zhong, Y.; Gu, C.; Ren, C. Adaptively random weighted cubature kalman filter for nonlinear systems. *Math. Probl. Eng.* **2019**, *2019*, 1–13. [CrossRef]
35. Gao, Z.; Gu, C.; Yang, J.; Gao, S.; Zhong, Y. Random weighting-based nonlinear gaussian filtering. *IEEE Access* **2020**, *8*, 19590–19605. [CrossRef]
36. Gao, S.; Zhong, Y.; Li, W. Random weighting method for multi-sensor data fusion. *IEEE Sens. J.* **2011**, *11*, 1955–1961. [CrossRef]
37. Wu, X.-Y.; Yang, Y.-N.; Zhao, L.-C. Approximation by random weighting method for M-test in linear models. *Sci. China Math.* **2007**, *50*, 87–99. [CrossRef]
38. Hu, G.; Gao, S.; Zhong, Y.; Gu, C. Asymptotic properties of random weighted empirical distribution function. *Commun. Stat.-Theory Methods* **2013**, *44*, 3812–3824. [CrossRef]
39. Gao, S.; Xue, L.; Zhong, Y.; Gu, C. Random weighting method for estimation of error characteristics in SINS/GPS/SAR integrated navigation system. *Aerosp. Sci. Technol.* **2015**, *46*, 22–29. [CrossRef]
40. Wei, W.; Gao, S.; Zhong, Y.; Gu, C.; Subic, A. Random weighting estimation for systematic error of observation model in dynamic vehicle navigation. *Int. J. Control Autom. Syst.* **2016**, *14*, 514–523. [CrossRef]

sensors

MDPI

Article

W-VSLAM: A Visual Mapping Algorithm for Indoor Inspection Robots

Dingji Luo [1,*], Yucan Huang [1], Xuchao Huang [1], Mingda Miao [2] and Xueshan Gao [1]

1 School of Mechatronical Engineering, Beijing Institute of Technology, Beijing 100081, China; 3120230155@bit.edu.cn (Y.H.); 221068359@stdmail.gxust.edu.cn (X.H.); xueshan.gao@bit.edu.cn (X.G.)

2 School of Mechanical Engineering College, Jiangsu Ocean University, Lianyungang 222005, China; 2023000090@jou.edu.cn

* Correspondence: 3120215083@bit.edu.cn; Tel.: +86-1881-087-5012

Abstract: In recent years, with the widespread application of indoor inspection robots, high-precision, robust environmental perception has become essential for robotic mapping. Addressing the issues of visual–inertial estimation inaccuracies due to redundant pose degrees of freedom and accelerometer drift during the planar motion of mobile robots in indoor environments, we propose a visual SLAM perception method that integrates wheel odometry information. First, the robot's body pose is parameterized in SE(2) and the corresponding camera pose is parameterized in SE(3). On this basis, we derive the visual constraint residuals and their Jacobian matrices for reprojection observations using the camera projection model. We employ the concept of pre-integration to derive pose-constraint residuals and their Jacobian matrices and utilize marginalization theory to derive the relative pose residuals and their Jacobians for loop closure constraints. This approach solves the nonlinear optimization problem to obtain the optimal pose and landmark points of the ground-moving robot. A comparison with the ORBSLAM3 algorithm reveals that, in the recorded indoor environment datasets, the proposed algorithm demonstrates significantly higher perception accuracy, with root mean square error (RMSE) improvements of 89.2% in translation and 98.5% in rotation for absolute trajectory error (ATE). The overall trajectory localization accuracy ranges between 5 and 17 cm, validating the effectiveness of the proposed algorithm. These findings can be applied to preliminary mapping for the autonomous navigation of indoor mobile robots and serve as a basis for path planning based on the mapping results.

Keywords: indoor inspection robots; V-SLAM; multi-sensor fusion

1. Introduction

In recent years, the rapid development of social productivity has led to a significant increase in the demand for security, especially with the continuous expansion of urban areas, which poses greater challenges for building and community security [1]. Additionally, there has been a notable rise in the need for round-the-clock patrol and surveillance in fields such as industrial parks, power facilities, and chemical enterprises. However, the security industry is facing challenges such as an aging workforce and rising labor costs, leading to an urgent need for efficient security robots to meet the demands of environmental perception in complex settings. In this context, visual SLAM technology, a multi-sensor fusion-based localization framework, has gained significant attention [2]. Its application in wheeled robots is particularly notable, as it enables autonomous perception in indoor environments lacking GPS signals, playing a critical role in the development of robot localization and mapping technology.

The most commonly used sensors for SLAM localization are laser and vision sensors. Laser SLAM technology is relatively mature, primarily based on TOF (time-of-flight) LiDAR. For research in laser SLAM, He et al. [3] successfully proposed a laser-inertial pose-estimation and map-construction method using B-spline curves, achieving robust registration steps. In the context of cooperative localization between drones and unmanned vehicles, Ryoga et al. [4] successfully achieved cooperative localization estimation by integrating drone point estimation results with the observation likelihood of unmanned vehicles. Bo et al. [5] combined the IESKF (Iterated Extended Kalman Filter) and factor graphs to propose a more precise local pose estimation method. By filtering LiDAR and IMU data with IESKF, they integrated the laser-inertial odometry with a back-end factor graph, offering new insights into the development of SLAM technology. To address challenges in different scenarios, researchers have proposed SLAM systems combining multi-line LiDAR and visual perception [6–9], offering diverse solutions for different environments. In the field of firefighting robots, a mapping system based on Cartographer SLAM theory provides autonomous patrol and automatic fire extinguishing functions in unknown indoor environments [10]. However, despite its notable advantages in meeting real-time demands, laser SLAM's loop closure detection strategies can still present issues with stability, particularly in geometrically symmetric environments. Overall, laser SLAM ensures real-time performance, but its loop closure detection strategies can lag behind, especially in some geometrically symmetric environments where they can lead to unstable loop closures.

The VINS (Visual–Inertial Navigation System) [11] for wheeled robots integrates low-frequency wheel encoder data to enable scale observation. Zhang et al. [12] designed a novel visual-based localization algorithm by combining camera, IMU, and wheel odometer data, providing more comprehensive information for ground robots. Liu et al. [13] tightly fused camera, IMU, and wheel encoder data, proposing a new extended visual–inertial odometry calculation method that significantly improves localization accuracy. In the field of ground vehicles, Lee et al. [14] effectively fused multi-modal visual, inertial, and 2D wheel odometry measurements in a visual–inertial-wheel odometry system using a sliding window filter approach, providing comprehensive information support for ground vehicles. Zheng et al. [15] improved visual estimation accuracy by directly parameterizing the pose of ground vehicles on SE(2) through an efficient pre-integration algorithm based on SE(2). Carlos et al. [16] developed a tightly integrated visual–inertial SLAM system, ORB-SLAM3, that is capable of performing visual, visual–inertial, and multi-map SLAM using monocular, binocular, and RGB-D cameras. Researchers have proposed various innovative methods across different scenarios, such as the planar PnP algorithm by Li J. et al. [17] and Zou's analysis and comparison of indoor navigation methods based on LiDAR SLAM. Zhu et al. [18] introduced a stereo visual–inertial fusion system based on VINS-Mono that closely integrates visual and inertial measurements. Additionally, Li J. et al. [19] proposed the Attention-SLAM method that combines a visual saliency model with traditional monocular visual SLAM to simulate human navigation patterns. In summary, pure visual SLAM can suffer from unobservable degrees of freedom under certain conditions, leading to estimation errors, particularly for wheeled robots during straight-line motion or constant local acceleration motion. In such cases, the IMU may not provide sufficient information.

In response to the planar movement of wheeled mobile robots in indoor environments, this paper proposes a multi-sensor visual SLAM perception method that integrates visual information with encoder-based wheel odometry information under the SE(2) framework. The high-precision relative pose observations provided by the wheel encoder speed data over short periods allow for accurate inter-frame estimations of visual keyframes. Additionally, the wheel odometry offers reliable pose estimations when visual signals fail (e.g., due to low texture or pure rotation). The integrated visual estimation information also helps to eliminate accumulated pose estimation errors from the wheel encoders, ultimately enabling accurate perception of the pose of the wheeled mobile robot.

2. Method Construction

The overall system framework is illustrated in Figure 1. The perception system for the wheeled mobile robot consists primarily of three threads: tracking, local optimization, and global optimization.

The tracking thread runs in real time at a frequency higher than the camera frame rate. Camera information and wheel speed information are soft-synchronized, with the first frame serving as the temporal origin, where the origin of the world coordinate system coincides with the origin of the robot coordinate system, making the first image frame the starting keyframe. First, the wheel speed information with a rate higher than the frame rate is pre-integrated to estimate the initial pose of the mobile robot. Each new image frame undergoes ORB feature extraction, and RANSAC [20,21] is used to remove outliers from feature matches. The thread matches the features in the current frame with the reference keyframe and estimates the initial map points using the pre-integrated information between the feature pairs and image frames. If the current frame meets the criteria for becoming a keyframe, it is added as a new reference keyframe to the local map.

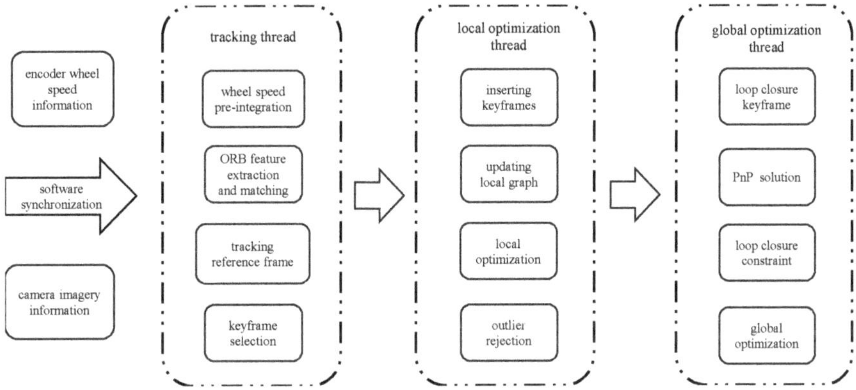

Figure 1. Block diagram of mobile robot visual SLAM with integrated wheel speed.

The local optimization thread maintains the local covisibility graph that is generated during the robot's movement, including keyframes and covisible map points with strong covisibility relationships. When a new keyframe is added to the local map, a search is conducted in the covisibility graph up to three levels deep to find keyframes with a covisibility relationship to the latest keyframe, which are considered local keyframes. Covisible map points associated with each local keyframe are treated as local map points. Local covisibility graph optimization is then performed using the nonlinear optimization library g2o [22] to update the poses of keyframes and the coordinates of map points. After optimization, map points with errors exceeding the threshold are considered outliers and are removed.

The global optimization thread maintains all keyframes and map points within the entire map. Keyframes that have undergone local optimization enter the global optimization thread, where similar keyframes are identified from the entire map using the "bag of words" approach [23] as loop closure candidate frames. Once loop closure is confirmed, the current keyframe, the loop closure keyframe, and their covisible map points form a subgraph. Optimization of the poses of the two keyframes and map points leads to an inter-frame loop closure constraint between the poses of the two frames. Visual constraints, wheel speed pre-integration constraints, and loop closure constraints are then incorporated into the global map to optimize the pose of the mobile robot and global map points, thereby eliminating accumulated errors caused by local optimization.

2.1. Coordinate System and Parameter Representation

As shown in Figure 2, the perception problem for wheeled mobile robots primarily involves three coordinate systems: the world coordinate system w, the robot coordinate system b, and the camera coordinate system c. The robot coordinate system can be any frame of reference fixed to the robot body, but in this context, it is defined as the odometry coordinate system. The origin, O_1, is located at the midpoint of the line connecting the wheel axes, with the positive x-axis pointing in the forward direction of the robot. The y-axis is aligned with the wheel axis line and positive to the left, while the positive z-axis points vertically upward from the origin. The camera coordinate system is centered at the camera's optical center O_2, with the positive z-axis pointing outwards from the camera lens, and the x and y axes aligned with the u and v directions of the pixel plane, respectively.

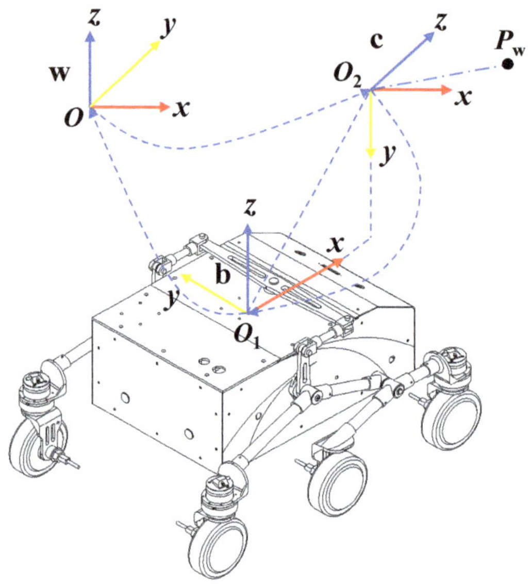

Figure 2. Schematic diagram of mobile robot coordinate system.

The Lie groups SO(3) and SE(3) are used to represent rotation R and pose T in three-dimensional space, respectively:

$$\mathrm{SO}(3) = \left\{ R \in \mathbb{R}^{3\times3} \middle| RR^{\mathrm{T}} = \mathbf{I}, \det(\mathbf{R}) = 1 \right\} \tag{1}$$

$$\mathrm{SE}(3) = \left\{ T = \begin{bmatrix} R & t \\ 0^{\mathrm{T}} & 1 \end{bmatrix} \in \mathbb{R}^{4\times4} \middle| R \in \mathrm{SO}(3), t \in \mathbb{R}^3 \right\} \tag{2}$$

Since the Lie group space does not support addition, making it impossible to perform small iterations on optimized variables, Lie algebra $\xi = [\phi,\ r]^{\mathrm{T}} \in \mathbb{R}^6$ is introduced to enable rotations and poses to be mapped as follows:

$$R = \exp(\phi^{\wedge}) \tag{3}$$

$$T = \exp(\xi^{\wedge}) \tag{4}$$

In Equation (3), the Lie algebra $\phi \in \mathbb{R}^3$ of the rotation matrix corresponds to the rotation vector, which can also be obtained using the Rodrigues formula:

$$t = J_l(\phi)r \tag{5}$$

The left Jacobian $J_l(\phi)$ is expressed as follows:

$$J_l(\phi) = \frac{\sin|\phi|}{|\phi|}I + (\frac{|\phi| - \sin|\phi|}{|\phi|^3})\phi\phi^{\mathrm{T}} + \frac{1-\cos|\phi|}{|\phi|^2}\phi^{\wedge} \tag{6}$$

Similarly, the pose on SE(2) corresponding to the three-dimensional vector $\varphi = [\nu, \psi]^{\mathrm{T}}$ can be expressed as follows:

$$\vartheta = \begin{bmatrix} \Phi & \nu \\ 0^{\mathrm{T}} & 1 \end{bmatrix} \tag{7}$$

where

$$\Phi = \exp(\psi)^{\wedge} = \begin{bmatrix} \cos\psi & -\sin\psi \\ \sin\psi & \cos\psi \end{bmatrix} \tag{8}$$

The correspondence between the Lie algebra se(3) and se(2) can be expressed as follows:

$$\mathrm{se}(3) \mapsto \mathrm{se}(2) : \varphi = \left[r_x, r_y, \phi_z\right]^{\mathrm{T}} \tag{9}$$

$$\mathrm{se}(2) \mapsto \mathrm{se}(3) : \xi = [\nu_1, \nu_2, 0, 0, 0, \psi]^{\mathrm{T}} \tag{10}$$

In general, the transformation matrix T_{AB} represents the transformation from coordinate system B to coordinate system A. The matrix includes a rotation matrix R_{AB} and a translation vector t_{AB}. The pose of a ground mobile robot is represented in SE(2) and denoted as T_{wb}, while the pose of the camera is represented in SE(3) and denoted as T_{cw}. The external calibration transformation matrix between the robot and camera coordinate systems is also represented in SE(3) and denoted as T_{bc}. A 3D map point is expressed in the world coordinate system as P_{w}, and the map point observed in the camera coordinate system is denoted as P_{c}. The transformation relationship between these coordinates is expressed as follows:

$$P_{\mathrm{c}} = T_{\mathrm{cw}} P_{\mathrm{w}} \tag{11}$$

2.2. Visual SLAM Perception Algorithm with Wheel Speed Fusion

Since optimization-based SLAM back-end methods can iteratively achieve more accurate state estimates, the sensing problem of wheeled mobile robots moving on indoor planes is modeled here as a maximum a posteriori estimation problem (MAP). Given the initial pose T_{wb_0} of the wheeled mobile robot, the camera intrinsic matrix K, and the external calibration matrix T_{cb}, and considering a series of continuous input images c_i ($i = 1, 2, \cdots$) and wheeled odometer speed information ω_j ($j = 1, 2, \cdots$), the goal is to estimate a series of poses T_{wb_i} ($i = 1, 2, \cdots$) of the wheeled mobile robot on SE(2) and the corresponding map points P_{w}^k ($k = 1, 2, \cdots$) throughout the observation process. Next, we construct visual observation constraints, wheeled pre-integration constraints, and loop closure observation constraints to form the final MAP problem.

2.3. Visual Observation Constraint

As shown in Figure 3, the camera uses a pinhole model to map the 3D coordinate point P_{w} in the world coordinate system to the 2D coordinate $P_{uv}(u, v)$ on the pixel plane.

In the camera coordinate system, the relationship between the 3D point P_{c} and the 2D point P_{uv} on the pixel plane is as follows:

$$\begin{bmatrix} u \\ v \end{bmatrix} = \frac{1}{z}\begin{bmatrix} f_x & 0 & c_x \\ 0 & f_y & c_y \end{bmatrix} P_{\mathrm{c}} \tag{12}$$

In the formula, z is the z-axis component of the 3D coordinate point. The parameters f_x and f_y represent the camera focal lengths along the x and y axes, respectively. The parameters c_x and c_y denote the coordinates of the image center point on the pixel plane.

This can be abbreviated as the projection function:

$$P_{uv} = \pi(P_c) \tag{13}$$

Given that visual measurements may be affected by noise, under the assumption of a Gaussian model, the measurement equation can be expressed as follows:

$$P_{uv} = \pi(P_c) + \eta_{uv} \tag{14}$$

where $\eta_{uv} \sim \mathcal{N}(0, \sigma_{uv}^2)$.

To facilitate the nonlinear optimization solution of the SLAM backend, we differentiate the previous equation with respect to the 3D point P_c in the camera coordinate system to obtain the Jacobian matrix of the pixel observation with respect to the camera coordinate point P_c [24]:

$$J_\pi(P_c) = \frac{1}{z}\begin{bmatrix} f_x & 0 & -x\frac{f_x}{z} \\ 0 & f_y & -y\frac{f_y}{z} \end{bmatrix} \tag{15}$$

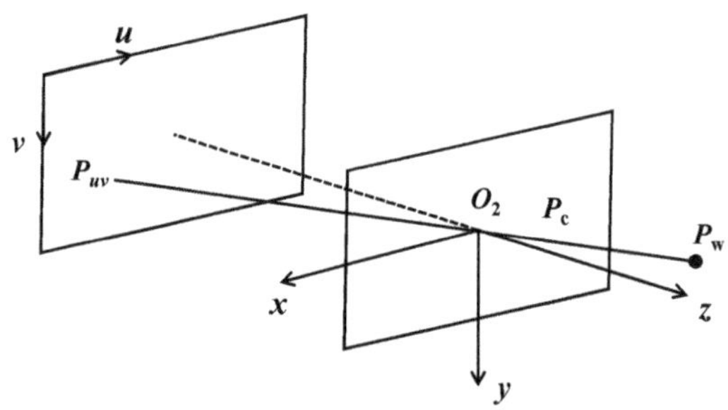

Figure 3. Camera projection model.

When the i-th keyframe observes a landmark point P_w^j and the camera extrinsic parameter rotation matrix R_{cb} and translation vector t_{cb} are known, the observation model in the camera coordinate system is given by

$$P_{uv}(T_{wb_i}, P_w^j) = \pi(R_{cb}R_{wb_i}^T(P_w^j - P_{wb_i}) + t_{cb}) + \eta_{uv} \tag{16}$$

In the equation, R_{wb_i} and P_{wb_i} represent the SE(3) expressions corresponding to the robot's pose parameterized on SE(2), while $\pi(\cdot)$ denotes the camera projection equation.

Unlike the unconstrained state of SE(3), a mobile robot undergoing planar motion is parameterized as a pose on SE(2). Given the minor vibrations that occur during the motion of the robot [15], perturbations in SE(3) are added to the SE(2) pose of the robot to account for the existing visual measurement noise. Assuming the translational perturbation experienced by the robot's pose is $\eta_t \sim \mathcal{N}(0_{3\times1}, \sigma_t^2)$ and the rotational perturbation is $\eta_\theta \sim \mathcal{N}(0_{3\times1}, \sigma_\theta^2)$, the actual pose of the robot should be represented as

$$R_{wb_i} \leftarrow \exp(\eta_\theta)R_{wb_i} \tag{17}$$

$$P_{wb_i} \leftarrow P_{wb_i} + \eta_t \tag{18}$$

Therefore, the observation equation in Equation (16) is modified as follows:

$$P_{uv}(T_{wb_i}, P_c^j) = \pi(R_{cb}\exp(\eta_\theta)R_{wb_i}^T(P_c^j - P_{wb_i} - \eta_t) + t_{cb}) + \eta_{uv} \tag{19}$$

From the above Equation (19), it is evident that the pixel observation is influenced by three types of noise: η_θ, η_t, and η_{uv}. To calculate the covariance of the camera observation, a first-order Taylor expansion is performed:

$$P_{uv}(T_{\mathrm{wb}_i}, P_c^j) \approx \pi(T_{\mathrm{wb}_i}, P_c^j) + J_{\eta_\theta}\eta_\theta + J_z\eta_t + \eta_{uv} \tag{20}$$

The Jacobian matrix of the camera observation with respect to the rotational perturbation η_θ and translational perturbation η_t is given by

$$J_{\eta_\theta} = J^\pi(T_{\mathrm{wb}_i}, P_{\mathrm{c}}^j)R_{\mathrm{cb}}R_{\mathrm{wb}_i}^{\mathrm{T}}(P_{\mathrm{c}}^j - P_{\mathrm{wb}_i})^\wedge \tag{21}$$

$$J_z\eta_t = -J^\pi(\rho_i, l_j)R_{\mathrm{cb}}R_{\mathrm{wb}_i}^{\mathrm{T}} \tag{22}$$

The term J^π refers to the Jacobian matrix of the camera projection equation, as shown in Equation (15).

Thus, we have

$$\delta\eta = J_{\eta_\theta}\eta_\theta + J_t\eta_t\delta + \eta_{uv} \tag{23}$$

Therefore, the covariance propagation equation for camera observation information under perturbation can be derived as follows:

$$\Sigma_{ij} = J_{\eta_\theta}\Sigma_\theta J_{\eta_\theta}^{\mathrm{T}} + J_{\eta_t}\Sigma_t J_{\eta_t}^{\mathrm{T}} + \Sigma_{uv} \tag{24}$$

Subsequently, under the aforementioned constraints, the camera's reprojection observation error equation can be constructed as follows:

$$e_{ij} = \pi(T_{\mathrm{wb}_i}, P_c^j) - P_{uv}(T_{\mathrm{wb}_i}, P_c^j) \tag{25}$$

The Jacobian matrices of the reprojection observation error with respect to the rotational and translational components of the robot's pose are, respectively,

$$\frac{\partial e_{ij}}{T_{\mathrm{wb}_i}} = \begin{bmatrix} \frac{\partial e_{ij}}{\partial \theta_i} & \frac{\partial e_{ij}}{\partial v_i} \end{bmatrix} \tag{26}$$

$$\frac{\partial e_{ij}}{\partial \theta_i} = -J^\pi(T_{\mathrm{wb}_i}, l_j)R_{\mathrm{cb}}R_{\mathrm{wb}_i}^{\mathrm{T}}\Lambda_{12} \tag{27}$$

$$\frac{\partial e_{ij}}{\partial v_i} = J^\pi(T_{\mathrm{wb}_i}, l_j)R_{\mathrm{cb}}R_{\mathrm{wb}_i}^{\mathrm{T}}(P_{\mathrm{c}}^j - P_{\mathrm{wb}_i})^\wedge e_3 \tag{28}$$

The Jacobian matrix of the reprojection error with respect to the landmark point is:

$$\frac{\partial e_{ij}}{\partial P_{\mathrm{c}}^j} = J^\pi(T_{\mathrm{wb}_i}, P_{\mathrm{c}}^j)R_{\mathrm{cb}}R_{\mathrm{wb}_i}^{\mathrm{T}} \tag{29}$$

At this point, we have derived the camera reprojection observation error, its corresponding covariance, and the Jacobian matrix of the error with respect to the variables to be optimized.

2.4. Wheel Speed Pre-Integration Constraint

In wheeled mobile robots, encoder information is typically generated at a higher frequency than visual data, resulting in each frame of visual data corresponding to multiple instances of encoder data. Consequently, updating the camera pose at each iteration of back-end optimization, alongside the robot poses represented by the multiple sets of encoder data, can either lead to redundant computations or affect the accuracy of perception.

Therefore, the concept of pre-integration is employed [25], as shown in Figure 4, to construct constraints between frames of visual information using pre-integrated encoder data.

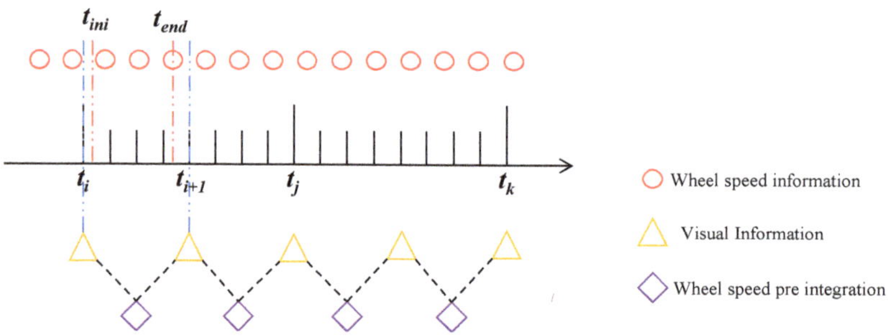

Figure 4. Schematic diagram of wheel speed information pre-integration.

First, the linear velocity v and angular velocity ω of a wheeled mobile robot can be derived from the left and right wheel speeds, as well as the robot's intrinsic parameters.

$$\begin{bmatrix} v \\ \omega \end{bmatrix} = \frac{1}{2B}\begin{bmatrix} B & B \\ 1 & -1 \end{bmatrix}\begin{bmatrix} v_r \\ v_l \end{bmatrix} + n_{v\omega} \tag{30}$$

In the equations, the linear velocity and angular velocity are subject to Gaussian noise, $n_{vw} \sim \mathcal{N}(0, \sigma_{vw}^2)$. To mitigate errors from manual measurements, accurate intrinsic parameters of the robot should be obtained through joint calibration with the camera [26].

Then, the discrete integration model for wheel speed at two consecutive time points can be derived, with a sampling interval of T.

$$\theta_{k+1} = \theta_k + \omega_k T \tag{31}$$

$$p_{k+1} = p_k + R(\theta_k)\begin{bmatrix} 0 \\ v_k T \end{bmatrix} \tag{32}$$

where $R(\theta_k) = \begin{bmatrix} \cos\theta_k & -\sin\theta_k \\ \sin\theta_k & \cos\theta_k \end{bmatrix}$.

Given $\varpi_k = [v_k, \omega_k]^{\mathrm{T}}$, the above equation can be expressed in a compact form as

$$q_{k+1} = f(q_k, \varpi_k) \tag{33}$$

Performing a Taylor expansion on the equation yields the error state equation as follows:

$$\delta q = \frac{\partial f}{\partial q_k} q_k + \frac{\partial f}{\partial \varpi_k}\varpi_k \tag{34}$$

$$\frac{\partial f}{\partial q_k} = \begin{bmatrix} 1 & 0 & -v_k \sin\theta_k T \\ 0 & 1 & v_k \cos\theta_k T \\ 0 & 0 & 1 \end{bmatrix} \tag{35}$$

where $\frac{\partial f}{\partial q_k} = \begin{bmatrix} 1 & 0 & -v_k \sin\theta_k T \\ 0 & 1 & v_k \cos\theta_k T \\ 0 & 0 & 1 \end{bmatrix}$, $\frac{\partial f}{\partial \varpi_k} = \begin{bmatrix} \cos\theta_k & 0 \\ \sin\theta_k & 0 \\ 0 & T \end{bmatrix}$.

Given $F_k = \frac{\partial f}{\partial q_k} G_k = \frac{\partial f}{\partial \omega_k}$, the error state equation between two consecutive wheel speed moments can be obtained as follows:

$$\begin{bmatrix} \delta\theta_{k+1} \\ \delta p_{k+1} \end{bmatrix} = F_k \begin{bmatrix} \delta\theta_k \\ \delta p_k \end{bmatrix} + G_k \begin{bmatrix} n_\omega \\ n_v \end{bmatrix} \tag{36}$$

In this way, the covariance propagation equation for wheel speed observations at two consecutive time intervals can be derived [27]:

$$P_{k+1} = F_k P_k F_k^{\mathrm{T}} + G_k Q_{v\omega} G_k^{\mathrm{T}} \tag{37}$$

The initial value of P_k is generally set as a zero matrix, and Q_{vw} is derived from the measurement noise n_{vw} of the encoder wheel speed:

$$Q_{v\omega} = \begin{bmatrix} \sigma_v^2 & 0 \\ 0 & \sigma_\omega^2 \end{bmatrix} \tag{38}$$

Assuming there are n instances of wheel speed information $\omega_{k+1}, \omega_{k+2}, \cdots, \omega_{k+n}$ between two consecutive keyframes c_i and c_{i+1}, as indicated by Equation (33), the wheel speed pre-integration observation between the two keyframes is as follows:

$$\Delta q_{i,i+1} = \sum_{i=k+1}^{k+n} f(q_i, \omega_i) \tag{39}$$

Accordingly, the covariance matrix of the pre-integrated quantities $\Sigma_{i,i+1}$ between two visual keyframes can be recursively obtained from Equation (37).

Then, let the pre-integrated observation be denoted as $\overline{T}_{i,i+1} = \Delta q_{i,i+1}$, and based on the pre-integrated quantities derived in Equation (39), the wheel speed measurement error equation can be constructed as follows:

$$e_{i,i+1} = \overline{T}_{i,i+1} T_{\mathrm{wb}_i} T_{\mathrm{wb}_{i+1}}^{-1} \tag{40}$$

By dividing the residuals into rotational and translational components, we obtain

$$\begin{bmatrix} e_{i,i+1}^{\theta} \\ e_{i,i+1}^{\rho} \end{bmatrix} = \begin{bmatrix} \bar{\theta}_{i,i+1} - (\theta_{i+1} - \theta_i) \\ \bar{p}_{i,i+1} - R^{\mathrm{T}}(\theta_i)(p_{i+1} - p_i) \end{bmatrix} \tag{41}$$

The Jacobian matrices of the wheel speed observation error with respect to the robot poses at two different time points are as follows:

$$\frac{\partial e_{i,i+1}}{\partial q_i} = \begin{bmatrix} 1 & 0 \\ R^{\mathrm{T}}(p_{i+1} - p_i)^{\wedge} & R^{\mathrm{T}} \end{bmatrix} \tag{42}$$

$$\frac{\partial e_{i,i+1}}{\partial q_{i+1}} = \begin{bmatrix} -1 & 0 \\ 0 & -R^{\mathrm{T}} \end{bmatrix} \tag{43}$$

The Jacobian matrix of the translational component of the error with respect to the rotational component of the i-th robot state is calculated using the right perturbation approach. Consequently, the observation error of the wheel speed pre-integration, its associated covariance, and the Jacobian matrix of the error with respect to the variables to be optimized have been obtained.

2.5. Loop Closure Co-Visibility Observation Constraint

In addition to the visual observation constraints and wheel speed pre-integration observation constraints introduced in the previous sections, there is also a loop closure constraint in the robot's movement. This constraint is important for eliminating cumulative errors that arise during the robot's state estimation process. When the robot's pose v_1 at a certain time forms a loop closure with the pose v_2 at a previous time, there will be a large number of co-visible landmarks between the two frames. This allows the system to construct a loop closure constraint to correct the robot's pose.

By taking the current frame, the loop closure frame, and the co-visible landmarks as nodes, and using the observation error of the landmarks in the camera coordinate system as edges, a subgraph optimization problem can be constructed. For instance, consider the observation of the k-th landmark $\overline{P}_k^{c_i}$ from the perspective of the current frame c_i (defined as c_1 for the current frame and c_2 for the loop closure frame). The measurement error can be expressed as

$$e_{c_i,k}(X_i) = \overline{P}_k^{c_i} - T_{c_i w} P_k \tag{44}$$

The Jacobian matrix of the measurement error $e_{c_i,k}$ with respect to the optimization variable X is

$$\begin{aligned} \frac{\partial e_{i,k}}{\partial P_k} &= -R_{c_i w} \\ \frac{\partial e_{i,k}}{\partial \delta\psi} &= \begin{bmatrix} -I & P_k^{\wedge} \end{bmatrix} \end{aligned} \tag{45}$$

Let $G_k = \begin{bmatrix} \frac{\partial e_{i,k}}{\partial P_k} & \frac{\partial e_{i,k}}{\partial \delta\psi} \end{bmatrix}$. By applying a first-order Taylor expansion, Equation (44) can be expressed as

$$e_{c_i,k}(X) \simeq e_{c_i,k}(X^*) + G_{i,k}\Delta X \tag{46}$$

Therefore, the error function to be optimized is

$$J(X) = \frac{1}{2}\sum_{i=1}^{2}\sum_{k=1}^{N} (e_{c_i,k}(X^*) + G_{i,k}\Delta X)^{\mathrm{T}} \Sigma_k^{-1} (e_{c_i,k}(X^*) + G_{i,k}\Delta X) \tag{47}$$

After the iteration converges, the optimized poses of the two frames can be obtained as poses $T_{c_1 w}^*$ and $T_{c_2 w}^*$.

In the next step, to obtain the covariance of the relative observations, we need to calculate the covariance of each pose individually. Under the assumption of a Gaussian distribution, let the current pose to be optimized and the loop closure pose be variables $X = [X_1,\ X_2]^{\mathrm{T}}$, respectively, while the shared 3D point is denoted as P. The joint probability distribution of (X, P) can be represented as the product of the marginal and conditional probabilities [28], which can be expressed as follows:

$$P(X,P) = P(X)P(P|X) \propto \exp\left(\frac{1}{2}\begin{bmatrix} X \\ P \end{bmatrix}^{\mathrm{T}} \begin{bmatrix} \Sigma_{XX} & \Sigma_{XP} \\ \Sigma_{PX} & \Sigma_{PP} \end{bmatrix}^{-1} \begin{bmatrix} X \\ P \end{bmatrix}\right) \tag{48}$$

After iterative optimization, the Hessian matrix used for solving the optimization problem in the graph can be obtained. This matrix is denoted as H:

$$H = \begin{bmatrix} H_{ii} & H_{ik} \\ H_{ki} & H_{kk} \end{bmatrix} \tag{49}$$

The matrix block $H_{ii}(12 \times 12)$ in the top left corner corresponds to the current frame and loop closure frame, while the matrix block $H_{kk}(3k \times 3k)$ in the bottom right corner corresponds to the landmark points.

For the likelihood distribution, the information matrix is the expected Hessian matrix of the negative log-likelihood problem [29]. Given the information matrix of the joint distribution, the information matrix of the marginal probability can be derived through decomposition [30]. By marginalizing the landmark points using the Schur complement, the information matrix for the marginal probability $P(X)$ can be obtained as follows:

$$H'_{ii} = H_{ii} - H_{ik} H_{ik}^{-1} H_{ki} \tag{50}$$

Next, we rewrite H'_{ii} in the following form:

$$H'_{ii} = \begin{bmatrix} H'_{11} & H'_{12} \\ H'_{21} & H'_{22} \end{bmatrix} \tag{51}$$

Assuming the current frame pose and loop closure frame pose are mutually independent, their information matrices correspond to the respective blocks in H'_{ii}. Since the covariance matrix is the inverse of the information matrix, the information matrices for the current frame and the loop closure frame can be obtained as follows:

$$\Sigma_{11}^{-1} = H'_{11} \tag{52}$$

$$\Sigma_{22}^{-1} = H'_{22} \tag{53}$$

Consequently, the perturbed poses of the current frame and the loop closure frame can be represented as follows:

$$\overline{T}_{\mathrm{c_1w}} = \exp(\epsilon_1^{\wedge}) T_{\mathrm{c_1w}}^{*} \tag{54}$$

In the context of this statement, $\epsilon_1 \in \mathbb{R}^6, \epsilon_1 \sim \mathcal{N}(0, \Sigma_{11})$ refers to the left perturbation in small quantities.

Similarly:

$$\overline{T}_{\mathrm{c_2w}} = T_{\mathrm{c_2w}}^{*} \exp(\epsilon_2^{\wedge}) \tag{55}$$

In the context of this statement, $\epsilon_1 \in \mathbb{R}^6, \epsilon_1 \sim \mathcal{N}(0, \Sigma_{11})$ refers to the right perturbation in small quantities. Both the left and the right perturbations are equivalent; however, we choose the right perturbation in this context for convenience in subsequent calculations.

Therefore, the relative observation between the current frame and the loop closure frame is expressed as follows:

$$\overline{T}_{12} = \overline{T}_{\mathrm{c_1w}} * \overline{T}_{\mathrm{c_2w}}^{-1} \tag{56}$$

Given the relative observation perturbation as ϵ, the perturbation mechanism leads to the following equation:

$$\exp(\epsilon^{\wedge})\overline{T}_{12} = \exp(\epsilon_1^{\wedge}) T_{\mathrm{c_1w}} (T_{\mathrm{c_2w}} \exp(\epsilon_2^{\wedge}))^{-1} \tag{57}$$

From identity,

$$\exp((\mathcal{T}\mathbf{x})^{\wedge}) = \mathbf{T} \exp(\mathbf{x}^{\wedge}) \mathbf{T}^{-1} \tag{58}$$

We obtain

$$\exp(\epsilon^{\wedge})\overline{T}_{12} = \exp(\epsilon_1^{\wedge}) T_{\mathrm{c_1w}} \exp\left((-\mathcal{T}_{\mathrm{c_1w}}\epsilon_2)^{\wedge}\right) T_{\mathrm{c_1w}} T_{\mathrm{c_2w}} \tag{59}$$

which is

$$\exp(\epsilon^{\wedge}) = \exp(\epsilon_1^{\wedge}) \exp\left((-\mathcal{T}_{\mathrm{c_1w}}\epsilon_2)^{\wedge}\right) \tag{60}$$

Given the formula $\epsilon'_2 = -\mathcal{T}_{\mathrm{c_1w}}\epsilon_2$, the covariance of ϵ (to the second-order term only) can be derived using the Baker–Campbell–Hausdorff (BCH) formula:

$$\overline{T}_{12} = \exp(\epsilon^{\wedge}) T_{12} \tag{61}$$

where $\epsilon \sim \mathcal{N}(0, \Sigma_{12})$.

Therefore, the observation error between the loop closure frame and the current frame can be expressed as

$$e_{12} = \log\left(\overline{T}_{12} T_{\mathrm{wb_1}} T_{\mathrm{wb_2}}^{-1}\right)^{\vee} \tag{62}$$

Similar to Equation (40), the transformation can be expressed in terms of the rotation and translation components on SE(2) as follows:

$$\begin{bmatrix} e_{12}^{\theta} \\ e_{12}^{\rho} \end{bmatrix} = \begin{bmatrix} \bar{\theta}_{12} - (\theta_1 - \theta_2) \\ \bar{\rho}_{12} - R^{\mathrm{T}}(\theta_1)(p_2 - p_1) \end{bmatrix} \tag{63}$$

The corresponding Jacobian matrix is similar to that in Equations (42) and (43), so it will not be repeated here. Thus, the loop closure covisual observation error, covariance matrix, and Jacobian matrix relative to the optimization variables have been obtained.

2.6. Formulation of the Maximum A Posteriori Estimation Problem

The essence of robot state estimation is maximum a posteriori estimation given known observations; when ignoring unknown prior information, this can be converted to maximum likelihood estimation. Under the assumption that the probability density function of the state variables follows a Gaussian distribution, the problem can be further transformed into a least squares estimation problem through negative logarithmic transformation. Thus, by treating the mobile robot pose and map points as state variables $\chi = (T_{\mathrm{wb}_1}, \cdots, T_{\mathrm{wb}_n}, P_1^{\mathrm{w}}, \cdots, P_m^{\mathrm{w}})$, and based on the aforementioned reprojection error Equation (25), wheel speed pre-integration error Equation (40), and loop closure co-visibility error Equation (62), a nonlinear error function can be constructed as follows:

$$F(\chi) = \sum_{in,jm} e_{ij}^{\mathrm{T}} \Sigma_{ij}^{-1} e_{ij} + \sum_{in-1} e_{i,i+1}^{\mathrm{T}} \Sigma_{i,i+1}^{-1} e_{i,i+1} + \lambda \sum_{ik} e_{12}^{\mathrm{T}} \Sigma_{12}^{-1} e_{12} \tag{64}$$

In the equation, λ takes a value of 0 or 1, where it is set to 1 when loop closure and global optimization are needed.

The optimal state vector χ^* is obtained by minimizing the objective function $F(\chi)$:

$$\chi^* = \arg\min_{\chi} F(\chi) \tag{65}$$

The error function can be linearized at the operating point χ using the second-order Taylor expansion, which yields

$$F(\chi + \delta\chi) \approx F(\chi) + J\delta\chi + \frac{1}{2}\delta\chi^{\mathrm{T}} H \delta\chi \tag{66}$$

To minimize the objective function, the function is differentiated with respect to $\delta\chi$ and set to zero. This results in the optimal perturbation value of the state vector, which is the solution to the following linear system:

$$H\delta\chi^* = -J \tag{67}$$

The optimal solution for the state variables is the best estimate of the pose of the wheeled mobile robot and the observed 3D map points:

$$\chi^* \leftarrow \chi^* \boxplus \delta\chi^* \tag{68}$$

In this context, $\boxplus$ represents the generalized summation symbol.

3. Experimental Comparison and Analysis

To verify the state estimation performance of the proposed Wheel Speed Visual SLAM (W-VSLAM) method—henceforth referred to as W-VSLAM in the text—for planar robot motion in an indoor environment without GPS, a cross-comparison with the latest visual SLAM algorithms in the field of visual methods was conducted. This comparison includes qualitative visual analysis and quantitative analysis based on general evaluation metrics, aiming to comprehensively validate the superiority of the proposed method. This section begins by detailing the experimental dataset, including the necessary parameters for different forms of data, and introduces two commonly used trajectory accuracy evaluation metrics in the SLAM field. Next, it provides an in-depth explanation of two sets of comparison experiments to validate the superiority of the W-VSLAM method. Finally, three groups of accuracy evaluation experiments are conducted to verify the effectiveness of the proposed W-VSLAM method.

3.1. Data Collection

Given that existing public datasets do not include indoor SLAM data with robot speed information or camera-encoder extrinsic parameters, a dataset was collected using a self-developed mobile robot platform in an indoor environment of an academic building. To facilitate comparison with open-source visual algorithms, the camera was mounted on top of the robot body in an upward-facing orientation (front-facing camera orientation rendered the latest visual SLAM algorithms completely ineffective). The encoders, coaxial with the drive motors, provided feedback on the wheel rotation speed of the robot platform at 1024 lines. The remotely controlled robot platform navigated an indoor environment to form a looped trajectory. During this motion, data from a 2D lidar, mounted on the front of the robot and operating at a frequency of 10 Hz, were recorded. In addition, grayscale images were captured at a frequency of 25 Hz with a resolution of 752 × 480 pixels. Wheel speed data, calculated at the base layer, were collected at a frequency of 100 Hz and measured in RPM (revolutions per minute). The coefficients of the radial–tangential distortion model for the camera are shown in Table 1. The intrinsic parameters matrix K for the pinhole camera model are as follows:

$$K = \begin{bmatrix} 366.9826923 & 0 & 361.35791565 \\ 0 & 366.7935063 & 246.71152313 \\ 0 & 0 & 1 \end{bmatrix} \tag{69}$$

Table 1. Performance index data of different segmentation algorithms.

Parameters	Radial Distortion k_1	Radial Distortion k_2	Tangential Distortion p_1	Tangential Distortion p_2
Value	−0.02209723	−0.00543498	0.00179311	−0.00069370

3.2. Quantitative Evaluation Metrics

To assess the performance of the proposed vision algorithm, it must be compared with the ground truth trajectory, which is typically acquired using more accurate sensors. Common trajectory evaluation metrics include Absolute Trajectory Error (ATE) and Relative Pose Error (RPE). Let T_k represent the pose of the trajectory being evaluated at time k, and let $T_{gt,k}$ represent the pose of the ground truth trajectory at the same time:

$$\mathrm{ATE}_k = \log\left(T_{gt,k}^{-1} T_k\right)^{\vee} \tag{70}$$

In the equation, $\log(\cdot)^{\vee}$ represents the operator that maps a transformation matrix (Lie group) to a vector (Lie algebra).

After obtaining the N trajectory errors for the two trajectories, statistical calculations can be performed to obtain common SLAM trajectory evaluation metrics, such as root mean square error (RMSE):

$$\mathrm{ATE}_{\mathrm{RMSE}} = \sqrt{\frac{1}{N}\sum_{k=1}^{N}\left\|\log\left(T_{gt,k}^{-1} T_k\right)^{\vee}\right\|^2} \tag{71}$$

After obtaining the absolute trajectory error, it is also necessary to evaluate the relative trajectory error. This can be conducted by considering the trajectory from time k to time $k + \Delta t$ and calculating the relative trajectory error as follows:

$$\mathrm{RPE}_k = \log\left(\left(T_{gt,k}^{-1} T_{gt,k+\Delta t}\right)^{-1} \left(T_k^{-1} T_{k+\Delta t}\right)\right)^{\vee} \tag{72}$$

Similarly, the root mean square error (RMSE) for the relative trajectory error can be calculated as follows:

$$\mathrm{RPE}_{\mathrm{RMSE}} = \sqrt{\frac{1}{N} \sum_{k=1}^{N} \left\| \log\left(\left(T_{gt,k}^{-1} T_{gt,k+\Delta t}\right)^{-1} \left(T_k^{-1} T_{k+\Delta t}\right)\right)^{\vee} \right\|^2} \tag{73}$$

In addition to evaluating the trajectory as a whole, one can decompose the pose into translation and rotation components and evaluate each separately. The calculation method is similar to the overall evaluation but involves extracting the corresponding translation or rotation components and calculating the differences. In the subsequent experiments, this separate evaluation approach is used, calculating the maximum error, minimum error, and root mean square error (RMSE) for each component. This allows for a detailed comparison of the estimation performance of the algorithm.

Due to the indoor environment without GPS, the robot's true trajectory is obtained using the common laser HectorSLAM method with a 2D lidar. Then, using the ATE and RPE evaluation metrics discussed earlier, the proposed method, the latest open-source visual algorithm ORBSLAM3, and the wheel odometry trajectories are compared in terms of error accuracy relative to the true trajectory. This allows for a cross-comparison of the methods.

It should be noted that the other well-known open-source algorithms, VINS-Mono and VINS-Fusion, are not included in the comparison because the accuracy of the VINS algorithms relies significantly on precise initialization. For wheeled mobile robots in planar motion, it is difficult to fully excite the accelerometer during initialization, which leads to the divergence of the estimated robot trajectory. This renders the algorithm ineffective and not suitable for comparison. Therefore, wheel odometry trajectories are used as a substitute.

3.3. *Indoor Comparative Experimental Results and Analysis*

To validate the superiority of the proposed method for the state estimation of planar motion robots in indoor environments, a cross-comparison was conducted in the visual SLAM field using two sets of experiments. The comparison dataset was obtained by manually controlling a robot to follow a square trajectory in an indoor environment, as depicted in Figure 5. During the movement, there were lighting changes, and at the corners, there were pure rotations of 90 degrees. The first set of data involved running two laps at a speed of 0.08 m/s, while the second set of data consisted of one lap at a speed of 0.04 m/s (the two sets of data were not meant to be controlled for; the number of laps was chosen randomly for ease of operation, so one set involved one lap, and the other involved two laps). Figure 6 shows the feature points tracked during the algorithm's execution. Matched feature points in the reference frame are indicated by red circles, while unmatched points are shown in blue circles. Matched feature points in the current frame are represented by green circles, while unmatched points are represented by red circles. The results of running the proposed SLAM algorithm are shown in Figures 7 and 8. Black points represent landmark points, black boxes represent keyframes, blue lines represent the co-visibility between keyframes, and red represents the current orientation and landmark points.

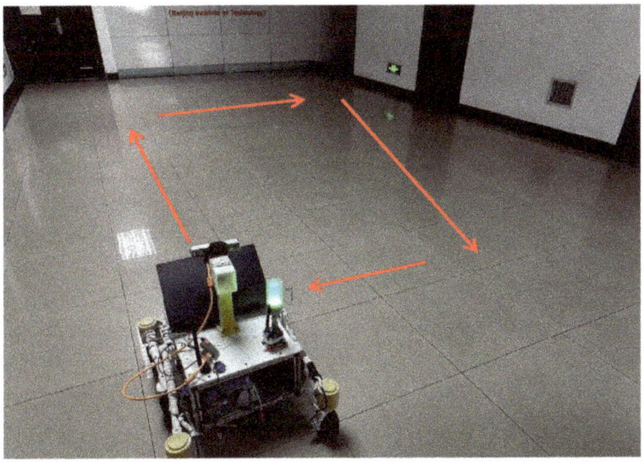

Figure 5. Schematic diagram of square movement in the comparative experiment.

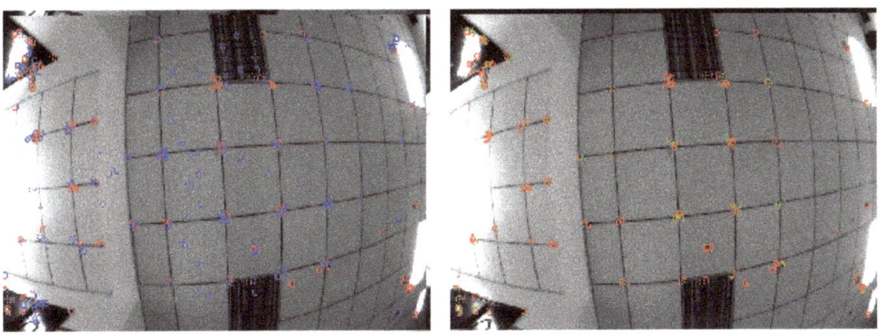

(a) (b)

Figure 6. Reference keyframe and current frame in previous tracking, (**a**) reference keyframe of previous tracking and (**b**) current frame of previous tracking.

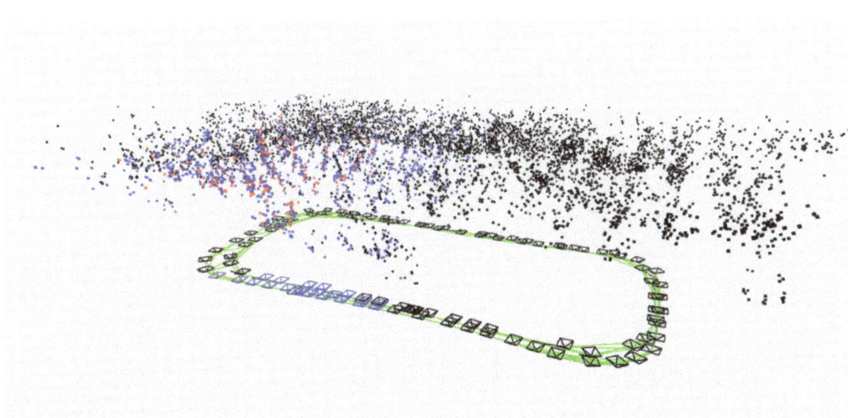

Figure 7. Comparative Experiment One: robot poses and environmental map points obtained by W-VSLAM.

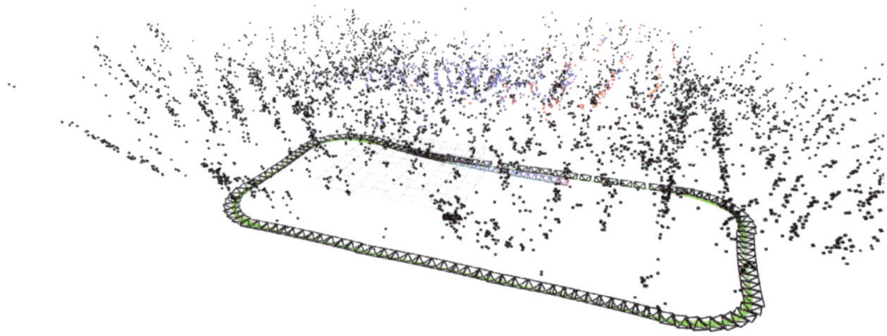

Figure 8. Comparative Experiment Two: robot poses and environmental map points obtained by W-VSLAM.

The trajectory obtained by Laser HectorSLAM is considered the ground truth. The trajectories of the proposed W-VSLAM method, the latest visual open-source algorithm ORBSLAM3 keyframe trajectory, and the robot's odometry trajectory were compared. The trajectory comparisons for the two sets of experiments are shown in Figures 9 and 10. Given the inconsistency in time frequencies across the different trajectories, the maximum time error is set at 0.01 s. Additionally, due to the external parameter transformation between different sensors, the results were aligned using the visualization tool.

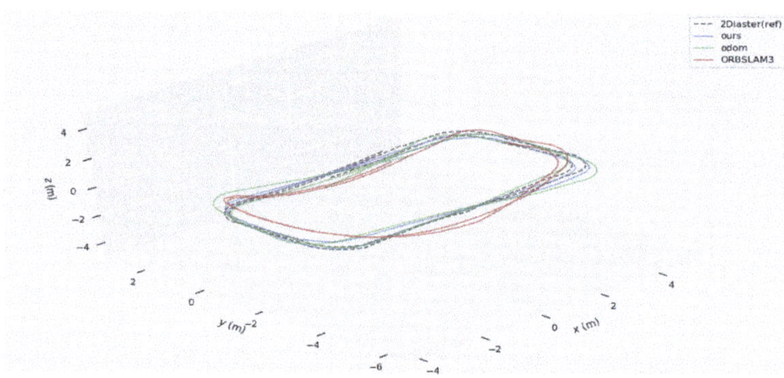

Figure 9. Comparative Experiment One: trajectory comparison chart of different algorithms.

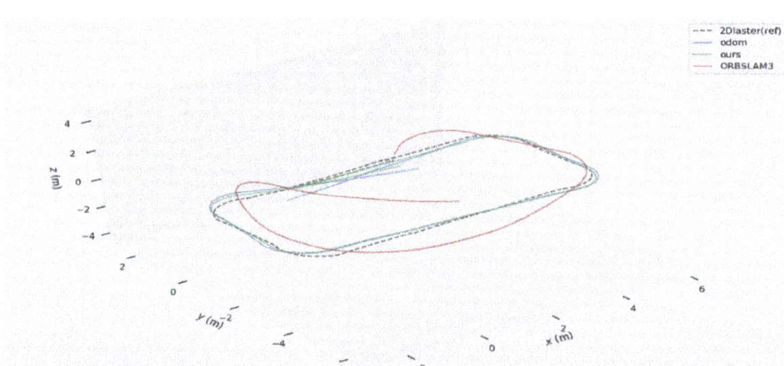

Figure 10. Comparative Experiment Two: trajectory comparison chart of different algorithms.

In Figure 9, the 2DLaser (ref) represents the reference trajectory obtained by 2DLaser and is shown as a dashed line. Ours represents the proposed W-VSLAM method and is depicted as a blue solid line. Odom represents the trajectory obtained from odometry and is depicted as a green solid line. ORBSLAM3 represents the latest open-source algorithm and is depicted as a solid red line. In Figure 10, the 2DLaser (ref) represents the reference trajectory obtained by Laser SLAM and is shown as a dashed line. Odom represents the trajectory obtained from odometry and is depicted as a solid blue line. Ours represents the proposed W-VSLAM method and is depicted as a solid green line. ORBSLAM3 represents the latest open-source algorithm and is depicted as a solid red line. From the two figures, it is visually apparent that the trajectory obtained using the ORBSLAM3 algorithm has significant errors in the z-axis direction, whereas the proposed visual algorithm aligns more closely with the reference trajectory.

More specifically, in Experiment One, the comparison of trajectories obtained by the four methods in terms of translation is illustrated in Figure 11. In the x and y axes, the odometry trajectory experiences drift due to accumulated error from integration. Similarly, the ORBSLAM3 algorithm requires initialization, resulting in a missing segment at the beginning and also showing accumulated error, though less than the odometry error. In contrast, the proposed W-VSLAM method aligns more closely with the reference trajectory, displaying a smaller error. In the z-axis direction, the proposed algorithm uses a 3-degree-of-freedom representation for the robot's pose, ensuring an accurate estimation of planar motion. Meanwhile, the ORBSLAM3 algorithm uses a six-degree-of-freedom parameterization for the robot's pose during planar motion, leading to erroneous estimation. This algorithm also exhibits significant errors in the pitch and roll directions, which are not presented here. The results of Experiment Two are similar and are not included.

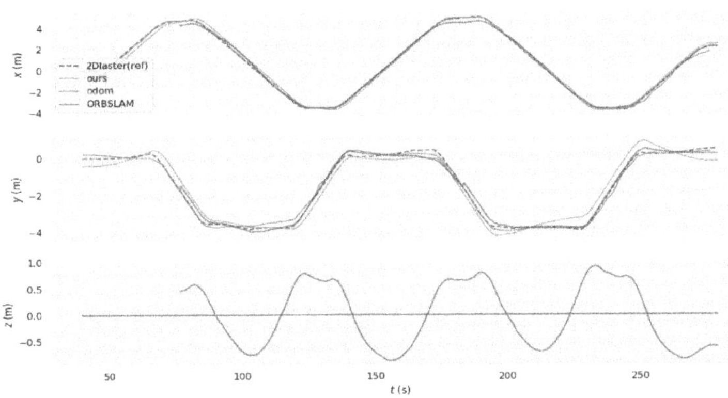

Figure 11. Comparative test one: translational component comparison chart of trajectories from different algorithms.

After the qualitative analysis presented above, a quantitative analysis of the trajectory accuracy of the various algorithms in the two comparison experiments was conducted using two evaluation metrics: absolute trajectory error (ATE) and relative pose error (RPE). The results for Experiment One are shown in Tables 2 and 3, while the results for Experiment Two are presented in Tables 4 and 5. For relative pose error (RPE), considering the variation in time frequencies of the different trajectories, the relative error was calculated over a relative increment of 1 m in the translation direction and a relative increment of 1 degree in the rotation estimation.

Table 2. Absolute trajectory error (ATE) estimates of different algorithms for two loop trajectories.

Error Type	ORBSLAM3 Algorithm	Wheel Odometer	The Proposed Algorithm
Maximum Translational Error (m)	0.996424	0.663076	0.135999
Minimum Translational Error (m)	0.053682	0.191564	0.005725
Translational Root Mean Square Error (m)	0.649618	0.411756	0.070133
Maximum Rotational Error (rad)	1.798408	0.249882	0.073018
Minimum Translational Error (rad)	1.499492	0.000036	0.000234
Rotational Root Mean Square Error (rad)	1.674251	0.138140	0.024885

Table 3. Relative trajectory error estimates of different algorithms for two loop trajectories.

Error Type	ORBSLAM3 Algorithm	Wheel Odometer	The Proposed Algorithm
Maximum Translational Error (m)	2.258283	0.072512	0.083033
Minimum Translational Error (m)	1.679300	0.004918	0.005750
Translational Root Mean Square Error (m)	1.934542	0.030060	0.035044
Maximum Rotational Error (rad)	0.267859	0.015824	0.042052
Minimum Translational Error (rad)	0.004864	0.000011	0.000113
Rotational Root Mean Square Error (rad)	0.080078	0.002707	0.013080

Table 4. Absolute trajectory error (ATE) estimates of different algorithms for one loop trajectory.

Error Type	ORBSLAM3 Algorithm	Wheel Odometer	The Proposed Algorithm
Maximum Translational Error (m)	2.448048	0.877545	0.265917
Minimum Translational Error (m)	0.270060	0.130613	0.008439
Translational Root Mean Square Error (m)	1.263682	0.384537	0.160804
Maximum Rotational Error (rad)	2.242332	0.288801	0.255336
Minimum Translational Error (rad)	1.169881	0.001404	0.001039
Rotational Root Mean Square Error (rad)	1.844953	0.157194	0.101190

Table 5. Relative trajectory error (RPE) estimates of different algorithms for one loop trajectory.

Error Type	ORBSLAM3 Algorithm	Wheel Odometer	The Proposed Algorithm
Maximum Translational Error (m)	1.798007	0.100730	0.120928
Minimum Translational Error (m)	1.286840	0.064568	0.066626
Translational Root Mean Square Error (m)	1.445999	0.083673	0.094710
Maximum Rotational Error (rad)	0.375543	0.026001	0.037211
Minimum Translational Error (rad)	0.007266	0.000009	0.000061
Rotational Root Mean Square Error (rad)	0.069250	0.001821	0.010742

Absolute trajectory error reflects the algorithm's accuracy over the entire trajectory. In Experiment One, as shown in Table 2, the proposed algorithm achieved a root mean square error of 0.135999 m in the translation direction, which is significantly better than the other two algorithms. The root mean square error in rotation estimation was 0.024885 rad, also outperforming the other two algorithms.

In Experiment Two, as shown in Table 4, the root mean square error in translation estimation was 0.160804 m, while the root mean square error in rotation estimation was 0.101190 rad. In horizontal comparisons, the proposed algorithm outperformed the other two algorithms, demonstrating its ability to integrate two relatively imprecise sensor data streams to achieve more accurate pose estimation and validate the effectiveness of the algorithm.

Moreover, a vertical comparison of Tables 2 and 4 reveals that the translation estimation in Experiment One is superior to that in Experiment Two, as is the rotation estimation. This is likely because the trajectory in Experiment One had more loop closures, resulting in more accurate pose estimation for the robot. This finding further supports the effectiveness of the proposed algorithm. Relative trajectory error reflects the algorithm's performance on local trajectories.

In Experiment One, as shown in Table 3, the proposed algorithm achieved a root mean square error of 0.035044 m in translation estimation and 0.013080 rad in rotation estimation. Experiment Two showed similar results, with a root mean square error of 0.094710 m in translation estimation and 0.010742 rad in rotation estimation. In horizontal comparisons, these results are slightly below the best performance, but the differences are minimal. It is notable that while the wheel odometry method accumulates errors quickly due to high-frequency integration, resulting in inaccurate global trajectory estimation, it is relatively accurate in local relative pose estimation. The proposed visual SLAM algorithm, which integrates wheel odometry, leverages this aspect. By using pre-integration to obtain accurate inter-camera frame estimations as constraints, and by shortening the wheel odometry integration interval to reduce cumulative errors, the algorithm improves the relative pose estimation performance of the camera. This ensures that the proposed algorithm can achieve more accurate global pose estimation.

In summary, for planar motion robot state estimation under significant lighting variations, the proposed method demonstrates superior performance in both global and local accuracy compared to the latest visual algorithms in the field. When compared to wheel odometry, the proposed method also exhibits superior performance globally, while achieving nearly comparable, albeit slightly lower, performance at the local level.

3.4. Results and Analysis of Multiple Indoor Experiments

Additionally, three experimental sets were conducted multiple times to validate the robustness and perceptual performance of the proposed W-VSLAM method, mitigating potential environmental factors. The comparative results with the latest visual SLAM algorithms have been discussed in the previous section and will not be repeated here. Instead, the trajectories provided by laser HectorSLAM are used as reference true values to evaluate the perceptual accuracy of the proposed visual SLAM method. The results of the robot's environmental perception across different motion trajectories, as well as comparisons with the reference trajectory, are shown in Figures 12–18. The RMSE accuracy of the W-VSLAM method across different trajectories is presented in Table 6. In the first dataset, the robot initially follows a small square trajectory and then expands its movement range. In the second dataset, the robot traverses two adjacent closed square trajectories. The third dataset involves a back-and-forth motion pattern. The first two datasets mainly highlight the impact of lighting changes during the robot's movement on the perception algorithm, while the third emphasizes the effect of a 360-degree pure rotation at the midpoint of a straight trajectory.

(a) (b)

Figure 12. (**a**) Perception results of robot poses and map points in Experiment One. (**b**) Comparison between Experiment One and reference trajectory.

(a) (b)

Figure 13. (**a**) Perception results of robot poses and map points in Experiment Two. (**b**) Comparison between Experiment Two and reference trajectory.

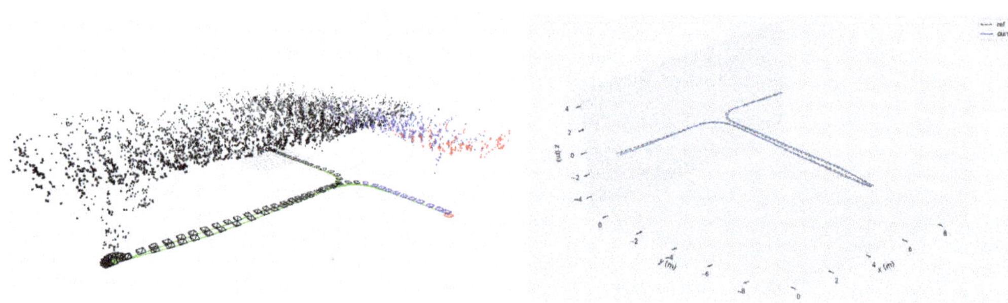

(a) (b)

Figure 14. (**a**) Perception results of robot poses and map points in Experiment Three. (**b**) Comparison between Experiment Three and reference trajectory.

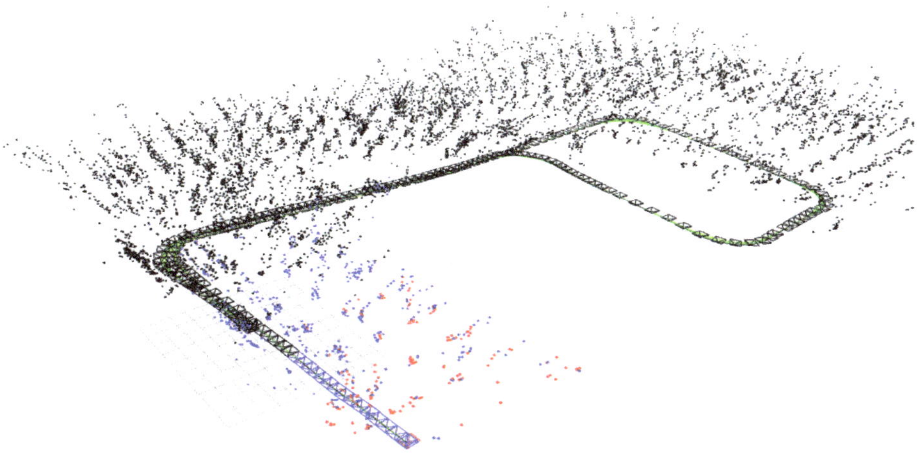

Figure 15. Indoor long corridor environment trajectory; Rivz result diagram.

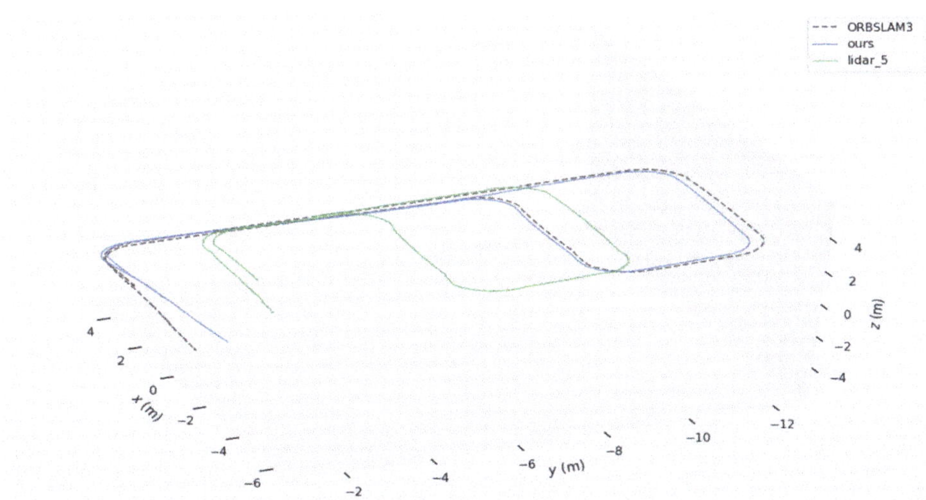

Figure 16. Comparison chart of trajectories in the indoor long corridor environment.

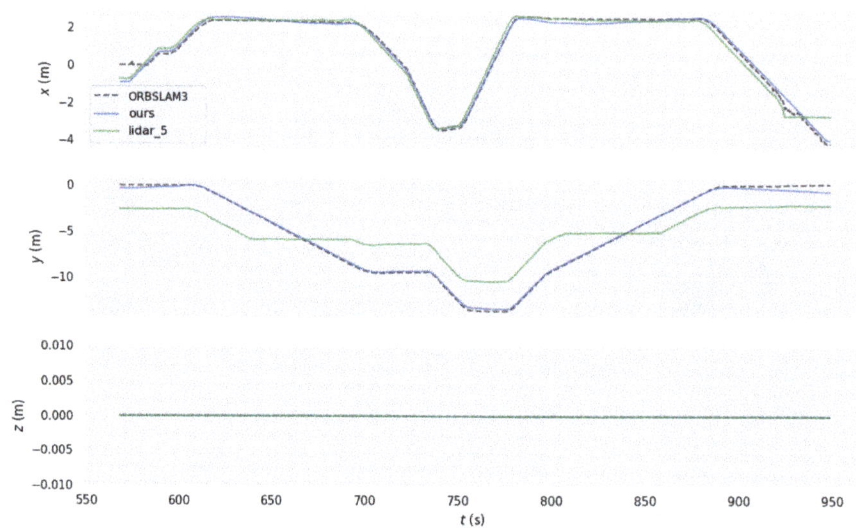

Figure 17. Comparison chart of translational components of trajectories in the indoor long corridor environment.

Table 6. Results of trajectory accuracy evaluation from multiple experiments.

Category	Experiment 1	Experiment 2	Experiment 3
Absolute Translational RMSE (m)	0.069569	0.174467	0.083839
Absolute Rotation RMSE (rad)	0.079908	0.103241	0.096109
Relative Translation RMSE (m)	0.110006	0.106630	0.107273
Relative Rotation RMSE (rad)	0.012823	0.015159	0.025358

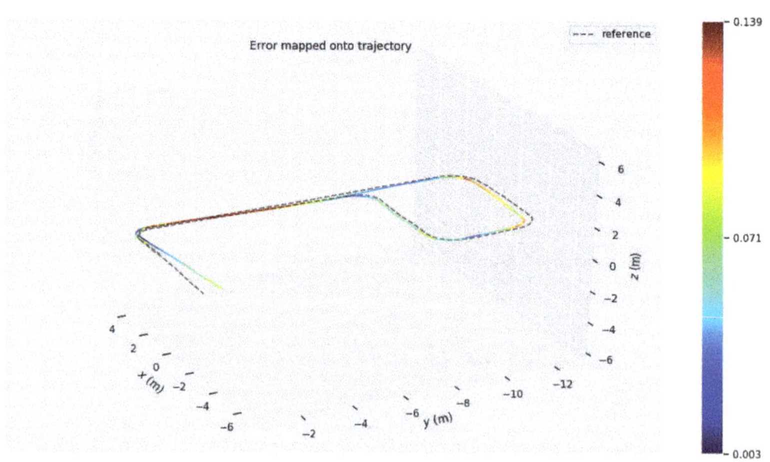

Figure 18. Absolute accuracy of rotational estimation in the indoor long corridor environment.

The results from the three different experimental sets indicate that the proposed W-VSLAM perception algorithm can robustly perceive robot poses and map points even under conditions of strong lighting changes or 360-degree pure rotational motion. The trajectory localization accuracy ranges from 5 cm to 17 cm, aligning with the current perception levels of visual SLAM methods and further verifying the effectiveness of the proposed approach.

3.5. The Results and Analysis of the Indoor Long Corridor Environment Experiment

The indoor long corridor environment has always posed challenges for laser SLAM perception. Due to the limited range or angle of sensors, the information perceived by the sensors does not significantly change during movement, leading to inaccurate state estimation. In this experiment, a robot was controlled to navigate through a long corridor within an indoor environment, returning through the corridor to form a closed-loop trajectory. Throughout the movement, significant changes in illumination intensity were observed. The results of the proposed W-VSLAM method in this long corridor environment are shown in Figure 15.

In the aforementioned corridor environment, we operated HectorSLAM using a 2D LiDAR with a scan range parameter set to 5 m. The resulting trajectory was compared with the trajectories obtained using the proposed algorithm and ORBSLAM3. The visual comparisons of the trajectories are illustrated in Figures 16 and 17. Based on the conclusions from the quantitative accuracy comparison in the previous subsection, the trajectory from the wheel speed odometer was neither included in the comparison nor reported. Observing the trajectory obtained by HectorSLAM with a 5 m detection range, it is evident that when the sensor's detection distance is insufficient to cover the corridor length, the trajectory generated by HectorSLAM becomes distorted. This distortion occurs because the point cloud observations within the detection range do not change as the robot moves through the corridor, leading to a lack of translational estimation (in the y-axis direction in the world coordinate system). In contrast, the proposed algorithm can still accurately perceive the environment in the long corridor, demonstrating the advantages of visual SLAM methods in specific settings. Additionally, as seen in the translational component comparison in Figure 17, the trajectory obtained by ORBSLAM3 is generally accurate in the r-axis direction (perpendicular to the corridor direction), except at the end of the trajectory. The proposed algorithm's trajectory closely aligns with the reference trajectory, except for a noticeable deviation at the trajectory's end due to visual odometry estimation errors. The trajectory generated by the proposed method is closer to the actual path, further showcasing its advantages in environments where LiDAR fails.

Although the translational estimation of HectorSLAM's reference trajectory is inaccurate and thus cannot be analyzed in detail, its rotational estimation can still be qualitatively assessed. The absolute and relative trajectory accuracies for rotational estimation, as achieved by the proposed algorithm compared to the reference trajectory, are shown in Figures 18 and 19, respectively. In addition to the maximum and minimum errors illustrated in the figures, the root mean square error (RMSE) of the absolute rotational error is 0.0726080 rad, and the RMSE of the relative rotational error is 0.022059 rad, which is consistent with the results from the previous subsection. Overall, the comparison with the HectorSLAM laser algorithm demonstrates that the proposed visual SLAM algorithm can achieve more accurate perception in indoor long corridors or diffuse reflection environments where 2D LiDAR fails.

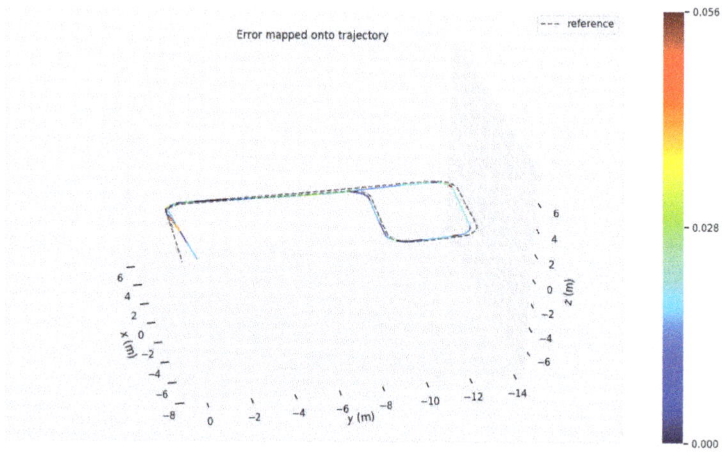

Figure 19. Relative accuracy of rotational estimation in the indoor long corridor environment (with a 1° increment).

4. Conclusions

To address the estimation inaccuracies in indoor environments without GPS signals for ground mobile robots, caused by excess pose degrees of freedom and the lack of excitation in accelerometers, a visual SLAM perception method that integrates wheel odometry information is proposed. The method parameterizes the robot pose on SE(2) and the camera pose on SE(3). It then derives visual constraint residuals and Jacobian matrices for reprojection observations using the camera projection model and uses the pre-integration concept to derive pose-constraint residuals and Jacobian matrices from wheel odometry pre-integration quantities. Loop closure observation constraints are derived using marginalization theory to calculate relative pose residuals and Jacobians. These are used to solve the nonlinear optimization problem and obtain the optimal poses and landmark points for the ground mobile robot. Comparisons with the ORBSLAM3 algorithm and wheel odometry in experimental tests indicate that the proposed visual SLAM method integrating wheel odometry provides higher perception accuracy in planar motion estimation, particularly in indoor environments with significant lighting variations or pure rotational motion. The method's superiority is demonstrated with accuracy in a range of 5–17 cm across multiple experiments, further confirming its effectiveness. It is worth noting that the proposed visual SLAM perception method with integrated wheel odometry is not limited to wheeled mobile platforms; it can also be applied to tracked mobile platforms.

Author Contributions: Conceptualization, D.L.; methodology, D.L. and X.G.; software, D.L. and X.H.; validation, X.G., D.L. and Y.H.; formal analysis, D.L. and M.M.; investigation, D.L. and X.H.; resources, X.G.; data curation, D.L.; writing—original draft preparation, X.H. and D.L.; writing—review and editing, D.L. and X.G.; visualization, D.L. and Y.H.; supervision, X.G.; project administration, X.G.; funding acquisition, X.G. All authors have read and agreed to the published version of the manuscript.

Funding: This research received no external funding.

Institutional Review Board Statement: Not applicable.

Informed Consent Statement: Not applicable.

Data Availability Statement: Data are contained within the article.

Conflicts of Interest: The authors declare no conflicts of interest.

References

1. Gao, Y.; Li, S.; Wang X. A patrol mobile robot for power transformer substations based on ROS. In Proceedings of the 2018 Chinese Control and Decision Conference (CCDC), Shenyang, China, 9–11 June 2018.
2. Brossard, M.; Bonnabel, S.; Barrau, A. Unscented Kalman filter on Lie groups for visual inertial odometry. In Proceedings of the 2018 IEEE/RSJ International Conference on Intelligent Robots and Systems (IROS), Madrid, Spain, 1–5 October 2018.
3. He, G.; Liu, Y.; Li, C. Tightly coupled laser-inertial pose estimation and map building based on B-spline curves. *Meas. Sci. Technol.* **2023**, *34*, 125130. [CrossRef]
4. Shimizu, R.; Sugita, Y. Accuracy improvement of cooperative localization using UAV and UGV. *Adv. Robot.* **2023**, *37*, 999–1011. [CrossRef]
5. You, B.; Zhong, G.; Chen, C. A Simultaneous Localization and Map System Using the Iterative Error State Kalman Filter Judgment Algorithm for Global Navigation Satellite System. *Sensors* **2023**, *23*, 6000. [CrossRef] [PubMed]
6. Su, Z.; Zhou, J.; Dai, J. Optimization Design and Experimental Study of Gmap Algorithm. In Proceedings of the 2020 Chinese Control and Decision Conference (CCDC), Hefei, China, 22–24 August 2020.
7. Gao, H.; Zhang, X.; Wen, J. Autonomous indoor exploration via polygon map construction and graph-based SLAM using directional endpoint features. *IEEE Trans. Autom. Sci. Eng.* **2018**, *16*, 1531–1542. [CrossRef]
8. Yang, A.; Cao, Y.; Liu, Y. AGV robot for laser-SLAM based method testing in automated container terminal. *IEEE Trans. Image Process.* **2023**, *50*, 969–980. [CrossRef]
9. Wang, X.; Ma, X.; Li, Z. Research on SLAM and Path Planning Method of Inspection Robot in Complex Scenarios. *Electronics* **2023**, *12*, 2178. [CrossRef]
10. Li, S.; Yun, J.; Feng, C. An Indoor Autonomous Inspection and Firefighting Robot Based on SLAM and Flame Image Recognition. *Fire* **2023**, *6*, 93. [CrossRef]
11. Wu, J.; Guo, X.; Georgiou G. Vins on wheels. In Proceedings of the 2017 IEEE International Conference on Robotics and Automation (ICRA), Singapore, 29 May–3 June 2017.
12. Zhang, M.; Chen, Y.; Li, M. Vision-aided localization for ground robots. In Proceedings of the 2019 IEEE/RSJ International Conference on Intelligent Robots and Systems (IROS), Macau, China, 3–8 November 2019.
13. Liu, J.; Gao, W.; Hu, Z. Visual-inertial odometry tightly coupled with wheel encoder adopting robust initialization and online extrinsic calibration. In Proceedings of the 2019 IEEE/RSJ International Conference on Intelligent Robots and Systems (IROS), Macau, China, 3–8 November 2019.
14. Lee, W.; Eckenhoff, K.; Yang, Y. Visual-inertial-wheel odometry with online calibration. In Proceedings of the 2020 IEEE/RSJ International Conference on Intelligent Robots and Systems (IROS), Las Vegas, NV, USA, 25–29 October 2020.
15. Zheng, F.; Liu, H. Visual-odometric localization and map for ground vehicles using SE(2)-XYZ constraints. In Proceedings of the 2019 International Conference on Robotics and Automation (ICRA), Montreal, QC, Canada, 20–24 May 2019.
16. Campos, C.; Elvira, R.; Rodríguez, G. Orb-slam3: An accurate open-source library for visual, visual–inertial, and multimap slam. *IEEE Trans. Robot.* **2021**, *37*, 1874–1890. [CrossRef]
17. Li, J.; Yang, B.; Huang, K. Robust and efficient visual-inertial odometry with multi-plane priors. In Proceedings of the 2nd Chinese Conference on Pattern Recognition and Computer Vision, PRCV 2019, Xi'an, China, 8–11 November 2019.
18. Zhu, J.; Yao, C.; Janschek, K. Stereo Visual-Inertial Fusion for UAV State Estimation. *IFAC-PapersOnLine* **2020**, *53*, 9420–9425. [CrossRef]
19. Li, J.; Pei, L.; Zou, D. Attention-SLAM: A visual monocular SLAM learning from human gaze. *IEEE Sens. J.* **2020**, *21*, 6408–6420. [CrossRef]
20. Derpanis, G. Overview of the RANSAC Algorithm. *Image Rochester NY* **2010**, *4*, 2–3. Available online: https://scholar.google.com/scholar_lookup?title=Overview+of+the+RANSAC+Algorithm&author=Derpanis,+K.G.&publication_year=2010&journal=Image+Rochester+NY&volume=4&pages=2%E2%80%933 (accessed on 26 August 2024).
21. Miguel, C.; Domingo, M.; Andrés, C. An Efficient Point-Matching Method Based on Multiple Geometrical Hypotheses. *Electronics* **2021**, *10*, 246. [CrossRef]
22. Kummerle, R.; Grisetti, G.; Strasdat, H. g2o: A general framework for graph optimization. In Proceedings of the 2011 IEEE International Conference on Robotics and Automatio, Shanghai, China, 9–13 May 2011.
23. Galvez-Lpez, D. Tardos.Bags of Binary Words for Fast Place Recognition in Image Sequences. *IEEE Trans. Robot.* **2012**, *28*, 1188–1197. [CrossRef]
24. Zhu, Y.; An, H.; Wang, H. RC-SLAM: Road Constrained Stereo Visual SLAM System Based on Graph Optimization. *Sensors* **2024**, *24*, 536. [CrossRef] [PubMed]
25. Forster, C.; Carlone, L.; Dellaert, F. On-manifold preintegration for real-time visual–inertial odometry. *IEEE Trans. Robot.* **2016**, 33, 1–21. [CrossRef]
26. Heng, L.; Li, B.; Pollefeys, M. Camodocal: Automatic intrinsic and extrinsic calibration of a rig withmultiple generic cameras and odometry. In Proceedings of the 2013 IEEE/RSJ International Conference on Intelligent Robots and Systems, Tokyo, Japan, 3–7 November 2013.
27. He, Y.; Chai, Z.; Liu, X. Tightly-coupled vision-gyro-wheel odometry for ground vehicle with online extrinsic calibration. In Proceedings of the 2020 3rd International Conference on Intelligent Autonomous Systems (ICoIAS), Singapore, 26–29 February 2020.

28. Dou, H.; Wang, Z.; Wang, C. Immediate Pose Recovery Method for Untracked Frames in Feature-Based SLAM. *Sensors* **2024**, *24*, 3. [CrossRef] [PubMed]
29. Ly, A.; Marsman, M.; Verhagen, J. A tutorial on Fisher information. *J. Math. Psychol.* **2017**, *80*, 40–55. [CrossRef]
30. Ralph, H. Schur type comparison theorems for affine curves. *J. Geom.* **2023**, *115*, 1. [CrossRef]

sensors

MDPI

Article

Event-Driven Maximum Correntropy Filter Based on Cauchy Kernel for Spatial Orientation Using Gyros/Star Sensor Integration

Kai Cui [1,2,3,*], Zhaohui Liu [1,2,3], Junfeng Han [1,2,3], Yuke Ma [4], Peng Liu [1,2,3] and Bingbing Gao [4,5,*]

1 University of Chinese Academy of Sciences, Beijing 100049, China; lzh@opt.ac.cn (Z.L.); hanjunfeng@opt.ac.cn (J.H.); liupeng1@opt.ac.cn (P.L.)
2 Xi'an Institute of Optics and Precision Mechanics, Xi'an 710119, China
3 Key Laboratory of Space Precision Measurement, Chinese Academy of Sciences, Xi'an 710119, China
4 School of Automation, Northwestern Polytechnical University, Xi'an 710072, China; myk@mail.nwpu.edu.cn
5 Research & Development Institute, Northwestern Polytechnical University in Shenzhen, Shenzhen 518063, China
* Correspondence: cuikai@opt.ac.cn (K.C.); bingbing.gao@nwpu.edu.cn (B.G.)

Abstract: Gyros/star sensor integration provides a potential method to obtain high-accuracy spatial orientation for turntable structures. However, it is subjected to the problem of accuracy loss when the measurement noises become non-Gaussian due to the complex spatial environment. This paper presents an event-driven maximum correntropy filter based on Cauchy kernel to handle the above problem. In this method, a direct installation mode of gyros/star sensor integration is established and the associated mathematical model is derived to improve the turntable's control stability. Based on this, a Cauchy kernel-based maximum correntropy filter is developed to curb the influence of non-Gaussian measurement noise for enhancing the gyros/star sensor integration's robustness. Subsequently, an event-driven mechanism is constructed based on the filter's innovation information for further reducing the unnecessary computational cost to optimize the real-time performance. The effectiveness of the proposed method has been validated by simulations of the gyros/star sensor integration for spatial orientation. This shows that the proposed filtering methodology not only has strong robustness to deal with the influence of non-Gaussian measurement noise but can also achieve superior real-time spatial applications with a small computational cost, leading to enhanced performance for the turntable's spatial orientation using gyros/star sensor integration.

Keywords: maximum correntropy filter; spatial orientation; gyros/star sensor integration; non-Gaussian noise; Cauchy kernel

1. Introduction

The turntable structure can realize the orientation motion of the payload (camera, radar and so on) independently of the spacecraft platform. It plays a vital role for the space tracking and detection system in the aerospace field [1,2]. A two-dimensional turntable is one of the most widely used structures in the above systems. It consists of two rotating platforms, in which their shafting is perpendicular to each other [1,3]. The highly precise orientation measurement of the two-dimensional turntable is an important prerequisite for achieving high stability and enhancing the stability of the space tracking and detection system.

Spatial orientation accuracy refers to the error between the ideal spatial orientation and the actual spatial orientation associated with the inner frame of the turntable after the rotation and transformation of each axis. Due to the inevitable control error, the installation error and other factors in the practical application of the system, there is a deviation between the actual orientation and the ideal orientation of the turntable, which significantly affects the control stability of the turntable [4,5]. Therefore, highly precise orientation measurements of the turntable are particularly important.

Currently, in spatial orientation applications, the star sensor is a commonly used measurement device that can provide high-precision attitude and orientation information [6,7]. However, the operating frequency of the star sensor is quite low and it is easily limited by the time and environmental conditions [7,8]. In contrast, gyros, as an autonomous inertial measurement method, can continuously output high-frequency angular/attitude information with good real-time performance [9,10]. However, gyros have the weakness of error drifts, leading to poor long-term attitude/orientation accuracy [11]. It can be seen that the above two sensors have a good complementary characteristic in terms of attributes, so the gyros/star sensor integration scheme can use the high-precision attitude or orientation information obtained from the star sensor to compensate for the drift of the gyros, achieving high-precision turntable orientation information [12,13].

When the gyros/star sensor integration is implemented for the two-dimensional turntable's spatial orientation, there are usually two schemes: one is the strap-down installation mode and the other is the direct installation mode [13,14]. In the former, both the gyros and star sensor are installed on the fixed base of the turntable, while in the latter, the gyros are installed in the inner frame of the two-dimensional turntable and firmly connected with the load camera/radar, and the star sensor is installed on the fixed base of the turntable. It should be noted that in many applications, due to the limited space, the star sensor is hardly installed in the inner frame of the turntable because of its large size. In the strap-down installation mode, the turntable's control stability is easily affected by the turntable error since the gyros/star sensor integration cannot directly perceive the turntable angle. Thus, the direct installation mode has gained increasing attention by scholars in recent years. The research of this paper is also based on the direct installation mode of gyros/star sensor integration.

Nowadays, the data fusion of gyros/star sensor integration is mostly achieved based on Kalman filtering [15,16]. Under the assumption of linear and Gaussian noise models, Kalman filtering can obtain the optimal estimation of a system state [15,16]. However, due to the influence of the spatial light environment, the measurement of star sensors will be contaminated. This makes the measurement noise of gyros/star sensor integration no longer follow a Gaussian distribution, resulting in non-Gaussian noise. In this situation, the performance of Kalman filtering will be significantly deteriorated [17,18].

Research efforts have been dedicated to dealing with non-Gaussian noise involved in the dynamic system. Farahmand et al. proposed a doubly robust smoothing method to provide double robustness that enhances the accuracy and reliability of estimations, particularly in the presence of bias measurement noise covariance and model misspecification [18]. However, the smoothing method is not suitable for a real-time application. Further, Soken et al. established a robust Kalman filter by using a scalar factor to adjust the measurement noise covariance for curbing the influences of inaccurate measurement noise covariances [19]. However, its improvement is very limited since the scalar factor is chosen empirically. Particle filter (PF) and Gaussian sum filter (GSF) are also two typical forms of non-Gaussian filtering [20,21]. Kim et al. developed PF to further extend the non-Gaussian capabilities of filtering by employing sampling and resampling techniques to approximate posterior distributions [20]. Sorenson et al. constructed a GSF by using random sampling to obtain the global approximation of the system state's posterior probability density, which is not limited to Gaussian noise [21]. However, global approximation requires a large amount of computation, making the above filters difficult to apply for the spatial orientation [22].

As an emerging filtering theory, information theoretic learning has garnered attention for its optimization criteria, which rely on information theoretic measures such as maximum correntropy criterion (MCC) derived directly from data, instead of conventional second-order statistical metrics [23,24]. It exhibits an excellent robustness to handle non-Gaussian noise since it uses the signals' high-order moments rather than the second-order moments in the minimum mean square error principle. Chen et al. [23] first established the maximum correntropy Kalman filter by using the maximum correntropy criterion. It has superior performance to resist the influences of outliers and non-Gaussian noise for a linear system. Compared to PF and GSF, the filter based on the MCC shows a quite low computational burden. Further, Mohammadi et al. [24] presented a maximum–minimum correntropy criterion for robust and stable gene selection. Gao et al. [25] developed the theory of MCC to handle outliers and non-Gaussian noise to improve the performance of spectral redshift estimation for spectral redshift navigation. In the traditional maximum correntropy (MC) filter, the Gaussian kernel is always used to define the distance between distinct vectors. Nevertheless, it may not always be an optimal selection for the kernel function. An issue arises when the measurement is affected by multi-dimensional non-Gaussian noise, such as heavy-tailed noise in real-world applications. In such cases, the aforementioned MC filters may fail due to the occurrence of singular matrices, causing certain characteristics of the noise to be overlooked and resulting in reduced estimation accuracy [26,27].

To address the drawbacks of Gaussian kernel-based MC filters, a novel Cauchy kernel-based MCC filter is proposed [27,28]. It uses the Cauchy kernel instead of the original Gaussian kernel to describe the distance between different vectors. The Cauchy kernel originated from the Cauchy distribution. Its function curve has a long tail, indicating that the kernel function is very broad and can be used for data with high dimensions. Because of the above reason, the Cauchy kernel-based MC filter shows a superior performance when handling multi-dimensional non-Gaussian noise. It also behaves in a consistently stable manner when a different Cauchy kernel bandwidth is selected [28,29]. Unfortunately, the existing MC filters or Cauchy kernel-based MC filters mostly utilize the fixed-point iteration method to update the estimates of the state, which will burden the computational efficiency [27,30]. For the application of spatial orientation, the real-time performance is a very important property index. Thus, if the MCC is used for the whole filtering process of gyros/star sensor integration, it will increase the unnecessary computational burden, leading to a deteriorated real-time performance. This is because in the time point without measurement fault, i.e., non-Gaussian measurement noise does not exist, the Kalman filter can obtain the optimal estimation result with a lower computational cost than MC filters, while MC filters need to use the fixed-point iteration to update the estimates of the state, requiring a large computational cost.

To address the above problem for a spatial orientation application, this paper presents a novel method of a Cauchy kernel-based maximum correntropy filter that is event-driven to curb the influence of non-Gaussian measurement noise on the solution of gyros/star sensor integration. It first establishes a direct installation mode of gyros/star sensor integration to achieve a high turntable control stability and then derives its associated mathematical model. Further, an event-driven maximum correntropy filter based on the Cauchy kernel is proposed to handle the non-Gaussian noise involved in the measurement of gyros/star sensor integration caused by the complex spatial environment. It also constructs an event-driven mechanism according to the innovation information to reduce the redundant computational cost for the optimization of the filter's real-time performance. The proposed methodology can achieve robustness to curb the influence of non-Gaussian measurement noise with a superior computational cost compared to the existing MC filters. Simulations and a comparison analysis based on the gyros/star sensor integration for a two-dimensional turntable's spatial orientation have been carried out to comprehensively evaluate the effectiveness of the proposed filtering method.

2. Model of Gyros/Star Sensor Integration for Spatial Orientation

The direct installation mode is used for the integration of the gyros and star sensor, which is described in Figure 1. It can be seen from Figure 1 that the gyros are installed in the inner frame of the two-dimensional turntable and firmly connected with the load camera/radar, and the star sensor is installed on the fixed base of the turntable.

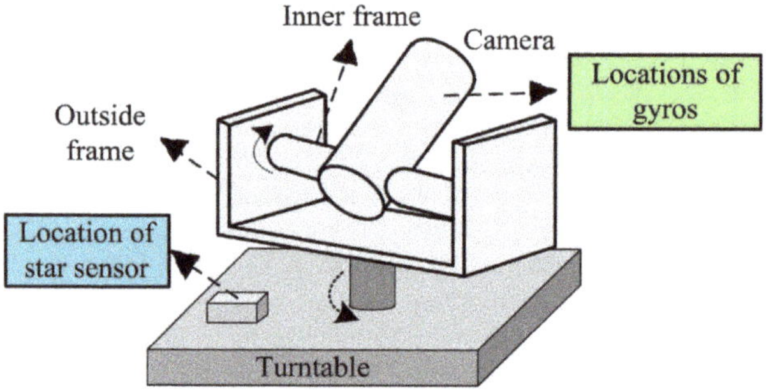

Figure 1. The direct installation mode of the gyros/star sensor integration.

Based on the direct installation mode, the framework of the gyros/star sensor integration used for the spatial orientation is illustrated in Figure 2, which has two loops: a filtering loop and error calculation loop. In the filtering loop, the gyros' measurement information, the turntable control quantities, and the attitude of the star sensor serve as the input of filter for the error state estimation of gyros/star sensor integration. Then, the estimated gyros' attitude error is fed back to correct the gyros and the gyros' attitude is updated for the next time step. In the error calculation loop, the orientation error between the tracking vector (the normalized vector between the spacecraft and target) and the attitude reference of the spacecraft is calculated by using the turntable's control quantities and the spacecraft's attitude at each time step to reflect the tracking accuracy.

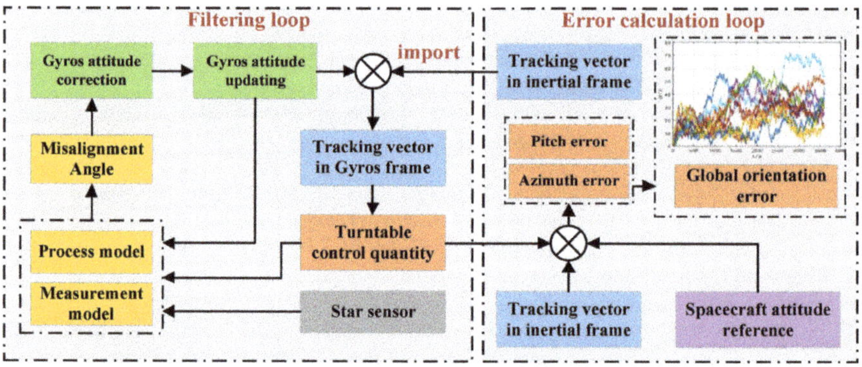

Figure 2. Framework of gyros/star sensor integration for spatial orientation. The different colored lines in the global orientation error plot represent the orientation error curves of multiple repeat tests.

It should be noted that as shown in Figures 1 and 2, the star sensor realizes the three-dimensional attitude information of turntable base. Since the rotation angles of the turntable's inner frame and outside frame are known through the pitch and azimuth code discs, the three-dimensional attitude information of the turntable base can be transformed into the turntable's inner frame (E-frame) and inertial frame (I-frame). Based on this attitude information, the measurements and measurement model for the gyros/star sensor integration are established. Further, combining the process model with the measurement model, gyros' platform misalignment angle can be estimated through the filtering process. This estimation will be fed back to the gyros to correct the attitude error. Thus, the star sensor can restrain and correct the gyros' drift and further enhance the turntable's spatial orientation accuracy.

Based on the above, the model of gyros/star sensor integration can be further derived and established, which includes the process and measurement models.

2.1. Process Model

Denote the inertial frame as the I-frame, body frame as the B-frame, turntable's base frame as the G-frame, turntable's outside frame as the A-frame, turntable's inner frame as the E-frame and gyros' composite frame as the P-frame.

The I-frame is chosen as the navigation frame; thus, the angular relation between the I-frame and P-frame can be expressed by the attitude matrix [31,32]

$$\dot{C}_P^I = C_P^I\left(\omega_{IP}^P \times\right) \tag{1}$$

where C_P^I is the attitude matrix from the P-frame and I-frame, $\dot{C}_P^I$ is the differential form of C_P^I, ω_{IP}^P represents the gyros' output, and $(\omega_{IP}^P\times)$ stands for the skew symmetric matrix of ω_{IP}^P, which can be calculated as [33]

$$\left(\omega_{IP}^P \times\right) = \begin{bmatrix} 0 & -\omega_{IPz}^P & \omega_{IPy}^P \\ \omega_{IPz}^P & 0 & -\omega_{IPx}^P \\ -\omega_{IPy}^P & \omega_{IPx}^P & 0 \end{bmatrix} \tag{2}$$

where ω_{IPi}^P $(i = x, y, z)$ represents the components of ω_{IP}^P in the directions of x, y and z.

Considering the measurement error, the above attitude matrix should be written as $C_P^{I'}$, and the I'-frame and I-frame have a platform misalignment angle. Thus, we have the following relationship [32,34]

$$C_P^{I'} = C_I^{I'} C_P^I = \left(\mathbf{I} - \phi^I \times\right) C_P^I \tag{3}$$

where ϕ^I stands for the platform misalignment angle.

Further, according to the definition of platform misalignment angle ϕ^I in [35], it has a differential relation with the gyros' zero-bias error, which can be described as follows. The specifics can be seen in [35].

$$\dot{\phi}^I = -C_P^I \varepsilon^P \tag{4}$$

where ε^P is the gyros' zero-bias error.

It can be seen that the attitude update of the gyros diverges with time due to the existence of gyros' zero-bias error; thus, we can use the star sensor's output to estimate and correct the gyros' attitude errors.

Denote the system state of gyros/star sensor integration as

$$X = \left[(\phi^I)^{\mathrm{T}} \quad (\varepsilon^P)^{\mathrm{T}} \right]^{\mathrm{T}} \tag{5}$$

Based on (5), the process model of gyros/star sensor integration can be described as

$$\dot{X} = \begin{bmatrix} 0_{3\times3} & -C_P^I \\ 0_{3\times3} & 0_{3\times3} \end{bmatrix} X + \begin{bmatrix} w_P \\ 0_{3\times1} \end{bmatrix} \tag{6}$$

Discretizing (6), we can obtain

$$X_k = \begin{bmatrix} \phi_k^I \\ \varepsilon_k^P \end{bmatrix} = \begin{bmatrix} I_{3\times3} & -C_P^I \\ 0_{3\times3} & I_{3\times3} \end{bmatrix} \begin{bmatrix} \phi_{k-1}^I \\ \varepsilon_{k-1}^P \end{bmatrix} + \begin{bmatrix} I_{3\times3} \\ 0_{3\times3} \end{bmatrix} W_{k-1} \tag{7}$$

2.2. Measurement Model

In the direct installation mode, there is a turntable angle between the gyros and the star sensor. Thus, according to Figure 1, the error matrix can be constructed based on the output of the star sensor

$$M_{Zb} = \tilde{C}_I^{P_i} \tilde{C}_B^I C_G^B C_A^G C_E^A = \tilde{C}_E^{P_i} \tag{8}$$

where C is the rotation matrix from one frame to another frame.

In the real application, since the installation angle can be calibrated to compensate for its error, the E-frame can be seen coincident with the P-frame. Further, by using the Rodrigues formula [34], (8) can be further written as

$$\begin{aligned} M_{Zb} &= \widetilde{C}_E^{P_i} = \widetilde{C}_P^{P_i} = (\widetilde{C}_{P_i}^P)^{\mathrm{T}} \\ &= \left(I + \frac{\sin\phi^P}{\phi^P}(\boldsymbol{\phi}^P\times) + \frac{1-\cos\phi^P}{(\phi^P)^2}(\boldsymbol{\phi}^P)^2\right)^{\mathrm{T}} \approx (I + (\boldsymbol{\phi}^P\times))^{\mathrm{T}} = (I - (\boldsymbol{\phi}^P\times)) \end{aligned} \tag{9}$$

where $\boldsymbol{\phi}^P = [\phi_x^P \quad \phi_y^P \quad \phi_z^P]^{\mathrm{T}}$ is the platform misalignment angle between the P-frame and P_i-frame computed by the output of star sensor, which can be seen as a small amount; $\phi^P = |\boldsymbol{\phi}^P|$ represents its modulus.

According to the definition of the skew symmetric matrix [33], we have

$$\left(\boldsymbol{\phi}^P\times\right) = \begin{bmatrix} 0 & -\phi_z^P & \phi_y^P \\ \phi_z^P & 0 & -\phi_x^P \\ -\phi_y^P & \phi_x^P & 0 \end{bmatrix} \tag{10}$$

Thus, expanding (9) yields

$$M_{Zb} = I - \begin{bmatrix} 0 & -\phi_z^P & \phi_y^P \\ \phi_z^P & 0 & -\phi_x^P \\ -\phi_y^P & \phi_x^P & 0 \end{bmatrix} = \begin{bmatrix} 1 & \phi_z^P & -\phi_y^P \\ -\phi_z^P & 1 & \phi_x^P \\ \phi_y^P & -\phi_x^P & 1 \end{bmatrix} \tag{11}$$

Then, the measurement of gyros/star sensor integration can be constructed as

$$Z_k = -\frac{1}{2}\begin{bmatrix} M_{32} - M_{23} \\ M_{13} - M_{31} \\ M_{21} - M_{12} \end{bmatrix} = \begin{bmatrix} \phi_x^P \\ \phi_y^P \\ \phi_z^P \end{bmatrix} = \boldsymbol{\phi}^P \tag{12}$$

where M_{ij} represents the ith row and jth column element of the matrix M_{Zb}.

Using the relationship $\boldsymbol{\phi}^P = C_I^P \boldsymbol{\phi}^I$, the measurement model can be established by

$$Z_k = \frac{1}{2}\begin{bmatrix} M_{32} - M_{23} \\ M_{13} - M_{31} \\ M_{21} - M_{12} \end{bmatrix} = \begin{bmatrix} C_I^P & 0_{3\times 3} \end{bmatrix} X_k + V_k \tag{13}$$

where V_k is the measurement noise.

Thus, the model of gyros/star sensor integration for spatial orientation, including the process and measurement models, is achieved.

3. Event-Driven Maximum Correntropy Filter Based on Cauchy Kernel

As discussed previously, the star sensor's measurement is subject to the spatial environment's interference due to low light intensity. This will make the measurement noise of gyros/star sensor integration in (12) no longer adhere to a Gaussian distribution. In this situation, the traditional Kalman filtering has no ability to deal with the non-Gaussian measurement noise, leading to divergent filtering results. This may cause ineffective correction for the gyros by using the star sensor. Thus, this section establishes an event-driven maximum correntropy filter based on the Cauchy kernel to address the above problem.

3.1. Maximum Correntropy Filter Based on Cauchy Kernel

3.1.1. Maximum Correntropy Criterion

In information theory, correlation entropy is often defined as a statistic that describes the similarity between two random variables. Cross-correlation entropy (correntropy) is an extension of the correlation between random variables. It is not only used to measure the second-order information of the joint probability density function but can also measure higher-order moments.

Firstly, the correntropy between the random variables X and Y is defined as [23]

$$C(X,Y) = E[G_\sigma(X,Y)] = \int G_\sigma(X,Y)dF_{XY}(x,y) \tag{14}$$

where $E[\cdot]$ represents the calculation of expectation, $G_\sigma(\cdot)$ is a shift-invariant Mercer kernel, $F_{XY}(x,y)$ stands for the joint probability distribution of variables X and Y, and $\int \cdot$ represents the integral operation.

The Gaussian kernel is the most widely used kernel function in Mercer theory, and it is described as [29]

$$G_\sigma(X,Y) = G_\sigma(X-Y) = \exp\left(-\frac{\|X-Y\|^2}{2\sigma^2}\right) \tag{15}$$

where $\sigma > 0$ represents the bandwidth of the Gaussian kernel.

However, the Gaussian kernel may not always be an optimal selection due to the appearance of singular matrices, making the filter break down. In this paper, we use the Cauchy kernel instead of the Gaussian kernel as the kernel function for the correntropy to better capture the heavy-tailed features in the noise. The Cauchy kernel has merits such as being less sensitive to kernel bandwidth and having better stability compared to the Gaussian kernel. It is defined as [27]

$$G_\sigma(X,Y) = G_\sigma(X-Y) = \frac{1}{1+\|X-Y\|^2/\sigma} \tag{16}$$

where $\sigma > 0$ is the bandwidth of the Cauchy kernel, which is used to adjust the degree of influence of the kernel function on non-Gaussian noise.

In real applications, the true joint distribution function of random variables is usually inaccessible; thus, the sample mean estimation of correntropy is often calculated to approximate the true joint distribution function [28]

$$\hat{C}(X,Y) = \frac{1}{N}\sum_{i=1}^{N} G_\sigma(x_i, y_i) \tag{17}$$

where N is the sample number, and x_i and y_i are the sample data of the random variables X and Y.

Correntropy can capture the higher-order moment information of the observed signal, which can be described by the Taylor series expansion of the kernel function [28]

$$\hat{C}(X,Y) = \sum_{n=0}^{\infty} \frac{(-1)^n}{\sigma^n} E\left[(X-Y)^{2n}\right] \tag{18}$$

It can be seen from (15) and (16) that when $X = Y$, the correntropy of two random variables reaches its maximum, indicating the strongest similarity between them, which is the core principle of the maximum correntropy criterion. In addition, it can be seen from (18) that the correntropy is a weighted sum of the even-order moments of $X - Y$. Unlike the minimum mean square error criterion in Kalman filtering, only the second-order moment is considered. Thus, the maximum correntropy criterion can be used to handle the influence of non-Gaussian noises.

Further, from (15) and (16), we can see that the Gaussian kernel gets exponentially closer to 0 as $X - Y$ increases, which leads to an increase in the possibility of matrix singularity. As a comparison, the Cauchy kernel approaches zero much more slowly. Thus, it can effectively reduce the probability of matrix singularity. This is the reason why the Cauchy kernel instead of the Gaussian kernel is used as the kernel function for the correntropy in this paper.

3.1.2. Process of the Maximum Correntropy Filter Algorithm Based on Cauchy Kernel

According to the above-described maximum entropy criterion based on the Cauchy kernel in Section 3.1.1, we can further combine it with Kalman filtering to establish the maximum correntropy filter based on the Cauchy kernel.

As shown in (7) and (13), the system process and measurement models of gyros/star sensor integration can be summarized as the following form

$$\begin{cases} \boldsymbol{X}_k = \boldsymbol{F}_{k/k-1}\boldsymbol{X}_{k-1} + \boldsymbol{W}_k \\ \boldsymbol{Z}_k = \boldsymbol{H}_k\boldsymbol{X}_k + \boldsymbol{V}_k \end{cases} \tag{19}$$

where

$$\boldsymbol{F}_{k/k-1} = \begin{bmatrix} \boldsymbol{I}_{3\times3} & -\boldsymbol{C}_P^I \\ \boldsymbol{0}_{3\times3} & \boldsymbol{I}_{3\times3} \end{bmatrix} \tag{20}$$

The calculation procedure of traditional Kalman filter can be described as

Step 1: Initialization parameter.

$$\begin{cases} \hat{\boldsymbol{X}}_0 = E[\boldsymbol{X}_0] \\ \boldsymbol{P}_0 = E[(\boldsymbol{X}_0 - \hat{\boldsymbol{X}}_0)(\boldsymbol{X}_0 - \hat{\boldsymbol{X}}_0)^{\mathrm{T}}] \end{cases} \tag{21}$$

Step 2: Time update. Using the previous state estimate and its error covariance matrix in time $k-1$, the state prediction and its error covariance matrix are calculated as [17]

$$\hat{\boldsymbol{X}}_{k/k-1} = \boldsymbol{F}_{k/k-1}\hat{\boldsymbol{X}}_{k-1} \tag{22}$$

$$\boldsymbol{P}_{k/k-1} = \boldsymbol{F}_{k/k-1}\boldsymbol{P}_{k-1}\boldsymbol{F}_{k/k-1}^{\mathrm{T}} + \boldsymbol{Q}_{k-1} \tag{23}$$

where $\boldsymbol{F}_{k/k-1}$ is the state transition matrix described by (20); $\hat{\boldsymbol{X}}_{k/k-1}$ and $\boldsymbol{P}_{k/k-1}$ are the state prediction and its error covariance at time k; $\hat{\boldsymbol{X}}_{k-1}$ and $\boldsymbol{P}_{k-1}$ are the state estimation and its error covariance at the previous time $k-1$; and $\boldsymbol{Q}_{k-1}$ is the process noise covariance.

Step 3: Measurement update. The processes for measurement updates are as follows [17]

$$\boldsymbol{K}^k = \boldsymbol{P}_{k/k-1}\boldsymbol{H}_k^{\mathrm{T}}(\boldsymbol{H}_k\boldsymbol{P}_{k/k-1}\boldsymbol{H}_k^{\mathrm{T}} + \boldsymbol{R}_k)^{-1} \tag{24}$$

$$\hat{\boldsymbol{X}}_k = \hat{\boldsymbol{X}}_{k/k-1} + \boldsymbol{K}_k[\boldsymbol{Z}_k - \boldsymbol{H}_k\hat{\boldsymbol{X}}_{k/k-1}] \tag{25}$$

$$\boldsymbol{P}_k = (\boldsymbol{I} - \boldsymbol{K}_k\boldsymbol{H}_k)\boldsymbol{P}_{k/k-1} \tag{26}$$

where $\boldsymbol{R}_k$ is the measurement noise covariance; $\boldsymbol{K}_k$ is the filtering gain; and $\boldsymbol{H}_k$ is the measurement matrix.

However, due to the weak light environment in space, the star sensor's measurement information is subject to be disturbed, leading to the occurrence of non-Gaussian measurement noise. This further reduces the estimation accuracy of the Kalman filtering and may even cause divergence, resulting in the loss of its ability to correct the gyros.

According to the maximum correntropy criterion based on the Cauchy kernel, we construct a new filtering method to modify the measurement update process of Kalman filtering to handle the influence of non-Gaussian measurement noise. Firstly, based on the system model shown in (19), we have

$$\begin{bmatrix} \hat{\boldsymbol{X}}_{k/k-1} \\ \boldsymbol{Z}_k \end{bmatrix} = \begin{bmatrix} \boldsymbol{I} \\ \boldsymbol{H}_k \end{bmatrix}\boldsymbol{X}_k + \boldsymbol{A}_k \tag{27}$$

where

$$A_k = \begin{bmatrix} \hat{X}_{k/k-1} - X_k \\ V_k \end{bmatrix}, \text{ and } E\left[A_k A_k^T\right] = \begin{bmatrix} P_{k/k-1} & 0 \\ 0 & R_k \end{bmatrix} = B_k B_k^{\mathrm{T}} \tag{28}$$

where B_k is the Cholesky decomposition of $E\left[A_k A_k^T\right]$.

Multiplying both sides of (27) by B_k^{-1} yields

$$Y_k = M_k X_k + E_k \tag{29}$$

where $Y_k = B_k^{-1}\begin{bmatrix} \hat{X}_{k/k-1} \\ Z_k \end{bmatrix}$, $M_k = B_k^{-1}\begin{bmatrix} I \\ H_k \end{bmatrix}$, and $E_k = B_k^{-1} A_k$.

Then, a correntropy cost function can be established as

$$J = \frac{1}{L}\sum_{i=1}^{L} G_\sigma\left(E_k^i\right) = \frac{1}{L}\sum_{i=1}^{L} G_\sigma\left(y_k^i - m_k^i X_k\right) \tag{30}$$

where $L = n + m$ represents the dimension of Y_k, and m and n are the dimensions of Z_k and X_k, respectively; E_k^i and y_k^i represent the ith elements of E_k and Y_k; and m_k^i is the ith row of M_k.

The optimal state estimation can be obtained using the following equation.

$$\frac{\partial J}{\partial X_k} = \frac{2}{\sigma}\frac{1}{L}\sum_{i=1}^{L} G_\sigma(E_k^i)^2 m_k^{i,T}\left(y_k^i - m_k^i X_k\right) = 0 \tag{31}$$

After simplification, it is evident that the following result can be obtained [36]

$$\begin{aligned} X_k &= \left(\sum_{i=1}^{L} G_\sigma(E_k^i)^2 m_k^{i,T} m_k^i\right)^{-1}\left(\sum_{i=1}^{L} G_\sigma(E_k^i)^2 m_k^{i,T} y_k^i\right) \\ &= \left(M_k^T D_k M_k\right)^{-1}\left(M_k^T D_k Y_k\right) \end{aligned} \tag{32}$$

where $D_k = \begin{bmatrix} D_k^x & 0 \\ 0 & D_k^z \end{bmatrix}$ with

$$\begin{aligned} D_k^x &= diag\left(G_\sigma(E_k^1)^2 \quad ,\ldots, \quad G_\sigma(E_k^n)^2 \right) \\ &= diag\left(\left(\frac{1}{1+\left(\hat{X}_{k/k-1}^1 - m_k^1 \hat{X}_k\right)^2/\sigma}\right)^2 \quad ,\ldots, \quad \left(\frac{1}{1+\left(\hat{X}_{k/k-1}^n - m_k^n \hat{X}_k\right)^2/\sigma}\right)^2 \right) \end{aligned} \tag{33}$$

$$\begin{aligned} D_k^z &= diag\left(G_\sigma(E_k^{n+1})^2 \quad ,\ldots, \quad G_\sigma(E_k^{n+m})^2 \right) \\ &= diag\left(\left(\frac{1}{1+\left(Z_k^1 - m_k^{n+1} \hat{X}_k\right)^2/\sigma}\right)^2 \quad ,\ldots, \quad \left(\frac{1}{1+\left(Z_k^m - m_k^{n+m} \hat{X}_k\right)^2/\sigma}\right)^2 \right) \end{aligned} \tag{34}$$

where $\hat{X}_{k/k-1}^i (i = 1, 2, \cdots, n)$ and $Z_k^i (i = 1, 2, \cdots, m)$ are the ith elements of $\hat{X}_{k/k-1}$ and Z_k.

By expanding (32), we can obtain

$$\begin{aligned} X_k &= \left(M_k^{\mathrm{T}} D_k M_k\right)^{-1}\left(M_k^{\mathrm{T}} D_k Y_k\right) \\ &= \left(\left(B_k^{P,\mathrm{T}}\right)^{-1} D_k^x \left(B_k^P\right)^{-1} + H_k^{\mathrm{T}}\left(B_k^{R,\mathrm{T}}\right)^{-1} D_k^z \left(B_k^R\right)^{-1} H_k\right)^{-1} \\ &\quad \left(\left(B_k^{P,\mathrm{T}}\right)^{-1} D_k^x \left(B_k^P\right)^{-1} \hat{X}_{k/k-1} + H_k^{\mathrm{T}}\left(B_k^{R,\mathrm{T}}\right)^{-1} D_k^z \left(B_k^R\right)^{-1} Z_k\right) \end{aligned} \tag{35}$$

where B_k^P and B_k^R are the Cholesky decomposition of $P_{k/k-1}$ and R_k.

Applying the matrix inversion lemma ($(A+BD^{-1}C)^{-1} = A^{-1} - A^{-1}B(D+CA^{-1}B)^{-1}CA^{-1}$), (35) can be become

$$\hat{X}_k = \hat{X}_{k/k-1} + \overline{K}_k\left[Z_k - H_k\hat{X}_{k/k-1}\right] \tag{36}$$

where

$$\overline{K}_k = \overline{P}_{k/k-1}H_k^{\mathrm{T}}(H_k\overline{P}_{k/k-1}H_k^{\mathrm{T}} + \overline{R}_k)^{-1} \tag{37}$$

where $\overline{P}_{k/k-1} = B_k^P(D_k^x)^{-1}B_k^{P,\mathrm{T}}, \overline{R}_k = B_k^R(D_k^z)^{-1}B_k^{R,\mathrm{T}}$. The specific derivation from (35)–(37) can be found in [36].

To estimate the system state through (36), it is necessary to compute the filter gain $\widetilde{K}_k$ in (35) firstly. The calculation of the gain inherently relies on the posterior state estimate $\hat{X}_k$, which is involved in (33) and (34). This clearly indicates an iterative process for state estimation. Therefore, we outline the measurement update process of the maximum correntropy filter based on the Cauchy kernel as follows:

(i) Let the iteration index $t = 1$ and $\hat{X}_k^1 = \hat{X}_{k/k-1}$, $\hat{X}_k^{t+1}$ can be achieved by the following process

$$\hat{X}_k^{t+1} = \hat{X}_k^t + \widetilde{K}_k\left[Z_k - H_k\hat{X}_k^t\right] \tag{38}$$

$$\widetilde{K}_k = \widetilde{P}_{k/k-1}H_k^{\mathrm{T}}(H_k\widetilde{P}_{k/k-1}H_k^{\mathrm{T}} + \widetilde{R}_k)^{-1} \tag{39}$$

$$\widetilde{P}_{k/k-1} = B_k^P\left(\widetilde{D}_k^x\right)^{-1}B_k^{P,T} \tag{40}$$

$$\widetilde{R}_k = B_k^R\left(\widetilde{D}_k^z\right)^{-1}B_k^{R,T} \tag{41}$$

where $\widetilde{D}_k^x = diag(G_\sigma(\widetilde{E}_k^1)^2, \ldots, G_\sigma(\widetilde{E}_k^n)^2)$ and $\widetilde{D}_k^z = diag(G_\sigma(\widetilde{E}_k^{n+1})^2, \ldots, G_\sigma(\widetilde{E}_k^{n+m})^2)$; $\widetilde{E}_k^i$ is the element of $\widetilde{E}_k$ and $\widetilde{E}_k = Y_k - M_k\hat{X}_k^t$.

(ii) Compare the posterior estimation from the current iteration $\hat{X}_k^{t+1}$ with those from the previous iteration $\hat{X}_k^t$. If it satisfies the following condition, the iterative process is terminated; otherwise, set the final state estimation $\hat{X}_k = \hat{X}_k^{t+1}$ and return to Step (i).

$$\frac{\left\|\hat{X}_k^{t+1} - \hat{X}_k^t\right\|}{\left\|\hat{X}_k^t\right\|} \leq \varepsilon \tag{42}$$

where ε is a relatively small number.

(iii) Compute the posterior error covariance matrix as

$$P_k = (I - \widetilde{K}_kH_k)P_{k/k-1} \tag{43}$$

At this time, the measurement update process at time k of the maximum correntropy filter based on the Cauchy kernel can be terminated.

It can be seen from (33) to (37) that as the bandwidth $\sigma \to \infty$, D_k^x and D_k^z will tend to unity matrices. In this case, the maximum correntropy filter based on the Cauchy kernel will be degraded to the Kalman filter and its robustness will gradually vanish. If $\sigma \to 0$, D_k^x and D_k^z will become $\mathbf{0}$, the calculation of filter gain $\widetilde{K}_k$ will be out of operation and the filter will be disabled. Thus, an appropriate kernel bandwidth is vital to affect the performance of the maximum correntropy filter based on the Cauchy kernel. Currently, it is usually determined by prior information or a multiple test computation prior to the filtering solution.

3.2. Event-Driven Conditions

From the above filtering algorithm process, it can be seen that multiple iterations are involved. This means it needs a large computational cost for the filtering calculation at each time. Thus, in the times without non-Gaussian measurement noise, the maximum correntropy filter based on the Cauchy kernel will involve too many redundant calculations. Due to this reason, this section establishes an adaptive event-driven mechanism to effectively recognize the non-Gaussian measurement noise, which can significantly improve the computational efficiency of the maximum correntropy filter based on the Cauchy kernel.

The constructed event-driven conditions are related to the filter's innovation vector information, which is calculated as [37]

$$\widetilde{\boldsymbol{Z}}_k = \boldsymbol{Z}_k - \boldsymbol{H}_k \hat{\boldsymbol{X}}_{k/k-1} \tag{44}$$

In the normal case, the innovation $\widetilde{\boldsymbol{Z}}_k$ should adhere to a Gaussian distribution and its covariance is [37]

$$\boldsymbol{P}_{ZZ,k/k-1} = \boldsymbol{H}_k \boldsymbol{P}_{k/k-1} \boldsymbol{H}_k^{\mathrm{T}} + \boldsymbol{R}_k \tag{45}$$

Consider that $\boldsymbol{P}_{ZZ,k/k-1}$ is a positive matrix, we can construct a unitary matrix $\boldsymbol{L}_k$ to make the following relationship hold

$$\boldsymbol{L}_k^{\mathrm{T}} \boldsymbol{P}_{ZZ,k/k-1} \boldsymbol{L}_k = \Sigma_k \tag{46}$$

where $\Sigma_k = diag\{\lambda_{k1}, \lambda_{k2}, \ldots, \lambda_{kn_i}\} \in n_i \times n_i$, $\lambda_{k1}, \lambda_{k2}, \ldots, \lambda_{kn_i}$ are the eigenvalues of $\boldsymbol{P}_{ZZ,k/k-1}$ and n_i represents the number of eigenvalues.

Further, define $\boldsymbol{F}_k = \boldsymbol{L}_k \Sigma_k^{-1/2}$, the innovation vector can be normalized as

$$\overline{\boldsymbol{Z}}_k = \boldsymbol{F}_k^{\mathrm{T}} \widetilde{\boldsymbol{Z}}_k \tag{47}$$

Equation (47) is called Mahalanobis transformation. The elements of $\overline{\boldsymbol{Z}}_k$ should be uncorrelated and obey the standard Gaussian distribution [22,37]. Thus, the event-driven conditions for the detection of the non-Gaussian measurement noise can be constructed as

$$\tau_k = \begin{cases} 1 & \kappa_\beta \leq \left\| \overline{\boldsymbol{Z}}_k \right\|_\infty \leq \kappa_\alpha \\ 0 & Otherwise \end{cases} \tag{48}$$

where $\left\| \overline{\boldsymbol{Z}}_k \right\|_\infty$ represents the infinity norm of $\overline{\boldsymbol{Z}}_k$, i.e., $\left\| \overline{\boldsymbol{Z}}_k \right\|_\infty = \max\left\{ \overline{\boldsymbol{Z}}_{k,1}, \overline{\boldsymbol{Z}}_{k,2}, \ldots, \overline{\boldsymbol{Z}}_{k,q} \right\}$; κ_β and κ_α are the lower threshold and the higher threshold, respectively.

Based on (48), there are three cases that will appear:

Case 1: $\left\| \overline{\boldsymbol{Z}}_k \right\|_\infty < \kappa_\beta$ and $\tau_k = 0$. This case considers that the integration of the gyros/star sensor can be predictable and the measurement does not need to be applied to update the estimation. The purpose of this case is to reduce the communication burden between the gyros and star sensor so as to decrease the whole system's computational cost.

Case 2: $\kappa_\beta \leq \left\| \overline{\boldsymbol{Z}}_k \right\|_\infty \leq \kappa_\alpha$ and $\tau_k = 1$. This case considers that the measurement is normal and the non-Gaussian noise is not involved;

Case 3: $\left\| \overline{\boldsymbol{Z}}_k \right\|_\infty > \kappa_\alpha$ and $\tau_k = 0$. This case considers that the measurement is deviant and the non-Gaussian noise is involved.

It should be noted that the selection of thresholds κ_β and κ_α is very important for the detection of the non-Gaussian measurement noise. κ_β is chosen to balance the communication burden and estimation performance, which can be set according to the required communication rate. κ_α is always set according to the confidence level. Through the simulation test, a 97.5% confidence level can obtain a relatively satisfactory detection performance.

3.3. Algorithm Description

Based on the maximum correntropy Kalman filter based on the Cauchy kernel described in Section 3.1 and event-driven conditions constructed in Section 3.2, we can establish the algorithm of the event-driven maximum correntropy Kalman filter based on the Cauchy kernel, whose computational process is illustrated in Algorithm 1. It can balance the computational efficiency and robustness of the filter, reducing unnecessary computational load while ensuring the algorithm's anti-interference performance.

Algorithm 1. Event-driven- maximum correntropy filter based on Cauchy kernel (ED-MCFCK)

1 **Initialization:** $\hat{x}_0$, P_0
2 **For** k = 1,2,. . .
{
3 Compute $\hat{X}_{k/k-1}$ and $P_{k/k-1}$ as (22) and (23) of Kalman filtering procedures;
4 Calculate filter's innovation vector as well as its covariance as (44) and (45), and further construct the normalized innovation vector as (47).
5 Compute and conduct the event driven condition by (48).
6 **If** $\|\overline{Z}_k\|_\infty < \kappa_\beta$ **and** $\tau_k = 0$, set the state prediction $\hat{X}_{k/k-1}$ and $P_{k/k-1}$ as the final state estimation.
7 **If** $\kappa_\beta \le \|\overline{Z}_k\|_\infty \le \kappa_\alpha$ and $\tau_k = 1$, execute (24)–(26) to obtain system state esti-mation $\hat{X}_k$ and P_k.
8 **If** $\|\overline{Z}_k\|_\infty > \kappa_\alpha$ **and** $\tau_k = 0$, (indicating the presence of non-Gaussian noise in the measurement information), the MCFCK is driven.
{
9 Let the iteration index t = 1 and $\hat{X}_k^1 = \hat{X}_{k/k-1}$;
10 Compute $\widetilde{P}_{k/k-1}$ and $\widetilde{R}_k$ according to (40) and (41), and further calcu late filter gain $\widetilde{K}_k$ by (39).
11 Compute $\hat{X}_k^t$ by (38) and test the *Condition* (42);
12 **If** *Condition* $\ge \varepsilon$
13 Let $t = t + 1$ (iterations) and $\hat{X}_k = \hat{X}_k^{t+1}$, and conduct the next iteration.
14 **Else**, the iterative process is terminated and $\hat{X}_k = \hat{X}_k^t$.
}
15 Compute the posterior error covariance matrix by (43).
}

4. Simulation Evaluations and Discussions

Simulation evaluations have been conducted to comprehensively validate the efficacy of the proposed event-driven maximum correntropy filter based on the Cauchy kernel (ED-MCFCK) for a two-dimensional turntable's spatial orientation using the gyros/star sensor integration. A comparison analysis of the proposed ED-MCFCK with the Kalman filter and MCF based on the Gaussian kernel (abbreviated as MCF) as well as MCFCK is also conducted in this section for the handling of non-Gaussian noise involved in the measurements.

As shown in Figure 3, a trajectory of spacecraft is designed according to its actual flight process, which is a classical parabolic curve motion. Further, the attitude of the spacecraft at every time step, including yaw, pitch, and roll is depicted in Figure 4. The initial attitude of the spacecraft was $(142.16^\circ, -89.78^\circ, 63.16^\circ)$ and the total simulation time was 3600 s. Accordingly, a two-dimensional turntable was installed on the spacecraft, and the tracking vector between the spacecraft and the target star is also shown in Figure 3.

For the filtering loop shown in Figure 2, the gyros' sampling rates are both 20 Hz, the star sensor's sampling rate is 1 Hz, and the other simulation parameters are listed in Table 1. The initial state and its error covariance, as well as the initial process noise covariance, are set as

$$\begin{cases} X_0 = [0_{6\times1}]^{\mathrm{T}} \\ P_0 = diag[\ (1rad)^2 \ \ (1rad)^2 \ \ (1rad)^2 \ \ (1'')^2 \ \ (1'')^2 \ \ (1'')^2\ ] \\ Q_0 = diag[\ (0.01^\circ/\sqrt{h})^2 \ \ (0.01^\circ/\sqrt{h})^2 \ \ (0.01^\circ/\sqrt{h})^2\ ] \end{cases} \quad (49)$$

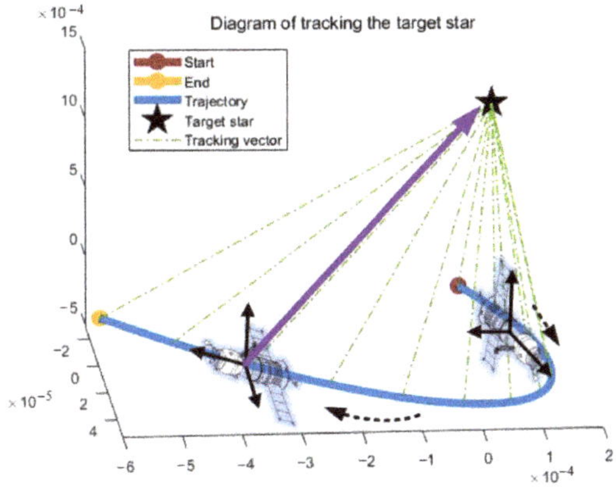

Figure 3. Trajectory and tracking vector of the spacecraft in *I*-frame. The spacecraft's trajectory is plotted using the endpoint of the tracking vector, so this trajectory is unitless. The black solid arrows represent the spacecraft's attitude, the green dashed and purple solid arrows represent the tracking vector, and the black dashed arrows represent the spacecraft's direction of motion.

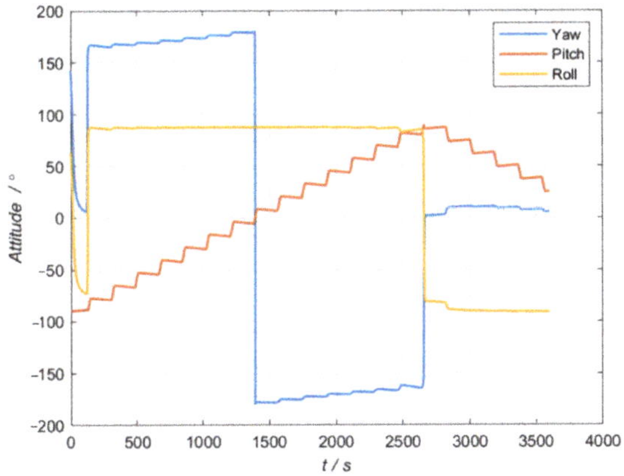

Figure 4. Attitude change of the spacecraft.

Table 1. Simulation parameters of gyros/star sensor integration.

Information Sources	Simulation Parameters	Values
Gyros	Gyros' constant drift	$0.1°/\mathrm{h}$
	Gyros' random walk coefficient	$0.01°/\sqrt{h}$
	Scale factor error	100 PPM
	Misalignment error	$5''$
Star sensor	Goniometric error	$5''$ (3σ)

A. Accuracy Evaluation and Analysis

To verify the accuracy and robustness of the proposed ED-MCFCK in the presence of non-Gaussian measurement noise, two typical non-Gaussian measurement noise scenarios were considered, i.e., non-Gaussian noise with outliers in measurements and α stable noise in measurements.

By using the trial and error method [38], the bandwidth parameters σ in the above-mentioned maximum correntropy filters were determined to be 13. For the designed event-driven conditions, the confidence level is set as 97.5% (α = 0.025) with 3 degrees of freedom (m = 3).

(1) Non-Gaussian noise with outliers in measurements

To simulate this situation, a random noise with Gaussian distribution was directly injected into the measurement during the time interval (1500 s, 2500 s) with a probability of $\rho = 0.5$. Thus, the measurement becomes the following

$$\mathbf{Z}_k = \begin{cases} \mathbf{Z}_k, & others \\ (1-\rho)\mathbf{Z}_k + \rho[\mathbf{Z}_k + N(0, (4 \times 10^{-4} rad)^2)], & (1500s, 2500s) \end{cases} \tag{50}$$

It can be seen that the measurement noise in (50) is mixed Gaussian, which is a non-Gaussian and strong impulsive noise (outliers).

For the case of non-Gaussian noise with outliers in measurements, Figures 5 and 6 illustrate the orientation errors in azimuth and pitch achieved by the above-mentioned Kalman filter, MCF, MCFCK and ED-MCFCK, respectively. Further, the corresponding 3σ error boundary for the ED-MCFCK, which is calculated based on the 10 runs of Monte Carlo simulations as in [39], is also provided in Figures 5 and 6. As shown in Figures 5 and 6, during the time interval (1500 s, 2500 s) with non-Gaussian noise, the Kalman filter has no ability and fails to handle the abnormal non-Gaussian noise. This further deteriorates the obtained state estimate during the filtering measurement update process, resulting in a significant degradation in orientation accuracy. The orientation errors in azimuth and pitch exhibit severe fluctuations. In contrast, MCF, MCFCK and ED-MCFCK can leverage the robustness of the Gaussian/Cauchy kernel to effectively handle the non-Gaussian noise information, leading to stable orientation errors in azimuth and pitch.

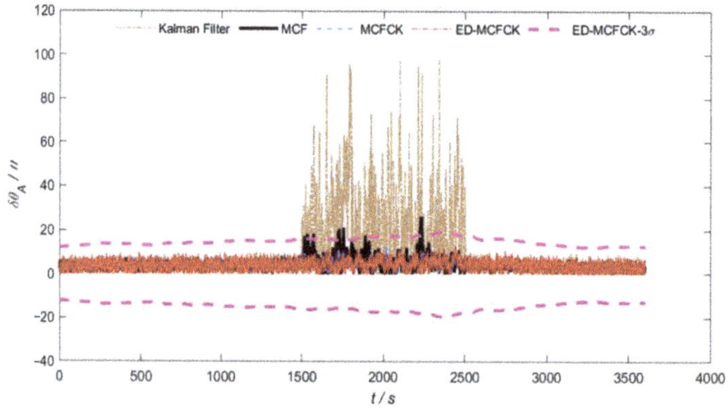

Figure 5. Orientation errors in azimuth and 3σ error boundary for the case of non-Gaussian noise with outliers.

The global orientation accuracy and the corresponding Root Mean Square Error (RMSE) are further used to evaluate the orientation accuracy, where the global orientation error is defined as

$$\|\delta\theta\| = \sqrt{\delta\theta_A^2 + \delta\theta_E^2} \tag{51}$$

where $\delta\theta_A$ and $\delta\theta_E$ are the azimuth error and pitch error of the turntable, respectively.

The RMSE at time k is defined as

$$RMSE = \sqrt{\frac{1}{N}\sum_{i=1}^{N}|\delta\theta_k|^2} \tag{52}$$

where N is the total simulation time, and $\delta\theta_k$ represents the orientation error at time k.

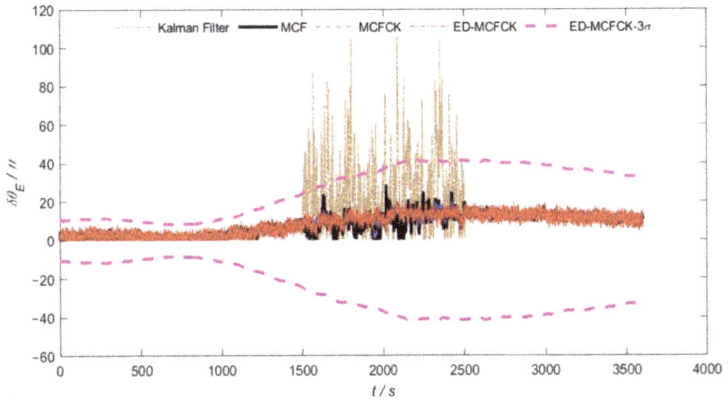

Figure 6. Orientation errors in pitch and 3σ error boundary for the case of non-Gaussian noise with outliers.

Table 2 shows the mean RMSEs of global orientation error as well as the azimuth and pitch errors for each filter. Due to the ability of robustness on non-Gaussian noise, the mean RMSEs of global orientation error as well as the azimuth and pitch errors achieved by the MCF are 55.8%, 67.76% and 48.46% lower than those of the Kalman filter. Meanwhile, the mean RMSEs of global orientation error as well as the azimuth and pitch errors achieved by the MCFCK are 56.26%, 70.35% and 47.97% lower than those of the Kalman filter. Further, the mean RMSEs of global orientation error as well as the azimuth and pitch errors achieved by the ED-MCFCK are 56.91%, 70.47% and 48.84% lower than those of the Kalman filter. These indicate that the MCF, MCFCK and ED-MCFCK exhibit a similar estimation performance for the case of non-Gaussian noise with outliers in measurements and they have a better sensitivity to the non-Gaussian measurement noises compared to the Kalman filter.

Table 2. Mean RMSEs of the orientation errors by Kalman filter, MCF, MCFCK and ED-MCFCK for the case of non-Gaussian noise with outliers.

Filtering Methods	Global Orientation Error (″)	Azimuth Error (″)	Pitch Error (″)
Kalman Filter	24.60	16.22	18.49
MCF	10.87	5.23	9.53
MCFCK	10.76	4.81	9.62
ED-MCFCK	10.6	4.79	9.46

(2) α stable noise in measurements

The α stable noise is a significant type of non-Gaussian random distribution, and its two most important features are the stability of the probability distribution and the heavy-tailed probability density function. The characteristic function of the α stable noise is given by [40]

$$\varphi(t) = \exp\left\{j\delta t - \alpha|t|^{\beta}[1 + j\zeta \mathrm{sgn}(t)\omega(t,\beta)]\right\} \tag{53}$$

with

$$\omega(t,\beta) = \begin{cases} \tan(\frac{\beta\pi}{2}), \beta \neq 1 \\ \frac{2}{\pi}\log|t|, \beta = 1 \end{cases} \tag{54}$$

$$\operatorname{sgn}(t) = \begin{cases} 1, & t > 0 \\ 0, & t = 0 \\ -1, & t < 0 \end{cases} \tag{55}$$

where $\beta \in (0, 2]$ is the characteristic index, $\zeta \in [-1, 1]$ is the symmetry parameter, δ is the location factor and $\alpha > 0$ is the dispersion factor.

To test the performance of the four filters under α stable noise, the above noise form was injected into the measurement during the time intervals (1000 s, 1500 s) and (2500 s, 3000 s), where $(\beta, \zeta, \alpha, \delta) = (1.8, 0, 1.6, 0)$ is set in this simulation.

Figures 7 and 8 provide the orientation errors in azimuth and pitch achieved by the Kalman filter, MCF, MCFCK and ED-MCFCK as well as the 3σ error boundary of ED-MCFCK for the case of α stable noise in measurement. It can be seen that as stated previously, MCF is unable to handle heavy-tailed noise anomalies due to the limitation of the Gaussian kernel function; thus, its estimation error is divergent. In contrast, MCFCK and ED-MCFCK can effectively handle the influence of α stable noise with a heavy-tailed characteristic through the robustness of the Cauchy kernel. They show similar performance with the case of non-Gaussian noise with outliers in measurements and have a stronger robustness than the Kalman filter. Table 3 shows the mean RMSEs of global orientation error as well as the azimuth and pitch errors by the above four filters. Due to the robustness of the Cauchy kernel of MCFCK and ED-MCFCK, the mean RMSEs of the global orientation error as well as the azimuth and pitch errors achieved by the MCFCK are 63.19%, 78.65% and 53.46% lower than those of the Kalman filter, and the mean RMSEs of ED-MCFCK are 66.43%, 81.97% and 56.98% lower than those of the Kalman filter. However, due to the limitation of the Gaussian kernel in handling the heavy-tailed noise anomalies, the mean RMSEs of global orientation error as well as the azimuth and pitch errors by the MCF increased by 2008.05%, 3599.69% and 1625.61% compared the ED-MCFCK. Therefore, it is evident that the proposed ED-MCFCK has a superior robust performance to curb heavy-tailed non-Gaussian noise such as α stable noise, leading to improved orientation accuracy in the complex space environment.

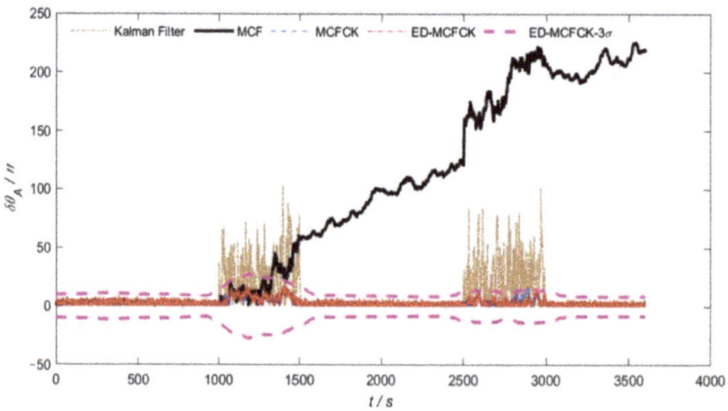

Figure 7. Orientation errors in azimuth and 3σ error boundary for the case of σ stable noise.

Table 3. Mean RMSEs of the orientation errors by Kalman filter, MCF, MCFCK and ED-MCFCK for the case of σ stable noise.

Filtering Methods	Global Orientation Error ($''$)	Azimuth Error ($''$)	Pitch Error ($''$)
Kalman Filter	26.27	18.08	19.06
MCF	185.93	120.61	141.5
MCFCK	9.67	3.86	8.87
ED-MCFCK	8.82	3.26	8.20

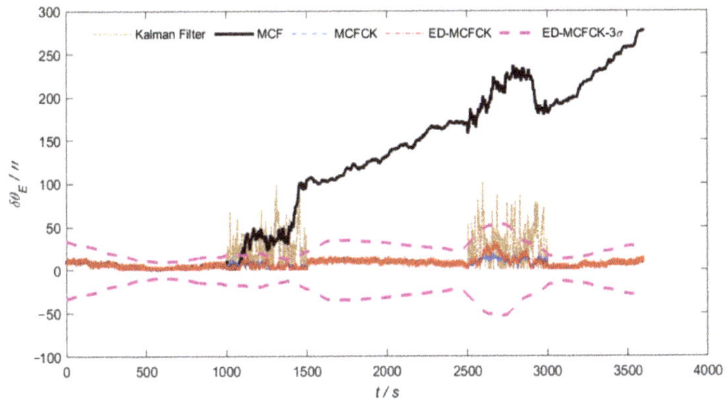

Figure 8. Orientation errors in pitch and 3σ error boundary for the case of σ stable noise.

B. Computational Performance Evaluation

The computational efficiency of the proposed ED-MCFCK is also discussed by comparison with the Kalman filter, MCF and MCFCK via a Monte Carlo simulation. It was conducted 10 times on an Intel(R) Core(TM) i5-9300H 2.40 GHz PC with 8 GB memory to study the computational performance of the proposed ED-MCFCK based on the above two types of non-Gaussian measurement noise scenarios. The computational times for each Monte Carlo run of the Kalman filter, MCF, MCFCK and ED-MCFCK are listed in Table 4. Further, the relative efficiencies of all the filters compared to the ED-MCFCK are also computed and shown in Figure 9.

Table 4. The computational times for each Monte Carlo run of the Kalman Filter, MCF, MCFCK and ED-MCFCK based on the simulation cases.

Filtering Methods	Non-Gaussian Noise with Outliers	α Stable Noise
	Computational Time	Computational Time
Kalman Filter	2.78 s	2.80 s
MCF	5.22 s	6.98 s
MCFCK	5.14 s	5.62 s
ED-MCFCK	2.84 s	2.89 s

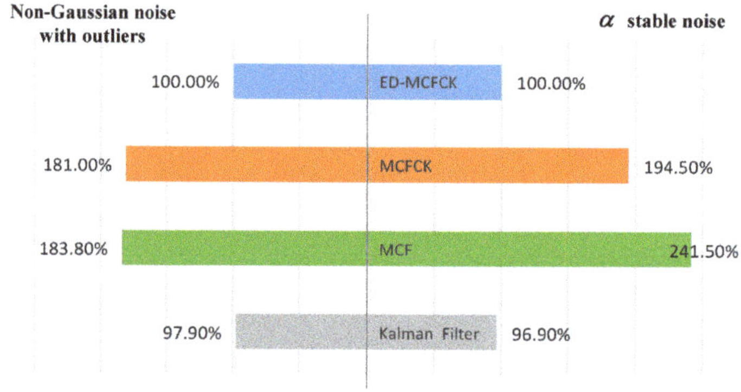

Figure 9. Relative efficiencies of all the filters compared to the ED-MCFCK based on the simulation cases. The blue, orange, green, and gray histograms respectively represent the computational efficiency of the ED-MCFCK, MCFCK, MCF, and Kalman Filter algorithms.

It can be seen from Table 4 and Figure 9 that the Kalman filter has optimal real-time performance and its computational time for each Monte Carlo run is smallest among all the four filters. This is the reason why it is widely used in engineering practices. Further, the computational times of MCF and MCFCK are significantly longer than those of the Kalman filter. The filtering computational time of MCF is 87.8% (for the first case) and 149.3% (for the second case) longer than that of the Kalman filter. The filtering computational time of MCFCK is 84.9% (for the first case) and 100.7% (for the second case) longer than that of the Kalman filter. They have a similar computational performance since multiple iterations are involved in filtering every time and result in unnecessary redundant calculations. Both MCF and MCFCK's real-time performance are poor; thus, they are not suitable for the practice applications. As a comparison, the proposed ED-MCFCK establishes a flexible event-driven mechanism to detect and process the case of non-Gaussian measurement noise. This effectively avoids the unnecessary redundant calculations and significantly improves the computational efficiencies of MCF and MCFCK. The ED-MCFCK has an approximated computational efficiency compared to the Kalman filter (97.9% for the first case and 96.9% for the second case). Thus, through the designed event-driven mechanism, ED-MCFCK can simultaneously obtain a good robustness on non-Gaussian measurement noise and a strong real-time performance in practical engineering.

The results of the above simulation evaluations and analysis demonstrate that the proposed ED-MCFCK not only has a superior ability to deal with the influence of non-Gaussian measurement noise but can also achieve superior real-time spatial applications by avoiding unnecessary computational costs in time points without non-Gaussian measurement noise, leading to improved spatial orientation performance for the gyros/star sensor integration used in the two-dimensional turntable structure.

5. Conclusions

This paper proposes an event-driven maximum correntropy filter based on the Cauchy kernel to improve the performance of gyros/star sensor integration used for spatial orientation under the environment of non-Gaussian measurement noise. The main contributions of this paper can be concluded as follows: (i) it establishes and derives a direct installation mode of gyros/star sensor integration as well as its mathematical model to ensure the turntable control stability; (ii) it further proposes an event-driven maximum correntropy filter based on the Cauchy kernel to handle the influence of non-Gaussian measurement noise on the above gyros/star sensor integration, in which an event-driven mechanism is also constructed based on the innovation information to reduce the unnecessary computational costs involved in the filtering solution. The results of the simulations and comparison analysis indicate that the proposed methodology can effectively curb the non-Gaussian measurement noise through the maximum correntropy criterion based on the Cauchy kernel, and it also can avoid the unnecessary computational cost of MC filters such as MCF and MCFCK, leading to improved robustness for the gyros/star sensor integration of spatial orientation in comparison with the conventional Kalman filter and MC filters.

Future research will focus on the combination of the proposed ED-MCFCK with advanced intelligent methods such as neural networks or deep learning to improve its robustness by adaptively adjusting the kernel bandwidth in real-time to accommodate the complex spatial environment. Further, it will also concentrate on the practical/extension application of the ED-MCFCK in the fields of vehicle navigation or target tracking.

Author Contributions: Conceptualization, K.C., Z.L. and B.G.; methodology, K.C., Z.L. and J.H.; software, K.C., Z.L. and J.H.; validation, K.C., Y.M. and B.G.; formal analysis, Z.L. and J.H.; investigation, K.C., J.H. and P.L.; resources, K.C., Z.L. and Y.M.; data curation, K.C., Z.L. and B.G.; writing—original draft preparation, K.C., Z.L., J.H. and Y.M.; writing—review and editing, K.C., Z.L., J.H., Y.M., P.L. and B.G.; funding acquisition, K.C. and B.G. All authors have read and agreed to the published version of the manuscript.

Funding: This research was funded by the Youth Innovation Promotion Association of the Chinese Academy of Sciences, grant number 2022405; the Aeronautical Science Foundation of China, grant number ASFC-20220019053003; the Key Research and Development Program of Shaanxi, grant number 2023-YBGY-375; the Guangdong Basic and Applied Basic Research Foundation, grant numbers 2024A1515030002 and 2023A1515011560; and the Shenzhen Science and Technology Program, grant number JCYJ20220530161606015.

Institutional Review Board Statement: Not applicable.

Informed Consent Statement: Not applicable.

Data Availability Statement: The original contributions presented in the study are included in the article; further inquiries can be directed to the corresponding author.

Conflicts of Interest: The authors declare no conflicts of interest.

Abbreviations

Abbreviations	Full Names
PF	Particle filter
GSF	Gaussian sum filter
MCC	maximum correntropy criterion
MC	maximum correntropy
ED-MCFCK	event-driven maximum correntropy filter based on Cauchy kernel
MCF	maximum correntropy filter
MCFCK	maximum correntropy filter based on Cauchy kernel
RMSE	Root Mean Square Error

References

1. Wang, Z.; Huang, M.; Qian, L.; Zhao, B. Near-earth space two-dimension opto-electronic turntable design. *Optik* **2020**, *200*, 163387. [CrossRef]
2. Yao, Q. Adaptive dynamic surface control for attitude tracking of spacecraft under measurement biases. *J. Aerosp. Eng.* **2020**, *35*, 04022099. [CrossRef]
3. Zhang, K.; Duan, G.; Ma, M. Adaptive sliding-mode control for spacecraft relative position tracking with maneuvering target. *Int. J. Robust. Nonlinear Control* **2018**, *28*, 5786–5810. [CrossRef]
4. Cai, X.; Zhong, K.; Fu, Y.; Chen, J.; Liu, Y.; Huang, C. Calibration method for the rotating axis in panoramic 3D shape measurement based on a turntable. *Meas. Sci. Technol.* **2020**, *32*, 035004. [CrossRef]
5. Wang, D.; Fu, F.; Li, W.; Tu, Y.; Liu, C.; Liu, W. A review of the diagnosability of control systems with applications to spacecraft. *Annu. Rev. Control* **2020**, *49*, 212–229. [CrossRef]
6. Ramachandran, M.P. Analytical derivation of star sensor accuracies due to centroid error. *Sādhanā* **2021**, *46*, 205. [CrossRef]
7. Zheng, X.; Huang, Y.; Mao, X.; He, F.; Ye, Z. Research status and key technologies of all-day star sensor. *J. Phys. Conf. Ser.* **2020**, *1510*, 012027. [CrossRef]

8. He, Y.; Wang, H.; Feng, L.; You, S. A novel method of eliminating stray light interference for star sensor. *IEEE Sens. J.* **2020**, *20*, 8586–8596. [CrossRef]
9. Kojima, H. Backstepping-based steering control for spacecraft attitude control using two-skewed control moment gyroscopes. *J. Guid. Control Dyn.* **2023**, *46*, 80–96. [CrossRef]
10. Mok, S.H.; Byeon, S.Y.; Bang, H.; Choi, Y. Performance comparison of gyro-based and gyroless attitude estimation for cubesats. *Int. J. Control Autom. Syst.* **2020**, *18*, 1150–1160. [CrossRef]
11. Gao, S.; Zhang, W.; He, X. Observer-based multiple faults diagnosis scheme for satellite attitude control system. *Asian J. Control* **2020**, *22*, 307–322. [CrossRef]
12. Wang, D.; Xu, T.; Li, M.; Shi, Y. Navigation performance analysis of Earth–Moon spacecraft using GNSS, INS, and star tracker. *Satell. Navig.* **2024**, *5*, 16. [CrossRef]
13. Tan, W.; Dai, D.; Wu, W.; Wang, X.; Qin, S. A comprehensive calibration method for a star tracker and gyroscope units integrated system. *Sensors* **2018**, *18*, 3106. [CrossRef]
14. Gou, B.; Cheng, Y.M. INS/CNS integrated navigation based on corrected infrared earth measurement. *IEEE Trans. Ins. Meas.* **2018**, *68*, 3358–3366. [CrossRef]
15. Xu, L.; Li, X.R.; Mallick, M.; Duan, Z.S. Modeling and State Estimation of Destination-Constrained Dynamic Systems. Part II: Uncertain Arrival Time. *IEEE Trans. Signal. Process*, 2024; *in press*. [CrossRef]
16. Henna, H.; Toubakh, H.; Kafi, M.; Sayed-Mouchaweh, M.; Djemai, M. Hybrid supervision scheme for satellite attitude control with sensor faults. *CEAS Space J.* **2024**, *16*, 753–767. [CrossRef]
17. Yang, B.; Huang, H.; Cao, L. Centered error entropy-based sigma-point Kalman filter for spacecraft state estimation with non-Gaussian noise. *Space Sci. Technol.* **2022**, *1*, 381–393. [CrossRef]
18. Farahmand, S.; Giannakis, G.; Angelosante, D. Doubly robust smoothing of dynamical processes via outlier sparsity constraints. *IEEE Trans. Signal. Process.* **2011**, *59*, 4529–4543. [CrossRef]
19. Soken, H.; Hajiyev, C.; Sakai, S. Robust Kalman filtering for small satellite attitude estimation in the presence of measurement faults. *Eur. J. Control* **2014**, *20*, 64–72. [CrossRef]
20. Kim, S.; Jang, M.; La, H. Development of a particle filter-based path tracking algorithm of autonomous trucks with a single steering and driving module using a monocular camera. *Sensors* **2023**, *23*, 3650. [CrossRef]
21. Sorenson, H.; Alspach, D. Recursive Bayesian estimation using Gaussian sums. *Automatica* **1971**, *7*, 465–479. [CrossRef]
22. Gao, B.; Hu, G.; Zhang, L.; Zhong, Y.; Zhu, X. Cubature Kalman filter with closed-loop covariance feedback control for integrated INS/GNSS navigation. *Chin. J. Aeronaut.* **2023**, *36*, 363–376. [CrossRef]
23. Chen, B.; Xing, L.; Liang, J.; Zheng, N.; Principe, J.C. Steady-state mean-square error analysis for adaptive filtering under the maximum correntropy criterion. *IEEE Signal Process. Lett.* **2014**, *21*, 880–884.
24. Mohammadi, M.; Noghabi, H.; Hodtani, G.; Mashhadi, H. Robust and stable gene selection via Maximum–Minimum Correntropy Criterion. *Genomics* **2016**, *107*, 83–87. [CrossRef] [PubMed]
25. Gao, G.; Zhong, Y.; Gao, Z.; Zong, H.; Gao, S. Maximum correntropy based spectral redshift estimation for spectral redshift navigation. *IEEE Trans. Ins. Meas.* **2023**, *72*, 8503110. [CrossRef]
26. Yi, C.; Li, M.; Li, S. Multi-Sensor Fusion Target Tracking Based on Maximum Mixture Correntropy in Non-Gaussian Noise Environments with Doppler Measurements. *Information* **2023**, *14*, 461. [CrossRef]
27. Wang, J.; Lyu, D.; He, Z.; Zhou, H.; Wang, D. Cauchy kernel-based maximum correntropy Kalman filter. *Int. J. Syst. Sci.* **2020**, *51*, 3523–3538. [CrossRef]
28. Ye, X.; Lu, S.; Wang, J.; Wu, D.; Zhang, Y. Robust State Estimation Using the Maximum Correntropy Cubature Kalman Filter with Adaptive Cauchy-Kernel Size. *Electronics* **2023**, *13*, 114. [CrossRef]
29. Shen, B.; Wang, X.; Zou, L. Maximum correntropy Kalman filtering for non-Gaussian systems with state saturations and stochastic nonlinearities. *IEEE/CAA J. Autom. Sin.* **2023**, *10*, 1223–1233. [CrossRef]
30. Chu, S.; Qian, H.; Ding, P. Improved maximum correntropy unscented Kalman filter for spacecraft attitude estimation. *Int. J. Control Autom. Syst.* **2023**, *21*, 2020–2030. [CrossRef]
31. Tang, H.; Xu, J.; Chang, L.; Shi, W.; He, H. Invariant error-based integrated solution for SINS/DVL in earth frame: Extension and comparison. *IEEE Trans. Ins. Meas.* **2022**, *72*, 9500617. [CrossRef]
32. Groves, P. *Principles of GNSS, Inertial, and Multisensor Integrated Navigation Systems*; Artech House: London, UK, 2008.
33. Geiger, P.; Knebel, J.; Frey, E. Topologically robust zero-sum games and Pfaffian orientation: How network topology determines the long-time dynamics of the antisymmetric Lotka-Volterra equation. *Phys. Rev. E* **2018**, *98*, 062316. [CrossRef]
34. Ritto-Corrêa, M.; Camotim, D. On the differentiation of the Rodrigues formula and its significance for the vector-like parameterization of Reissner–Simo beam theory. *Int. J. Numer. Methods Eng.* **2002**, *55*, 1005–1032. [CrossRef]
35. Seo, Y.; Yu, H.; Ryu, K.; Lee, I.; Lee, I.; Oh, J.; Kim, C.; Lee, S.; Park, C. Analysis of gyro bias depending on the position of inertial measurement unit in rotational inertial navigation systems. *Sensors* **2022**, *22*, 8355. [CrossRef] [PubMed]
36. Chen, B.; Liu, X.; Zhao, H.; Principe, J. Maximum correntropy Kalman filter. *Automatica* **2017**, *76*, 70–77. [CrossRef]
37. Chang, G. Kalman filter with both adaptivity and robustness. *J. Process. Control* **2014**, *24*, 81–87. [CrossRef]
38. Cho, S.; Kim, B. Adaptive IIR/FIR fusion filter and its application to the INS/GPS integrated system. *Automatica* **2008**, *44*, 2040–2047. [CrossRef]

39. Wang, M.; Wu, W.; Zhou, P.; He, X. State transformation extended Kalman filter for GPS/SINS tightly coupled integration. *GPS Solut.* **2018**, *22*, 112. [CrossRef]
40. Xing, J.; Jiang, T.; Li, Y. *q*-Rényi kernel functioned Kalman filter for land vehicle navigation. *IEEE Trans. Circuits Syst. II Express Briefs* **2022**, *69*, 4598–4602. [CrossRef]

sensors

MDPI

Article

YOLO-OD: Obstacle Detection for Visually Impaired Navigation Assistance

Wei Wang *, Bin Jing, Xiaoru Yu, Yan Sun, Liping Yang and Chunliang Wang

College of Computer Science and Technology, Changchun University, No. 6543, Satellite Road, Changchun 130022, China; 231502527@mails.ccu.edu.cn (B.J.); 231501508@mails.ccu.edu.cn (X.Y.); 221501464@mails.ccu.edu.cn (Y.S.); yangliping@ccu.edu.cn (L.Y.); 221501491@mails.ccu.edu.cn (C.W.)
* Correspondence: wangwei@ccu.edu.cn

Abstract: Visually impaired individuals frequently encounter difficulties in detecting and avoiding obstacles in the wild. To address this issue, we propose an obstacle detection method for visual navigation assistance, named YOLO-OD. To improve the ability to detect and differentiate between different sized obstacles in outdoor environments, we introduce the Feature Weighting Block (FWB), which improves feature importance discrimination. To address the challenges of detecting cluttered outdoor environments and handling occlusions, we introduce the Adaptive Bottleneck Block (ABB), which captures varying features across different scenes. To solve the problem of detecting relatively small obstacles in outdoor environments, we propose the Enhanced Feature Attention Head (EFAH). The proposed YOLO-OD achieves an average precision of 30.02% on a public dataset, making it a worth studying approach for blind and visually impaired navigation aids.Our study effectively addresses the navigation challenges faced by visually impaired individuals by improving model performance, thereby enhancing its practical values. The code for YOLO-OD has been made publicly available to ensure reproducibility and facilitate further research.

Keywords: obstacle detection; YOLO-OD; feature weighting block; adaptive bottleneck block; enhanced feature attention head

1. Introduction

According to the World Health Organization, the global population is approximately 7.7 billion, with around 246 million individuals experiencing visual impairments, including 39 million who are blind. A primary challenge faced by visually impaired individuals is mobility, as they struggle with recognizing and navigating obstacles, particularly in urban environments. These obstacles significantly impact their safety and independence. Advances in computer technology have led to a proliferation of computer-assisted devices, particularly in image processing and medical image analysis. However, visual detection tasks for visually impaired individuals are particularly hindered by challenges such as occlusion [1], dense crowds [2], and small object detection [3], especially in real-world settings with complex environments. These challenges limit the effectiveness of existing assistive technologies, making it difficult for visually impaired individuals to navigate safely. Occlusion complicates object localization, while small objects are challenging to detect due to their diminutive size and inconspicuous features. In densely populated scenes, overlapping objects blur boundaries, further complicating segmentation and recognition. Accurate and efficient object detection technology is crucial for visually impaired individuals. However, existing algorithms often struggle with precision and reliability in practical applications, particularly when faced with overlapping objects, varying lighting conditions, and small obstacle detection. These limitations make current solutions insufficient for ensuring the safety and autonomy of visually impaired individuals. Therefore, the research presented in this paper is essential for addressing these challenges.

Visually impaired individuals need to locate and avoid obstacles in strange environments [4]. Traditional techniques, such as guide dogs [5] and white canes [6], are limited in terms of speed, precision, and coverage. Guide dogs are effective navigation means, but they cannot detect every small obstacle and are also costly and require professional training. White canes, on the other hand, can only detect obstacles within a very limited range, making them insufficient for safe navigation in complex urban settings. Consequently, there is an urgent need for computer-aid approaches [7]. Numerous studies utilizing object detection aim to assist visually impaired individuals. For example, Said et al. [8] proposed a real-time intelligent navigation assistance system. Compared to traditional methods, data-driven object detection technologies offer higher precision, greater generalizability, and sustainable development potential. These advances have the potential to be integrated into assistive devices, providing visually impaired individuals with a greater sense of safety and independence in navigating urban and unfamiliar environments.

In the field of computer vision, object detection is a fundamental task that has been extensively researched. Object detectors can be roughly defined as one-stage and two-stage methods. One-stage object detectors [9–11] perform both object localization and classification by directly predict bounding boxes and class scores for each object from extracted features. These methods are generally faster as they perform detection and classification in a single network pass, which makes them well-suited for real-time applications. In contrast, two-stage object detectors [12–14] employ a two-step strategy. These detectors first generate region proposals in the first stage and then refine these proposals by classifying them and adjusting their bounding boxes in the second stage. These methods normally offer higher detection precision and generalizability, which makes them ideal for scenarios that demand high precision. The advent of universal object detectors has facilitated the deployment of object detection technology in assistive devices designed for individuals with visual impairments. By enabling real-time recognition of pedestrians, vehicles, and other obstacles, object detection helps visually impaired individuals navigate more effectively and improves their quality of life.

As illustrated in Figure 1, for visually impaired individuals, object detection in the real world often suffers from challenges such as occlusion, small object detection, and complex backgrounds, making safe navigation extremely challenging in dynamic environments. Specifically, occlusion results in an incomplete presentation of object features, which can impair semantic information and increase the difficulty of subsequent regression and classification, directly impacting the ability of visually impaired individuals to accurately detect obstacles in their surroundings, especially in busy environments. The high complexity of densely crowded scenes, such as crowded pedestrian crossings, poses a significant challenge to feature extractors because distinguishing between objects and backgrounds becomes difficult, leading to redundant information that interferes with object features. This significantly affects the ability of assistive technologies to provide timely and accurate information for visually impaired individuals. Overlapping object regions also present a challenge to detectors trying to distinguish between different object boundaries. In the detection of smaller obstacles, such as potholes or traffic cones, due to the lack of effective pixels representing these smaller obstacles, less useful information is extracted, which ultimately hampers the effectiveness of traditional detection methods. This issue is especially critical for visually impaired individuals who rely on precise detection of small obstacles to ensure their safety. This hampers the ability of feature extractors to accurately adapt to the features of objects across different scales, ultimately limiting the discriminative power and precision of the features. Our proposed methods aim to mitigate these challenges by enhancing feature extraction and improving robustness to occlusion and background complexity, ultimately ensuring that visually impaired individuals can receive reliable and accurate assistance in various real-world environments.

To address the aforementioned challenges, we propose YOLO-OD. Building on YOLOv8, we integrated an Adaptive Bottleneck Block (ABB) within the detection neck to enhance object detection in complex scenes, added a Feature Weighting Block (FWB) to process features initially extracted by the backbone, Enhanced Feature Attention Head (EFAH) is proposed, which incorporates Convolutional Block Attention Module (CBAM) [15] into the classification header to improve the classification precision, and introduces a transposed convolution layer and a convolutional layer in the regression header to improve the localization precision. These modifications collectively aim to enhance the robustness and precision of the model, thereby providing effective navigational assistance to visually impaired individuals.

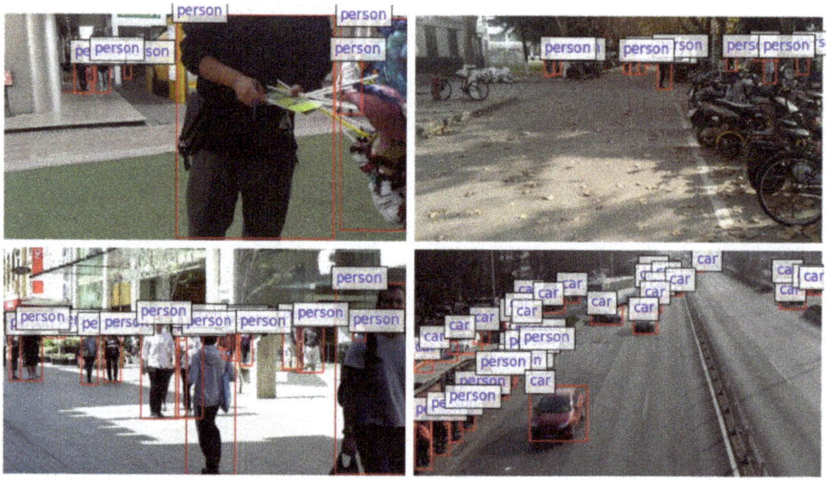

Figure 1. Visualization of picture samples from the real world.

To provide a concise overview of our contributions, we present them in five key aspects:

(1) We propose a new obstacle detection method that incorporates a feature-weighted block (FWB), an adaptive bottleneck block (ABB), and an enhanced feature attention head (EFAH) to improve usability in visual impairment detection.

(2) To enhance the detection and distinction of obstacles of varying sizes in outdoor environments, we introduce the Feature Weighting Block (FWB), designed to improve the discrimination of feature importance.

(3) To tackle the challenges of cluttered outdoor environments and occlusion, we present the Adaptive Bottleneck Block (ABB), which captures diverse features across different scenes.

(4) For the detection of relatively small obstacles, We propose the Enhanced Feature Attention Head (EFAH), which can better capture the detailed information of features through the transposed convolution layer.

(5) Our proposed method was validated on a publicly available dataset and achieved an average accuracy of 30.02%, making it a worthwhile research method for navigational aids for the blind and visually impaired.

2. Related Work

2.1. One-Stage Detectors

One-stage object detectors, exemplified by the YOLO [16] and SSD series, are optimized for real-time applications due to their high detection speeds. YOLOv3 [17] introduced a more complex network structure and multi-scale predictions, thereby improving precision while maintaining speed. YOLOv4 [18] further optimized the architecture with CSPNet and various training techniques, enhancing both speed and precision. YOLOv5 [19] was designed to enhance ease of use, with a focus on ensuring efficient implementation and deployment. One of the state of the art YOLO, YOLOv8 [20], introduces architectural refinements, enhanced anchor-free detection, dynamic task prioritization, and superior real-time performance.The Single Shot MultiBox Detector SSD [21] utilize multi-scale feature maps for efficient object detection, offering a balance between speed and accuracy without region proposal steps, ideal for real-time applications.

2.2. *Two-Stage Detectors*

Two-stage object detectors, particularly the R-CNN series [22], are distinguished by their precision in object detection. R-CNN (Region-CNN) [23] introduces the two-stage approach by generating region proposals using selective search and applying CNNs to classify each proposal. While effective, it was slow due to the need for individual CNN processing per region. Fast R-CNN [24] improves the efficiency of R-CNN by using shared convolutional feature maps and introducing RoI pooling. This allowed all region proposals to share the same feature map, greatly reducing computation time and maintaining high accuracy. Faster R-CNN [25] introduces the Region Proposal Network (RPN), integrating proposal generation directly into the network, which significantly increased speed and efficiency. By jointly predicting object bounds and scores, it streamlined the two-stage detection process.Cascade R-CNN [26] extendes the Faster R-CNN architecture by progressively refining region proposals across multiple stages, with each stage using increasingly higher IoU thresholds. This made the model more accurate for difficult detection cases, especially in handling high-quality proposals.

2.3. *Similar Works*

Many researchers have already done similar work on solving the problem of obstacle avoidance for the visually impaired [27,28]. Atitallah et al. [29] propose developing an obstacle detection system based on a modified YOLOv5 neural network architecture. The proposed system is capable of recognizing and locating a set of indoor and outdoor objects that are extremely useful for Blind and Visually Impaired (BVI) navigation aids. Woo et al. [30] propose a smart data driven mobility aid for VIP (SOMAVIP), enabling real-time scene perception. The ODOMOVIP comprises of responsive devices and IoT embedded smart city environments, deep machine learning, computer vision, and data processing methods. Although much research has been conducted, many challenges remain, such as the significant shortcomings in detecting small objects and complex backgrounds. Our proposed YOLO-OD addresses these issues by utilizing FWB for preliminary weighting of the extracted features, ABB to enhance the ability of model to detect objects in complex scenes, by adding multi-scale convolution, learnable normalization layer and learnable activation function. Finally, EFAH to improve the classification and regression of model by capturing more detailed information.

3. Methodology

3.1. *Architecture of the Proposed Method*

To address the challenges faced by visually impaired individuals in complex environments, such as small obstacles, occlusion, and background noise, we propose YOLO-OD, inspired by the architecture of YOLOv8. As illustrated in Figure 2, the proposed YOLO-OD leverages its advanced structure to improve detection performance, particularly in real-world tests involving urban obstacles such as potholes, traffic cones, and pedestrian congestion. These enhancements are crucial for improving the safety and independence of visually impaired individuals. The YOLO-OD consists of three main components: the backbone network (CSPDarknet-53), the detection neck (KAN Path Aggregation Feature Pyramid Network, KANPAFPN) [31], and the detection head (Enhanced Feature Attention Head).

CSPDarknet-53 is chosen as the backbone due to its effective utilization of cross-stage partial connections, which scientifically optimize feature extraction by reducing redundancy in the network. This structure allows for efficient feature reuse and enhances the ability of model to process high-dimensional data with minimal computational overhead. The multiscale feature extraction capabilities of backbone make it particularly well-suited for identifying both subtle and prominent obstacles, essential for assisting visually impaired individuals in dynamic environments.

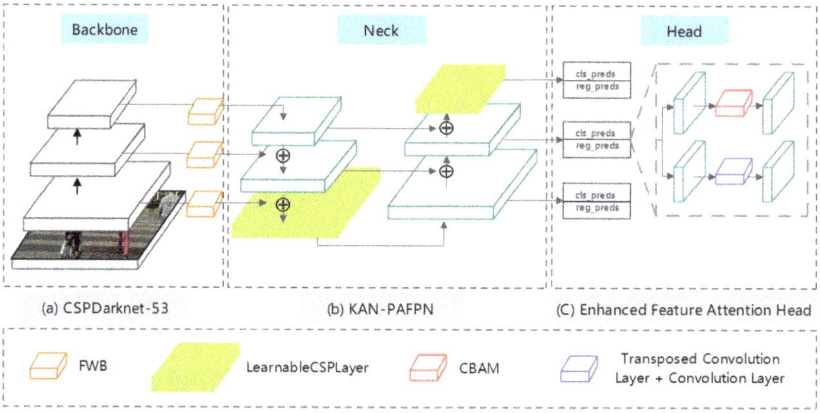

Figure 2. Architecture of the proposed method.

The KAN-PAFPN is employed for feature fusion, where integration is executed through top-down and bottom-up pathways to capture detailed and abstract features. However, the indiscriminate merging of features may cause significant information to be diluted, especially in urban environments where visually impaired individuals need precise obstacle detection. Subsequently, the extracted features from the obstacles are indiscriminately merged within the PAFPN, resulting in equal treatment of both significant and insignificant features, the model may struggle to prioritize key information, potentially leading to missed detections or inaccurate classifications. To address this challenge, the incorporation of the Feature Weighting Block (FWB) before the KAN-PAFPN allows for differential weighting of features, which significantly improved the ability of model to detect small and critical obstacles in real-world tests, such as potholes and narrow pathways, ensuring more reliable navigation for visually impaired individuals [32]. Furthermore, to address the complex environments faced by visually impaired individuals, we introduces the Adaptive Bottleneck. The module enhances object detection by capturing diverse features and adapting to various scenarios, thereby improving the overall precision and robustness of the system.

The EFAH consists of two parts: the classification head and the regression head. The classification head incorporates the CBAM, which selectively focuses on spatial and channel-wise information, helping to differentiate obstacles more effectively from complex backgrounds. To enhance performance in scenarios involving visually impaired individuals, several modifications have been implemented. The classification head incorporates the CBAM to enhance its ability to distinguish between obstacles. The regression head has been enhanced with a transposed convolution followed by a convolution layer, adds more detailed information, improves precision in predicting obstacle locations for more reliable navigation assistance [33].

3.2. Feature Weighting Block

As illustrated in Figure 3, to help visually impaired individuals navigate safely, particularly in urban environments with varying obstacles, we propose the FWB, which is a feature-weighted processing module. The specific computational process can be described as follows:

$$\begin{aligned} F &= FWB(f) \\ FWB(f) &= HSigmoid(LeakyReLU(Conv2d\_1 \times 1 \\ &\quad (AvgPool(f)))) \end{aligned} \tag{1}$$

where f represents the original features extracted by CSPDarknet-53, with dimensions $H \times W \times C$, where H, W, and C represent the height, width, and number of channels, respectively. $Conv2d\_1 \times 1$ represents a convolution layer with a feature kernel size of 1×1, mainly used to adjust the number of channels. Averagepooling (Avgpool) is a downsampling technique in convolutional neural networks that reduces the spatial dimensions of feature maps by taking the average of values within a sliding window, after which the features become $1 \times 1 \times C$, Following this, a convolutional layer adjusts the number of channels to 1. The LeakyReLU activation function is mainly used to apply a small non-zero slope to negative input values to ensure that all inputs contribute to the output, thus preventing neurons from becoming inactive during training [34]. The Hard Sigmoid activation function is a computationally efficient piecewise linear approximation of the standard sigmoid, designed to reduce complexity by capping and scaling inputs without the use of exponential computation, after which values between 0 and 1 are compressed, producing an output dimension of $1 \times 1 \times w$, where w is a specific value. F is the final feature representation. FWB assigns weights to said features based on their semantic importance, ensuring that more critical features related to obstacles are highlighted. This is crucial for improving detection accuracy in scenarios where small or partially occluded obstacles might be overlooked by conventional methods. This block allows the model to better differentiate between obstacles of different sizes, as demonstrated in our experiments with datasets containing varying-sized obstacles. In particular, the FWB showed a significant improvement in detecting smaller obstacles, which are often missed by conventional models. This enhancement is crucial in providing reliable navigational assistance to visually impaired individuals in complex and dynamic environments.

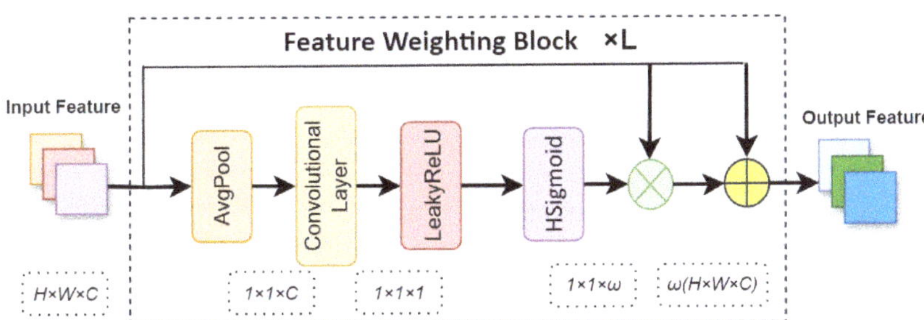

Figure 3. Feature Weighting Block.

3.3. *Adaptivebottleneck Block*

As illustrated in Figure 4, we propose the Adaptive Bottleneck Block in KAN-PAFPN, which aims to address the lack of flexibility when dealing with various scenarios and features. The Adaptive Bottleneck Block enhances the ability of model to detect high-contrast edges and fine textures in complex environments. This capability is crucial for providing safer and more reliable obstacle avoidance assistance for visually impaired individuals navigating urban environments. To achieve this, the module employs multi-scale convolutional kernels (1×1, 3×3, 5×5) [35] to capture multi-scale features. The use of multi-scale convolutional kernels allows the model to capture diverse features across different receptive fields, which is particularly important for detecting a wide range of obstacles, from large, easily visible barriers to small, less noticeable hazards that visually impaired individuals might encounter on urban streets. The multi-scale convolutional kernel is calculated as follows:

$$\begin{aligned} f_{1\times1} &= Conv2d_{1\times1}(f) \\ f_{3\times3} &= Conv2d_{3\times3}(f) \\ f_{5\times5} &= Conv2d_{5\times5}(f) \end{aligned} \tag{2}$$

$$f_{concat} = Concat(f_{1\times1}, f_{3\times3}, f_{5\times5}) \tag{3}$$

where $ConvLayer_{n\times n}$ refers to a convolution layer with a kernel size of $n \times n$. *Concat* means the features are concatenated along the channel dimension. After the convolutionallayers, the fixed BatchNormalizationlayer is replaced by the LearnableBatchNormalizationlayer, which includes additional trainable parameters that adaptively adjust the normalization process based on the input data distribution.

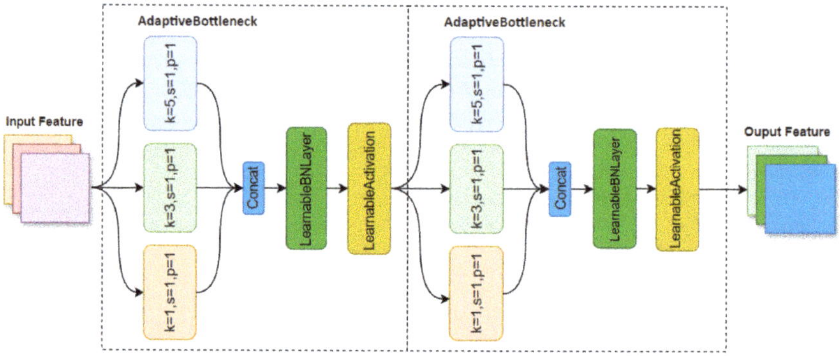

Figure 4. AdaptiveBottleneck Block.

In the standard Batch Normalization (BN) layer, the normalized output is computed as:

$$\hat{x} = \frac{x - \mu_{\text{batch}}}{\sqrt{\sigma^2_{\text{batch}} + \epsilon}}$$

where μ_{batch} and σ^2_{batch} are the mean and variance calculated across the batch, and ϵ is a small constant for numerical stability, $\hat{x}$ indicates the output eigenvalues after normalization. Here, the mean and variance are calculated statically from the current batch of data, limiting the model's ability to adapt to more complex data distributions over time.

Next, a linear transformation is applied:

$$y = \gamma\hat{x} + \beta$$

where y represents the output eigenvalue after normalization and linear transformation. γ and β are trainable parameters that scale and shift the normalized features. This allows the model to adjust the normalized features slightly, but the reliance on batch statistics for μ_{batch} and σ^2_{batch} makes it less flexible in handling diverse or complex input distributions.

However, in our proposed LearnableBatchNormlayer, we introduce additional flexibility by making the mean and variance learnable. This is achieved by introducing trainable parameters Δ_μ and Δ_σ, which allow the model to adjust the mean and variance during the training process, rather than relying solely on batch statistics. The normalization process is calculated as:

$$f_{norm} = LearnableBatchNormLayer\left(\gamma\frac{x - (\mu_{\text{batch}} + \Delta_\mu)}{\sqrt{\sigma^2_{\text{batch}} + \Delta_\sigma + \epsilon}} + \beta\right)$$

where Δ_μ and Δ_σ are trainable parameters that adjust the batch mean and variance, respectively. This modification allows the normalization process to dynamically adjust based on the varying input data distribution, such as different lighting conditions in urban areas or varying obstacle textures in crowded environments. Experiments show that this approach enhances the detection stability in these challenging scenarios, thereby improving the safety of visually impaired individuals.

By introducing these learnable adjustments, Δ_μ and Δ_σ, the Adaptive Batch Normalization module provides the flexibility to handle diverse input conditions more effectively than standard Batch Normalization. This capability is especially useful in visually impaired navigation scenarios where environmental conditions, such as lighting and obstacle types, can vary significantly. Specifically:

- Δ_μ acts as a learnable shift to the batch mean μ_{batch}, allowing the model to modify the mean in response to changes in data distribution.
- Δ_σ adjusts the batch variance σ^2_{batch}, allowing the model to fine-tune the spread of the data distribution.

Thus, the learnable parameters γ, β, Δ_μ, and Δ_σ enable the LearnableBatchNormlayer to adapt more effectively to diverse input conditions, making it particularly useful in complex and dynamic environments, such as obstacle detection for visually impaired individuals, where robustness is critical for system performance.

The learnable adjustments to mean and variance provide greater stability and adaptability, especially when working with small batch sizes or data with high variability, resulting in improved model robustness.

This adaptive normalization technique enhances robustness in a variety of situations, including different lighting conditions and complex backgrounds, and thus effectively helps visually impaired people with obstacle avoidance. In addition, the AdaptiveBottleneck module incorporates the LearnableActivation function. Unlike fixed activation functions such as $SiLU$ in the original pafpn, the LearnableActivation function is calculated as:

$$F = LearnableActivation(\omega(\frac{x}{1+e^{-x}} + \sum_i c_i B_i(x))) \tag{4}$$

where F represents the output feature map after processing through the Adaptive Bottleneck module, ω and c_i being trainable parameters, x represents the feature, and $B_i(x)$ representing basis functions such as B-splines.

This dynamic adjustment ensures optimal non-linear transformation for varying input features, enhancing the detection of high-contrast edge features and fine textures, which are critical for accurate obstacle detection.

3.4. Enhanced Feature Attention Head

As illustrated in Figure 5, in order to improve the detection accuracy of smaller obstacles in scenarios where visually impaired people are traveling, we propose the Enhanced Feature Attention Head, which is divided into a classification head and a regression head. In the classification stage, the input feature is first processed through a convolutional layer, which extracts preliminary features. Subsequently, the Convolutional Block Attention Module (CBAM) is integrated to refine the feature by focusing on crucial regions and suppressing irrelevant information. This attention mechanism enhances the information on space and channel, thereby improving classification accuracy. A subsequent convolutional layer operation processes these attended features, which are then utilized to calculate the classification loss. In the regression head, the input feature is firstly processed by a convolutional layer to resize and channel dimension. For precise localization, a transposed convolution module is introduced to expand the size of the features so that more detailed information about the features can be captured, thus helping to achieve finer object localization. This is followed by another convolutional layer that refines the upsampled features, resize features back to original size. A final convolutional layer processes these refined features, which are then used to output prediction results.

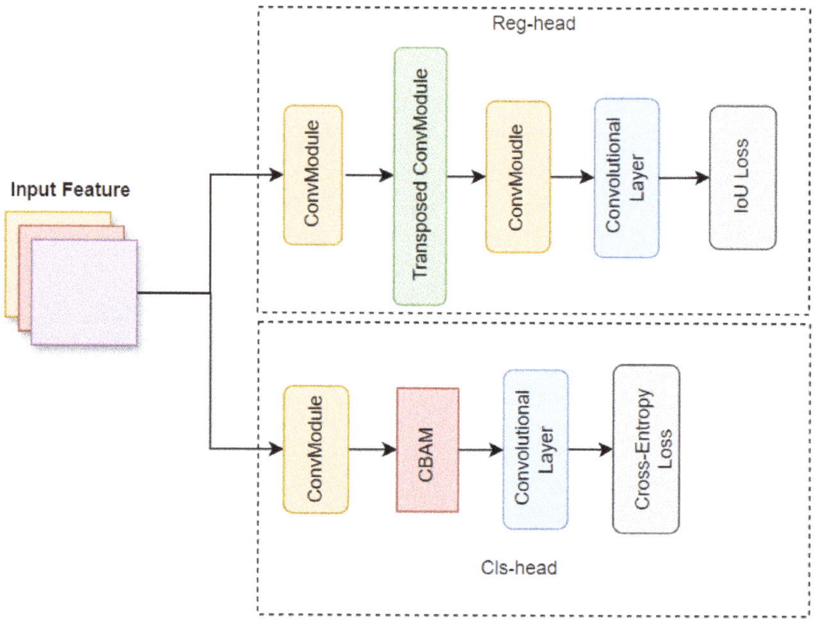

Figure 5. Enhanced Feature Attention Head.

4. Experiment

4.1. Datasets and Evaluation Metrics

The dataset used in this article is a road obstacle dataset in COCO format, available at the link [36]. This dataset contains a total of 6276 images, which has four categories, with the number of samples for each category as follows: car: 22,318, person: 53,756, traffic cone: 1176, and pothole: 1719.

We employ a set of metrics to evaluate the effectiveness of our model, including mAP (50, s, m, l), GFLOPS, and Params. mAP is the Average Precision (AP) across all classes. mAP50 represents the mean average precision at an IoU (Intersection over Union) threshold of 0.5. mAP_s quantifies the mean detection precision for objects with an area smaller than 32×32 pixels, mAP_m assesses the mean detection precision for objects with an area between 32×32 and 96×96 pixels, and mAP_l evaluates the mean detection precision for objects with an area larger than 96×96 pixels. GFLOPS is employed to evaluate the computational speed of a method, whereas Params is utilized to quantify the number of parameters within the method.

4.2. Implementation Details

The code for the proposed model and the comparison models was implemented using the MMYOLO and MMDetection frameworks. YOLO-OD was trained on two RTX 3090 GPUs for 150 epochs, while the other models used their default settings. MMDetection used the SGD optimizer with a learning rate of 0.02, while MMYOLO used the SGD optimizer with a learning rate of 0.01. For MMDetection, the input image size was 1333×800 pixels, and the test image size was also 1333×800 pixels. In contrast, MMYOLO used an input image size of 640×640 pixels. The batch size was set to 16 by default, and the confidence threshold for visualization was set to 0.5 by default.

4.3. Ablation Studies

Effect of the Feature Weighting Block (FWB). In Table 1, the addition of FWB to YOLOv8-s leads to a modest increase in computational cost, with the parameters rising from 11.13 to 11.14 and GFLOPs from 14.27 to 14.28. However, the benefits are evident through the improvement in mAP_s, which increases from 18.93 to 20.02. This gain in performance demonstrates the effectiveness of FWB in enhancing small obstacle detection capabilities, which is particularly crucial for visually impaired individuals navigating complex environments. Furthermore, there is also a significant improvement in overall mAP, increasing from 28.51 to 29.33, mAP_m from 28.93 to 30.42, and mAP_l from 34.01 to 34.72. These improvements indicate that FWB not only provides considerable enhancements in small object detection but also yields notable gains in the detection of medium and large objects. This comprehensive enhancement across different object sizes underscores the versatility of FWB in improving the overall detection performance of the model. Similarly, Table 2 demonstrates the impact of incorporating FWB into YOLOv5-s. The parameters increase slightly from 8.12 to 8.13, and GFLOPs rise from 12.33 to 12.34. Despite this minimal increase, there is a significant improvement in the mAP_s metric, which rises from 11.24 to 12.17. Additionally, the overall mAP improves from 14.23 to 15.22, mAP_m from 17.71 to 19.14, and mAP_l from 22.88 to 24.55. These improvements clearly demonstrate that the inclusion of FWB not only enhances the ability of model to detect small obstacles but also significantly improves its performance for medium and large objects. This further validates the effectiveness of FWB across different scenarios and object scales. The inclusion of FWB significantly enhances the feature representation capability of the model, thereby achieving substantial improvements in detection performance with a negligible impact on model complexity.

Table 1. Ablation experiments.

Models	mAP	mAP_{50}	mAP_s	mAP_m	mAP_l	Flops/G	Parameters/M
YOLOv8-s (baseline)	28.51	42.02	18.93	28.93	34.01	14.27	11.14
YOLOv8-s + FWB	29.33	42.54	20.02	30.42	34.72	14.28	11.14
YOLOv8-s + ABB	29.55	42.85	20.23	29.94	35.14	58.52	40.09
YOLOv8-s + EFAH	29.44	42.46	19.25	29.86	35.06	14.90	11.37
YOLOv8-s + FWB + ABB	30.00	42.33	20.87	30.41	35.74	58.53	48.09
YOLOv8-s + ABB + EFAH	29.78	42.86	19.98	30.05	35.37	59.14	48.32
YOLOv8-s + FWB + EFAH	29.72	42.72	19.92	30.35	35.08	14.90	11.37
YOLOv8-s + FWB + ABB + EFAH	30.02	43.32	20.96	31.07	35.58	59.15	48.32

Table 2. Ablation experiments on other one-stage detectors.

Models	mAP	mAP_{50}	mAP_s	mAP_m	mAP_l	Flops/G	Parameters/M
YOLOv5-s	14.23	28.52	10.65	17.71	12.83	8.12	12.33
YOLOv5-s + FWB	15.22	29.41	11.24	19.05	14.45	8.13	12.34
YOLOv5-s + EFAH	15.34	28.84	12.17	19.16	13.78	8.23	12.39
YOLOv7-tiny	21.91	37.61	13.02	24.01	25.92	6.56	6.02
YOLOv7-tiny + FWB	23.43	38.85	14.66	25.32	27.67	6.57	6.03
YOLOv7-tiny + EFAH	22.12	38.26	13.68	23.65	27.09	6.67	6.12
YOLO X-s	23.22	40.31	14.51	25.06	27.52	13.32	8.94
YOLO X-s + FWB	23.54	40.82	14.76	25.67	27.61	13.33	8.94
YOLO X-s + ABB	24.76	41.76	15.25	26.23	30.27	47.92	38.56
YOLO X-s + EFAH	23.91	41.12	14.65	25.42	28.47	15.80	9.83
SSD300	21.22	39.04	8.52	21.95	28.61	30.58	24.15
SSD300 + FWB	22.06	40.12	9.26	23.23	29.47	30.58	24.15
SSD300 + EFAH	21.63	39.37	8.63	22.67	29.09	44.72	56.23

Effect of the Adaptive Bottleneck Block (ABB). As shown in Table 1, adding the Adaptive Bottleneck Block (ABB) significantly improves YOLOv8-s in terms of both small and medium obstacle detection. Specifically, mAP_s increased from 18.93 to 20.23, and mAP_m rose from 28.93 to 29.94. Although the computational cost increased—with GFLOPs rising from 14.27 to 58.52 and parameters growing from 11.14 to 40.09—this trade-off is justified by the notable accuracy gains. Such improvements are crucial for accurately detecting overlapping and small obstacles in complex urban environments, thereby enhancing the safety of navigation systems for visually impaired individuals.Similarly, Table 2 shows that incorporating ABB into YOLO X-s increased the mAP from 23.22 to 24.76 and mAP_{50} from 40.31 to 41.76. While the GFLOPs increased from 13.32 to 47.92 and the parameters grew from 8.94 to 38.56, these performance improvements highlight effectiveness of ABB in capturing complex features. This enhanced capability is particularly beneficial in navigating intricate and cluttered environments, offering more reliable detection of potential obstacles, which is key to supporting visually impaired individuals.

Effect of the Enhanced Feature Attention Head (EFAH). As shown in Table 1, integrating the Enhanced Feature Attention Head (EFAH) into YOLOv8-s results in improved performance across various metrics, with mAP increasing from 28.51 to 29.44, mAP_s rising from 18.93 to 20.23, and only a slight increase in model complexity, as GFLOPs grow from 14.27 to 14.90 and parameters from 11.14 to 11.37. This indicates that EFAH effectively enhances the ability of model to focus on crucial features, significantly benefiting the detection of small obstacles, which is critical for navigation assistance for visually impaired individuals in crowded environments. In Table 3, applying EFAH to Deformable DETR yields similar advantages, with mAP_s improving from 15.92 to 16.73 and mAP_l from 30.36 to 31.65. Although GFLOPs and parameters increase from 208.00 to 231.00 and 41.36 to 41.63, respectively, this slight complexity boost is offset by the enhanced detection accuracy. Such improvements are essential in complex scenarios involving obstacles of various sizes, ensuring the model can capture fine details necessary for the safe and efficient navigation of visually impaired individuals. The consistent gains from EFAH across different models underscore its effectiveness in addressing the challenge of small-object detection, providing an essential enhancement for applications requiring precise obstacle recognition in dynamic and intricate environments.

Table 3. Ablation experiments on two-stage detectors.

Models	mAP	mAP_{50}	mAP_s	mAP_m	mAP_l	Flops/G	Parameters/M
Cascade R-CNN	27.23	40.95	17.12	27.26	34.01	14.27	11.14
Cascade R-CNN + FWB	27.12	41.85	18.06	27.62	34.72	14.28	11.14
Cascade R-CNN + EFAH	27.63	41.66	18.45	28.47	35.14	58.52	48.09
Faster R-CNN	25.90	41.61	17.72	25.94	33.04	236.00	69.16
Faster R-CNN + FWB	26.03	41.52	19.04	27.16	32.77	236.00	69.17
Faster R-CNN + EFAH	26.32	41.74	18.12	26.55	32.82	236.00	69.38
Deformable-DETR	27.92	44.75	15.92	29.93	30.36	208.00	41.36
Deformable-DETR + FWB	28.52	45.24	16.95	30.31	30.38	208.00	41.38
Deformable-DETR + EFAH	28.32	45.02	16.73	30.44	31.65	231.00	41.63

4.4. Comparisons

As illustrated in Table 4, the proposed method demonstrates a significant advantage over both the YOLO series models [20] and other mainstream object detectors, including the one-stage SSD [21] model and two-stage detectors like Faster R-CNN [25], Cascade R-CNN [26], and Deformable DETR [37]. This superiority is evident in both accuracy and efficiency, especially in detecting small objects—a critical feature for ensuring the safe navigation of visually impaired individuals in complex environments. Compared to the baseline YOLOv8-s model, our method achieves an increase in mAP from 28.51 to 30.01 and in mAP_s from 18.93 to 20.96, underscoring the enhanced ability of model to detect small, less conspicuous obstacles. Such improvements are crucial for applications requiring precise small-object detection, as they provide better obstacle awareness in environments where visual cues are limited or subtle. Additionally, while the two-stage Cascade R-CNN achieves a respectable mAP of 27.23 and mAP_s of 17.12, our model surpasses it with 30.01 and 20.96, respectively, while also being more computationally efficient. Specifically, our model has fewer parameters (48.32 vs. 69.16) and lower GFLOPs (59.15 vs. 236.00), making it more suitable for deployment on resource-limited assistive devices. The proposed method stands out as the most precise among both the two-stage detectors (Faster R-CNN, Cascade R-CNN, Deformable DETR) and the one-stage detectors (SSD, YOLOv5-s, YOLOv7-tiny, YOLOX-s, YOLOv8-s, YOLOv10-n). Its ability to excel in detecting small obstacles further substantiates its potential as an effective tool for assistive technology aimed at visually impaired individuals. Overall, the proposed model's high precision, combined with its efficient computation, positions it as a robust solution for developing assistive technology dedicated to enhancing the independence and safety of visually impaired users. The performance gains in small-object detection, in particular, highlight the capability of model to navigate complex urban environments, making it an invaluable aid in supporting visually impaired individuals in real-world scenarios.

Table 4. Comparison experiments.

Models	Backbone	mAP	mAP_{50}	mAP_s	mAP_m	mAP_l	Flops/G	Parameters/M
YOLOv5-s	YOLOv5CSPDarknet	14.23	28.52	10.65	17.71	12.83	8.12	12.33
YOLOv7-tiny	E-ELAN	21.91	37.61	13.02	24.01	25.92	6.56	6.02
YOLO X-s	YOLOXCSPDarknet	23.22	40.31	14.51	25.06	27.52	13.32	8.94
SSD300	SSDVGG	21.22	39.04	8.52	21.95	28.61	30.58	24.15
YOLOv8-s	YOLOv8CSPDarknet	28.51	42.02	18.93	28.93	34.01	14.27	11.14
YOLOv10-n	CSPDarknet	29.1	42.3	19.9	30.1	35.2	8.25	2.69
Faster R-CNN	ResNet-50	25.90	41.6	17.7	25.9	30.3	208.00	41.36
Cascade R-CNN	ResNet-50	27.23	40.95	17.12	27.26	33.04	236.00	69.16
Deformable DETR	ResNet-50	27.9	44.7	15.9	29.9	35.4	193.00	40.10
Ours	CSPDarknet	30.01	43.32	20.96	31.07	35.58	59.15	48.32

4.5. Visualization

As illustrated in Figure 6, the detection results of YOLOv8 (left) and our proposed YOLO-OD (right) are visualized. The red box represents the actual detected box, and the green box highlights the difference between the detection results of the two models. We can see that the red car in the first row on the left is not detected by YOLOv8, whereas it is accurately detected by YOLO-OD, reflecting the superiority of YOLO-OD in detecting small obstacles that are crucial for visually impaired individuals. Similarly, the pedestrians in the first set of images in the second row are not detected by YOLOv8 due to the occlusion of the railings and the complex background. However, they are accurately detected by YOLO-OD, demonstrating the ability of YOLO-OD to handle challenging conditions such as occlusion and complex environments. This ability is particularly critical for visually impaired individuals who need reliable detection to navigate safely through urban areas with many obstacles. These visualization results clearly illustrate the practical benefits of YOLO-OD in real-world scenarios, particularly for visually impaired individuals who face significant challenges when navigating through environments with multiple overlapping objects and varying background complexities.

Figure 6. Visualization of YOLOv8 (**left**) and YOLO-OD (**right**) detection results.

5. Discussion

Our work proposes an object detection method named YOLO-OD, specifically designed to help visually impaired individuals navigate safely through challenging environments, such as low-light or cluttered urban areas, by effectively detecting obstacles that may compromise their mobility and safety. YOLO-OD was developed to address the shortcomings of existing object detection methods, which often struggle with small or occluded obstacles in dynamic environments that visually impaired individuals face. We have demonstrated the superior detection performance of YOLO-OD through experiments on public obstacle detection datasets, emphasizing its capability to overcome these challenges. The experimental results show that YOLO-OD significantly outperforms other comparison models, particularly in detecting small and partially occluded obstacles which are critical for the safety of visually impaired individuals. We conducted ablation experiments on different models, including both one-stage and two-stage detectors, to understand the contribution of individual components, such as the Adaptive Bottleneck Block and Feature Weighting Block, to the improved performance of YOLO-OD in challenging environments. Our goal is to further improve the detection accuracy of our method while optimizing inference speed. Below are the directions for improvement that we plan to explore:

1. Addressing the specific challenges of obstacle detection for visually impaired individuals in low-light conditions to further refine the model.

2. Upgrading the model using techniques such as wavelet convolution and StarNet to enhance both detection accuracy and inference speed.

Through these improvements, we expect that YOLO-OD will achieve even more outstanding performance in obstacle detection tasks for visually impaired individuals, providing stronger support for those with visual impairments.

Author Contributions: Conceptualization, W.W.; methodology, B.J.; software, W.W.; validation, X.Y.; formal analysis, Y.S.; investigation, L.Y.; resources, W.W.; data curation, C.W.; writing—original draft preparation, B.J.; writing—review and editing, W.W.; visualization, X.Y.; supervision, W.W.; project administration, L.Y.; funding acquisition, W.W. All authors have read and agreed to the published version of the manuscript.

Funding: This work was supported by the Jilin Provincial Education Department's Science and Technology Project (Project No. JJKH20240749KJ) and by Changchun Yaxin Technology Co., Ltd. through the Jilin Provincial Science and Technology Development Program (Project No. 2023JBH05L54).

Institutional Review Board Statement: Not applicable

Informed Consent Statement: Informed consent was obtained from all subjects involved in the study.

Data Availability Statement: You can access and obtain the data through the following links: Code repository for YOLO-OD: https://github.com/jjking00/YOLO-OD (accessed on 21 November 2024). Dataset: https://aistudio.baidu.com/datasetdetail/198589 (accessed on 21 November 2024). We confirm that the code and dataset will remain publicly accessible to ensure reproducibility and verification of the experimental results.

Conflicts of Interest: The authors declare no conflicts of interest.

Abbreviations

The following abbreviations are used in this manuscript:

MDPI	Multidisciplinary Digital Publishing Institute
DOAJ	Directory of open access journals
TLA	Three letter acronym
LD	Linear dichroism

References

1. Li, F.; Li, X.; Liu, Q.; Li, Z. Occlusion handling and multi-scale pedestrian detection based on deep learning: A review. *IEEE Access* **2022**, *10*, 19937–19957. [CrossRef]
2. Khan, S.D.; Salih, Y.; Zafar, B.; Noorwali, A. A deep-fusion network for crowd counting in high-density crowded scenes. *Int. J. Comput. Intell. Syst.* **2021**, *14*, 168.
3. Liu, Y.; Sun, P.; Wergeles, N.; Shang, Y. A survey and performance evaluation of deep learning methods for small object detection. *Expert Syst. Appl.* **2021**, *172*, 114602. [CrossRef]

4. Real, S.; Araujo, A. Navigation systems for the blind and visually impaired: Past work, challenges, and open problems. *Sensors* **2019**, *19*, 3404. [CrossRef] [PubMed]
5. Saegusa, S.; Yasuda, Y.; Uratani, Y.; Tanaka, E.; Makino, T.; Chang, J.Y.J. Development of a guide-dog robot: Leading and recognizing a visually-handicapped person using a LRF. *J. Adv. Mech. Des. Syst. Manuf.* **2010**, *4*, 194–205. [CrossRef]
6. dos Santos, A.D.P.; Medola, F.O.; Cinelli, M.J.; Garcia Ramirez, A.R.; Sandnes, F.E. Are electronic white canes better than traditional canes? A comparative study with blind and blindfolded participants. *Univers. Access Inf. Soc.* **2021**, *20*, 93–103. [CrossRef]
7. Arakeri, M.P.; Keerthana, N.; Madhura, M.; Sankar, A.; Munnavar, T. Assistive technology for the visually impaired using computer vision. In Proceedings of the 2018 International Conference on Advances in Computing, Communications and Informatics (ICACCI), Bangalore, India, 19–22 September 2018; IEEE: Piscataway, NY, USA, 2018; pp. 1725–1730.
8. Said, Y.; Atri, M.; Albahar, M.A.; Ben Atitallah, A.; Alsariera, Y.A. Obstacle detection system for navigation assistance of visually impaired people based on deep learning techniques. *Sensors* **2023**, *23*, 5262. [CrossRef]
9. Jiang, Z.; Zhao, L.; Li, S.; Jia, Y. Real-time object detection method based on improved YOLOv4-tiny. *arXiv* **2020**, arXiv:2011.04244.
10. Wang, C.Y.; Bochkovskiy, A.; Liao, H.Y.M. YOLOv7: Trainable bag-of-freebies sets new state-of-the-art for real-time object detectors. In Proceedings of the IEEE/CVF Conference on Computer Vision and Pattern Recognition, Vancouver, BC, Canada, 17–24 June 2023; pp. 7464–7475.
11. Li, C.; Li, L.; Jiang, H.; Weng, K.; Geng, Y.; Li, L.; Ke, Z.; Li, Q.; Cheng, M.; Nie, W.; et al. YOLOv6: A single-stage object detection framework for industrial applications. *arXiv* **2022**, arXiv:2209.02976.
12. Wang, X.; Shrivastava, A.; Gupta, A. A-fast-rcnn: Hard positive generation via adversary for object detection. In Proceedings of the IEEE Conference on Computer Vision and Pattern Recognition, Honolulu, HI, USA, 21–26 July 2017; pp. 2606–2615.
13. Ren, S.; He, K.; Girshick, R.; Sun, J. Faster r-cnn: Towards real-time object detection with region proposal networks. *Adv. Neural Inf. Process. Syst.* **2015**, *28*, 1137–1149. [CrossRef]
14. He, K.; Gkioxari, G.; Dollár, P.; Girshick, R. Mask r-cnn. In Proceedings of the IEEE International Conference on Computer Vision, Venice, Italy, 22–29 October 2017; pp. 2961–2969.
15. Woo, S.; Park, J.; Lee, J.Y.; Kweon, I.S. Cbam: Convolutional block attention module. In Proceedings of the European Conference on Computer Vision (ECCV), Munich, Germany, 8–14 September 2018; pp. 3–19.
16. Diwan, T.; Anirudh, G.; Tembhurne, J.V. Object detection using YOLO: Challenges, architectural successors, datasets and applications. *Multimed. Tools Appl.* **2023**, *82*, 9243–9275. [CrossRef] [PubMed]
17. Redmon, J.; Farhadi, A. Yolov3: An incremental improvement. *arXiv* **2018**, arXiv:1804.02767.
18. Bochkovskiy, A.; Wang, C.Y.; Liao, H.Y.M. Yolov4: Optimal speed and accuracy of object detection. *arXiv* **2020**, arXiv:2004.10934.
19. Liu, G.; Hu, Y.; Chen, Z.; Guo, J.; Ni, P. Lightweight object detection algorithm for robots with improved YOLOv5. *Eng. Appl. Artif. Intell.* **2023**, *123*, 106217. [CrossRef]
20. Terven, J.; Córdova-Esparza, D.M.; Romero-González, J.A. A comprehensive review of yolo architectures in computer vision: From yolov1 to yolov8 and yolo-nas. *Mach. Learn. Knowl. Extr.* **2023**, *5*, 1680–1716. [CrossRef]
21. Liu, W.; Anguelov, D.; Erhan, D.; Szegedy, C.; Reed, S.; Fu, C.Y.; Berg, A.C. Ssd: Single shot multibox detector. In Proceedings of the Computer Vision–ECCV 2016: 14th European Conference, Amsterdam, The Netherlands, 11–14 October 2016; Springer: Cham, Switzerland, 2016; pp. 21–37.
22. Cheng, B.; Wei, Y.; Shi, H.; Feris, R.; Xiong, J.; Huang, T. Revisiting rcnn: On awakening the classification power of faster rcnn. In Proceedings of the European Conference on Computer Vision (ECCV), Munich, Germany, 8–14 September 2018; pp. 453–468.
23. Ren, Y.; Zhu, C.; Xiao, S. Object detection based on fast/faster RCNN employing fully convolutional architectures. *Math. Probl. Eng.* **2018**, *2018*, 3598316. [CrossRef]
24. Girshick, R. Fast r-cnn. In Proceedings of the IEEE International Conference on Computer Vision, Santiago, Chile, 7–13 December 2015; pp. 1440–1448.
25. Liu, B.; Zhao, W.; Sun, Q. Study of object detection based on Faster R-CNN. In Proceedings of the 2017 Chinese Automation Congress (CAC), Jinan, China, 20–22 October 2017; IEEE: Piscataway, NY, USA, 2017; pp. 6233–6236.
26. Cai, Z.; Vasconcelos, N. Cascade r-cnn: Delving into high quality object detection. In Proceedings of the IEEE Conference on Computer Vision and Pattern Recognition, Salt Lake City, UT, USA, 18–23 June 2018; pp. 6154–6162.
27. Chandna, S.; Singhal, A. Towards outdoor navigation system for visually impaired people using YOLOv5. In Proceedings of the 2022 12th International Conference on Cloud Computing, Data Science & Engineering (Confluence), Virtual, 27–28 January 2022; IEEE: Piscataway, NY, USA, 2022; pp. 617–622.
28. Elgendy, M.; Sik-Lanyi, C.; Kelemen, A. A novel marker detection system for people with visual impairment using the improved tiny-yolov3 model. *Comput. Methods Programs Biomed.* **2021**, *205*, 106112. [CrossRef]
29. Atitallah, A.B.; Said, Y.; Atitallah, M.A.B.; Albekairi, M.; Kaaniche, K.; Boubaker, S. An effective obstacle detection system using deep learning advantages to aid blind and visually impaired navigation. *Ain Shams Eng. J.* **2024**, *15*, 102387. [CrossRef]
30. Khan, W.; Hussain, A.; Khan, B.M.; Crockett, K. Outdoor mobility aid for people with visual impairment: Obstacle detection and responsive framework for the scene perception during the outdoor mobility of people with visual impairment. *Expert Syst. Appl.* **2023**, *228*, 120464. [CrossRef]
31. Liu, Z.; Wang, Y.; Vaidya, S.; Ruehle, F.; Halverson, J.; Soljačić, M.; Hou, T.Y.; Tegmark, M. Kan: Kolmogorov-Arnold networks. *arXiv* **2024**, arXiv:2404.19756.

32. Dai, X.; Chen, Y.; Xiao, B.; Chen, D.; Liu, M.; Yuan, L.; Zhang, L. Dynamic head: Unifying object detection heads with attentions. In Proceedings of the IEEE/CVF Conference on Computer Vision and Pattern Recognition, Nashville, TN, USA, 20–25 June 2021; pp. 7373–7382.
33. Zeiler, M.D.; Krishnan, D.; Taylor, G.W.; Fergus, R. Deconvolutional networks. In Proceedings of the 2010 IEEE Computer Society Conference on Computer Vision and Pattern Recognition, San Francisco, CA, USA, 13–18 June 2010; IEEE: Piscataway, NY, USA, 2010; pp. 2528–2535.
34. Xu, J.; Li, Z.; Du, B.; Zhang, M.; Liu, J. Reluplex made more practical: Leaky ReLU. In Proceedings of the 2020 IEEE Symposium on Computers and communications (ISCC), Rennes, France, 7–10 June 2020; IEEE: Piscataway, NY, USA, 2020; pp. 1–7.
35. Cui, Z.; Chen, W.; Chen, Y. Multi-scale convolutional neural networks for time series classification. *arXiv* **2016**, arXiv:1603.06995.
36. Zyh. Road Obstacle Dataset in COCO Format. *Baidu AI Studio*, 2023. Available online: https://aistudio.baidu.com/datasetdetail/198589 (accessed on 13 October 2024).
37. Zhu, X.; Su, W.; Lu, L.; Li, B.; Wang, X.; Dai, J. Deformable detr: Deformable transformers for end-to-end object detection. *arXiv* **2020**, arXiv:2010.04159.

sensors

MDPI

Article

Empowering People with Disabilities in Smart Homes Using Predictive Informing †

Marko Periša *, Petra Teskera, Ivan Cvitić and Ivan Grgurević *

University of Zagreb, Faculty of Transport and Traffic Sciences, Vukelićeva 4, 10000 Zagreb, Croatia; petra.teskera@fpz.unizg.hr (P.T.); ivan.cvitic@fpz.unizg.hr (I.C.)
* Correspondence: marko.perisa@fpz.unizg.hr (M.P.); ivan.grgurevic@fpz.unizg.hr (I.G.)
† This paper is an extended version of our paper published in the 7th EAI International Conference on Management of Manufacturing Systems, Bratislava, Slovakia, 24–26 October 2023.

Abstract: The possibilities of the Ambient Assisted Living (AAL)/Enhanced Living Environments (ELE) concept in the environment of a smart home were investigated to improve accessibility and improve the quality of life of a person with disabilities. This paper focuses on the concept of predictive information for a person with disabilities in a smart home environment concept where artificial intelligence (AI) and machine learning (ML) systems use data on the user's preferences, habits, and possible incident situations. A conceptual mathematical model is proposed, the purpose of which is to provide predictive user information from defined data sets. This paper defines the taxonomy of communication technologies, devices, and sensors in the environment of the user's smart home and shows the interaction of all elements in the environment of the smart home. Through the integration of assistive technologies, it is possible to adapt the home to users with diverse types of disabilities and needs. The smart home environment with diverse types of sensors whose data are part of sets defined by a mathematical model is also evaluated. The significance of establishing data sets as a foundation for future research, the development of ML models, and the utilization of AI is highlighted in this paper.

Keywords: accessibility; artificial intelligence; assistive technology; ambient assisted living; navigation

1. Introduction

According to the World Health Organization (WHO), an estimated 2.2 billion people have some form of vision impairment (distance or near), while 1.3 billion people have a significant vision disability, representing 16% of the world's population. Vision loss can affect a person's ability to perform daily activities and affects people of all ages. However, the majority of people with vision impairment and blindness are over 50 years old [1]. The implementation of assistive technologies can range from physical products such as wheelchairs, glasses, prosthetic limbs, white canes, and hearing aids to modern IT solutions and services such as speech recognition, smart home management, and the like. With the aging of the global population and the increase in non-communicable diseases, it is estimated that around 3.5 billion people will need assistive technology by 2050. Statistical data predicts that the global number of smart home users in the smart home market will continuously increase between 2023 and 2028 by a total of 424.5 million users (+117.69%), which is an indicator of a trend that can certainly affect the possibility of applying the smart home environment for people with visual impairments [2]. Everyday needs are an integral part of a person's life that affect its quality. The home and its environment,

https://doi.org/10.3390/s25010284

and the contents found in it certainly have a significant impact on the quality of life. The development of information and communication technologies (ICT) and services opens the field of creating a smart home, which with its functionalities can raise the level of quality of life of the person living in it. Many years of work, as well as numerous pieces of research conducted in the domain of assistive technologies and their adaptation to people with visual impairments and other types of impairments, have shown that this field requires the development of services tailored to the requirements of users. The above is based on the fact that not all people have the same knowledge about the use of assistive technologies, therefore, the individual adaptation of services is the basis for better future application.

Using the methods of analysis and synthesis, information and data from previous research were collected and a taxonomy of communication technologies and services was defined. Also, using the above methods, data sets were defined that are the basis for the work of the mathematical model. The observation method was used to collect information (phenomena and processes) surrounding a blind and visually impaired person who is in a smart home environment.

The aim of this paper is to highlight the complex interdependence of the different elements of a smart home system and underscore the importance of defining data sets and creating user profiles depending on the type of impairment of the user.

2. Methodology

Today's technological development in the field of assistive technology and the environment of people with disabilities brings significant changes in the way people with disabilities live and participate in everyday life [3]. Figure 1 shows the research methodology in this paper. The research hypothesis is as follows: based on defined data sets, it is possible to create user profiles of user behaviour for the purpose of predictive information. For this purpose, the literature covering the development and application of assistive technologies (devices and services) was researched, the technical characteristics of communication technologies that can work in the IoT environment of a smart home were investigated, and user needs depending on the type of user impairment were also investigated.

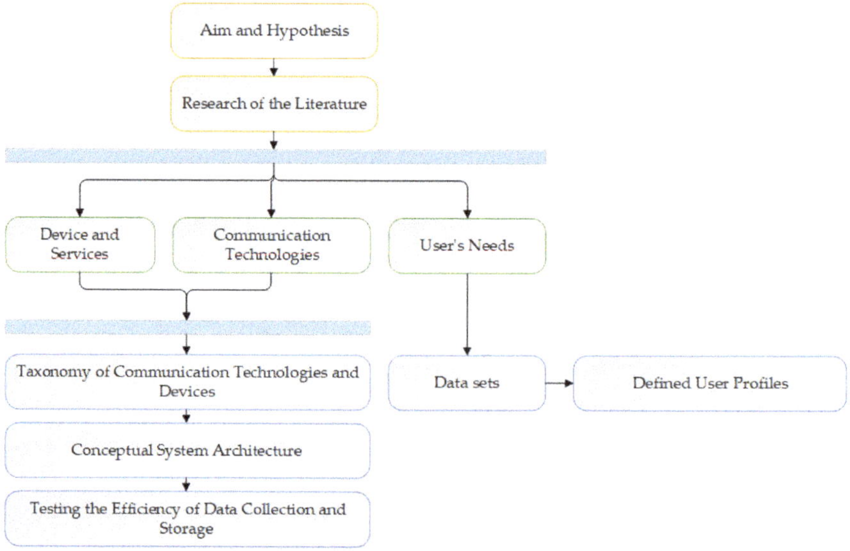

Figure 1. Research methodology.

Based on the methods used in the research, a conceptual architecture of the system will be presented, the aim of which is to show the interaction of data collection, processing, and storage. The defined data sets are presented mathematically and represent the basis for the development of machine learning (ML) models and the creation of a user knowledge base. Based on this, it is possible to develop user profiles necessary for predictive user information using artificial intelligence (AI). Predictive information works in such a way that data and information about user needs are collected daily in the smart home environment and personalized user information is created using AI.

2.1. Literature Review

Assistive technologies are not representation of only information and communication devices and systems, but also services aimed at improving the functional abilities of people with diverse types of disabilities [4]. Having a disability can impact a person's ability to conduct daily tasks and fully participate in various environments, including their own home. However, everyday activities within their home such as cooking, cleaning, safe movement around the home, or health monitoring can be integrated into a smart home environment where users can control individual devices [5].

A smart home can be defined as the interaction of information and communication technologies and services with the user through the home network to create better living conditions and environments for the user [6]. Today's development of smart homes and indoor environments intended for a person with disabilities is based on the concepts of Ambient Assisted Living (AAL) and Enhanced Living Environments (ELE). The technologies and working principles used in the mentioned concepts are based on modern information and communication technologies such as Cloud Computing (CC) (Fog, Edge, Dew), the Internet of Things (IoT), Machine to Machine (M2M), and diverse types of sensor networks and corresponding sensors [7]. Several different communication devices and systems are currently in daily use, and in most cases, the devices are used for multiple purposes and solve more than just one need [8]. The diversity of technologies available to support the autonomy and independence of people with disabilities offers the possibility of choosing a particular technological solution or set of solutions that are most suitable for their functional abilities [9]. According to the data on the quantity of IoT devices used in smart environments until the year 2025, there are 71 billion installed IoT devices expected, whereas in smartphone environments 12.86 billion devices are expected [10]. Following the development trends in the field of AAL and ELE concepts, the need for the application of diverse types of sensors, robotics, the IoT, and voice assistants to manage them is increasingly emerging [11,12]. The context of the use of AAL and ELE concepts is presented through a literature review, where the authors stated the possibility of use and purpose in closed spaces and the importance of network heterogeneity [13]. The AAL concept, whose aim is to help and provide real-time information to elderly people and persons with disabilities, is also increasingly applied in the design of new services based on decision support systems [14]. The older group of users is becoming increasingly represented in numerous studies, based on the fact that the number of users aged 65+ is increasing. This is also supported by data from the United Nations, where in 2021 there were 761 million elderly people, and it is predicted that this number will reach 1.6 billion in 2050 [15]. The aforementioned fact prompted the authors to design and evaluate the SecurHome TV service, which is based on the interaction of elderly people with their TV receivers as it is one of the most frequently used devices in everyday life [16]. SecurHome TV represents an ecosystem, with a special emphasis on the algorithm for warning users and detecting dangerous situations, contributing to monitoring the activities of an older group of users. For modeling IoT services and delivering real-time information to persons with disabilities, it is possible to use a conceptual model of system architecture applicable to an IoT environment [17]. According to a research paper [18], incorporating IoT technology in a smart home can greatly benefit visually impaired individuals by facilitating their daily activities. This, in turn, enhances the overall autonomy, safety, and comfort of users in their own home environment.

The development of technology such as AI provides the ability to detect and monitor the daily activities of people with disabilities (users) in their homes. In the user's smart home environment, the following system components are important: appropriate hardware equipment for data collection (sensors and actuators), a communication network (sensor autonomous network), a set of applications and tools for data processing and management, and data storage equipment [5]. The e-care@home system defines certain data sets and the modularity of the smart home system, which enables use regardless of the type of impairment the user has [19]. Also, the application of AI technology and the 6G network is applied in medicine as a closed user space where appropriate IoT services (IoMT) are applied [20]. The application of autonomous and intelligent networks and systems is today also increasingly used indoors, mostly in the field of eHealth and in providing services for the senior persons and persons with disabilities [21,22].

In addition to the application of AI technology and the environment of a smart home for people with disabilities, it is also possible to use it for elderly users who also have certain health problems, such as dementia and an increased risk of falling when moving. For this purpose, the authors propose the improvement of data sets with the aim of a more efficient system for detecting user falls and informing caregivers [23]. AI requires the definition and use of certain types of data sets and thus the creation of user forms for which it is also necessary to define user needs. Daily activities depend on the distinctive characteristics or types of the user's impairment and an individual approach to solving daily needs is frequently required. For this purpose, to model innovative services in a sustainable ecosystem, it is necessary to clearly define data sets, all to deliver accurate information to the end user and the possibility of predictive information.

User needs are the basis for defining data sets, but also for designing a smart home according to user requirements [8,24–26]. There are 624,019 persons with disabilities in the Republic of Croatia, of which 353,550 are male (56.7%), and 270,469 are female (43.3%), thus persons with disabilities make up about 16.0% of the total population. Groups of users and their number in the Republic of Croatia, according to the type of impairment that is the subject of this research, are locomotor system impairment 177,547, speech-voice communication impairment 51,149, visual impairment 20,526, and hearing impairment 17,679 [27].

2.2. Defining User Needs in a Smart Home Environment

Research [28] has explored the lifetime home standards (LHTS) of individuals with disabilities. The conclusions are mainly focused on the architectural adaptation of the environment in which the user is located. The research also highlighted the importance of incorporating technological solutions to enhance their living conditions. However, high-tech solutions such as electronic doors and windows remain financially inaccessible to most users. Nonetheless, users expressed that implementing such solutions would improve their quality of life. Safety within the home was also emphasized, particularly the need to detect electric shocks and gas leaks as well as the importance of lighting management. Another piece of research has examined the benefits of home automation systems for visually impaired individuals [29]. The research results point out that there is a strong interest in home automation to increase the autonomy of visually impaired individuals, with lighting control and heating systems being the most prominent areas of focus.

To design the conceptual system architecture, the smart home environment for the visually impaired can be considered through three segments: user needs, the home environment or architecture, as well as the technology used in the home [5]. This approach can also be applied to other groups of people with disabilities, where it is important to show the characteristics of impairment as shown in Table 1.

Table 1. User characteristics according to the type of impairment.

Type of Impairment	Characteristics
Blind and partially sighted	• slow and uncertain movement • difficulties with planning, organizing time, materials, and tasks • "fear" of light • inability to distinguish colours or shades of colours and letter sizes • spatial disorientation • limited living space
Locomotive system	• physical and architectural barriers • free space for wheelchairs to turn around in the room • limited living space
Speech-voice communication	• speech impairments • voice impairment • the problem of reading, writing, and arithmetic
Hearing	• limited and poor communication • problem using speech • limited living space • avoiding social activities • concentration difficulties • weaker ability to maintain balance • weaker motor speed and coordination

Based on the type of user impairment, Table 2 defines increased risks depending on the smart home usage scenario (kitchen—K, living room—LR, bedroom—SR, bathroom/toilet—TWC, terrace/balcony—TB, and daily activities—DA).

Table 2. Display of increased risks depending on the type of impairment and the scenario of using the smart home service.

Type of Impairment	Usage Scenario	Increased Risks
Blind and partially sighted	K	• fire during food preparation • uncontrolled gas leakage • injuries and burns • flood • electric shock
	LR	• obstacles and possible injuries • inability to locate individual elements • fall

Table 2. *Cont.*

Type of Impairment	Usage Scenario	Increased Risks
Blind and partially sighted	SR	• obstacles and possible injuries • inability to locate individual elements
	TWC	• flood • electric shock • injuries and falls
	TB	• injuries and falls
	DA	• health status monitoring • incorrect information using the information service
Locomotive system	K	• uncontrolled gas leakage • injuries and burns • flood • electric shock • unavailability of elements and injuries
	LR	• obstacles and possible injuries • fall • poor availability of elements
	SR	• obstacles and possible injuries • fall
	TWC	• injuries and falls • electric shock • flood
	TB	• injuries and falls
	DA	• health status monitoring • incorrect information using the information service
Speech-voice communication	All scenarios	• panic and poor reaction to possible incident situations
Hearing	All scenarios	• concentration

2.3. *Defining Data Sets in a Smart Home Environment*

Data sets will be defined as elements of the smart home management system based on the above-mentioned increased risks for certain groups of users and usage scenarios. It is also important to emphasize the communications segment that is necessary for the operation of an automated system based on generative AI. The sensors and function tags used in the smart home environment are defined through three subsets, shown in Table 3.

Table 3. The range of functional dependencies required to deliver information at the user's request.

Range of Function Dependence	User Health (S_{hi})	Incident Information (S_{ii})	User Environment (S_{ei})
Sensors and function tags	Pressure (S_p) Pulse (S_{pu}) Sugar (S_s) User's temperature (S_t) Breathing (S_b) Burn (S_{bu}) Fall (S_{fa}) Electric shock (S_{el}) Air quality (external) (S_{ao}) Dust (S_d)	Fire (S_f) Gas (S_g) Fall (S_{fa}) Flood (S_{fl}) Electric shock (S_{el}) Earthquake (S_{er}) Burn (S_{bu})	Object detection (S_{od}) Air quality (indoor) (S_{ai}) Food quality (refrigerator) (S_{fq}) Light intensity (S_l) Temperature (S_{te}) Humidity (S_h) Dust (S_d) Air quality (outside) (S_{ao}) Open/close door (S_{ocd}) Fire (S_f) Flood (S_{fl}) Electric shock (S_{el}) Fall (S_{fa}) Earthquake (S_{er})

Within each subset, defined sensors collect the data needed to deliver the information requested by the user. Sensors are defined as variables and have their labels, and they are shown with a Venn diagram and a mathematical model (Figure 2) [17]. The conceptual mathematical model for providing information (predictive information) to the user is presented as a union of defined sets, i.e., data collected from around the smart home.

$$S_{shome} = S_{hi} \cup S_{ii} \cup S_{ei} \text{ or if it is an expansion with additional subsystems} \quad (1)$$

$$S_{shome} = \bigcup_{i=1}^{S} S_i \quad (2)$$

from which:

$$S_{hi} = \left\{ S_p,\ S_{pu}, S_s, S_t, S_b, S_{bu},\ S_{fa}, S_{el}, S_d, S_{ao} \right\} \quad (3)$$

$$S_{ii} = \left\{ S_f,\ S_g, S_{fa}, S_{fl},\ S_{el},\ S_{er}, S_{bu} \right\} \quad (4)$$

$$S_{ei} = \left\{ S_{od},\ S_{ai},\ S_{fg},\ S_l,\ S_{te}\ ,\ S_h\ , S_d, S_{ao}\ , S_{ocd}, S_f, S_{fl}, S_{el},\ S_{fa},\ S_{er} \right\}. \quad (5)$$

Dependence of individual variables in several subsystems is as follows:

$$S_{hi} \bigcap S_{ii} \bigcap S_{ei} = \left\{ S_{fa}, S_{el} \right\} \quad (6)$$

$$S_{hi} \bigcap S_{ii} = \{ S_{bu} \} \quad (7)$$

$$S_{ii} \bigcap S_{ei} = \left\{ S_f, S_{fl},\ S_{er} \right\} \quad (8)$$

$$S_{hi} \bigcap S_{ei} = \{ S_{ao}, S_d \} \quad (9)$$

The graphical representation in Figure 2 shows the inclusion of individual data (variables) from all these subsystems. Fall detection (S_{fa}) and electric shock (S_{el}) are connected to all three subsystems because the delivery of information about the user's health (state in which they are and type of impairment) and location must be real-time according to the emergency services or the caregivers if this has been previously defined. Burn detection (S_{bu}) on the user is identified by sensors from the subset of user health (S_{hi}) and incident information (S_{ii}) to call emergency services. When a flood (S_{fl}), fire (S_f), and earthquake (S_{er}) are detected from the environment where the user is located, a call to the emergency services or the caregiver is also made. To preserve the user's health (allergic reactions), the sensor of dust (S_d) and air quality in the external environment (S_{ao}) informs the user about this. All defined variables represent a set of user data necessary for the creation of real-time information and the possibility of predictive user information.

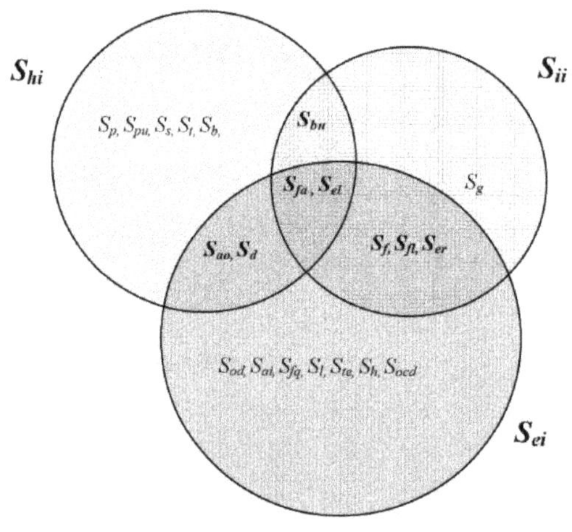

Figure 2. Representation of variables in subsets.

The concept of a smart home for person with disabilities and associated sensors is shown in Figure 3. The figure shows all the sensors listed in Table 3. Inside the smart home, the user can be informed by voice about the time (hours and minutes), reminder management, and environmental detection based on data on incidents such as vibrations (earthquake), flood, etc. The device can be used to control TV and radio devices, refrigerators, vacuum cleaners, air conditioners, front doors, blinds, cameras, and other available IoT devices. It is also possible to provide information to the user about health, during which it is possible to monitor the user's pulse, pressure, temperature, and the amount of sugar and oxygen in the blood. For this purpose, the user should have a smart bracelet that is equipped with the appropriate sensors. The smart bracelet, with its characteristics, can be modular, i.e., adaptable to the user's type of impairment. Depending on the type of impairment, appropriate sensors are activated and the format of providing information to the user is adjusted [30]. The connection of the bracelet with the smart home system is also required to create a user base.

Inside the smart home, there are sensors for a safer and more comfortable execution of everyday activities. The virtual assistant provides voice control of smart IoT devices and provides the user with information collected from sensors. Sensor information is delivered upon user request. For the mentioned concept, it is important to use an application solution made according to the guidelines of universal design. The idea of the presented concept is to enable the creation of a user base of the user's needs to provide accurate and real-time information and predictive user information.

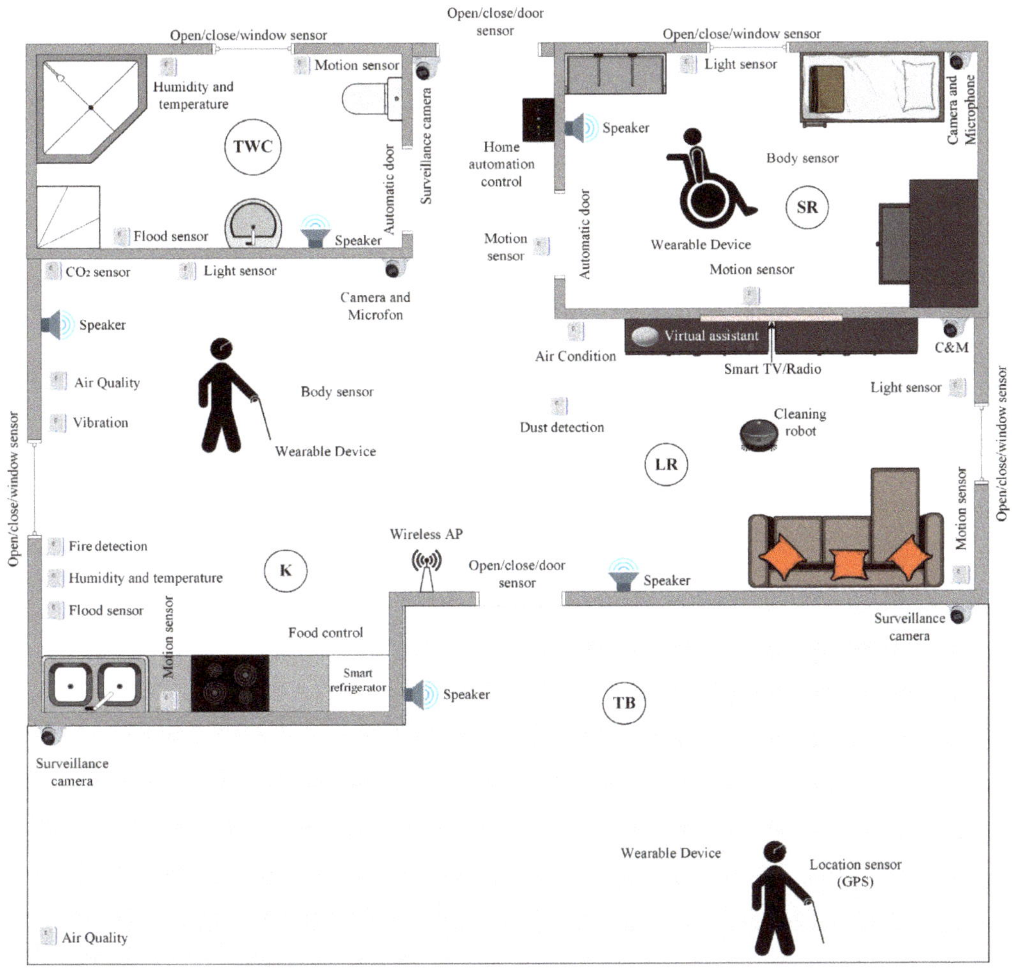

Figure 3. Concept of a smart home for a person with disabilities.

2.4. Taxonomy of Communication Technologies and Devices

To apply AI technology, i.e., perception of the environment in which the user is located and for the provision of relevant information, a taxonomy of communication technologies and devices shown in Figure 4 is proposed. The purpose of using AI is to ensure, based on collected data from sensors that monitor everyday situations and movements, that the user's behaviour and obligations are created by appropriate user patterns as a basis for predictive user information. User forms creation (depending on the type of impairment) defines the basis for modeling IoT services based on AI technology, all to develop a predictive model for information delivery.

For service delivery to the user, the smart home environment is divided into four layers of automation: Perception and Actuations, Network/Transmission, Preprocessing and Middleware, and Applications and Devices. The Perception and Actuations layer represents physical objects that serve to collect information from the user's environment. The user's environment is made up of defined scenarios in which they find themself every day (kitchen, living room, bedroom, toilet, and outdoor balcony/terrace), as well as the user's health and daily needs.

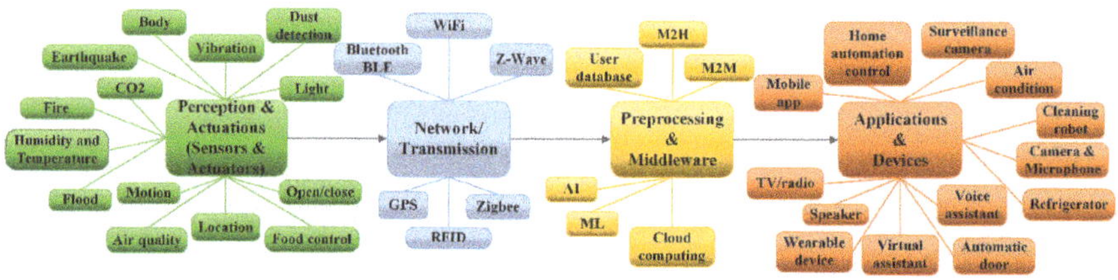

Figure 4. Taxonomy of the used technology and devices.

The user's health condition is monitored via a smart bracelet equipped with body sensors. Body sensors are a group of sensors that collect data on heart rate, pressure, sugar, temperature, and breathing. Also, to monitor the user's health, a sensor is used to detect burns, electric shock, and sensors that can indicate possible allergic reactions (dust sensor and air quality in the external environment) are used. Sensors that are used to warn about the possible occurrence of an accident and belong to the group for providing incident information have the role of detecting the following: a fire, gas (CO_2), user fall, a flood, electric shock, an earthquake, and burns. To monitor user activity in all scenarios, sensors are used for objects/obstacles detection, the monitoring of air quality in the indoor environment, food quality, the detection of light intensity, temperature measurements in the home, and measurements of the amount of moisture and dust.

The Network layer or Transmission layer is responsible for the secure transmission of data obtained from the sensor devices to the processing system, i.e., to the Preprocessing and Middleware layer through various network technologies. It consists of an access sublayer that has the role of collecting data from the Perception layer and of sending it to the Internet sublayer. In the access sublayer, which is in the part of the core network, the key role is played by the technologies Z-Wave, Zigbee, Bluetooth BLE, GPS, and RFID. Wi-Fi technology is used to transfer the obtained information from the access sublayer to the next layer, which is Preprocessing and Middleware [5]. GPS technology has the function of determining the user's position if the user is in the external environment of the smart home or if the user moves away from the home in which they are located (shown in Figure 3). The above is important if the user has some form of dementia or similar impairments.

The Preprocessing and Middleware layer is a software layer that can directly synchronize the services with the appropriate requirements and has a connection to the database. It performs processes such as preparing the information for further processing, data cleaning, standardization, values of data within a dataset, balancing, intelligent routing, and translating network addresses. It is based on CC because of the great benefits it provides, such as scalability and flexibility. In the proposal of this paper, the user data are saved in the user database that is stored in the CC infrastructure. The preparation of information for delivery in the appropriate form is enabled using M2H (Machine–to–Human) and M2M technologies. The ML (machine learning) algorithms enable the intelligent adaptation of applications to changes in the environment, and the AI technology helps to collect data from the devices, predict the user behaviour within the smart home environment, and help to improve data security [5].

Based on the processed data in the Preprocessing and Middleware layer, the Application and Devices layer is responsible for providing end-user services and the proper operation of applications and devices. The identification, processing, and storage of data are performed in CC models depending on the level of service used. In addition to sensors and actuators (S_{n+1} and A_{n+1}), devices (D_{n+1}) are also used, which by their characteristics belong to the group of IoT devices and have been integrated into the smart home system.

Devices include smart robots, air conditioners, automatic doors at the home and room entrances, a smart TV, a refrigerator, cameras, a virtual assistant, and smart home management systems. Application solutions on smartphones and smart home management systems are also used. An API (Application Programming Interface) is used to provide the appropriate form of information that is defined through the user profile (degree of impairment, etc.).

3. Conceptual System Architecture

According to the defined sensors and the communication network used in the smart home environment, and the defined scenarios of the application of the smart home service, Figure 5 shows the conceptual architecture of the data collection and processing system. The communication of the elements is based on the AAL/ELE concept, the main purpose of which is to work in closed spaces adapted to user needs [12,13,31].

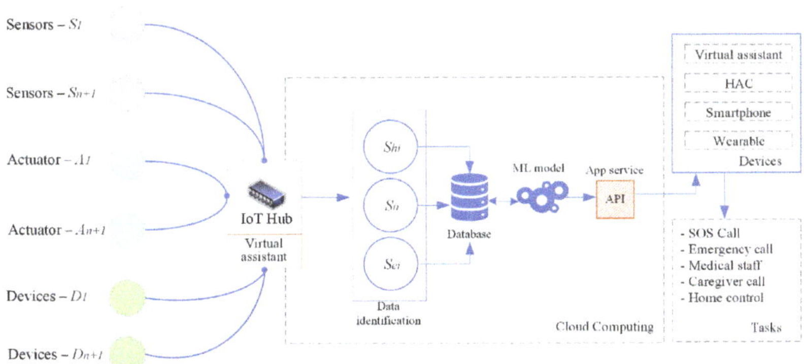

Figure 5. Conceptual architecture of the data collection and processing system.

The role of sensors and actuators (S_{n+1} and A_{n+1}) is to collect data from the user's environment depending on the scenario in which the smart home service is used. Devices (D_{n+1}) used in the smart home environment also generate data sets in the IoT hub and forward them to the CC, whose role is to collect, identify, and process data. The interaction between all levels of communication towards the CC depends on the service-level agreement (SLA), which must satisfy the corresponding level of service. The data are stored in the database depending on the identification of the associated set (function dependencies—S_{hi}, S_{ii}, and S_{ei}). An example of detecting an elevated pressure level in a user and performing tasks is shown in Figure 6.

When detecting pressure, it is important to set the pressure levels (normal–high). The mentioned levels also represent a weighting factor that is crucial for the execution of tasks and activities. One of the tasks called emergency call consists of several activities (activities such as contacting emergency services in case of fire, flood, fall, injury) that can be performed in different usage scenarios (K, LR, SR, TWC, TB, DA) depending on the type of user impairment. At the same time, the activities represent the way and the possibility of delivering information to the user or one of the defined stakeholders, and their sets are defined as tasks that are performed through specific devices.

Through the Application Programming Interface (API), information is displayed to the user/stakeholder in a specific format. Each piece of information is adapted to the user (depending on the type of impairment) using M2H communication in an acceptable form. To collect data on users, it is possible to use a separate database where the user data needed to define user profiles (type of impairment and method of data delivery) are located. The content is adjusted based on the user's characteristics and the functionalities selected by the individual user and is presented in Table 4.

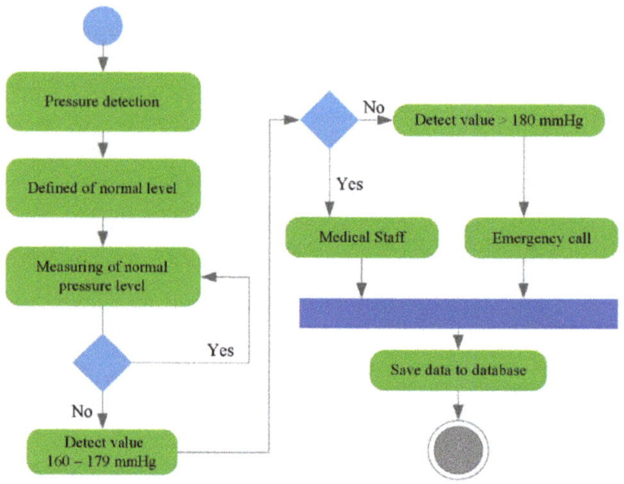

Figure 6. The process of detecting an increased level of pressure in the user.

Table 4. Form of providing information depending on the type of disability.

Type of Impairment	Form of Providing Information
Visually impaired person	Easy to use Designed according to universal design guidelines Using accessibility options for text and background customization Adaptable for TXT option Enabled access to use voice control Recognition of objects through the camera SOS option High accuracy in voice recognition in text
Voice–speech communication difficulties	Adaptable content for visual information Through video visual content, Possibility of using sign language Adjusting the intensity of the speaking speed Easy to use Designed according to universal design guidelines Easy to provide technical assistance SOS option
Hearing-impaired person	Adaptable content for visual information Through video visual content Possibility of using sign language Easy to use Designed according to universal design guidelines Adjusted for the possibility of using light effects SOS option

For persons with locomotor impairments, it is important to consider all the above-mentioned guidelines if it is an additional type of impairment; however, for persons who move with the help of wheelchairs, the aspect of providing accurate information and communication with the caregiver or emergency services is important. When identifying an incident situation, a person must have the simplest viable way out of the home environment. People who do not have speech or hearing problems but have a visual impairment or are physically impaired and use a wheelchair can use the services of a voice assistant in a smart home environment. Its role is the ability to manage IoT devices in the home and predictive user information.

4. Defining User Profiles

Based on the defined data sets and depending on the type of user impairment, user profiles are defined as shown in Table 5. The information is crucial for the design and development of personalized user services with the aim of providing accurate and real-time information in the user's home environment.

Table 5. User profile elements.

User Characteristics	User Needs	Data Collection Methods	Type of Personalized Messages
• Type of impairment • Level of support (partial or full assistance) • Level of knowledge about using technology • Health information • Additional specific information depending on the type of impairment	• Daily activities • Use of medical personnel • Prone to falls or • other forms of risk • Need for different forms of • communicators • Mobility and navigation • Accessibility	• Activity tracking sensors • Behaviour tracking • Health monitoring capabilities	• Personalized reminders • Automated devices and elements in the home environment • Anticipation of potential • problems (falls, etc.) • Notifications about planned events • Warning messages about • potential dangers • Voice and visual messages • Communication with a • caregiver or another person

The user profiles shown in Table 5 are divided into four categories: user characteristics, user needs, methods of collecting data and information, and the possibility of personalized messages. In the group of user characteristics, it is important to know what type of impairment is involved and what level it is (e.g., 75% vision impairment or less with no colour recognition). In this group, it is also important if, for example, it is a case of a deaf–blind person for whom it is necessary to additionally provide communication modules. Also important is the user's level of IT literacy, how much they use today's smartphones, etc. In this section, the required level of assistance support is also important, whether it is fully automated or partial, e.g., for full automation in a smart form, voice, visual, and tactile information are required, while for partial automation, the user could perform certain activities. Additional specific information relates, for example, if the user has a problem holding the device or speaks more slowly, so that adaptation is also required in that area.

Information about user needs is focused on monitoring daily activities (morning, day, night) in a 24 h time interval or at a specific time, the possibility of using medical staff or a caregiver requested by the user themself and communicating with them, and monitoring the increased risk of other forms of injury (falls or something similar). If the user requires the ability to use diverse types of communicators, for example sign language, this is also essential information in creating the profile. The user's level of mobility, navigation, and accessible environment are information that increase the user's autonomy in their home.

Data and information can be collected from the sensors shown in this paper and depending on the example in the smart home at the Figure 3. In addition, it is possible to collect data on user behaviour and use data from health records if available.

For creating predictive information, the way of personalized messages is important. Personalized reminders aim to provide information based on the user's daily routines, e.g., taking medication, rest, exercise, etc. Working with automated devices implies the possibility of voice or some other form to be used with devices in the home to provide information about planned events, e.g., if a trip to the doctor is planned, etc., with the user receiving a notification about it.

When delivering information in real time, it is necessary to consider the mechanisms for providing notifications, and to enable:

- The possibility of synchronizing the information delivery system, which must be consistent with the environment in which the user is located;
- The possibility of providing information (notifications) in different modalities depending on the type of user impairment (visual, audio, or haptic);
- That the user information service is designed according to the principles of universal design, which allows avoiding straining the user with excessive or distracting notifications.

The importance of using individual mechanisms in the system of providing information to users is also stated by the authors in the paper [32]. In the paper, an analysis of user perceptions regarding the provision of information used by users in different scenarios was made, during which various measurement variables were used (engagement, distraction, awareness, and emotional effort). With the approach, the authors tried to create a balanced user experience in the observed scenario (delivery of synchronized information related to TV content, and informing users via visual, audio, and haptic stimuli).

5. Testing and Performance Results of Sensors in a Smart Home Environment

To verify the effectiveness of the proposed mathematical model, which aims to define information for users that are in smart home environment, specific sensors and IoT equipment were utilized. The data were collected and processed in the Thingspeak cloud environment for data storage and continuous monitoring of information from the user's environment. Figure 7 shows the Thingspeak environment and the data collection time in laboratory conditions. Data were collected at average intervals of 15 [s]. Each record includes a unique identifier (id), time interval between measurements, sensor value, latitude, longitude, elevation, and status (2023-08-29T16:14:05+02:00, 15, 28.00000, 82.39999, 40.00000).

The collected values are shown in Figure 7, for example, the temperature value from 26 °C to 28 °C, the humidity amount from 20% to 40%, fall detection (motion), and recognition of objects at a certain distance from 3 [m] to 15 [m]. The equipment used for simulation and testing is shown in Table 6. The test was conducted in the Laboratory of Development and Research of Information and Communication Assistive Technology at the Faculty of Transport and Traffic Sciences, University of Zagreb.

The sensors collected various values including fire, burn, CO_2, earthquake, object, temperature, humidity, and fall. These values were then stored in the Thingspeak cloud environment and later in the MYSQL database, as shown in Figure 8. The database was created based on the mathematical model that described the type of data and values. The tables that collected the sensor data were linked to the table responsible for storing and delivering information to the end user.

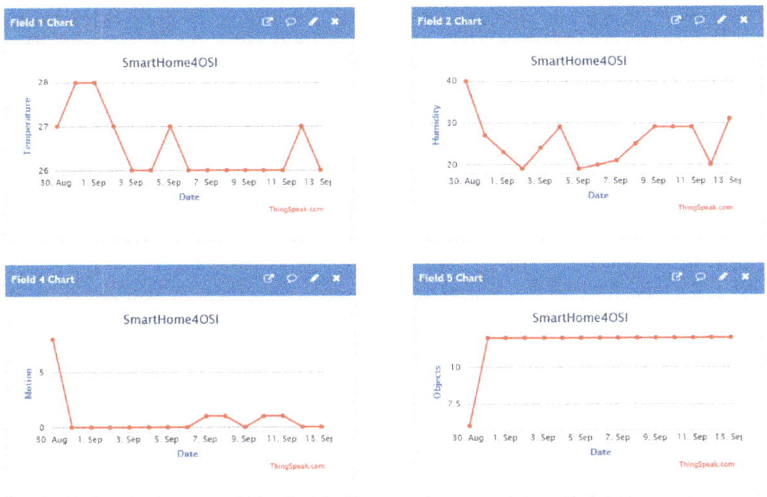

Figure 7. Thingspeak environment for monitoring collected data from sensors.

Table 6. Display of equipment used for the purpose of simulating data collection in a smart home.

Equipment	The Role
Sensor LM393—Adafruit Industries, New York, NY, USA	Detection of fire in the apartment or burns
Sensor MQ-2—Arduino, Monza, Italy	CO_2 gas level detection
Sensor SW-420—Arduino, Monza, Italy	Earthquake detection
Sensor HC-SR04—Arduino, Monza, Italy	Object detection
Sensor DHT11—Arduino, Monza, Italy	Measurement of temperature and humidity levels
Sensor PI-HCSR501—Arduino, Monza, Italy	Fall detection
Arduino Mega 2560Rev3—Arduino, Monza, Italy	Microcontroller for connecting sensors and computers
Arduino Ethernet Shield—Arduino, Monza, Italy	A component for connecting and sending data to the Cloud
Thingspeak—The MathWorks, Inc., Natick, MA, USA	Data collection, storage, and real-time monitoring

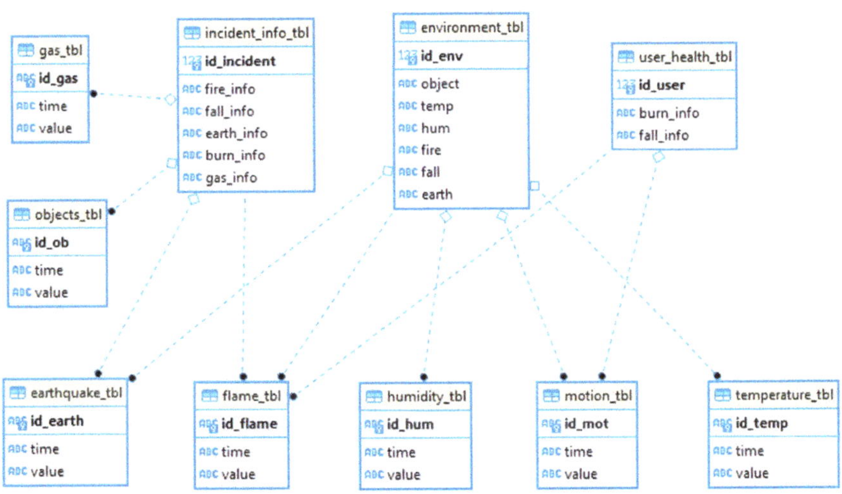

Figure 8. Relations within the database.

To deliver information in a suitable format for a specific user, such as a blind or partially sighted person with unique characteristics, an appropriate API must be defined in the middleware layer. In this case, an API can be used to receive data and send configuration values to an API that delivers information in the appropriate format for the user's device, regardless of the device type.

Testing of sensor performance and storage of collected data according to defined sets S_{hi}, S_{ii}, and S_{ei} showed the importance of storing individual values in precisely defined sets. These sets are essential for defining user profiles, which are necessary for the development of ML models. For instance, in a 15 day period, 80,398 data points were collected. Using a conceptual mathematical model and defined variables, the value of generating information about the fire is shown:

$$S_{info\_fire} = \{S_{ii}, S_{ei}, S_{hi}\}. \tag{10}$$

Based on the information above, when reporting a fire, incident information (S_{ii}) is provided along with the ability to call emergency services. The user's environment (S_{ei}) also provides information about the user's location within different scenarios (K, LR, SR, TWC, TB, DA) and whether the user is in danger (S_{hi}). This method has been found to be highly successful in storing data based on distinct variables and sets, enabling efficient delivery to the user and possible interest groups (stakeholders).

Based on the defined mathematical model for providing predictive information, data collection and storage testing according to the proposed model was conducted. In addition, an in-depth analysis of the papers presented through the literature review was also conducted (Table 7). At the end of the table, the contribution of the conducted research and testing to this paper is presented.

Table 7. In-depth analysis of research methods.

Papers	Research Topic	Data Collection Method	User Data Identification	Importance of API Development	Database Structure Overview
[3]	Overview and categorization of existing devices (AT)	Analysis of previous research results	Yes (partially)	No	No
[6]	Analysis of device operation and smart home functionality	Analysis of previous research results	Yes (partially)	No	No
[8]	Analysis of communication technologies for the smart home	Analysis of previous research results	Yes (no details)	No	No
[9]	Analysis of the application of assistive technologies in a smart home	Analysis of previous research results	Yes (no details)	Yes	No
[11]	Development of a service for the rehabilitation of users	Collecting data from sensors	Yes	Yes	No
[12]	Overview of architectures for assisted living	Analysis of previous research results	Yes (no details)	Yes	No
[13]	Development of AAL environment and presentation of technologies	Collecting data from sensors	Yes	Yes	No
[16]	Development of a notification service in a smart home	Data collection via forms	Yes	Yes	No
[18]	Exploring the challenges and opportunities of a smart home for the visually impaired	Analysis of previous research results	Yes (no details)	Yes	No
[19]	Development of an e-care system for the elderly	Collecting data from sensors	Yes	Yes	Yes
[20]	Development of a framework for the application of fog computing in eHealth	IoT device data collection	No	No	No
[23]	Activity classification for elderly person using ML	Data collection from radar sensors	No	No	No

Table 7. *Cont.*

Papers	Research Topic	Data Collection Method	User Data Identification	Importance of API Development	Database Structure Overview
[24]	Analysis of user needs in nursing homes	Analysis of previous research results	No	No	No
[25]	Online infrastructure design for collecting data on people with disabilities	Conducting surveys and interviews	Yes	No	No
[28]	Analysis of the lifestyle of visually impaired people at home	Conducting interviews	Yes (no details)	No	No
[29]	Research of technologies in the function of home automation for visually impaired people	Online survey and interviews	Yes	No	No
Our Research	Defining user profiles for the purpose of predictive information in a smart home	Analysis of the results of previous research and collection from sensors	Yes	Yes	Yes

The analyzed papers show the following categories: data collection methods, user data identification, the need for API development, and the existence of a database structure. The analyzed categories are important from the aspect of creating user profiles, all with the aim of creating a sustainable ecosystem in which the service of predictive informing of smart home users is developed. It is evident from the analyzed papers that some papers do not have clearly defined data necessary for the development of the service, or at least data sets, and only some works emphasize the importance of using APIs, which have a clear role in providing relevant information. It is also evident that only one paper has a presentation of the database structure. Also, as shown in Table 7, the article [19] has similarity to this research in all categories. However, the contribution of this paper is manifested through the following:

- The literature review enabled the identification of insights and methodological approaches that are the basis for understanding the issues of data collection and processing. The above is crucial for defining the user profiles and data sets presented in the paper;
- The data sets use relevant data from the user's environment, and the data are collected from a larger sensor database, which is shown by a mathematical model;
- The mathematical model allows the display of all data regardless of the type of user (a person with disabilities, an older user group, etc.), which enables the modularity of the service's operation to provide predictive information to the end user.

The above contributions can provide flexibility and long-term scalability to the system and show how the proposed architecture allows for rapid adaptation to different scenarios and end-user needs. Data collection in this paper was performed with the aim of ensuring the accuracy and reliability of the system, and the focus is on validation in laboratory conditions, which is shown by testing the operation of the sensors and the database. This approach allows for the identification and resolution of potential technical issues before implementation in real-world conditions and creates a solid foundation for later experiments in real-world conditions. The use of APIs in this research is the basis for providing predictive information depending on the type of user. The database must contain key data about user needs (shown in Table 1) in which the provision of real-time information is delivered via the API interface.

6. Conclusions

This paper proposes a method of data collection as a basis for creating a user knowledge base for the purpose of predictive user information. From the presented data collection testing, it is possible to store data in precisely defined databases, which is the basis for designing new innovative services. With an innovative approach, it is possible to create individual or personalized information depending on the user profiles presented in this paper.

The development of assistive technologies and the application of AI to help people with disabilities today can be said to be not just a type of device, but a set of methods, services, and an entire ecosystem that must be sustainable and accessible. Services based on IoT technology and AI can raise the user's level of quality of life, which increases the degree of mobility and involvement in all spheres of social life and everyday life. Designing and modeling such systems should be viewed according to user needs and based on that, to create a sustainable and accessible ecosystem. The possibilities provided today by information and communication networks, and technologies such as AI, ML, and autonomous systems, can create a faster and more efficient system for the user. Such possibilities provide predictive user information and interaction with the environment in which they are located.

Based on the proposed taxonomy from this paper, it is possible to create a smart home for a person with disabilities depending on the type and degree of impairment, which aims to achieve a more pleasant and safer everyday life. Based on the displayed interaction between all elements, it is possible to provide the user with all information from the home environment, information about the user's health (pulse, temperature, pressure, sugar, etc.), and the possibility of calling the emergency services or the caregiver in case of incident. Management of sensors, actuators, and devices in the home is possible through speech using a virtual assistant, which has the potential to improve the mood of the user in his home, if it is a person with impaired vision or a person with physical disabilities. People with voice–speech difficulties and people with a hearing impairment must receive adapted video–visual information to also provide them with a higher quality of life and independence.

The solution described in this paper is an assistive technology model that includes a device, user, service, and environment, providing a personalized smart home experience. With relevant devices and sensors gathering specific data sets, defined by a mathematical model, this solution serves as a vital sub-log for potential application of AI technologies. Laboratory testing has confirmed the significance of defining data sets in generating user-friendly information to the end user. In future development, an API interface should be defined for the sensor to segment and merge data based on user requirements. Future research will be focused on the development of ML models, verification of model effectiveness (simulation and validation), and the creation of a user knowledge base, all to model new IoT services for predictive user information. Future research should also contribute to creating a higher level of quality of life, and thus an improved mood, which can be proven by appropriate methods of measuring user satisfaction. By using different multi-criteria-decision-making approaches, it is possible to select the appropriate IoT components in a smart home and the personalized services themselves.

Author Contributions: Conceptualization, M.P. and P.T.; methodology, M.P.; formal analysis, I.G.; Investigation, P.T.; writing—original draft preparation, M.P. and P.T.; writing—review and editing, I.C. and I.G.; visualization, M.P.; supervision, I.C. and I.G. All authors have read and agreed to the published version of the manuscript.

Funding: This research is funded by the University of Zagreb through the grants for core financing of scientific and artistic activities of the University of Zagreb in academic year 2023/2024, under the project (210275) "Modeling Framework for the Development and Application of Assistive Technologies".

Institutional Review Board Statement: Not applicable.

Informed Consent Statement: Not applicable.

Data Availability Statement: Data are contained within the paper.

Conflicts of Interest: The authors declare no conflicts of interest.

References

1. World Health Organization. Blindness and Vision Impairment. Available online: https://www.who.int/news-room/fact-sheets/detail/blindness-and-visual-impairment (accessed on 4 December 2024).
2. Statista Market Insights. *Number of Users of Smart Homes Worldwide from 2019 to 2028*; Statista Research Department: New York, NY, USA, 2024.
3. Mashiata, M.; Ali, T.; Das, P.; Tasneem, Z.; Badal, M.F.R.; Sarker, S.K.; Hasan, M.M.; Abhi, S.H.; Islam, M.R.; Ali, M.F.; et al. Towards Assisting Visually Impaired Individuals: A Review on Current Status and Future Prospects. *Biosens. Bioelectron. X* **2022**, *12*, 100265. [CrossRef]
4. Periša, M.; Peraković, D.; Zorić, P. Challenges of Assistive Technologies Implementation into Industry 4.0: A Review. In Proceedings of the The Seventh International Conference Transport and Logistics, Niš, Serbia, 6 December 2019; Janošević, D., Ed.; University of Niš, Faculty of Mechanical Engineering: Niš, Serbia, 2019; pp. 5–10.
5. Periša, M.; Cvitić, I.; Zorić, P.; Grgurević, I. Concept, Architecture, and Performance Testing of a Smart Home Environment for the Visually Impaired Persons. In Proceedings of the EAI MMS 2022—7th EAI International Conference on Management of Manufacturing Systems, Krynica-Zdrój, Poland, 5–7 October 2022; Knapčíková, L., Peraković, D., Eds.; Springer: Cham, Switzerland, 2023; pp. 3–14.
6. Maswadi, K.; Ghani, N.B.A.; Hamid, S.B. Systematic Literature Review of Smart Home Monitoring Technologies Based on IoT for the Elderly. *IEEE Access* **2020**, *8*, 92244–92261. [CrossRef]
7. Dobre, C.; Mavromoustakis, C.X.; Garcia, N.M.; Mastorakis, G.; Goleva, R.I. Introduction to the AAL and ELE Systems. In *Ambient Assisted Living and Enhanced Living Environments: Principles, Technologies and Control*, 1st ed.; Dobre, C., Mavromoustakis, C.X., Garcia, N., Goleva, R.I., Mastorakis, G., Eds.; Butterworth-Heinemann: Newton, MA, USA, 2017; ISBN 978-0-12-805195-5.
8. Jamwal, R.; Jarman, H.K.; Roseingrave, E.; Douglas, J.; Winkler, D. Smart Home and Communication Technology for People with Disability: A Scoping Review. *Disabil. Rehabil. Assist. Technol.* **2022**, *17*, 624–644. [CrossRef] [PubMed]
9. Colnar, S.; Dimovski, V.; Grah, B.; Rogelj, V.; Bogataj, D. Smart Home Supporting Integrated Health and Care Services for Older Adults in the Community: Literature Review and Research Agenda. In Proceedings of the 2020 24th International Conference on System Theory, Control and Computing (ICSTCC), Sinaia, Romania, 8–10 October 2020; pp. 526–531.
10. Cvitić, I.; Peraković, D.; Periša, M.; Gupta, B. Ensemble Machine Learning Approach for Classification of IoT Devices in Smart Home. *Int. J. Mach. Learn. Cybern.* **2021**, *12*, 3179–3202. [CrossRef]
11. Bisio, I.; Garibotto, C.; Lavagetto, F.; Sciarrone, A. When EHealth Meets IoT: A Smart Wireless System for Post-Stroke Home Rehabilitation. *IEEE Wirel. Commun.* **2019**, *26*, 24–29. [CrossRef]
12. El Murabet, A.; Abtoy, A.; Touhafi, A.; Tahiri, A. Ambient Assisted Living System's Models and Architectures: A Survey of the State of the Art. *J. King Saud. Univ. Comput. Inf. Sci.* **2020**, *32*, 1–10. [CrossRef]
13. Cicirelli, G.; Marani, R.; Petitti, A.; Milella, A.; D'Orazio, T. Ambient Assisted Living: A Review of Technologies, Methodologies and Future Perspectives for Healthy Aging of Population. *Sensors* **2021**, *21*, 3549. [CrossRef] [PubMed]
14. Kunnappilly, A. Modeling and Formal Analysis of E-Health Systems. Ph.D. Thesis, Mälardalen University, Stockholm, Sweden, 2021.
15. United Nations. *World Social Report: Leaving No One behind in an Ageing World*; United Nations Publication: New York, NY, USA, 2023; ISBN 9789211304589.
16. Abreu, J.; Oliveira, R.; Garcia-Crespo, A.; Rodriguez-Goncalves, R. TV Interaction as a Non-Invasive Sensor for Monitoring Elderly Well-Being at Home. *Sensors* **2021**, *21*, 6897. [CrossRef] [PubMed]
17. Periša, M.; Marković, G.; Kolarovszki, P.; Madleňák, R. Proposal of a Conceptual Architecture System for Informing the User in the IoT Environment. *Promet. Traffic Transp.* **2019**, *31*, 37–47. [CrossRef]
18. De Oliveira, G.A.A.; Oliveira, O.D.F.; de Abreu, S.; de Bettio, R.W.; Freire, A.P. Opportunities and Accessibility Challenges for Open-Source General-Purpose Home Automation Mobile Applications for Visually Disabled Users. *Multimed. Tools Appl.* **2022**, *81*, 10695–10722. [CrossRef]
19. Köckemann, U.; Alirezaie, M.; Renoux, J.; Tsiftes, N.; Ahmed, M.U.; Morberg, D.; Lindén, M.; Loutfi, A. Open-Source Data Collection and Data Sets for Activity Recognition in Smart Homes. *Sensors* **2020**, *20*, 879. [CrossRef] [PubMed]
20. Sodhro, A.H.; Zahid, N. AI-Enabled Framework for Fog Computing Driven E-Healthcare Applications. *Sensors* **2021**, *21*, 8039. [CrossRef] [PubMed]

21. Donnici, R.; Coronato, A.; Naeem, M. A Self-Learning Autonomous and Intelligent System for the Reduction of Medication Errors in Home Treatments. In Proceedings of the Intelligent Environments 2021: Workshop Proceedings of the 17th International Conference on Intelligent Environments, Dubai, United Arab Emirates, 21–24 June 2021; Volume 29, pp. 156–166.
22. Aski, V.J.; Dhaka, V.S.; Kumar, S.; Verma, S.; Rawat, D.B. Advances on Networked EHealth Information Access and Sharing: Status, Challenges and Prospects. *Comput. Netw.* **2022**, *204*, 108687. [CrossRef]
23. Taylor, W.; Dashtipour, K.; Shah, S.A.; Hussain, A.; Abbasi, Q.H.; Imran, M.A. Radar Sensing for Activity Classification in Elderly People Exploiting Micro-Doppler Signatures Using Machine Learning. *Sensors* **2021**, *21*, 3881. [CrossRef] [PubMed]
24. Zimmermann, G.; Ableitner, T.; Strobbe, C. User Needs and Wishes in Smart Homes: What Can Artificial Intelligence Contribute? In Proceedings of the 2017 14th International Symposium on Pervasive Systems, Algorithms and Networks & 2017 11th International Conference on Frontier of Computer Science and Technology & 2017 Third International Symposium of Creative Computing (ISPAN-FCST-ISCC), Exeter, UK, 21–23 June 2017; pp. 449–453.
25. Park, J.S.; Bragg, D.; Kamar, E.; Morris, M.R. Designing an Online Infrastructure for Collecting AI Data from People With Disabilities. In Proceedings of the 2021 ACM Conference on Fairness, Accountability, and Transparency, Virtual Event, 3–10 March 2021; ACM: New York, NY, USA, 2021; pp. 52–63.
26. Samim, A. A New Paradigm of Artificial Intelligence to Disabilities. *Int. J. Sci. Res.* **2023**, *12*, 478–482. [CrossRef]
27. Benjak, T. *Izvješće o Osobama s Invaliditetom u Republici Hrvatskoj*; Hrvatski Zavod za Javno Zdravstvo: Zagreb, Croatia, 2021; Volume 1.
28. Rooney, C.; Hadjri, K.; Mcallister, K.; Rooney, M.; Faith, V.; Craig, C. Experiencing Visual Impairment in a Lifetime Home: An Interpretative Phenomenological Inquiry. *J. Hous. Built Environ.* **2018**, *33*, 45–67. [CrossRef] [PubMed]
29. Leporini, B.; Buzzi, M. Home Automation for an Independent Living. In Proceedings of the Proceedings of the 15th International Web for All Conference, New York, NY, USA, 23–25 April 2018; ACM: New York, NY, USA, 2018; pp. 1–9.
30. Peraković, D.; Periša, M.; Sente, R.E.; Bijelica, N.; Zorić, P.; Brletić, L.; Bucak, B.; Ignjatić, A.; Mišić, V.; Papac, A.; et al. Information and Communication System for Informing Users in Traffic Environment—SAforA. In Proceedings of the 2thEAI International Summit, Smart City 360°, Bratislava, Slovakia, 22–24 November 2016; Balog, M., Knapcikova, L., Soviar, J., Dorcak, P., Pollak, F., Caganova, D., Fazio, P., Aydin, K., Eds.; EAI: Bratislava, Slovakia, 2016.
31. Goleva, R.I.; Garcia, N.M.; Mavromoustakis, C.X.; Dobre, C.; Mastorakis, G.; Stainov, R.; Chorbev, I.; Trajkovik, V. AAL and ELE Platform Architecture. In *Ambient Assisted Living and Enhanced Living Environments*; Elsevier: Amsterdam, The Netherlands, 2017; pp. 171–209.
32. Almeida, P.; Abreu, J.; Silva, T.; Duro, L.; Aresta, M.; Oliveira, R. Notification Mechanisms In Second-Screen Scenarios—Towards a Balanced User Experience. In Proceedings of the 7th International Conference on Intelligent Technologies for Interactive Entertainment, Torino, Italy, 10–12 June 2015; pp. 95–99.

sensors

MDPI

Article

Autonomous Mission Planning for Fixed-Wing Unmanned Aerial Vehicles in Multiscenario Reconnaissance

Bei Chen [1,2,3], Jiaxin Yan [1], Zebo Zhou [1,3,*], Rui Lai [2] and Jiejian Lin [1]

[1] School of Aeronautics and Astronautics, University of Electronic Science and Technology of China, Chengdu 611731, China; 202112100618@std.uestc.edu.cn (B.C.); 202411100611@std.uestc.edu.cn (J.Y.); o1n1n7@gmail.com (J.L.)
[2] AVIC (Chengdu) Unmanned Aerial Vehicle Systems Co., Ltd., Chengdu 611743, China; lairuihit@163.com
[3] System of Systems and Artificial Intelligence Laboratory, Chendu 610041, China
* Correspondence: klinsmann.zhou@gmail.com

Abstract: Before a fixed-wing UAV executes target tracking missions, it is essential to identify targets through reconnaissance mission areas using onboard payloads. This paper presents an autonomous mission planning method designed for such reconnaissance operations, enabling effective target identification prior to tracking. Existing planning methods primarily focus on flight performance, energy consumption, and obstacle avoidance, with less attention to integrating payload. Our proposed method emphasizes the combination of two key functions: flight path planning and payload mission planning. In terms of path planning, we introduce a method based on the Hierarchical Traveling Salesman Problem (HTSP), which utilizes the nearest neighbor algorithm to find the optimal visit sequence and entry points for area targets. When dealing with area targets containing no-fly zones, HTSP quickly calculates a set of waypoints required for coverage path planning (CPP) based on the Generalized Traveling Salesman Problem (GTSP), ensuring thorough and effective reconnaissance coverage. In terms of payload mission planning, our proposed method fully considers payload characteristics such as scan resolution, imaging width, and operating modes to generate predefined mission instruction sets. By meticulously analyzing payload constraints, we further optimized the path planning results, ensuring that each instruction meets the payload performance requirements. Finally, simulations validated the effectiveness and superiority of the proposed autonomous mission planning method in reconnaissance tasks.

Keywords: fixed-wing UAV; mission planning; coverage path planning; payload characteristics

1. Introduction

In recent years, fixed-wing UAVs have played an increasingly important role in various reconnaissance mission scenarios. Their unique flight performance and long endurance capabilities have demonstrated significant application potential in both military reconnaissance and civilian monitoring fields [1–3]. One crucial application is target tracking [4–6]. Before a fixed-wing UAV executes target tracking missions, it is crucial to first identify targets within reconnaissance mission areas using onboard payloads. This paper presents an autonomous mission planning method designed for these reconnaissance operations, enabling effective target identification prior to initiating tracking tasks. We believe that UAVs completing reconnaissance missions must address two main aspects: path planning and payload planning. However, traditional planning methods are often limited to single-objective optimization, such as focusing on flight performance, obstacle avoidance, and energy consumption [7–9], while neglecting the deep integration of payload characteristics.

Currently, UAV path planning works are primarily focused on applications with single scenarios, lacking consideration of different scenarios. Cui [10] formulated the problem of multiple single-point reconnaissance targets as a Dynamic Constrained Traveling Salesman Problem with Neighborhoods (DCTSPN) and proposed a hierarchical algorithm based on deep reinforcement learning to solve it. Although the effectiveness of this algorithm was validated, it is well known that learning-based methods are not interpretable. Zhang et al. [11] proposed a bi-level hybridization-based metaheuristic algorithm for fixed-wing UAVs used in forest fire monitoring. While this algorithm can reduce data collection cycles and energy consumption in most cases, it is limited to single-point targets, restricting its applicability to other scenarios. At the same time, the coverage path planning (CPP) problem has garnered increasing attention [12–15], especially in missions requiring UAVs to scan entire areas. The CPP problem involves planning a path that covers all Regions of Interest (ROI) while avoiding obstacles or no-fly zones [16–18]. Different CPP methods have their own advantages and limitations. Selecting an appropriate CPP method requires the consideration of specific application requirements and environments [19]. Barrientos et al. [20] proposed a Breadth First Search (BFS)-based coverage planner, which can be applied to grid-based workspaces to generate paths with minimal turning maneuvers. However, search-based methods have high computational complexity, particularly in large-scale environments where they require substantial computational resources, making them unsuitable for long-endurance fixed-wing UAVs. In [21], a random walk-based CPP algorithm for robots was introduced. This method is simple and has low memory requirements, making it easy to deploy. However, random walk paths are only suitable for smaller environments and struggle to cover ROI with obstacles. Additionally, the robot may traverse the same path multiple times, leading to overall inefficiency in the coverage path. Bähnemann et al. [22] proposed a decomposition-based CPP algorithm for quadrotor, which extends the Boustrophedon Cell Decompositon (BCD) [23–25] by optimizing different sweep combinations to find the optimal sweep path and accounting for obstacles between the decomposed cells. This method is computationally efficient, easy to implement, and effectively handles complex areas.

Although the above works have achieved significant results in reconnaissance path planning for single-point or ROI targets, few works have been able to combine these two typical scenarios. Additionally, these studies have not adequately considered the role of payloads in the planning process. We need to clearly recognize that path planning without effective payload used for reconnaissance integration cannot ensure the successful execution of flight missions [26–29]. We introduce a reconnaissance fixed-wing UAV, typically operating at altitudes above 5000m, equipped with Synthetic Aperture Radar (SAR) systems for aerial reconnaissance [30,31]. SAR systems have two operating modes: beam forming [32] and strip modes [33]. One of the most challenging aspects of mission planning is adjusting the UAV's path based on the payload. This complexity arises from the need to combine the payload's system response time, imaging width, and others with the existing flight plan. Such integration is crucial for meeting mission requirements.

Therefore, this paper takes this as a motivation to explore how to achieve autonomous mission planning for fixed-wing UAVs in multiscenario reconnaissance. The innovations of this paper include the following: Firstly, in Section 2, we designed a path planning algorithm featuring the HTSP. This algorithm initially solves the TSP problem using the nearest neighbor algorithm to determine the reconnaissance sequence for different mission targets. Additionally, by incorporating the BCD technique, it addresses the coverage problem for area targets. We transformed the GTSP problem into a graph structure, ensuring the accuracy of the planning results. Fixed-wing UAVs need to switch flexibly between different mission scenarios. In Section 3, the proposed payload planning module considers the characteristics, workflow, and instruction parameters of the payload. It adjusts the pre-planned waypoint set according to the payload's operating mode. Furthermore, to ensure the safety of UAV transitions between distinct mission scenarios, a visibility graph obstacle avoidance algorithm was integrated to refine the waypoint sets. Finally, in Section 4, we developed a user-friendly UI interface that supports visualization and adjustment, and validated the effectiveness of the autonomous mission planning module in multiscenario situtations.

2. HTSP Path Planning

In multiscenario situations, UAVs must visit multiple point targets, which are specific locations, and area targets, which are ROI with defined boundaries. This problem can be modeled as an HTSP. The main challenge of HTSP is to determine the sequence of visiting each target to ensure the shortest flight path, as well as to plan the coverage flight path within each area target.

To address the HTSP problem, we adopt a two-stage strategy:

1. Sequence Planning: This stage determines the order of visiting different scenarios. By optimizing the sequence, the UAV can minimize the overall flight range.
2. Coverage Path Planning: The lower level addresses the coverage of area targets using BCD. This ensures that the UAV efficiently covers each area while adhering to constraints such as no-fly zones.

2.1. Sequence Planning

The inputs for the autonomous mission planning module include the entry point **S**, the exit point **E**, the set of no-fly zones or obstacle **Obs**, the set of area target **F**, and the set of point targets **M** .

$$\begin{cases} \mathbf{S} = \{x_S, y_S, h_S\} \\ \mathbf{E} = \{x_E, y_E, h_E\} \\ \mathbf{Obs} = \{\mathbf{obs}_1, \mathbf{obs}_2, \ldots, \mathbf{obs}_i\} \\ \mathbf{obs}_i = \{\mathbf{op}_{i1}, \mathbf{op}_{i2}, \ldots, \mathbf{op}_{in}\} \\ \mathbf{op}_{in} = \{x_{in}, y_{in}, h_{in}\} \\ \mathbf{F} = \{\mathbf{f}_1, \mathbf{f}_2, \ldots, \mathbf{f}_j\} \\ \mathbf{f}_j = \{\{\mathbf{fv}_{j1}, \mathbf{fv}_{j2}, \ldots, \mathbf{fv}_{jn}\}, mode, iw\} \\ \mathbf{fv}_{jn} = \{x_{jn}, y_{jn}, h_{jn}\} \\ \mathbf{M} = \{\mathbf{m}_1, \mathbf{m}_2, \ldots, \mathbf{m}_k\} \\ \mathbf{m}_k = \{x_k, y_k, h_k, mode, iw\} \end{cases} \tag{1}$$

In (1), x, y, and h represent the two-dimensional coordinates and altitude, respectively. $\mathbf{obs}_i$ denotes the no-fly zones constructed as polygons, with their vertices represented by $\mathbf{op}_{in}$. $\mathbf{f}_j$ represents the area targets, also constructed as polygons, with their vertices represented by $\mathbf{fv}_{jn}$. $\mathbf{m}_k$ refers to the point targets. *mode* and *iw* are integer variables indicating the payload operating mode and imaging width, respectively.

Sequence planning uses the nearest neighbor algorithm not only to determine the traversal order but also to further output the entry points for the coverage paths of polygonal area targets. Identifying the entry points for area targets helps minimize the overall path length. When covering polygons, choosing entry points judiciously can prevent redundant paths, thereby reducing the total flight range. The point set V is constructed as follows:

$$V = \left(\bigcup_{n=1}^{m} \mathbf{fp}_{jn} \right) \cup \left(\bigcup_{k=1}^{l} \mathbf{M}_k \right) \cup \{\mathbf{S}, \mathbf{E}\} \tag{2}$$

where m represents the number of area targets and l represents the number of point targets. To compute the Euclidean distance between any two points in the set V, we obtain the distance matrix D:

$$D_{ij} = \sqrt{(x_i - x_j)^2 + (y_i - y_j)^2} \tag{3}$$

where (x_i, y_i) and (x_j, y_j) are the coordinates of points V_i and V_j, respectively. Using the nearest neighbor algorithm, we solve for the path from the entry point S to the exit point E that passes through all points, denoted as T_{seq}. We initialize the current node as $current = v_0$ and the starting point as $v_0 = S$. We initialize a boolean array $visited$ to mark the visit status of all nodes, where all nodes are initially unvisited: $visited = \{\text{false}, \text{false}, \ldots, \text{false}\}$. We mark the starting node v_1 as visited: $visited[0] = \text{true}$). We initialize the sequence T_{seq} and add the starting node to the sequence: $T_{\text{seq}} = [v_1]$.

We loop from the starting node to the second-to-last node, executing the following iterative steps: For each node i from 1 to $n-2$, we select the node $next$ closest to node $current$ among all unvisited nodes, and thus we have:

$$next = \arg \min_{j \notin \text{visited}} D_{current,j} \tag{4}$$

We add $next$ to the sequence:

$$T_{seq} = T_{seq} \cup \{next\} \tag{5}$$

We update the current node.

$$current = next \tag{6}$$

We mark node $next$ as visited.

$$visited[\text{next}] = \text{true} \tag{7}$$

Finally, we add the exit point $v_{n-1} = E$ to the sequence.

$$T_{seq} = T_{seq} \cup \{v_{n-1}\} \tag{8}$$

From Equations (4)–(8), we use the nearest neighbor algorithm to obtain the shortest sequence of point set T_{seq} from **S** to **E**.

Finally, we determine the entry point of each polygon based on T_{seq}. We initialize the entry point array for each polygon $entry\_points$:

$$entry\_points = \{(0,0)\}_{i=1}^{m} \tag{9}$$

We initialize the polygon visit array $poly\_visited$:

$$poly\_visited = \{\text{false}\}_{i=1}^{m} \tag{10}$$

We iterate over each element V_i in the sequence T_{seq}. If V_i belongs to the i-th polygon and this polygon has not yet been visited, then we set V_i to:

$$entry\_points[i] = V_i \tag{11}$$

We mark the i-th polygon as visited.

$$polygon\_visited[i] = true \tag{12}$$

Through the formalization shown above, we can comprehensively describe the nearest neighbor algorithm for solving the approximate TSP problem, ensuring that the shortest sequence of target visits and entry points is determined from the start **S** to the end **E**.

2.2. Coverage Path Planning with Area Target

For UAVs equipped with SAR systems, continuous scanning is performed along the coverage path during a single pass over the area target. The data collected are processed by algorithms to generate high-resolution images. However, suboptimal flight paths can significantly degrade image quality. To ensure optimal scanning, the planning of a UAV's coverage path becomes even more critical.

To address the coverage path planning problem, the module employs the BCD algorithm to divide area targets with no-fly zones into multiple cells. This decomposition ensures that each cell is a simple polygon free of no-fly zones.

BCD is a method used to divide an AOI into smaller, simpler cells that can be covered systematically. The main goal is to ensure the complete coverage of the area while avoiding obstacles, ensuring efficient traversal and minimizing the total path length.

Let us consider an AOI $\mathbf{f}_j$ represented as a polygon with vertices$\{\mathbf{fv}_{j1}, \mathbf{fv}_{j2}, \ldots, \mathbf{fv}_{jn}\}$. The AOI may contain no-fly zones $\mathbf{Obs}$, each represented as a polygonal region. The task is to cover the area $\mathbf{f}_j$ while avoiding these no-fly zones. The algorithm identifies critical points on the boundaries of $\mathbf{f}_j$ and $\mathbf{Obs}$, which are points where the sweep line direction changes. This could be the vertices of $\mathbf{f}_j$ and $\mathbf{Obs}$. A sweep line moves from left to right (or top to bottom), dividing the area into cells whenever it encounters a critical point. The cells are simple polygons, free from any no-fly zones.

Each cell C_jk is formed as a result of the sweep line passing through critical points. Mathematically, a cell C_jk is represented as:

$$C_jk = \mathbf{f}_j \setminus \bigcup_{i=1}^{m} \mathbf{obs}_i \tag{13}$$

where C_jk is a simple polygon without holes, and m is the number of obstacles in the AOI.

Within each cell, a systematic back-and-forth (Boustrophedon) motion is used to cover the entire area. This motion can be described mathematically by generating parallel lines inside the cell, ensuring that they cover the cell fully without overlapping.

We choose a direction for the parallel lines. This is often chosen as perpendicular to the longest edge or the edge that aligns best with the general shape of the cell to minimize the number of turns. Let this direction be denoted by a unit vector $\mathbf{d}$.

The coverage path within C_jk is generated as a set of parallel lines l_ki that are spaced apart by the offset Δf. These lines are perpendicular to the chosen direction $\mathbf{d}$. Mathematically, l_ki can be represented as:

$$l_ki : \mathbf{r} = \mathbf{r}_0 + i\Delta f \mathbf{d}^{\perp}, \quad i = 0, 1, 2, \ldots, n \tag{14}$$

where $\mathbf{r}$ is the position vector of a point on the line. $\mathbf{r}_0$ is the position vector of the starting point of the first line L_0, typically taken as the intersection of the first line with the boundary of C_jk. These lines are separated by a distance called offset, denoted as Δf. $\mathbf{d}^{\perp}$ is a unit vector perpendicular to $\mathbf{d}$. n is the number of parallel lines required to cover the entire cell.

For each line segment l_ki, we identify the start and end points:

$$\mathbf{P}_{ki} = \left\{\mathbf{p}_{ki}^{start}, \mathbf{p}_{ki}^{end}\right\}, \quad i = 0, 1, 2, \ldots, n \tag{15}$$

Consequently, we obtain a set of all waypoints for the UAV to cover C_jk, which can be represented as:

$$\mathbf{P}_jk = \{\mathbf{P}_{k1}, \mathbf{P}_{k2}, \ldots\} \quad (16)$$

To minimize the total path length of the UAV, the coverage paths between adjacent cells are optimized. The objective is to find the optimal sequence for visiting cells, ensuring that each cell is visited exactly once. This problem is typically addressed using graph-based algorithms, and it is modeled as a GTSP.

The GTSP formulation [34,35] for UAV coverage path planning is defined as follows:

$$\text{Min} \sum_{i=1}^{m} \sum_{j=1}^{m} d_{ij} x_{ij} \quad (17)$$

where d_{ij} represents the travel distance between cell i and cell j. x_{ij} is a binary decision variable indicating whether the UAV travels directly from cell i to cell j. N is the set of all cells.

The optimization is subject to the following constraints:

Constraints:

1. Binary decision variables:

$$x_{ij} \in \{0, 1\} \quad \forall i, j \quad (18)$$

2. We set the start O to the cell where the entry point E obtained in Section 2.1 is located.

$$\sum_{i \in V} x_{Oi} = 1 \quad (19)$$

V is the set of cells except O.

3. We ensure that each cell is entered and left at most once:

$$\sum_{i \in N} x_{ij} \leq 1 \quad \forall j \in V \quad (20)$$

$$\sum_{j \in N} x_{ij} \leq 1 \quad \forall i \in V \quad (21)$$

4. Subtour elimination constraints (Miller–Tucker–Zemlin formulation):

$$u_i - u_j + m \cdot x_{ij} \leq m - 1 \quad \forall i, j \in V, i \neq j \quad (22)$$

$$1 \leq u_i \leq m - 1 \quad \forall i \in V \quad (23)$$

where u_i is an auxiliary variable which keeps the number of visited cells until cell i in the solution. These constraints ensure that the solution forms a single continuous path.

We solve the GTSP using GLKH (http://webhotel4.ruc.dk/~keld/research/GLKH/, accessed on 5 November 2024) as an open source solver [36]. After solving the GTSP, the coverage waypoints P_j of visiting the area target $\mathbf{f}_j$ are obtained. From the sequence planning in the previous section, we obtain the access order of the target T_{seq}. Let $T_{seq} = \{T_1, T_2, \ldots, T_n\}$ represent the sequence in which the targets are accessed, where each T_i is either a point target or an area target. P_T represents the waypoints associated with target T. For point target $\mathbf{m}_k$, the waypoint set P_k may consist of a single waypoint. For area target $\mathbf{f}_j$, the waypoint set P_j consists of multiple waypoints that cover the entire surface area.

In summary, the pre-planned waypoints P of the HTSP are obtained by connecting the waypoints of each target T_i in the sequence T_{seq}:

$$P = \bigcup_{i=1}^{n} P_{T_i} \tag{24}$$

3. Payload Mission Planning

In reconnaissance missions, the performance and efficiency of the UAV's payload are paramount. The payload mission planning process must therefore account for the specific operational characteristics and constraints. This section integrates payload characteristics to adjust the pre-planned waypoints P. Additionally, obstacle-avoidance waypoints during UAV cruising are generated based on a visibility graph method. The computational workflow for this section is designed to optimize the integration of payload characteristics and waypoint adjustments, as outlined in Figure 1.

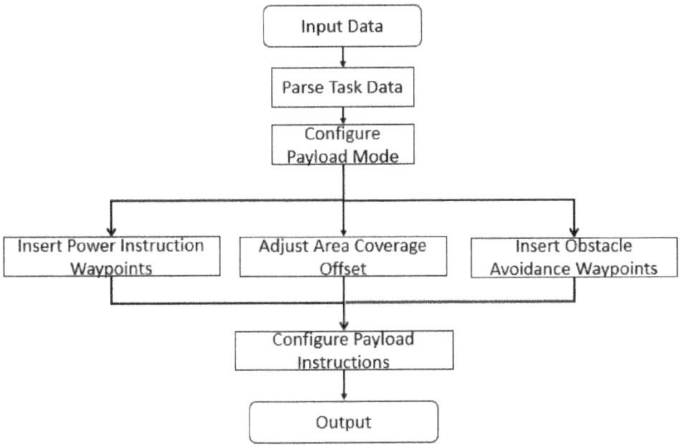

Figure 1. Computational workflow for UAV payload mission planning.

Equation (1) is the input for the autonomous mission planning module. It will provide the required payload *mode* (e.g., SAR Beam Focusing or SAR Strip) and *iw* for each target. We will complete mission planning based on this information.

3.1. Power Instruction

The first step in payload mission planning is managing the power state of the payload, which includes turning it on, standby, etc., depending on the mission's requirements. The goal here is to ensure that the payload is fully operational when the UAV reaches the target area, thereby avoiding any delays or inefficiencies in data acquisition.

Assuming that the payload's power-on time is t_{on} with the input *mode* in (1) and the UAV's speed is v, we calculate the distance d_{on}^i from the target T_i where the power-on command should be issued:

$$\|\mathbf{p}_{on} - \mathbf{p}_i\| \geq d_{on} = t_{on} \times v \tag{25}$$

In this formulation, $\mathbf{p}_{on}$ is the waypoint that needs to be inserted, and $\mathbf{p}_i$ is the position of T_i, which may be the point target position or the entry position of an area target. The updated set of waypoints P after inserting $\mathbf{p}_{on}$ can be expressed as:

$$P' = \{P_{T_1}, P_{T_2}, \ldots, \mathbf{p}_{on}, P_{T_i}, \ldots, P_{T_n}\} \tag{26}$$

Similarly, other kinds of commands for the payload can be inserted into the pre-planned waypoint set in the same manner, which will not be elaborated further here. By ensuring that the power command is issued at the correct waypoint, the UAV's payload will be in the appropriate operational state upon reaching the target.

3.2. Coverage Path Optimization with Payload

Different operating modes of the UAV's payload have different imaging widths based on factors like resolution and scan mode. By adjusting the offset of the coverage path to match the specific imaging width of the payload in its current mode, we ensure that each pass of the UAV maximizes its coverage area. This minimizes the number of required passes, thereby reducing mission time and energy consumption while ensuring complete coverage without gaps or excessive overlaps.

To adjust the offset of the full coverage path of $\mathbf{f}_j$, we need to replace the offset Δf in Equation (14) with the effective iw of the payload.

We adjust Δf to iw, so the updated formula for the coverage path lines l_{ki} becomes:

$$l_{ki} : \mathbf{r} = \mathbf{r}_0 + iw_{iw}\mathbf{d}^{\perp}, \quad i = 0, 1, 2, \ldots, n \tag{27}$$

Based on Equations (14)–(24), we can obtain the adjusted coverage waypoint set $P_{T_j}{}'$ for $\mathbf{f}_j$. The modified coverage waypoints are then inserted into the pre-planned waypoint set P, as shown in the following equation:

$$P' = \{P_{T_1}, P_{T_2}, \ldots, P_{T_j}{}', P_{T_j+1}, \ldots, P_{T_n}\} \tag{28}$$

The payload mission planning process described in this section plays a crucial role in the overall mission planning for reconnaissance fixed-wing UAVs. By carefully managing the power and optimizing the coverage path based on the payload's characteristics, a UAV can achieve optimal reconnaissance performance, ensuring that all mission objectives are met efficiently and effectively.

3.3. Obstacle Avoidance Based on Visibility Graph

Regardless of whether the target is a point or an area, obstacles may still exist in the region between two targets. Therefore, an obstacle avoidance method based on visibility graphs is adopted. The key steps of the visibility graph algorithm are as shown in Figure 2:

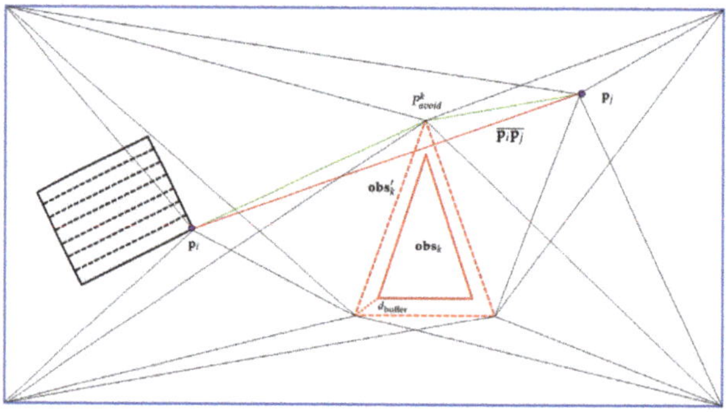

Figure 2. Visualization of the visibility graph algorithm for UAV obstacle avoidance. The algorithm identifies potential intersections between the planned path segment $\overline{\mathbf{p}_i\mathbf{p}_j}$, represented by the red line segment, and obstacle polygons. Obstacles are expanded by a buffer distance d_{buffer} to ensure safety, and alternate paths, such as through $\mathbf{p}^k_{\text{avoid}}$, are generated to bypass obstacles, as illustrated by the green line.

Firstly, each obstacle is represented as a convex polygon. For non-convex obstacles, convex decomposition is performed to divide them into multiple convex components. Subsequently, the vertices of the obstacle polygons are expanded by a safety buffer distance d_{buffer} to account for the UAV's size and maneuvering limitations:

$$\mathbf{obs}'_k = \{\mathbf{op}_{kn} + d_{\text{buffer}}\mathbf{n}_{kn} \mid \mathbf{op}_{kn} \in \mathbf{obs}_k\}$$

Here, $\mathbf{n}_{kn}$ represents the outward normal vector at vertex $\mathbf{op}_{kn}$.

Next, for the last waypoint $\mathbf{p}_i$ of the preceding target and the first waypoint $\mathbf{p}_j$ of the next target in the planned path, we examine whether the line segment $\overline{\mathbf{p}_i\mathbf{p}_j}$ intersects with any obstacles. The intersection condition can be expressed as:

$$\overline{\mathbf{p}_i\mathbf{p}_j} \cap \mathbf{obs}'_k \neq \emptyset \quad \forall k$$

The visibility graph $G = (V, E)$ is constructed where V contains all obstacle vertices, $\mathbf{p}_i$ and $\mathbf{p}_j$. E contains edges between vertices that are mutually visible.

Two vertices $\mathbf{v}_i$ and $\mathbf{v}_j$ are mutually visible if:

$$\overline{\mathbf{v}_i\mathbf{v}_j} \cap \mathbf{obs}'_k = \emptyset \quad \forall k \tag{29}$$

Then, using Dijkstra's algorithm on the visibility graph, we find the shortest collision-free path between $\mathbf{p}_i$ and $\mathbf{p}_j$:

$$P^k_{avoid} = \arg\min_{p \in G} \sum_{i=1}^{n-1} \|\mathbf{v}_{i+1} - \mathbf{v}_i\| \tag{30}$$

Finally, the obstacle avoidance waypoints are integrated into the final path:

$$P_{final} = \{P_{T_1}, P_{T_2}, \ldots, P_{T_k}, P^k_{avoid}, P_{T_k+1}, \ldots, P_{T_n}\} \tag{31}$$

This approach ensures safe navigation while maintaining optimal path efficiency. The visibility graph method is particularly effective in environments with multiple convex obstacles, providing guaranteed collision-free paths with minimal computational overhead.

4. Experiments and Discussions

The experimental platform for algorithm validation consisted of a high-performance computing workstation, assembled by Lenovo in Beijing, China. The workstation was equipped with an AMD Ryzen 9 7945HX CPU (16 cores, 2.5 GHz base frequency) and 16 GB DDR4 RAM. The computational environment was implemented in Visual Studio 2022 running on Windows 11. All algorithms were implemented using the Computational Geometry Algorithms Library [37] (CGAL, https://doc.cgal.org/latest/Manual/index.html, accessed on 5 November 2024) in C++, which provides robust implementations of geometric algorithms including polygon operations, convex hulls, and visibility graphs that are essential for our path planning and obstacle avoidance computations. The user interface was developed using the QT framework, providing an interactive visualization environment for mission planning and results analysis. All data were executed on this platform to ensure consistent and reliable performance metrics.

4.1. UI Design

We developed a user-friendly interface for the autonomous mission planning module, as shown in Figure 3, organized into three functional areas from right to left. The right section features input controls for defining mission areas, targets, and no-fly zones, along with options to manage mission parameters such as speed and altitude, and to import/export planning data in XML format. The central section displays a dynamic map with latitude and longitude axes, visually representing the mission area, targets, no-fly zones, and the UAV's planned waypoints. The left section presents a detailed table of the mission's waypoints, including their sequence, coordinates, UAV arrival modes, and payload control instructions. This interface streamlines mission setup, visualization, and result analysis.

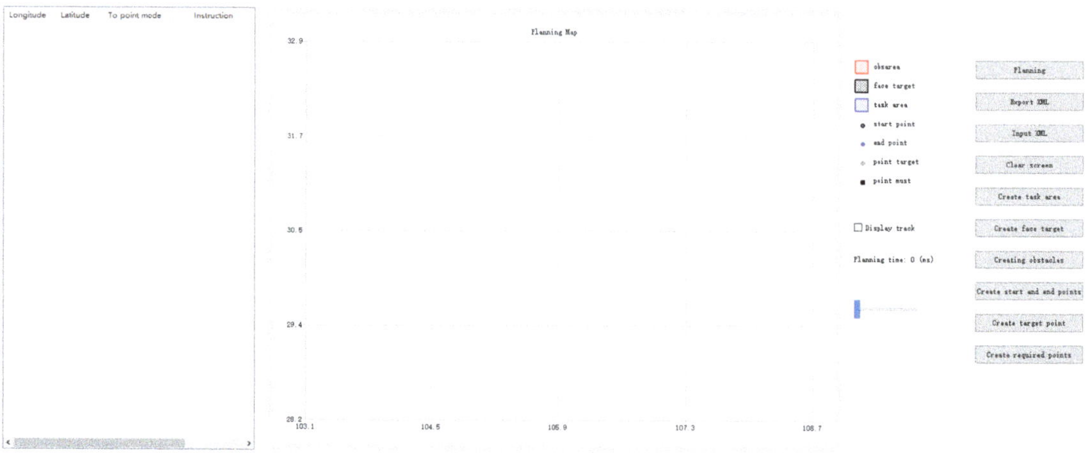

Figure 3. UI for autonomous mission planning module.

4.2. Autonomous Mission Planning

BCD stands out for its ability to ensure complete coverage of the area while simplifying the path planning problem. By dividing the workspace into simple, non-overlapping cells, BCD transforms the complex environment into a set of smaller polygons, each free of obstacles. This decomposition not only reduces computational complexity but also enhances the efficiency of path planning algorithms, ensuring that the UAV can systematically cover the entire area with minimal redundant traversal. Figure 4 visually represents this process, highlighting how the region is decomposed into discrete cells around no-fly zones.

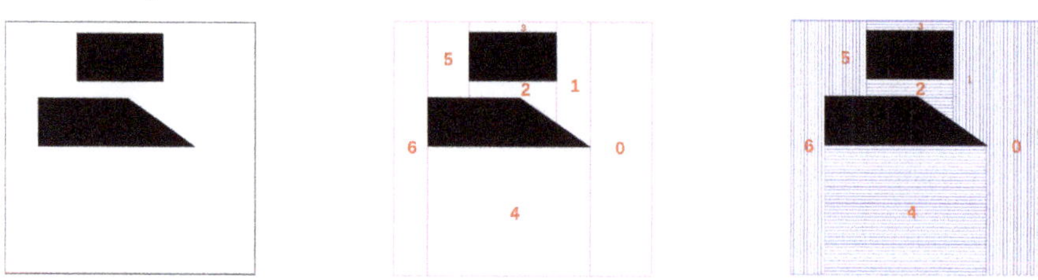

(**a**) (**b**) (**c**)

Figure 4. Area target with no-fly zones and its Boustrophedon Cell Decomposition. (**a**) An area target with no-fly zones. (**b**) Boustrophedon Cell Decomposition of an area target with no-fly zones.The area target is decomposed into 6 cells without no-fly zones, marked from 0 to 6 respectively. (**c**) Results of coverage path planning with an area target.

In the experimental results depicted in Figure 4a, the BCD method was applied to an area target defined by the vertices $(50,50)$, $(50,800)$, $(800,800)$, and $(800,50)$, which encompasses two polygonal no-fly zones with vertices $(148,275)$, $(418,275)$, $(621,148)$, $(148,418)$ and $(270,82)$, $(523,82)$, $(523,224)$, $(270,224)$. These no-fly zones are represented as black polygons within the diagram. The BCD method successfully divided the complex area into seven distinct cells (marked from left to right), each free of no-fly zones.

The decomposition process strategically navigated around the no-fly zones. This not only maintains the integrity of the no-fly zones but also maximizes the coverage of the area target by creating well-defined, accessible regions for the UAV to traverse.

Assuming the entry point of the area target in Figure 4b is the top left vertex, the optimal order for accessing these cells based on GTSP is as follows: $6 \rightarrow 4 \rightarrow 0 \rightarrow 1 \rightarrow 2 \rightarrow 5 \rightarrow 3$. Assuming $\Delta f = 10$, the result of the CPP based on BCD is shown in Figure 4c. Through the above numerical experiments, we have validated its effectiveness.

In the validation phase of the autonomous mission planning module, let us assume a fixed-wing UAV speed of 150 km/h and a flight altitude of 5000 m. The reconnaissance payload requires 70 s from power-on to operational state. The payload has two modes: an SAR Beam Focusing mode and an SAR Strip mode. For the SAR Beam Focusing mode, there are three resolution levels corresponding to three imaging widths, as shown in Table 1. Similarly, the characteristics of the SAR Strip mode are shown in Table 2. To better demonstrate the obstacle avoidance capability of the module in a large-span mission area, we set the d_{buffer} to 2000 m. The input target information includes the required reconnaissance payload mode and resolution for the mission. All position information is represented in latitude and longitude, and the coordinate transformation process during planning will not be elaborated here.

Table 1. The characteristics of the SAR Beam Focusing mode.

Default Work Altitude	Imaging Resolution	Range	Imaging Width
5000 m	0.5 m	16–80 km	5 km × 5 km
	0.3 m	12–60 km	3 km × 3 km
	0.1 m	12–30 km	1 km × 1 km

Table 2. The characteristics of the SAR Strip mode.

Default Work Altitude	Imaging Resolution	Range	Imaging Width
5000 m	5 m	72–120 km	405 km
	1 m	60–100 km	203 km
	0.5 m	48–80 km	12 km

As shown in Figure 5, in the planning map, within a 300 km × 300 km mission area (blue polygon), we input five area targets (black polygons), three no-fly zones (red polygons), five point targets (gray points), the starting point (green point), and the endpoint (blue point).

The proposed module was experimentally validated using the SAR Strip mode.The experiment was divided into two different scenarios. Figure 6 demonstrates the planning results. Firstly, sequence planning calculates the optimal access order, then mission planning optimizes the coverage path based on payload characteristics, and, finally, power instructions are also considered in the final mission instruction set. The following text will analyze these experimental results.

In the first area target scenario, the module successfully adjusted the UAV's path to accommodate the different resolution requirements of five area targets. For targets 1, 3, and 5, which required high-resolution imaging (0.5), the UAV followed a finer, closer-offset path to ensure detailed coverage. Target 2, requiring medium resolution (1.0), had a lower path density. For target 4, with the lowest resolution requirement (3.0), the module optimized the path for broader coverage with minimal overlap. In the second scenario, the experiment focused on processing point target scenes, where the resolution requirement for all five point targets was uniformly set to 0.5. This scenario was mainly used to verify the module's switching in various mission scenarios. Figure 7a illustrates the module output path. For the area target scenario, the module successfully adapts to varying resolution requirements, while the point target scenario demonstrates the module's effective switching between scenarios. Figure 7b illustrates an obstacle avoidance method based on visibility graphs. By expanding obstacle vertices according to the d_{buffer}, a visibility graph is constructed with waypoint No. 73 (lon: −59.5076; lat: −34.7782) as the starting point and waypoint No. 75 (lon: −58.8323; lat: −37.0782) as the endpoint. Utilizing Dijkstra's algorithm, the shortest collision-free path is identified, successfully generating waypoint No. 74 (lon: −59.3972; lat: −36.1817) as the obstacle avoidance waypoint. The distance between this waypoint and the closest vertex of the rectangular obstacle (longitude: −59.368118; latitude: −36.192308) satisfies the requirement d_{buffer}. This method ensures safe and efficient navigation.

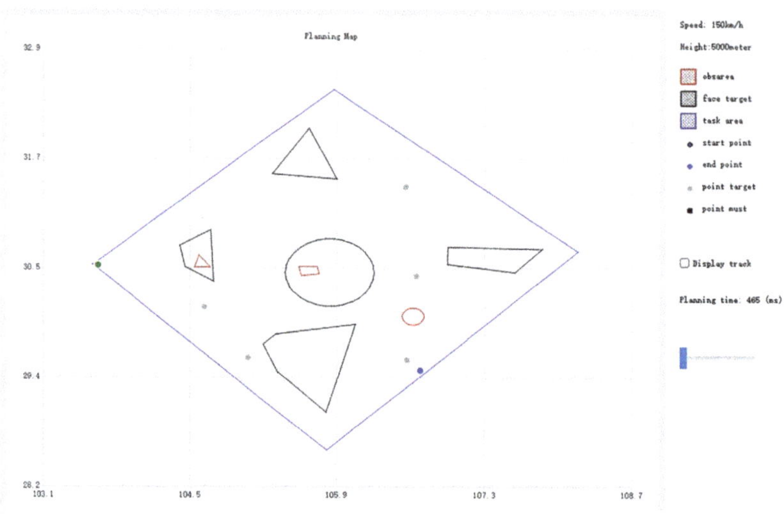

Figure 5. Planning map.

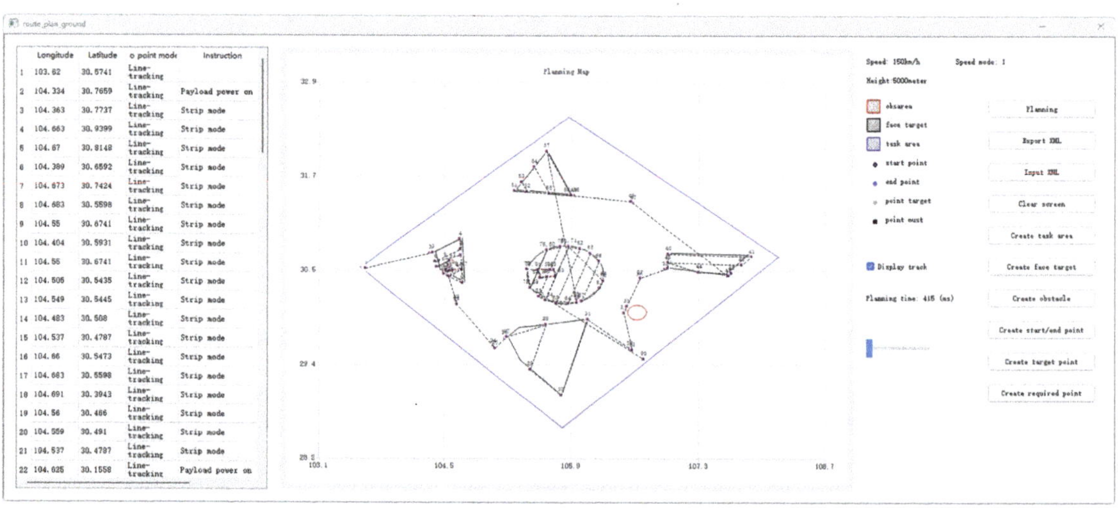

Figure 6. Planning results of autonomous mission planning module.

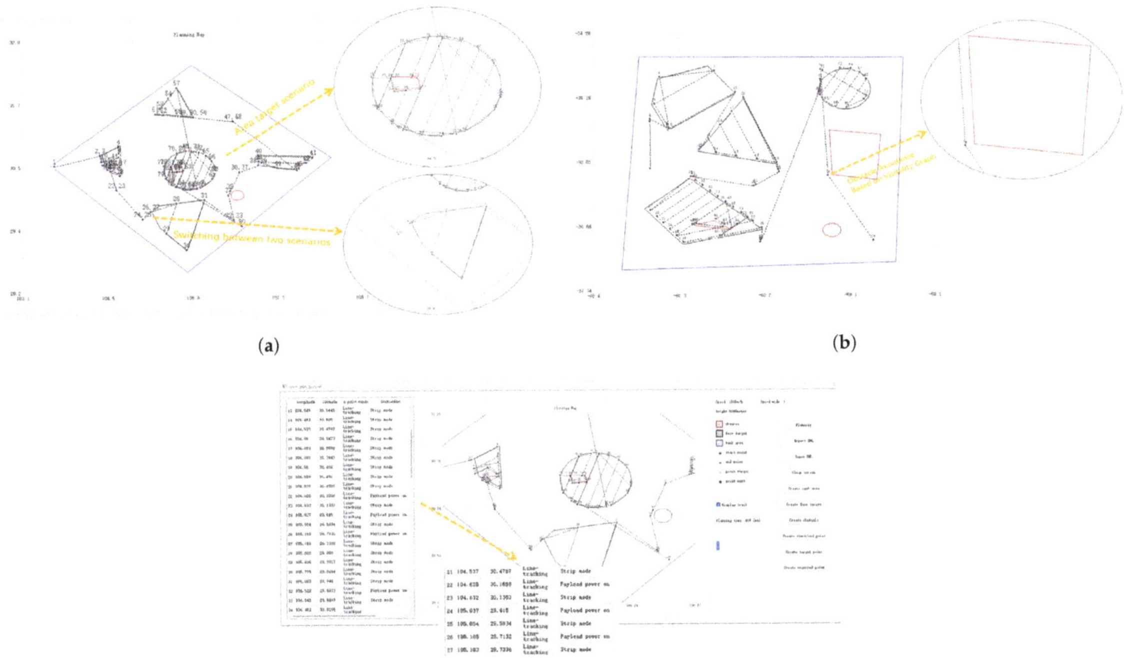

(**a**) (**b**)

(**c**)

Figure 7. Path planning analysis, visibility graph-based obstacle avoidance, and payload power instruction output. (**a**) Path planning for different target scenarios, illustrating the planned waypoint set under varying conditions. (**b**) Visibility graph-based obstacle avoidance, demonstrating the generated path that navigates around no-fly zones and obstacles. (**c**) The output of payload power instructions, showing the power-on and work instructions of the payload during the mission. (**a**) Path planning for different targets. (**b**) Visibility graph-based obstacle avoidance for path planning. (**c**) The output of payload power instructions.

As shown in the mission instruction list on the left side of Figure 7c, the module issued instruction No. 25 for the reconnaissance. Notably, the module inserted instruction No. 24 to power on the payload in advance, ensuring that it was fully operational upon reaching the target. To validate the computational efficiency of the proposed algorithm, we conducted multiple groups of test experiments. Each dataset consisted of 10 targets and three obstacles distributed across a 300 km × 300 km mission area. As shown in Figure 8, the experimental results demonstrate that the proposed method consistently completed calculations within 0.5 s across all test scenarios. The computation time, consistently below 500 ms, ensures compliance with the stringent operational requirements of fixed-wing UAVs.

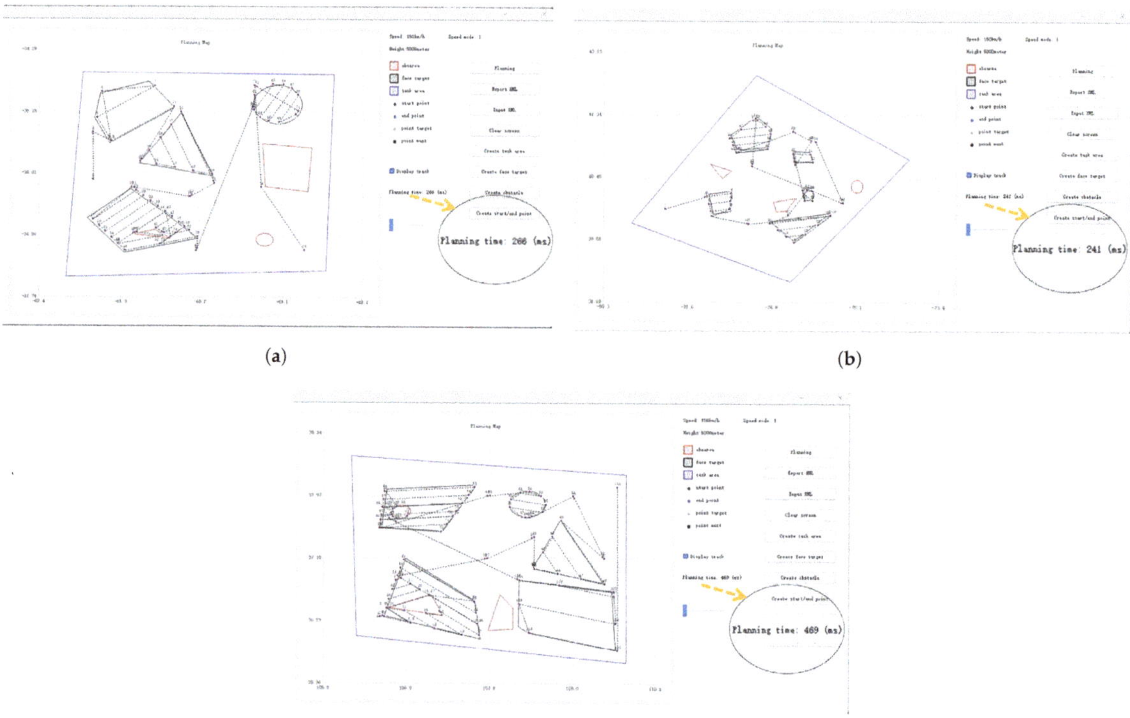

(**a**) (**b**) (**c**)

Figure 8. Comparison of algorithm execution times under three different test datasets. The results demonstrate that the computation time is within 0.5 s. (**a**) Time consumption results of the first set of test data. (**b**) Time consumption results of the second set of test data. (**c**) Time consumption results of the third set of test data.

5. Conclusions

This paper presented a comprehensive approach to autonomous mission planning for fixed-wing UAVs, specifically tailored to multiscenario reconnaissance missions. By addressing the limitations of existing UAV planning methods, which often overlook the integration of payload characteristics, we developed an advanced autonomous mission planning module that combines flight path planning and payload planning.

The proposed path planning strategy, grounded in the HTSP, effectively determines the optimal sequence for accessing reconnaissance targets and planning entry waypoints. The GTSP further enhances the coverage path planning, ensuring thorough reconnaissance even in complex environments with no-fly zones.

Our payload mission planning methodology meticulously considers key payload parameters such as scan resolution, imaging width, and modes. This ensures that each mission instruction is optimized to meet the payload's performance requirements, thereby maximizing reconnaissance effectiveness.

Experiments demonstrated the practical applicability of the module, allowing for intuitive verification and validation of its effectiveness. Overall, the results affirm the superiority of our integrated mission planning module, providing a robust solution for efficient and effective UAV reconnaissance in diverse scenarios. Future work could extend this framework to accommodate additional mission constraints and enhance real-time adaptability in dynamic environments.

Author Contributions: Conceptualization, B.C., J.Y. and Z.Z.; methodology, B.C. and J.Y.; software, J.Y., J.L. and R.L.; writing—review and editing, B.C. and J.Y.; visualization, J.L.; supervision, Z.Z.; project administration, R.L. and Z.Z.; funding acquisition, Z.Z. All authors have read and agreed to the published version of the manuscript.

Funding: This work was supported by the National Natural Science Foundation of China under Grant No. (42474017; 42074038) and, in part, by the System of Systems and Artificial Intelligence Laboratory pioneer fund grant.

Institutional Review Board Statement: Not applicable.

Informed Consent Statement: Not applicable.

Data Availability Statement: Data are contained within the article.

Conflicts of Interest: Authors Bei Chen and Rui Lai were employed by the company AVIC (Chengdu) Unmanned Aerial Vehicle Systems Co., Ltd. The remaining authors declare that the research was conducted in the absence of any commercial or financial relationships that could be construed as a potential conflict of interest.

References

1. Sonkar, S.; Kumar, P.; Philip, D.; Ghosh, A.K. Low-cost smart surveillance and reconnaissance using VTOL fixed wing UAV. In Proceedings of the 2020 IEEE Aerospace conference, Big Sky, MT, USA, 7 March 2020; pp. 1–7.
2. Zhang, J.; Huang, H. Occlusion-aware UAV path planning for reconnaissance and surveillance. *Drones* **2021**, *5*, 98. [CrossRef]
3. Ranasinghe, N.D.; Gunawardana, W.A.D.L. Development of gasoline-electric hybrid propulsion surveillance and reconnaissance VTOL UAV. In Proceedings of the 2021 IEEE International Conference on Robotics, Automation and Artificial Intelligence, Hong Kong, China, 21 April 2021; pp. 63–68.
4. Liao, S.L.; Zhu, R.M.; Wu, N.Q.; Shaikh, T.A.; Sharaf, M.; Mostafa, A.M. Path planning for moving target tracking by fixed-wing UAV. *Def. Technol.* **2020**, *16*, 811–824. [CrossRef]
5. Yang, L.; Liu, Z.; Wang, X.; Yu, X.; Wang, G.; Shen, L. Image-based visual servo tracking control of a ground moving target for a fixed-wing unmanned aerial vehicle. *J. Intell. Robot. Syst.* **2021**, *102*, 1–20. [CrossRef]
6. Sun, Z.; Garcia de Marina, H.; Anderson, B.D.; Yu, C. Collaborative target-tracking control using multiple fixed-wing unmanned aerial vehicles with constant speeds. *J. Guid. Control. Dyn.* **2021**, *44*, 238–250. [CrossRef]
7. Yang, Y.; Leeghim, H.; Kim, D. Dubins Path-Oriented Rapidly Exploring Random Tree* for Three-Dimensional Path Planning of Unmanned Aerial Vehicles. *Electronics* **2022**, *11*, 2338. [CrossRef]
8. Aiello, G.; Valavanis, K.P.; Rizzo, A. Fixed-wing uav energy efficient 3d path planning in cluttered environments. *J. Intell. Robot. Syst.* **2022**, *105*, 60. [CrossRef]
9. Machmudah, A.; Shanmugavel, M.; Parman, S.; Manan, T.S.A.; Dutykh, D.; Beddu, S.; Rajabi, A. Flight trajectories optimization of fixed-wing UAV by bank-turn mechanism. *Drones* **2022**, *6*, 69. [CrossRef]
10. Cui, Q. Multi-target points path planning for fixed-wing unmanned aerial vehicle performing reconnaissance missions. In Proceedings of the 5th International Conference on Information Science, Electrical, and Automation Engineering, Wuhan, China, 24 March 2023; Volume 12748, pp. 713–723.
11. Ding, Z.; Huang, Y.; Yuan, H.; Dong, H. Introduction to reinforcement learning. In *Deep Reinforcement Learning: Fundamentals, Research and Applications*; Dong, H., Ding, Z., Zhang, S., Eds.; Springer: Singapore, 2020; pp. 47–123.
12. Chen, J.; Ling, F.; Zhang, Y.; You, T.; Liu, Y.; Du, X. Coverage path planning of heterogeneous unmanned aerial vehicles based on ant colony system. *Swarm Evol. Comput.* **2022**, *69*, 101005. [CrossRef]
13. Theile, M.; Bayerlein, H.; Nai, R.; Gesbert, D.; Caccamo, M. UAV coverage path planning under varying power constraints using deep reinforcement learning. In Proceedings of the 2020 IEEE/RSJ International Conference on Intelligent Robots and Systems, Las Vegas, NV, USA, 24 October 2020; pp. 1444–1449.
14. Jia, Y.; Zhou, S.; Zeng, Q.; Li, C.; Chen, D.; Zhang, K.; Liu, L.; Chen, Z. The UAV path coverage algorithm based on the greedy strategy and ant colony optimization. *Electronics* **2022**, *17*, 2667. [CrossRef]
15. Majeed, A.; Hwang, S.O. A multi-objective coverage path planning algorithm for UAVs to cover spatially distributed regions in urban environments. *Aerospace* **2021**, *11*, 343. [CrossRef]
16. Vazquez-Carmona, E.V.; Vasquez-Gomez, J.I.; Herrera-Lozada, J.C.; Antonio-Cruz, M. Coverage path planning for spraying drones. *Comput. Ind. Eng.* **2022**, *168*, 108125 [CrossRef] [PubMed]
17. Galceran, E.; Carreras, M. A survey on coverage path planning for robotics. *Robot. Auton. Syst.* **2013**, *61*, 1258–1276. [CrossRef]

18. Vasquez-Gomez, J.I.; Marciano-Melchor, M.; Valentin, L.; Herrera-Lozada, J.C. Coverage path planning for 2d convex regions. *J. Intell. Robot. Syst.* **2020**, *97*, 81–94. [CrossRef]
19. Tan, C.S.; Mohd-Mokhtar, R.; Arshad, M.R. A comprehensive review of coverage path planning in robotics using classical and heuristic algorithms. *IEEE Access* **2021**, *9*, 119310–119342. [CrossRef]
20. Barrientos, A.; Colorado, J.; Cerro, J.D.; Martinez, A.; Rossi, C.; Sanz, D.; Valente, J. Aerial remote sensing in agriculture: A practical approach to area coverage and path planning for fleets of mini aerial robots. *J. Field Robot.* **2011**, *28*, 667–689. [CrossRef]
21. Hasan, K.M.; Reza, K.J. Path planning algorithm development for autonomous vacuum cleaner robots. In Proceedings of the 2014 International Conference on Informatics, Electronics & Vision, Dhaka, Bangladesh, 23 May 2014; pp. 1–6.
22. Bähnemann, R.; Lawrance, N.; Chung, J.J.; Pantic, M.; Siegwart, R.; Nieto, J. Revisiting boustrophedon coverage path planning as a generalized traveling salesman problem. In Proceedings of the Field and Service Robotics: Results of the 12th International Conference, Tokyo, Japan, 29 August 2019; pp. 277–290.
23. Kumar, K.; Kumar, N. Region coverage-aware path planning for unmanned aerial vehicles: A systematic review. *Phys. Commun.* **2023**, *59*, 102073. [CrossRef]
24. Choset, H.; Pignon, P. Coverage path planning: The boustrophedon cellular decomposition. In Proceedings of the Field and Service Robotics, Canberra, Australia, 8–10 December 1997; pp. 203–209.
25. Choset, H. Coverage of known spaces: The boustrophedon cellular decomposition. *Auton. Robot.* **2000**, *9*, 247–253. [CrossRef]
26. Apostolidis, S.D.; Kapoutsis, P.C.; Kapoutsis, A.C.; Kosmatopoulos, E.B. Cooperative multi-UAV coverage mission planning platform for remote sensing applications. *Auton. Robot.* **2022**, *46*, 373–400. [CrossRef]
27. Stecz, W.; Gromada, K. UAV mission planning with SAR application. *Sensors* **2020**, *4*, 1080. [CrossRef]
28. Zhang, Y.; Chen, Z.; Yan, Y.; Jiang, F.; Gao, X. Research on single target cognitive electronic reconnaissance strategy for unmanned aerial vehicle. *IET Radar Sonar Navig.* **2023**, *11*, 1711–1727. [CrossRef]
29. Stecz, W.; Gromada, K. Determining UAV flight trajectory for target recognition using EO/IR and SAR. *Sensors* **2020**, *20*, 5712. [CrossRef] [PubMed]
30. Asiyabi, R.M.; Ghorbanian, A.; Tameh, S.N.; Amani, M.; Jin, S.; Mohammadzadeh, A. Synthetic aperture radar (SAR) for ocean: A review. *IEEE J. Sel. Top. Appl. Earth Obs. Remote. Sens.* **2023**, *16*, 9106–9138. [CrossRef]
31. Luomei, Y.; Xu, F. Segmental aperture imaging algorithm for multirotor UAV-borne MiniSAR. *IEEE Trans. Geosci. Remote Sens.* **2023**, *61*, 1–8. [CrossRef]
32. Zhou, Y.; Wang, W.; Chen, Z.; Wang, P.; Zhang, H.; Qiu, J.; Zhao, Q.; Deng, Y.; Zhang, Z.; Yu, W.; et al. Digital beamforming synthetic aperture radar (DBSAR): Experiments and performance analysis in support of 16-channel airborne X-band SAR data. *IEEE Trans. Geosci. Remote Sens.* **2020**, *59*, 6784–6798. [CrossRef]
33. García-Fernández, M.; Álvarez-Narciandi, G.; López, Y.; Las-Heras, F. Array-based ground penetrating synthetic aperture radar on board an unmanned aerial vehicle for enhanced buried threats detection. *IEEE Trans. Geosci. Remote Sens.* **2023**, *61*, 1–18. [CrossRef]
34. Yuan, Y.; Cattaruzza, D.; Ogier, M.; Semet, F. A branch-and-cut algorithm for the generalized traveling salesman problem with time windows. *Eur. J. Oper. Res.* **2020**, *286*, 849–866. [CrossRef]
35. Nourmohammadzadeh, A.; Sarhani, M.; Voß, S. A matheuristic approach for the family traveling salesman problem. *J. Heuristics* **2023**, *29*, 435–460. [CrossRef]
36. Helsgaun, K. Solving the equality generalized traveling salesman problem using the Lin–Kernighan–Helsgaun Algorithm. *Math. Prog. Comp.* **2015**, *7*, 269–287. [CrossRef]
37. Fabri, A.; Pion, S. CGAL: The computational geometry algorithms library. In Proceedings of the 17th ACM SIGSPATIAL International Conference on Advances in Geographic Information Systems, Seattle, WA, USA, 4–6 November 2009; pp. 538–539.

sensors

MDPI

Article

A Lossless Scalar Calibration Algorithm Used for Tri-Axial Magnetometer Cross Array and Its Effectiveness Validation

Lihua Wu [1], Yu Huang [1,2,*] and Xintong Chen [1]

1 College of Physics and Optoelectronic Engineering, Harbin Engineering University, Harbin 150001, China; wulihua@hrbeu.edu.cn (L.W.); chen.xintong@hrbeu.edu.cn (X.C.)
2 National Key Laboratory of Underwater Acoustic Technology, Harbin Engineering University, Harbin 150001, China
* Correspondence: huangyu@hrbeu.edu.cn

Abstract: The accuracy of a magnetic gradient tensor (MGT) measured by tri-axial magnetometer cross arrays (TAMCAs) is compromised by inherent errors in individual tri-axial magnetometers (TAMs) and inter-sensor misalignment angles (MAs), both of which degrade the resultant MGT data quality. This paper proposes a novel lossless scalar calibration algorithm that eliminates mathematical approximations while tracking the fluctuation of the reference magnetic intensity (MI). The calibration algorithm is developed to improve TAMCAs' measurement precision; however it is difficult to provide a completely accurate MGT by experiments. Therefore, we have designed a kind of validation experiment based on a constrained Euler localization to demonstrate the effectiveness of the calibration algorithm. The fundamental principles of the proposed lossless scalar calibration methodology are systematically presented, accompanied by a numerical analysis of relative errors calibrating TAMCA parameters. Key influencing factors are carefully investigated, including the TAM noise level quantified by standard deviation (STD), calibration dataset size, and STD of reference MI fluctuations. In the experiments, to validate the effectiveness of calibrating TAMCAs composed of four fluxgate TAMs (FTAMs), we measured the true geo-MI using a proton magnetometer and regarded an energized circular coil as the alternating current (AC) magnetic source of the constrained Euler localization, respectively. The results indicated that the lossless scalar calibration algorithm significantly improves the measurement accuracy of the geo-MI of the calibration site and MGT of the energized coil.

Keywords: tri-axial magnetometer cross array; error; lossless scalar calibration; magnetic gradient tensor; effectiveness validation

1. Introduction

As the spatial derivative tensor of magnetic vectors, MGT inherently contains five independent components, demonstrating superior spatial resolution and enhanced signal discrimination capability compared to conventional analytical signals and vector measurements [1,2]. The form of the tensor enables the comprehensive reconstruction of magnetic source architectures while simultaneously suppressing geomagnetic noise and eliminating reliance on fixed base stations. Crucially, MGT permits direct determination of magnetization parameters through its rotationally invariant properties, requiring neither prior knowledge of source geometry nor positional information, and can show more effective capability in shallow magnetic anomaly detection over the total field measurement [3]. Unlike magnetic vector measurements vulnerable to attitude-induced errors in the presence of geomagnetic fields, MGT maintains operational stability and exhibits heightened

 https://doi.org/10.3390/s25072164

sensitivity to magnetic anomaly signals, owing to its inherent insensitivity to both geomagnetic diurnal variations and spatial orientation uncertainties [4,5]. These characteristics render MGT ideally suited for mobile deployment platforms in challenging environments where traditional magnetic prospecting fails, including low-latitude regions and rugged terrains. Specific applications span concealed mineral exploration, unexploded ordnance localization, and marine detection systems.

The limitations inherent in sensor manufacturing processes and system installation accuracy introduce MAs between different TAMs, in addition instrumental errors within individual TAMs, collectively degrading the data quality of MGT systems and necessitating calibration procedures to mitigate such discrepancies. TAM calibration methods are broadly categorized into vector-based and scalar-based approaches. The vector calibration method employs high-precision magnetic vectors as reference fields to calibrate TAMs, typically requiring magnetically shielded environments or active coil compensation system to suppress external magnetic interference, thereby imposing stringent application constraints. While this method is particularly suitable for characterizing TAM frequency responses [6], its implementation complexity limits practical applicability. In contrast, scalar calibration methods utilize high-precision MIs as reference standards [7–10], with constrained optimization-based techniques further eliminating the need for scalar magnetometer measurements to achieve cost-effective and simplified implementations [11]. However, a critical limitation of the scalar calibration method lies in the assumption of invariant MI outputs during calibration, which fails to track fluctuations in the reference magnetic field, potentially compromising calibration accuracy under real-world conditions.

The calibration of MAs in MGT systems relies on aligning the measurement benchmarks of TAMs to a unified coordinate system, either defined by a single TAM or the platform itself. In principle, a non-ideal MGT system in a uniform magnetic field produces outputs dependent on the field's uniformity, as the theoretical magnetic gradient in such regions is zero. This property enables calibration by rotating the MGT system within a uniform field and applying least squares or linear constrained least squares fitting to synchronize MGT-measured data with reference magnetic vectors [12]. For online calibration, Ref. [13] introduced a generalized least squares algorithm tailored to towed airborne fluxgate MGT systems, leveraging static error models, fluxgate dynamic characteristics, and the near-zero geo-MGT at high altitudes, validated via helicopter flight tests. However, real-world geomagnetic environments rarely satisfy uniform-field conditions, driving researchers to explore tensor invariant-based calibration strategies. Ref. [14] proposed calibrating MGT systems using two tensor invariants as constraints, analogous to scalar TAM calibration. Ref. [15] further integrated tensor components and invariants into an objective function, employing an improved differential evolution algorithm to simultaneously correct TAM instrumental errors and MAs in a single step, demonstrating enhanced noise resistance over conventional least squares methods. Additionally, an indirect calibration approach utilizing the generalized Hilbert transform was developed for airborne systems, bypassing the need for direct reference fields by leveraging high-precision total field gradiometer data as a processed calibration benchmark [16]. Faced with the challenges brought by non-uniform fields and dynamic operating environments, these advancements have collectively improved the accuracy of MGT systems under practical conditions.

Ref. [17] applies both implemented and optimized calibration approaches to full MGT systems utilizing superconducting quantum interference devices (SQUIDs), emphasizing rapid data processing, yet omitting explicit solutions for calibrating MAs. For tetrahedral MGT systems, Ref. [18] introduces a calibration algorithm based on constraint relationships among the nine components of the MGT, improving accuracy through inherent physical interdependencies. However, these constraints cannot directly extend to TAMCAs measuring only five independent tensor components. Ref. [19] proposes a linear calibration method for TAMCAs relying solely on vertical magnetic field constraints, which proves insufficient for comprehensive MA correction. Ref. [20] addresses this by constructing a nonlinear correction model within an artificial platform reference frame and employing the Levenberg–Marquardt algorithm for parameter fitting; however, the coupling between MAs and orthogonal error angles (OEAs) introduces interdependencies that challenge solution independence and accuracy. Ref. [21] enhances TAMCA calibration precision by adopting total least squares for ellipsoidal model parameter fitting, outperforming conventional least squares, yet fails to provide explicit sensor error parameter calculations. Meanwhile, MGT measurements are typically conducted via differential approximation, rotation modulation [22], or string vibration methods [23], with the differential approximation method favored for its simplicity and ability to capture full tensor information. For this method, centrally symmetric TAMCAs are preferred due to their minimized principle-based measurement errors [24]. Collectively, the calibration of MGT systems involves diverse methodologies with varying scopes and limitations.

This paper proposes a lossless scalar calibration algorithm for TAMCAs that eliminates mathematical approximations while dynamically tracking fluctuations in the reference MI. The algorithm derives an initial solution by linearizing the ellipsoid model equation, thereby accelerating computational efficiency. The structure of the paper is organized as follows. Section 2 details the principles of the proposed lossless scalar calibration algorithm for TAMCAs. Section 3 outlines validation strategies to assess the algorithm's effectiveness. In Section 4, numerical simulations analyze calibration accuracy under varying noise conditions, including dependencies on the STDs of TAM and scalar magnetometer noise, as well as calibration dataset size. Experimental validation is further conducted through geomagnetic field measurements and constrained Euler localization of an energized circular coil. Finally, Section 5 synthesizes key conclusions.

2. Calibration Methodology

2.1. Error Model of the TAMCA

The TAMCA comprises four TAMs positioned at vertices A, B, C, and D, labeled as nodes 1, 2, 3, and 4, respectively, as shown in Figure 1. Benefitting from the cross-shaped symmetric configuration of the array, the measured magnetic vector and MGT can effectively approximate to their true values at the geometric center P of the array. The baseline lengths of the array, defined as the distances L_x and L_y between the paired TAMs along the x- and y-axes, characterize the spatial resolution of the array in the x- and y-directions, respectively, served as critical parameters for spatial gradient calculations.

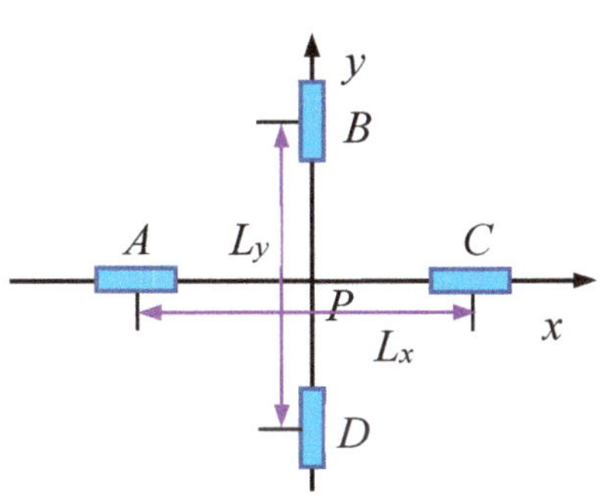

Figure 1. Schematic configuration of the TAMCA.

The measurement errors of the TAMCA stem from two primary sources: (1) intrinsic errors of the four TAMs including scale factor errors (SFEs), OEAs, and biases; and (2) MAs induced during system assembly. Calibrating the TAMCA as an integrated system, rather than individually calibrating each TAM, mitigates MAs caused by installation misalignments. As illustrated in Figure 2, the sensor coordinate system ($\xi_i\eta_i\zeta_i$, $i = 1,2,3,4$) and reference coordinate system ($x_iy_iz_i$) are defined such that their z-axes (ζ_i and z_i) are aligned. The OEAs of the i-th TAM are characterized by angles θ_i, φ_i, and ψ_i, where ψ_i represents the angle between the η_i-axis and y_i-axis, θ_i denotes the angle between the ξ_i-axis and its projection onto the x_iy_i-plane, and φ_i defines the angle between the projected l_i-axis and the x_i-axis. The MAs between the sensor coordinate system $x_jy_jz_j$ (for the j-th TAM, $j = 2,3,4$) and the reference system $x_1y_1z_1$ are further quantified by the angles γ_j, β_j, and α_j of right-handed sequential rotation (z-y-x), encapsulating the spatial misalignment inherent to the array configuration.

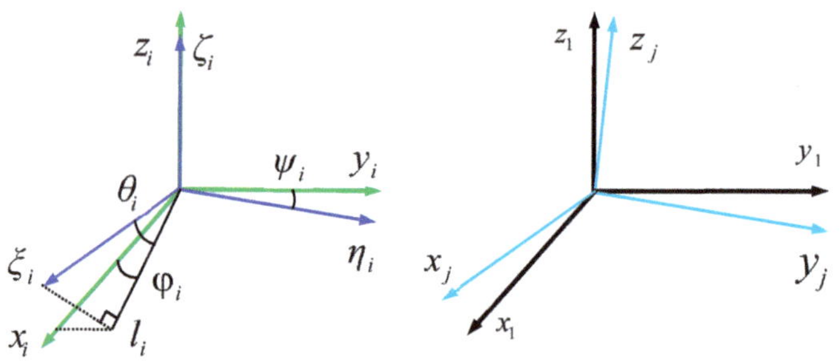

Figure 2. Spatial orientation of sensor's sensitive axes in the TAMCA.

B_i represents the true magnetic vector at the point of the i-th TAM in the coordinate system $x_1y_1z_1$, and B_i' is the expression of B_i in the coordinate system $x_iy_iz_i$, then

$$B_i' = \Lambda_i^{-1} B_i \tag{1}$$

where Λ_1 is an identity matrix with three ranks; namely, $B_1' = B_1$, $\Lambda_j = R_x(\alpha_j)R_y(\beta_j)R_z(\gamma_j)$, R_x, R_y, and R_z represent the rotation matrices around the x-axis, y-axis, and z-axis, respectively.

A uniform magnetic field is selected to calibrate the TAMCA, so the magnetic vectors B_i are all the same. The measurement output $\widetilde{B}_i$ of the i-th TAM is as follows [21]:

$$\widetilde{B}_i = D_i B_i' + B_{0i} + n_i \tag{2}$$

$$D_i = \begin{bmatrix} S_{xi}\cos\theta_i\cos\varphi_i & S_{xi}\cos\theta_i\sin\varphi_i & S_{xi}\sin\theta_i \\ 0 & S_{yi}\cos\psi_i & S_{yj}\sin\psi_i \\ 0 & 0 & S_{zi} \end{bmatrix} \tag{3}$$

where, D_i, B_{0i}, and n_i are the measurement matrix, bias, and noise of the i-th TAM, and S_{xi}, S_{yi}, and S_{zi} are the scalar factors (SFs) of the three sensitive axes, respectively. $\delta S_{xi} = S_{xi} - 1$, $\delta S_{yi} = S_{yi} - 1$, $\delta S_{zi} = S_{zi} - 1$ are, respectively, the SFEs. We can obtain from Equations (1) and (2) that

$$B_i' = \Lambda_i^{-1} B_i = \Omega_i(\widetilde{B}_i - B_{0i} - n_i) \tag{4}$$

where Ω_i is the inversed matrix of D_i.

Non-zero elements of Ω_i are, respectively, $\Omega_i(1,2) = p_{2i} = -\tan\varphi_i\sec\psi_i/S_{yi}$, $\Omega_i(2,3) = p_{5i} = -\tan\psi_i/S_{zi}$, $\Omega_i(2,2) = p_{4i} = \sec\psi_i/S_{yi}$, $\Omega_i(1,1) = p_{1i} = \sec\theta_i\sec\varphi_i/S_{xi}$,

$\Omega_i(3,3) = p_{6i} = S_{zi}^{-1}, \Omega_i(1,3) = p_{3i} = \sec\theta_i \sec\varphi_i \sec\psi_i(\cos\theta_i \sin\varphi_i \sin\psi_i - \sin\theta_i \cos\psi_i)/S_{zi}$. Given S_{xi}, S_{yi}, and S_{zi} approximate unity, while the small angles θ_i, φ_i, and ψ_i induce non-zero elements of Ω_i constrained to first-order contributions, the parameter identification accuracy is significantly enhanced. The expression of calculating the SFs and OEAs from $p_{ki}, k = 1, 2, \cdots, 6$ is as follows:

$$\begin{cases} S_{zi} = 1/p_{6i}, \quad \psi_i = -\arctan(S_{zi}p_{5i}) \\ S_{yi} = \sec\psi_i/p_{4i}, \quad \varphi_i = -\arctan(S_{yi}p_{2i}\cos\psi_i) \\ \theta_i = \arctan(\sin\varphi_i \tan\psi_i - S_{zi}p_{3i}\cos\varphi_i), \quad S_{xi} = \sec\theta_i \sec\varphi_i/p_{1i} \end{cases} \tag{5}$$

2.2. Principal of Lossless Scalar Calibration Algorithm

The ellipsoidal equation obtained from Equation (2) is as follows:

$$\widetilde{B}_i^{\mathrm{T}} A_i \widetilde{B}_i - 2B_{0i}^{\mathrm{T}} A_i \widetilde{B}_i + B_{0i}^{\mathrm{T}} A_i B_{0i} = 1 \tag{6}$$

where

$$B_{0i} = -A_i^{-1}[r_i s_i t_i]^{\mathrm{T}} \tag{7}$$

$||B_i||$ is the reference MI. Generally, $A_i = \left|\left|B_i\right|\right|^{-1}\Omega_i^{\mathrm{T}}\Omega_i$ is a symmetric positive definite matrix that can accurately track the fluctuations of $||B_i||$, and A_i is defined as follows:

$$A_i = \begin{bmatrix} a_i & d_i & e_i \\ d_i & b_i & f_i \\ e_i & f_i & c_i \end{bmatrix} \tag{8}$$

Let $u_i = [a_i, b_i, c_i, d_i, e_i, f_i, r_i, s_i, t_i]^{\mathrm{T}}$ and we can obtain the least square solution u_{0i} as expressed by Equation (9) through linearizing Equation (6).

$$u_{0i} = [K_i^{\mathrm{T}} K_i]^{-1} K_i^{\mathrm{T}} 1_{N\times 1} \tag{9}$$

$$K_i = \begin{bmatrix} \widetilde{B}_{xi1}^2 & \widetilde{B}_{yi1}^2 & \widetilde{B}_{zi1}^2 & 2\widetilde{B}_{xi1}\widetilde{B}_{yi1} & 2\widetilde{B}_{xi1}\widetilde{B}_{zi1} & 2\widetilde{B}_{yi1}\widetilde{B}_{zi1} & 2\widetilde{B}_{xi1} & 2\widetilde{B}_{yi1} & 2\widetilde{B}_{zi1} \\ \widetilde{B}_{xi2}^2 & \widetilde{B}_{yi2}^2 & \widetilde{B}_{zi2}^2 & 2\widetilde{B}_{xi2}\widetilde{B}_{yi2} & 2\widetilde{B}_{xi2}\widetilde{B}_{zi2} & 2\widetilde{B}_{yi2}\widetilde{B}_{zi2} & 2\widetilde{B}_{xi2} & 2\widetilde{B}_{yi2} & 2\widetilde{B}_{zi2} \\ \vdots & \vdots & \vdots & \vdots & \vdots & \vdots & \vdots & \vdots & \vdots \\ \widetilde{B}_{xiN}^2 & \widetilde{B}_{yiN}^2 & \widetilde{B}_{ziN}^2 & 2\widetilde{B}_{xiN}\widetilde{B}_{yiN} & 2\widetilde{B}_{xiN}\widetilde{B}_{ziN} & 2\widetilde{B}_{yiN}\widetilde{B}_{ziN} & 2\widetilde{B}_{xiN} & 2\widetilde{B}_{yiN} & 2\widetilde{B}_{ziN} \end{bmatrix} \tag{10}$$

What is more, the nonlinear equation system expressed by Equation (6) can be transformed into a constrained optimization problem denoted by Equation (11).

$$\begin{cases} \min f_i(u_i) \\ s.t. \quad u_{iL} < u_i < u_{iH} \end{cases} \tag{11}$$

where $f_i(u_i) = \sum_{n=1}^{N} y_i^2(n)$ and N is the calibration dataset size.

$$\begin{aligned} y_i = \; & \widetilde{B}_{xi}^2 u_{1i} + \widetilde{B}_{yi}^2 u_{2i} + \widetilde{B}_{zi}^2 u_{3i} + 2(\widetilde{B}_{xi}\widetilde{B}_{yi}u_{4i} + \widetilde{B}_{xi}\widetilde{B}_{zi}u_{5i} + \widetilde{B}_{yi}\widetilde{B}_{zi}u_{6i} - \widetilde{B}_{xi}u_{1i}u_{7i} \\ & -\widetilde{B}_{yi}u_{2i}u_{8i} - \widetilde{B}_{zi}u_{3i}u_{9i} - \widetilde{B}_{xi}u_{4i}u_{8i} - \widetilde{B}_{yi}u_{4i}u_{7i} - \widetilde{B}_{xi}u_{5i}u_{9i} - \widetilde{B}_{zi}u_{5i}u_{7i} - \widetilde{B}_{zi}u_{6i}u_{8i} \\ & -\widetilde{B}_{yi}u_{6i}u_{9i} + u_{4i}u_{7i}u_{8i} + u_{5i}u_{7i}u_{9i} + u_{6i}u_{8i}u_{9i}) + u_{1i}u_{7i}^2 + u_{2i}u_{8i}^2 + u_{3i}u_{9i}^2 - 1 \end{aligned} \tag{12}$$

The sequential quadratic programming algorithm is used to solve Equation (11) to obtain u_i, regarding u_{0i} as the initial solution, and then substitute u_i into Equation (7) to calculate out the biases $B_{0i} = [B_{0xi}, B_{0yi}, B_{0zi}]^{\mathrm{T}}$. Equation (12) is unrelated to $||B_i||$ but related to $\widetilde{B}_i$, so the solution of B_{0i} is not related to the noise of scalar magnetometer measuring $||B_i||$, but to the noise of the TAM.

After obtaining B_{0i}, we can derive the following from Equations (1) and (4):

$$\sum_{k=1}^{6} W_{ki} q_{ki} = ||B_i'||^2 = ||B_i||^2 \tag{13}$$

where $q_{1i} = p_{1i}^2$, $q_{2i} = p_{2i}^2 + p_{4i}^2$, $q_{3i} = p_{3i}^2 + p_{5i}^2 + p_{6i}^2$, $q_{4i} = p_{1i}p_{2i}$, $q_{5i} = p_{2i}p_{3i} + p_{4i}p_{5i}$, $q_{6i} = p_{1i}p_{3i}$, $W_{1i} = (\widetilde{B}_{xi} - B_{0xi})^2$, $W_{2i} = (\widetilde{B}_{yi} - B_{0yi})^2$, $W_{3i} = (\widetilde{B}_{zi} - B_{0zi})^2$, $W_{4i} = 2(\widetilde{B}_{xi} - B_{0xi})(\widetilde{B}_{yi} - B_{0yi})$, $W_{5i} = 2(\widetilde{B}_{yi} - B_{0yi})(\widetilde{B}_{zi} - B_{0zi})$, $W_{6i} = 2(\widetilde{B}_{xi} - B_{0xi})(\widetilde{B}_{zi} - B_{0zi})$.

Equation (13) is solved using linear least squares to obtain q_{ki}, and then to calculate p_{ki} using q_{ki}, while the SFs and OEAs can be calculated out substituting p_{ki} into Equation (5), which is called a nonlinear fitting transformation. As a result, B_i' can be calculated out from Equation (4) using the obtained biases, SFs, and OEAs. Equation (13) is related to both $\widetilde{B}_i$ and $||B_i||$; therefore, the calibration results of the SFEs and OEAs may depend on the noises of both the TAM and scalar magnetometer.

The identification of α_j, β_j, and γ_j can be regarded as the Procrustes problem, so the MAs are determined using Procrustes analysis on Equation (1), according to B_i'. Procrustes analysis is a method that directly estimates the translation and rotation parameters of coordinate transformation by performing singular value decomposition on the eigen matrix, which greatly simplifies the calculation.

In summary, there are no mathematical approximations for a lossless scalar calibration algorithm that can track fluctuations in the reference MI, and it is divided into three steps as shown in Figure 3. The first step is to obtain B_{0i} through solving Equation (11). The second step is to solve Equation (13) using linear least squares to calculate out q_{ki} and p_{ki}, and then the SFs and OEAs are all calculated through substituting p_{ki} into Equation (5). In the third step, all of the calculated biases, SFEs, and OEAs are substituted into Equation (4) to obtain B_i', and then α_j, β_j, and γ_j are obtained by Procrustes analysis.

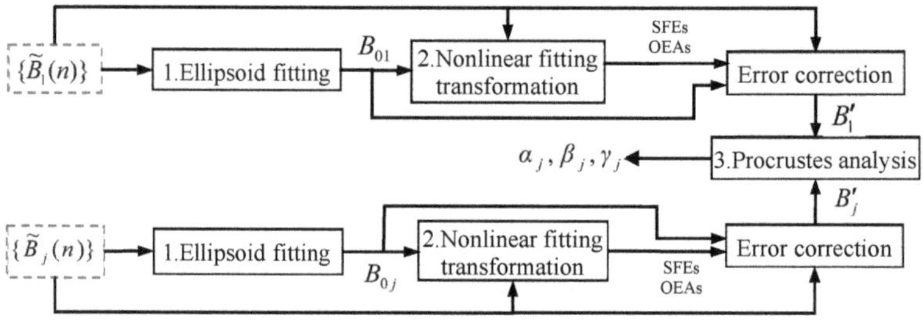

Figure 3. Flowchart of the lossless scalar calibration algorithm for the TAMCA.

3. Calibration Effectiveness Validation Method

3.1. Validation Using the Measurement of MI

The root mean square errors (RMSEs) of MI and MGT are often used as the parameters evaluating the performance of the lossless scalar calibration algorithm for the TAMCA. $\widetilde{e}_B$ and $\hat{e}_B$ are the RMSEs of the measured MI before and after calibrating the TAMCA, which are defined as follows, respectively:

$$\widetilde{e}_B = \sqrt{\frac{1}{N}\sum_{n=1}^{N}\left(\sqrt{\widetilde{B}_n^{\mathrm{T}}\widetilde{B}_n} - ||B_r||\right)^2} \tag{14}$$

$$\hat{e}_B = \sqrt{\frac{1}{N}\sum_{n=1}^{N}\left(\sqrt{\hat{B}_n^{\mathrm{T}}\hat{B}_n} - ||B_r||\right)^2} \tag{15}$$

where $\left\{\widetilde{B}_n\right\}$ and $\{\hat{B}_n\}, n = 1, 2, \cdots, N$ are the measurement data of the magnetic vector before and after calibrating the TAMCA, and B_r is the true value of magnetic vector.

$a = \partial B_x/\partial x$, $b = \partial B_y/\partial y$, $d = (\partial B_x/\partial y + \partial B_y/\partial x)/2$, $e = \partial B_z/\partial y$, and $f = \partial B_z/\partial x$ are five independent components of the MGT measured by the TAMCA [25]. An ideal field used to calibrate the TAMCA is a uniform magnetic field and, therefore, the true values of a, b, d, e, and f are all zero. $\widetilde{e}_T$ and $\hat{e}_T$ are the RMSEs of the MGT before and after calibration, whose expressions are as follows, respectively:

$$\widetilde{e}_T = \sqrt{\frac{1}{N}\sum_{n=1}^{N}\left(\widetilde{\delta}_{an}^2 + \widetilde{\delta}_{bn}^2 + \widetilde{\delta}_{dn}^2 + \widetilde{\delta}_{en}^2 + \widetilde{\delta}_{fn}^2\right)} \tag{16}$$

$$\hat{e}_T = \sqrt{\frac{1}{N}\sum_{n=1}^{N}\left(\hat{\delta}_{an}^2 + \hat{\delta}_{bn}^2 + \hat{\delta}_{dn}^2 + \hat{\delta}_{en}^2 + \hat{\delta}_{fn}^2\right)} \tag{17}$$

where $\left\{\widetilde{\delta}_{an}, \widetilde{\delta}_{bn}, \widetilde{\delta}_{dn}, \widetilde{\delta}_{en}, \widetilde{\delta}_{fn}\right\}$ and $\left\{\hat{\delta}_{an}, \hat{\delta}_{bn}, \hat{\delta}_{dn}, \hat{\delta}_{en}, \hat{\delta}_{fn}\right\}$ are the errors of a, b, d, e, and f before and after calibrating the TAMCA, respectively.

3.2. Validation Using Magnetic Source Localization

The MIs calculated via the four individual TAMs are independent of the misalignment errors between the TAMs and of the spatial orientations of the TAMCA. Thus, unlike $\hat{e}_T$, $\hat{e}_B$ cannot be used to determine the accuracy of the vectors measured by the TAMCA. Moreover, the true value of the MI is often unknown for lots of in situ calibrations, and geomagnetic reference fields probably contain gradients caused by the magnetic material located nearby the experimental environments. This means that in certain geomagnetic field conditions, it is difficult to accurately and conveniently verify the effectiveness of TAMCA calibration through MI and MGT. Instead, the relative position between the AC magnetic source that generates magnetic anomaly field and the TAMCA is easier to obtain. Furthermore, the constrained Euler localization results of the AC magnetic source are both related to its magnetic vector and MGT [25]. Therefore, it is feasible to verify the effectiveness of calibrating the TAMCA using the constrained Euler localization of AC magnetic sources.

4. Calibration Algorithm Testing

4.1. Numerical Simulations

In numerical simulation experiments of calibration, L_x and L_y are all 0.5 m, the horizontal component H and vertical component Z of the geomagnetic field are 25.56 μT and 48.65 μT, and the magnetic inclination angle I and magnetic declination angle D are 62°18′ and −9°55′, respectively. The noises of TAMs in the TAMCA are all zero-mean Gaussian random process, whose STD is denoted by σ_{VM}. The noise of scalar magnetometer measuring geo-MI is also a zero-mean Gaussian random process, whose STD is denoted by σ_{SM}. The error parameters of the TAMCA are shown in Table 1.

Two methods of rotating the TAMCA were used to simulate the generation of geomagnetic field data. One is to uniformly rotate around three axes of the TAMCA in turn, and the other is to randomly rotate around three axes of the TAMCA with a Gaussian distribution of Euler rotation angles. The simulation results of the relative calibration error obtained by the two methods are almost identical, so we adopted the random rotation in the simulation experiments. The parameters $\eta_{\delta S_{xi}}$, $\eta_{\delta S_{yi}}$, $\eta_{\delta S_{zi}}$, η_{θ_i}, η_{φ_i}, η_{ψ_i}, $\eta_{B_{0xi}}$, $\eta_{B_{0yi}}$, $\eta_{B_{0zi}}$, η_{α_j}, η_{β_j}, and η_{γ_j} represent the relative errors of δS_{xi}, δS_{yi}, δS_{zi}, θ_i, φ_i, ψ_i, B_{0xi}, B_{0yi}, B_{0zi}, α_j, β_j, and γ_j, respectively. All numerical simulations use 50 Monte Carlo experiments to represent the noise distribution.

Table 1. Error parameters of the TAMCA in numerical simulations.

Error Parameters	Unit	1st TAM	2nd TAM	3rd TAM	4th TAM
δS_x		0.03	0.02	−0.02	−0.05
δS_y		0.02	−0.04	0.01	0.03
δS_z		−0.02	0.05	0.03	−0.03
B_{0x}		36	−27	38	31
B_{0y}	nT	22	35	−31	−28
B_{0z}		−24	−41	24	−25
θ		0.14	−0.34	0.22	0.45
φ		−0.15	−0.26	−0.18	−0.38
ψ	°	0.21	0.35	0.31	0.32
α		0	−0.028	0.031	0.035
β		0	0.034	−0.028	−0.033
γ		0	−0.026	0.017	0.032

The size N of the calibration dataset is 360 and a geo-MI of 54955.83 nT is provided by the scalar magnetometer. Figure 4 shows the plotted curves between the all relative errors and σ_{VM}, where these relative errors exhibit linear growth trends with the increase in σ_{VM}, and the fluctuation of this growth is evident when σ_{VM} is greater than 1 nT.

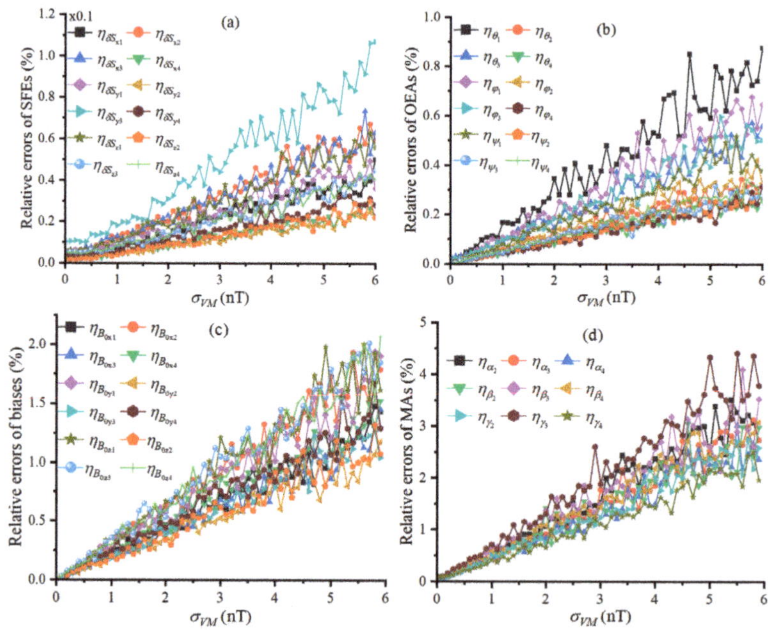

Figure 4. Relative error of calibrating TAMCA's error parameters, varying with σ_{VM}, and Figures (**a**–**d**) are the curves of the relative errors of SFEs, OEAs, biases and MAs under different STDs of TAM's noise.

$\widetilde{e}_{Bi}$ and $\hat{e}_{Bi}$ varying with σ_{VM} are shown in Figures 5a and 5b, and $\widetilde{e}_T$ and $\hat{e}_T$ varying with σ_{VM} are shown in Figures 5c and 5d, respectively. $\widetilde{e}_{Bi}$ and $\widetilde{e}_{Ti}$ are almost unaffected by the increase in σ_{VM}, while $\hat{e}_{Bi}$ and $\hat{e}_{Ti}$ show positive and linear correlation with σ_{VM}. This is because the lossless scalar calibration greatly reduces the measurement error induced by the error parameters of the TAMCA, highlighting the effect of TAMs' noise on the measurement accuracy of the geo-MI and MGT.

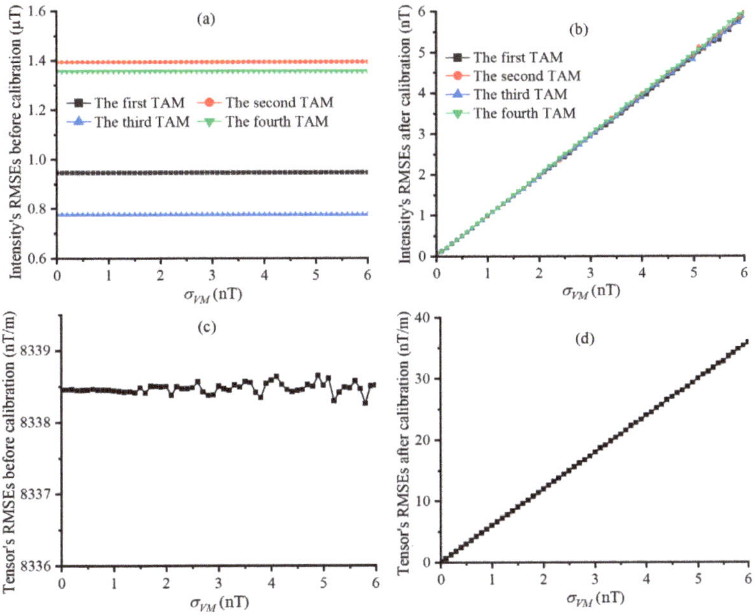

Figure 5. RMSEs of geo-MI and MGT, varying with σ_{VM}, Figures (**a**,**b**) are the curves of the intensity's RMSEs before and after calibration, and Figures (**c**,**d**) are the Tensor's RMSEs before and after calibration under different STDs of TAM's noise.

Similarly, the relationships between the relative errors and σ_{SM} under $\sigma_{VM} = 1\text{nT}$ and $N = 360$ are shown in Figure 6, where the relative errors of SFEs show linear growth trends with the increase in σ_{SM} and the other relative errors remain almost invariant. This indicates that σ_{SM} affects the calibration accuracy of δS_{xi}, δS_{yi}, and δS_{zi}, and it is the reverse for θ_i, φ_i, ψ_i, B_{0xi}, B_{0yi}, B_{0zi}, α_j, β_j, and γ_j.

$\widetilde{e}_{Bi}$, $\hat{e}_{Bi}$, $\widetilde{e}_T$, and $\hat{e}_T$ varying with σ_{SM} are shown in Figure 7, respectively. Geo-MI and MGT measured by the TAMCA before calibration are not affected by σ_{SM}, and it is because the measured values of the geo-MI and MGT are calculated from the TAMs' measurement data, unrelated to σ_{SM}. The comparison of Figure 7a with Figure 7b demonstrates that the calibrated TAMCA significantly reduces the measurement errors of the geo-MI, which tends to increase linearly with the increase in σ_{SM}, because it will lead to the increase in the measured reference geo-MI fluctuation. The comparison between Figure 7c,d proves that the calibrated TAMCA significantly reduces the measurement error of MGT, which is less affected by σ_{SM} because the geo-MI is much greater than σ_{SM}.

Figure 8 shows the relative errors of calibrating TAMCA's error parameters, as N changes when $\sigma_{VM} = 1\text{nT}$ and the scalar magnetometer provides a geo-MI of 54,955.83 nT. The relative errors all show decreasing trends with the increase in N; however, the calibration precision is less affected by N when it exceeds a certain threshold.

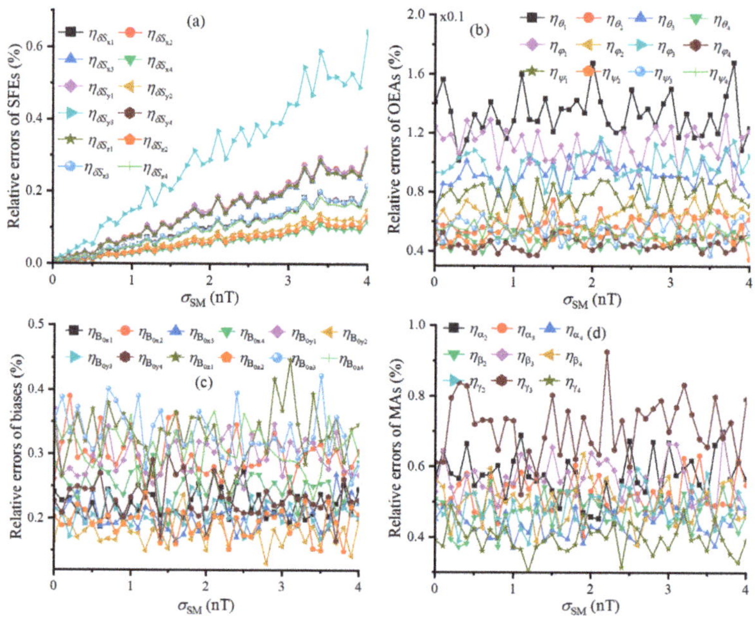

Figure 6. Relative errors of calibrating TAMCA's error parameters, varying with σ_{SM}, and Figures (**a–d**) are the curves of the relative errors of SFEs, OEAs, biases and MAs under different STDs of TAM's noise.

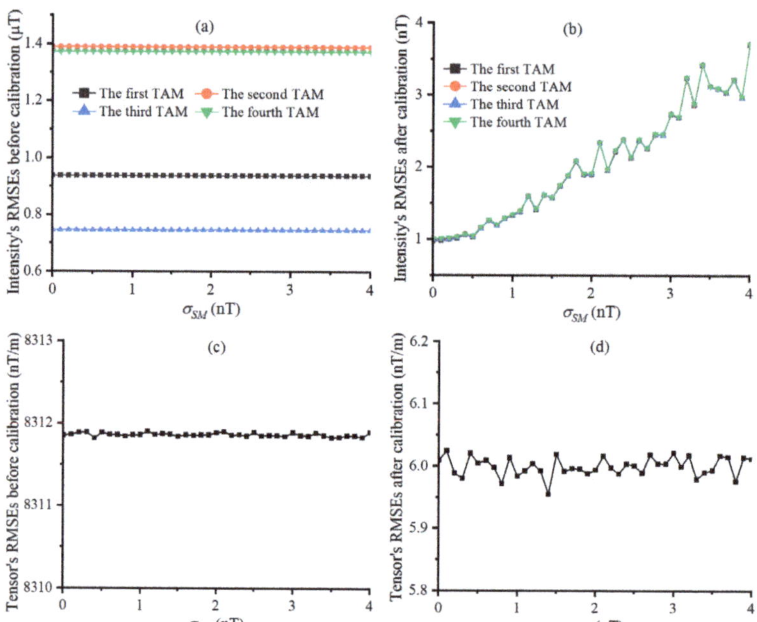

Figure 7. RMSEs of geo-MI and MGT, varying with σ_{SM}, Figures (**a**,**b**) are the curves of the intensity's RMSEs before and after calibration, and Figures (**c**,**d**) are the Tensor's RMSEs before and after calibration under different STDs of scalar magnetometer's noise.

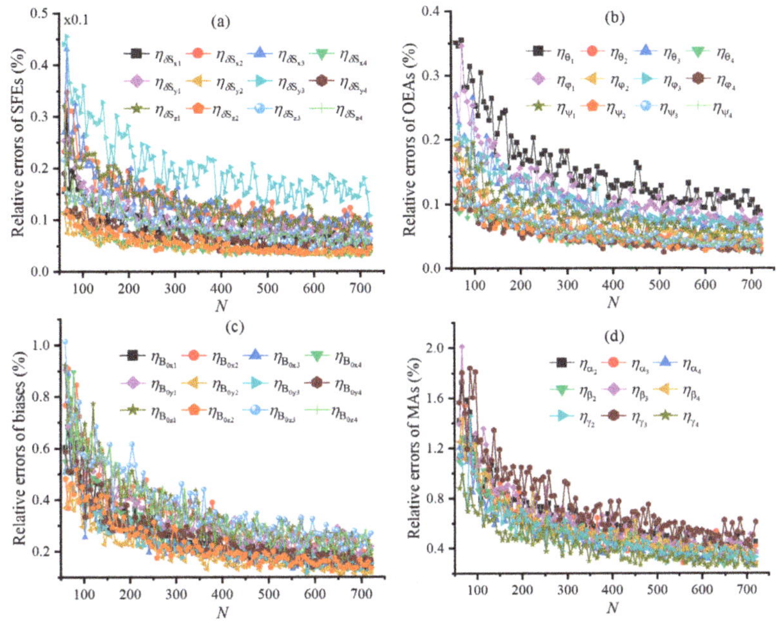

Figure 8. Relative error of calibrating TAMCA's error parameters, varying with N, and Figures (**a–d**) are the curves of the relative errors of SFEs, OEAs, biases and MAs under different sizes of the calibration dataset.

$\widetilde{e}_{Bi}$, $\hat{e}_{Bi}$, $\widetilde{e}_T$, and $\hat{e}_T$ varying with N are shown in Figure 9, respectively. $\widetilde{e}_{Bi}$ and $\widetilde{e}_T$ are all nearly unaffected by N and yet their fluctuation reduces with the increase in N, as shown in Figure 9a,c. The reason is that the STD of the TAMs' noise is much smaller than the measurement error of the geo-MI induced by the error parameters of the TAMCA. The comparison between Figure 9a,b demonstrates that the calibrated TAMCA significantly improves the measurement accuracy of the geo-MI, while it slightly increases and tends to stabilize with the increase in N. The comparison between Figure 9c,d reveals that the calibrated TAMCA significantly improves the measurement accuracy of the geo-MGT, while it slightly decreases and tends to stabilize with the increase in N. Their reasons are that the TAMs' noise with an STD of 1 nT has an observable effect on the geo-MI and geo-MGT measured by the calibrated TAMCA. This also indicates that increasing N alone cannot further improve the calibration accuracy of the TAMCA when N reaches a certain threshold.

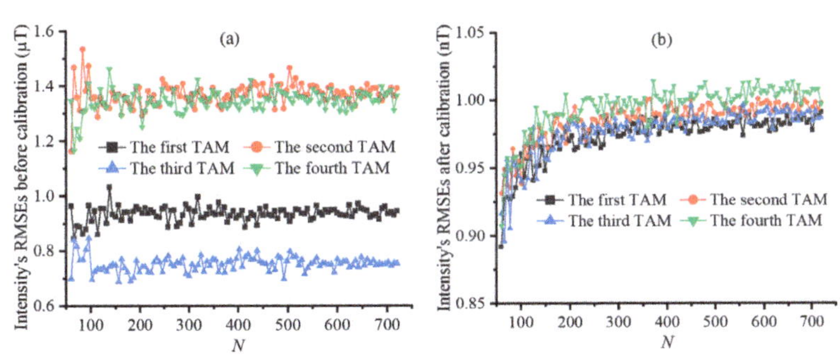

Figure 9. *Cont.*

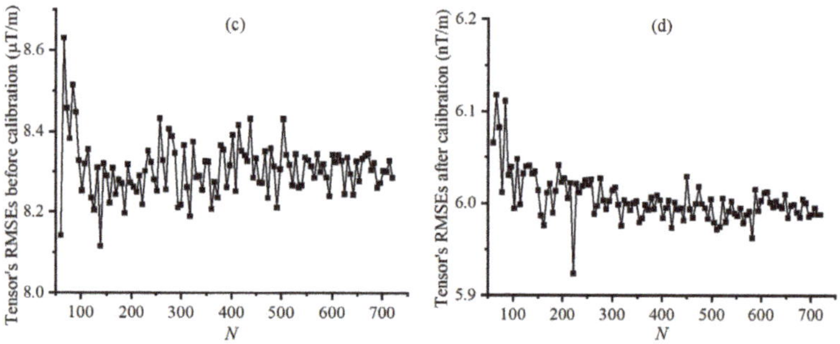

Figure 9. RMSEs of geo-MI and geo-MGT, varying with N, Figures (**a**,**b**) are the curves of the intensity's RMSEs before and after calibration, and Figures (**c**,**d**) are the Tensor's RMSEs before and after calibration under different sizes of the calibration dataset.

4.2. Experimental Tests

The lossless scalar calibration algorithm is tested using a cross-shaped and symmetric array consisting of four FTAMs under an outdoor environment. The fluxgate TAMCA (FTAMCA) with the baseline length of 0.21 m is placed on the table of a tri-axial non-magnetic turntable, as shown in Figure 10a. Figure 10b illustrates the experimental setup of testing the lossless scalar calibration algorithm through the measurement of geo-MI, where the true geo-MI is provided by a proton magnetometer. Figure 11c shows an energized circular coil, which is regarded as an AC magnetic source to test the calibration algorithm through its constrained Euler localization.

Figure 10. The experimental setup of testing the lossless scalar calibration algorithm: (**a**) is the FTAMCA placed on the table of the tri-axial non-magnetic turntable, (**b**) is the proton magnetometer placed on the table of the tri-axial non-magnetic turntable, and (**c**) is the energized circular coil; 1—FTAMCA (model: HSF133-2H3-AAB) with a noise level of no more than 15pTrms/Hz@1Hz and a measurement range of ±100 μT, 2—manual tri-axial non-magnetic turntable (model: HZT-500), 3—proton magnetometer probe (model: JHC-856T) with a resolution better than 0.1 nT, an accuracy better than 0.5 nT, and a measuring range from 0.01 μT to 100 μT, 4—the energized multi-turn circular coil.

The proton magnetometer with a sampling period of 5 s is used for in situ collection of the geo-MI. After smoothing and averaging the collected data, the geo-MI at the tri-axial non-magnetic turntable is 52185.0 nT. The initial angle of the turntable in the x-axis and y-axis is 0°, and the initial angle in the z-axis is −90°. Step 1: the turntable is rotated by one full circle around the x-axis with an interval of 60°. Step 2: the y-axis of the turntable is respectively adjusted to 60° and 120°, and then rotated around the x-axis following the way of the first step. Step 3: the z-axis of the turntable is adjusted to −60°, −30°, and 0° respectively; the rotation operations in both the first step and the second step are repeated. Therefore, the turntable has 72 sets of different orientations, and the corresponding dataset size is also 72. The duration of each set of data is approximately 20 s. The geomagnetic field of the calibration site of our university campus was measured by the proton magnetometer, and the STD of the geomagnetic field fluctuation was 8.26 nT, which could be regarded as magnetic noise or interference in the experiment.

Figure 11 illustrates the absolute errors of the geo-MI measured by four FTAMs of the FTAMCA before and after using the lossless scalar calibration algorithm, where black lines and red lines represent the absolute errors before calibration and after calibration, respectively. The absolute measurement errors of the geo-MI decrease from 339.2 nT, 277.2 nT, 283.2 nT, and 320.5 nT before calibration to 91.9 nT, 94.4 nT, 98.5 nT, and 97.1 nT after calibration, respectively. The results of calibrating the FTAMCA's error parameters are presented in Table 2, and they are all within the range of the error parameters' specification.

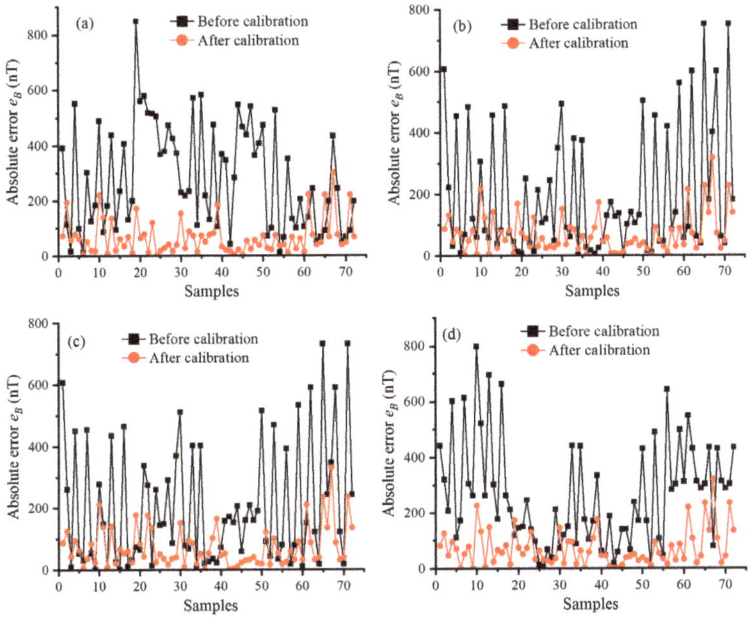

Figure 11. The absolute errors of geo-MI measured by the four FTAMs, Figure (**a–d**) represent respectively the absolute errors of geo-MI measured by the four FTAMs labeled by A, B, C and D before and after calibration.

The power amplifier (model: IT7624) generates a sine voltage signal with peak of 10 V and frequency of 10 Hz, which are used to drive a multi-turn circular coil with inner and outer diameters of 0.82 m and 0.96 m, respectively. The number of turns in the circular coil can be calculated out according to its center magnetic field of 10Gs@3.3A. The initial angles of the turntable in the x, y, and z directions are all 0°. The position coordinates of the coil's center relative to the FTAMCA's center in the x-axis, y-axis, and z-axis are 2.40 m, −1.00 m, and 0.46 m, respectively. By rotating around the z-axis of the turntable with an interval of 30°, we can obtain 12 sets of magnetic field data of the energized circular coil measured by FTAMCA under different orientations. After these magnetic field data are processed using a band-pass digital filter, the constrained Euler localization algorithm is used to locate the circular coil, and the localization results and their errors are obtained as shown in Figures 12a and 12b, respectively, where , , , and represent the results in the x-axis, y-axis, z-axis, and r-axis before calibration, and , , , and represent the results in the x-axis, y-axis, z-axis, and r-axis after calibration. The RMSEs of the localization in the x-axis, y-axis, z-axis, and r-axis are 0.2350 m, 0.3717 m, 0.0800 m, and 0.2924 m before calibration, and they are 0.2620 m, 0.2418 m, 0.0431 m, and 0.2516 m after calibration, respectively. Overall, the lossless scalar calibration of the FTAMCA improves the constrained Euler localization precision of the energized circular coil.

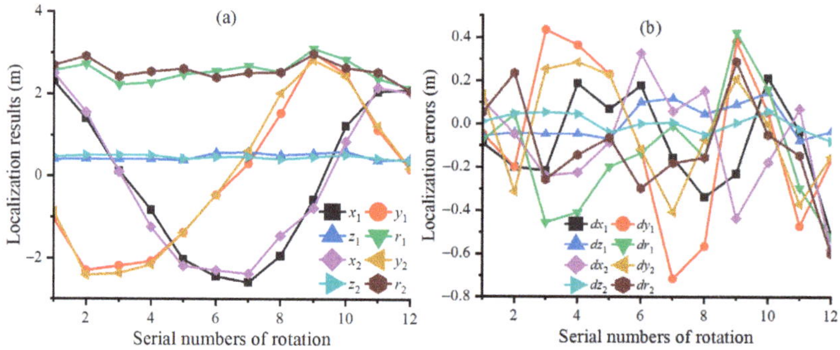

Figure 12. Localization results and their errors of the energized circular coil; the subscripts 1 and 2 denote, respectively, the values before and after calibrating the FTAMCA; (**a**,**b**) represent the localization results of the energized circular coil and the localization errors.

Table 2. The results of calibrating the FTAMCA's error parameters.

Error Parameter	Unit	1st FTAM	2nd FTAM	3rd FTAM	4th FTAM
δS_x		−0.0103	0.0062	0.0083	0.0023
δS_y		0.0076	−0.0129	−0.0124	0.0140
δS_z		0.0174	0.0038	−0.0004	−0.0143
B_{0x}		11.8467	−6.0092	−13.6844	−5.3428
B_{0y}	nT	26.1961	15.7185	10.3250	14.6912
B_{0z}		−17.8786	10.6258	29.7575	2.9363
θ		0.0387	0.0174	0.0253	0.0165
φ		0.0646	0.0981	0.0861	0.0987
ψ	°	−0.0622	−0.0683	−0.0650	−0.0715
α		0	0.0164	0.0074	0.0127
β		0	0.0118	0.0117	0.0145
γ		0	−0.0093	−0.0155	−0.0080

5. Conclusions

The four TAMs in the TAMCA are symmetrically arranged, with their intersection forming the measurement point of the TAMCA—a configuration that minimizes the principle-based measurement errors in the MGT compared to alternative designs. The paper proposes a lossless scalar calibration algorithm free from mathematical approximations, capable of tracking the reference MI fluctuations while precisely calibrating the SFEs, OEAs, and biases intrinsic to the four TAMs and the MAs between the TAMs. Simulation results of the relative errors calibrating the error parameters of the TAMCA demonstrate the calibration algorithm's good performance under various measurement noise STDs and calibration dataset sizes, which can be reasonably explained. An experimental scheme based on constrained Euler localization is designed to validate the effectiveness of the lossless scalar calibration algorithm for TAMCAs. Validation experiments of calibrating the FTAMCA were conducted using a proton magnetometer and an energized circular coil, and the results demonstrated that the proposed algorithm significantly enhanced the measurement accuracy of MIs and improved the precision in determining the coil's position coordinates relative to the FTAMCA. This underscores the calibration algorithm's capability to mitigate systematic errors of multi-sensor magnetic measurement devices.

Author Contributions: Funding acquisition and writing—original draft preparation, L.W.; methodology and writing—review and editing, Y.H.; experimental validation, X.C. All authors have read and agreed to the published version of the manuscript.

Funding: This research was funded by the National Nature Science Foundation of China under Grant Nos 61004130, the Natural Science Foundation of Heilongjiang under Grant Nos LH2020E072 and LH2023F023, the Stable Supporting Fund of Acoustic Science and Technology Laboratory under Grant Nos JCKYS2019604SSJS002, the Fundamental Research Funds for the Central Universities.

Institutional Review Board Statement: Not applicable.

Informed Consent Statement: Not applicable.

Data Availability Statement: The data that support the findings of this study are available on request from the corresponding author.

Acknowledgments: The authors would like to thank anonymous reviewers for their valuable comments and suggestions.

Conflicts of Interest: The authors declare no conflict of interest.

References

1. Ji, S.; Zhang, H.; Wang, Y.; Rong, L.; Shi, Y.; Chen, Y. Three-dimensional inversion of full magnetic gradient tensor data based on hybrid regularization method. *Geophys. Prospect.* **2019**, *67*, 226–261. [CrossRef]
2. Clark, D.A. New methods for interpretation of magnetic vector and gradient tensor data I: Eigenvector analysis and the normalised source strength. *Explor. Geophys.* **2012**, *43*, 267–282. [CrossRef]
3. Wang, Y.; Rong, L.; Qiu, L.; Lukyanenko, D.V.; Yagola, A.G. Magnetic susceptibility inversion method with full tensor gradient data using low-temperature SQUIDs. *Pet. Sci.* **2019**, *16*, 794–807. [CrossRef]
4. Heath, P.; Heinson, G.; Greenhalgh, S. Some comments on potential field tensor data. *Explor. Geophys.* **2003**, *34*, 57–62. [CrossRef]
5. Liu, K.; Sui, Y.; Cheng, H.; Wang, Z.; Clark, D.A. Magnetic Dipole Moment Determination Using Near-Field Magnetic Gradient Tensor Data. *IEEE Geosci. Remote Sens. Lett.* **2021**, *18*, 1169–1173. [CrossRef]
6. Pang, H.; Pan, M.; Wan, C.; Chen, J.; Zhu, X.; Luo, F. Integrated compensation of magnetometer array magnetic distortion field and improvement of magnetic object localization. *IEEE Trans. Geosci. Remote Sens.* **2014**, *52*, 5670–5676. [CrossRef]
7. Pang, H.; Luo, S.; Zhang, Q.; Li, J.; Chen, D.; Pan, M.; Luo, F. Calibration of a fluxgate magnetometer array and its application in magnetic object localization. *Meas. Sci. Technol.* **2013**, *24*, 075102. [CrossRef]
8. Hong, J.H.; Kang, D.; Kim, I.-J. A Unified Method for Robust Self-Calibration of 3-D Field Sensor Arrays. *IEEE Trans. Instrum. Meas.* **2021**, *70*, 1007211. [CrossRef]
9. Chen, W.; Zhang, H.; Zhang, X.; Lyu, W.; Li, X.; Sang, S. A correction method of magnetic gradient tensor system to improve magnet localization accuracy. *Sens. Actuators A* **2024**, *369*, 115194. [CrossRef]
10. Wang, C.; Qu, X.; Zhang, X.; Zhu, W.; Fang, G. A Fast Calibration Method for Magnetometer Array and the Application of Ferromagnetic Target Localization. *IEEE Trans. Instrum. Meas.* **2017**, *66*, 1743–1750. [CrossRef]
11. Huang, Y.; Wu, L.; Wang, Y.; Xu, Y. Scalar calibration of total instrument errors of tri-axial magnetometer using constrained optimization independent of magnetic field intensity. *IEEE Sens. J.* **2024**, *24*, 14352–14360. [CrossRef]
12. Keene, M.N.; Hubphrey, K.P.; Horton, T.J. Actively shielded, adaptively balanced SQUID gradiometer system for operation aboard moving platforms. *IEEE Trans. Appl. Supercond.* **2005**, *15*, 761–764.
13. Sui, Y.; Miao, H.; Wang, Y.; Luan, H.; Lin, J. Correction of a towed airborne fluxgate magnetic tensor gradiometer. *IEEE Geosci. Remote Sens. Lett.* **2016**, *13*, 1837–1841.
14. Sui, Y.; Liu, S.; Zhou, Z.; Wang, Y.; Cheng, D. Invariant Calibration of Magnetic Tensor Gradiometers. *IEEE Magn. Lett.* **2017**, *8*, 650105. [CrossRef]
15. Ding, X.; Li, Z.; Li, Y.; Liu, H. Calibration of magnetic gradient tensor system with differential evolution algorithm. *Chin. J. Geophys.* **2022**, *65*, 4930–4943. (In Chinese)

16. Liu, S.; Sui, Y.; Cheng, H.; Zhang, Z.; Clark, D. Accuracy evaluation of calibrating magnetic tensor gradiometer from total-field gradient measurements in aeromagnetic survey. *IEEE Trans. Instrum. Meas.* **2021**, *70*, 1007715. [CrossRef]
17. Schiffler, M.; Queitsch, M.; Stolz, R.; Chwala, A.; Krech, W.; Meyer, H.G.; Kukowski, N. Calibration of SQUID vector magnetometers in full tensor gradiometry systems. *Geophys. J. Int.* **2014**, *198*, 954–964. [CrossRef]
18. Lv, J.; Yu, Z.; Huang, J.; Zhou, J. The Compensation Method of Vehicle Magnetic Interference for the Magnetic Gradiometer. *Adv. Math. Phys.* **2013**, *2013*, 523164.
19. Tilbrook, D.L. Rotating magnetic tensor gradiometry and a superconducting implementation. *Supercond. Sci. Technol.* **2009**, *22*, 075002.
20. Yin, G.; Zhang, Y.; Fan, H.; Zhang, G.; Ren, G. Linear calibration method of magnetic gradient tensor systems. *Measurement* **2014**, *56*, 8–18.
21. Li, Q.; Li, Z.; Zhang, Y.; Yin, G. Artificial vector calibration method for differencing magnetic gradient tensor systems. *Sensors* **2018**, *18*, 361. [CrossRef] [PubMed]
22. Jia, W.; Lin, C.; Chen, C.; Zhai, G. Method of rotating-calculating magnetic gradient tensor. *J. Natl. Univ. Def. Technol.* **2018**, *40*, 183–188. (In Chinese)
23. Sunderland, A.; Ju, L.; Blair, D.G.; McRae, W.; Veryaskin, A.V. Optimizing a direct string magnetic gradiometer for geophysical exploration. *Rev. Sci. Instrum.* **2009**, *80*, 104705. [PubMed]
24. Huang, Y.; Hao, Y. Effect of three no-co-point axes in a magnetometer on underwater geomagnetic localization precision. *J. Huazhong Univ. Sci. Technol. (Nat. Sci. Ed.)* **2010**, *38*, 74–78. (In Chinese)
25. Wu, L.; Huang, Y.; Feng, D. Experimental investigation on the Euler-type location and orientation algorithm using magnetic field and its gradient tensor. *J. Magn. Magn. Mater.* **2022**, *546*, 168745.

sensors

MDPI

Article

An Improved Adaptive Monte Carlo Localization Algorithm Integrated with a Virtual Motion Model

Cili Zuo [1], Demin Xie [1], Lianghong Wu [1,2,*], Xiaolong Tang [1] and Hongqiang Zhang [1]

[1] School of Information and Electrical Engineering, Hunan University of Science and Technology, Xiangtan 411201, China; zuocl@hnust.edu.cn (C.Z.); 19967632798@163.com (X.T.); hongqiangzhang@hnust.edu.cn (H.Z.)

[2] School of Electrical and Engineering Hunan Industry Polytechnic, Changsha 410208, China

* Correspondence: lhwu@hnust.edu.cn

Abstract: Regarding the issue of high dependency on odometry in the adaptive Monte Carlo localization (AMCL) algorithm, an improved AMCL algorithm based on the normal distributions transform (NDT) and extended Kalman filter (EKF) is proposed. A virtual motion model is introduced into the AMCL framework to enable pose updates even when the robot has not moved. NDT is used for point cloud matching to estimate virtual displacement and calculate virtual control quantities, which are then fed into the motion model to predict and update particle states when the robot has not moved. Additionally, to avoid the negative impacts of encoder errors and wheel slippage on motion state estimation, the EKF algorithm integrates information from the wheel odometer and inertial measurement unit to estimate the robot's displacement, thereby improving localization accuracy and stability. The performance of the proposed algorithm was experimentally validated in both simulated and real environments and compared with other localization algorithms. Experimental results show that the proposed algorithm can effectively improve localization speed during the cold start phase and enhances localization accuracy and stability throughout the localization process. The proposed method is a potential method for improving the performance of mobile robot localization.

Keywords: mobile robot localization; adaptive Monte Carlo localization; normal distributions transform; extended Kalman filter

1. Introduction

With the rapid development of society, robots are playing an increasingly significant role in human production systems and daily life [1,2]. Among them, mobile robots, as one of the most widely used types of intelligent robots, are extensively employed in logistics, industrial applications, and other fields [3–5]. The process by which a mobile robot determines its pose within an environment based on sensor input is referred to as the localization problem [6]. Localization accuracy is a critical factor for achieving precise robot control [7]. For outdoor autonomous mobile robots, the Global Navigation Satellite System (GNSS) is typically used as the primary localization method, supplemented by local sensor systems. However, when satellite signals are obstructed by objects, localization errors can exceed several meters [8]. For indoor mobile robots, satellite signals are severely blocked, necessitating alternative localization methods. To address this, researchers have developed technologies such as infrared [9], Bluetooth [10], Wi-Fi [11], Ultra-Wide Band (UWB) [12], and pseudo-satellites [13]. Nevertheless, these indoor localization methods all have limitations. For example, infrared localization achieves high precision but suffers from weak penetration capabilities and short transmission distances [14]. Bluetooth has a limited range and relatively high costs [15], while Wi-Fi is constrained by coverage limitations, energy consumption, and susceptibility to interference [16]. Furthermore, these methods require the deployment of base stations or devices in the environment, and their signals are easily obstructed, degrading localization performance. In comparison, robot autonomous positioning is less constrained by infrastructure and has lower deployment costs, which has garnered extensive research [17].

Mobile robots can estimate their position and posture using motion state sensors carried by the robots. These sensors include motor encoders, Inertial Measurement Units (IMUs), etc., which capture the wheel speed, as well as the robot's acceleration, angular velocity, and other motion state information. By performing kinematic calculations, the robot's position and attitude are updated [18,19]. Although this method is cost-effective and does not rely on external devices, it has poor robustness, large cumulative errors, and generally only enables relative positioning [20]. Simultaneous Localization and Mapping (SLAM) technology allows robots to perceive their surroundings using laser radars or cameras and use the environmental information to draw maps. By referencing the maps, robots can autonomously localize within the environment, reducing the impact of cumulative errors [21]. Environmental information and map information can be matched using the grid method. By dividing the pose space into grids, sensor data are matched with existing grid maps using histogram filtering to update pose information [22]. However, in practical implementation, the performance of grid localization is often limited by the required grid granularity. Coarse grids, due to their high degree of discretization, may result in information loss, negatively impacting the filtering process and affecting localization accuracy and stability. Conversely, fine grids significantly increase computational load, leading to a considerable reduction in algorithm speed. To overcome the drawbacks of the above localization methods, researchers have proposed probabilistic localization methods based on the Markov model [23]. These methods divide the environment into grids and, by combining the robot's motion state model and observation model, use simple motion and observation models to reduce computational complexity, achieving robust robot pose estimation.

Monte Carlo localization (MCL) is a typical probability-based localization method that predicts the robot's movement displacement based on odometry data and updates the robot's pose estimation using sensed environment information and map data [24]. The MCL algorithm can efficiently solve pose tracking and global localization, but it has an issue with the "robot kidnapping" problem. To address this issue, adaptive Monte Carlo localization (AMCL) introduces random particles into the particle set under the assumption that the robot has a small probability of being "kidnapping". These random particles introduce stochasticity into the model, allowing the robot to recover from the kidnapped state. Even without "kidnapping", random particles enhance system robustness [25]. Currently, AMCL is a widely used and researched mobile robot localization method. Zhang L et al. treated robot pose as an unknown variable, used sensor-derived environmental data as observations, computed the conditional probability of poses given observations, and optimized this probability to achieve localization [26]. Hatem I et al. proposed generalized Monte Carlo localization (GMCL), which integrates three particle filters into AMCL to reduce pose tracking errors and improve success rates in global localization and kidnapped robot scenarios [27]. However, existing MCL research primarily focuses on improving particle sampling efficiency and localization success rates. The particle state update in MCL-based robot localization algorithms relies on the robot's movement displacement, meaning the robot must move to update its pose. When the movement displacement is small, the pose update efficiency is low. Furthermore, during the cold start phase, prolonged continuous movement is required to obtain a relatively accurate initial pose, leading to slow localization speed during the cold start phase.

In this paper, a new AMCL framework is proposed by introducing a virtual motion model. By using Normal Distributions Transform (NDT)-based point cloud matching, we estimate the deviation between the current pose and the actual pose, calculate the virtual displacement and virtual control quantities, and update the particle states based on the virtual control quantities. This allows for quick localization, even when the robot has not moved or has only moved a small distance, improving the localization speed during the cold start phase. Additionally, the EKF algorithm [28] is used to integrate IMU and encoder measurements of the robot's motion information as the robot odometry motion model. This addresses the issue of odometry model deviations caused by encoder errors and wheel slippage, effectively improving localization stability and accuracy.

2. AMCL Algorithm

Monte Carlo localization (MCL) is applicable to both local and global localization problems. It updates the pose by incorporating appropriate probabilistic motion and perception models into the particle filter algorithm. A particle filter is a robust, non-parametric method used to address problems with multimodal belief distributions. This technique originates from Bayesian filtering and builds the posterior distribution through a resampling process. Particles with higher importance weights are more likely to be retained, as higher weight values increase the probability of a particle being selected. The belief, or the estimate of the robot's current pose, is a probability density function. MCL uses a set of M particles $\chi_t = [\vec{x}_t^{[1]}, \vec{x}_t^{[2]}, \cdots \vec{x}_t^{[M]}]$ to represent the belief $\text{bel}(\vec{x}_t)$. Each particle is defined as a hypothetical pose $\vec{x}_t = [x_t^r, y_t^r, \theta_t^r]^T$ of the robot, representing the robot's position and yaw angle information in the reference coordinate system at time t. The weight of a particle represents the confidence level of the robot pose that the particle indicates. The initial confidence $\text{bel}(\vec{x}_0)$ is obtained by randomly generating M particles from a prior distribution $p(\vec{x}_0)$, while assigning the same importance factor M^{-1} to each particle. As shown in Equations (1)–(4), the next state of all particles is predictively updated through the motion model. Then, through the measurement model, weights are calculated for the updated particles. Finally, particles with low weights are discarded, and new particles are randomly generated around the existing particles for repeated iteration. During the particle discarding process, all particles near the correct pose may accidentally be discarded. When the number of particles is small and they are distributed over a large range, the probability of discarding the correct particles increases. This situation can lead to failure in global localization or the kidnapped robot problem, and the system may not be able to recover from global localization. To address this issue, random particles can be added to the particle set for mitigation.

$$\overline{bel}(\vec{x}_t) = \int p(\vec{x}_t \middle| u_t, \vec{x}_{t-1}) bel(\vec{x}_{t-1}) d\vec{x}_{t-1} \tag{1}$$

In the equation, u_t represents the control input of the robot at time t and $\overline{bel}(\vec{x}_t)$ represents the prior belief at time t, which is a prediction of the robot's pose at time t after it has moved under the control u_t at time t, based on the posterior belief that $bel(\vec{x}_{t-1})$ at time $t-1$. $bel(\vec{x}_{t-1})$ represents the posterior belief that at time $t-1$. $p(\vec{x}_t \middle| u_t, \vec{x}_{t-1})$ is the motion model.

$$bel(\vec{x}_{t-1}) = \eta p(z_t \middle| \vec{x}_t) \overline{bel}(\vec{x}_{t-1}) \tag{2}$$

In the equation, η is the normalization constant, whose value is chosen in a way that the $bel(\vec{x}_{t-1})$ sums up to one. $p(z_t \middle| \vec{x}_t)$ is a correction model used to update robot pose.

The model will generate a measurement probability $p(z_t|\vec{x}_t^{[n]})$ for each particle $\vec{x}_t^{[n]}$. z_t represents exteroceptive measurements.

MCL adds particles by monitoring the probability $p(z_t|z_{t-1}, u_t, m)$ of sensor measurements and comparing it with the average measurement probability. The average value can be noted as

$$\frac{1}{M}\sum_{n=1}^{M}\omega_t^{[n]} \approx p(z_t|z_{t-1}, u_t, m) \tag{3}$$

In the equation, m is the map; $\omega_t^{[n]}$ as the weights of particles represents random approximation of this probability.

$\omega_{avg,t}$ is the average weight of all particles at time t, which can be expressed as

$$\omega_{avg,t} = \omega_{avg,t-1} + \frac{1}{M}\sum_{n=1}^{M}\omega_t^{[n]} \tag{4}$$

As an improved algorithm of Monte Carlo localization, AMCL introduces random particles mathematically by assuming the robot is "kidnapped" with a small probability. The injection of these random particles creates a degree of random states in the motion model, and their presence can enhance the robustness of the system even when the robot is not actually kidnapped [25].

The AMCL algorithm determines if the robot is "kidnapped" based on the long-term estimated weight $\omega_{slow,t}$ and the short-term estimated weight $\omega_{fast,t}$ at time t. If $\omega_{slow,t}$ is greater than $\omega_{fast,t}$, it indicates that the robot is "kidnapped". By adding random particles during resampling, the robot can recover from being "kidnapped", which is expressed as follows.

$$\begin{cases} \omega_{slow,t} = \omega_{slow,t-1} + \alpha_{slow}(\omega_{avg,t} - \omega_{slow,t-1}) \\ \omega_{fast,t} = \omega_{fast,t-1} + \alpha_{fast}(\omega_{avg,t} - \omega_{fast,t-1}) \end{cases} \tag{5}$$

In the equation, $\omega_{avg,t}$ represents the average weight of all particles. The parameters α_{slow} and α_{fast} represent the decay rates of the exponential filters for estimating long-term and short-term averages, respectively, and $0 \leq \alpha_{slow} \leq \alpha_{fast}$ is defined.

Additionally, AMCL uses Kullback–Leibler distance (KLD) [24] for adaptive adjustment of the number of particles during the resampling phase, which is expressed as follows:

$$M_\chi = \frac{k-1}{2\varepsilon}\left[1 - \frac{2}{9(k-1)} + \sqrt{\frac{2}{9(k-1)}}z_{1-\delta}\right]^3 \tag{6}$$

In the equation, ε represents the error between the true distribution and the estimated distribution, $z_{1-\delta}$ denotes the $1-\delta$ quantile on the standard normal distribution, and k represents the non-empty bins of the histogram. The upper limit of particles, M_χ, has a linear relationship with the non-empty bins k.

In the initial phase of global localization, all bins are non-empty, and the value of k increases with each new resampling, further leading to an increase in the upper limit of particles. After some time, once global localization is completed, the particles converge near the true pose, and the k value decreases; then, M_χ is reduced. This dynamic adjustment of the number of particles improves the computational efficiency of the algorithm. The AMCL algorithm framework is shown in Figure 1. At the start of the algorithm, the particle population is initialized, and the motion model is used to predictively update the population particles, representing the change in the robot's pose after movement due to control input. Then, the measurement model is used to calculate the weights of the new population particles; higher weight values indicate that the particles are closer to the robot's true pose, while particles with low weights will be removed from the population. Finally, new particles are introduced into the population through resampling, and the motion model is used to predictively update the new population particles, continuing the cycle.

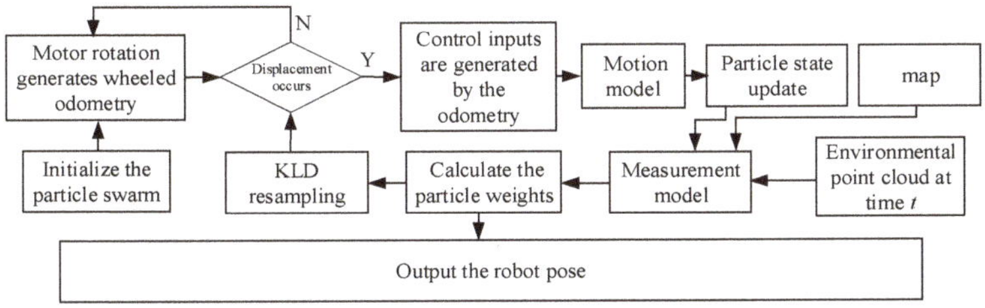

Figure 1. AMCL algorithm framework.

However, the AMCL algorithm has the following issues:

1. During the robot's localization process, the robot must be displaced through motion control for the AMCL algorithm to use the motion model for pose predictive updating.
2. Wheel slippage and drift can cause deviations in the motion information collected by the motor encoders, resulting in odometry measurement errors. Such errors can affect the predictive accuracy of the robot's motion in the AMCL algorithm, leading to a decrease in localization accuracy.

3. Improved AMCL Algorithm

To address the issue with the AMCL algorithm where particle states can only be updated through robot displacement via motion control, a Normal Distribution Transform (NDT) [29] algorithm is introduced to optimize the AMCL algorithm. The EKF algorithm [30] is used to integrate the motion information of the robot measured by the IMU and encoders as the robot's odometry motion model, to address the issue of deviations in the odometry model caused by encoder errors and wheel slippage. The framework of the improved algorithm is shown in Figure 2. First, particles are initialized to distribute across the map space. The EKF is used to integrate IMU information and motor rotation data to generate fused odometry information. When the robot moves, the particle states are updated using the real displacement information. When the robot's displacement is minimal, the NDT algorithm is used to generate virtual displacement, and the particle states are predictively updated based on this virtual displacement. The measurement model then calculates the particle weights, selecting particles that are close to the true pose. Finally, new particles are injected through KLD, and the above filtering process is repeated.

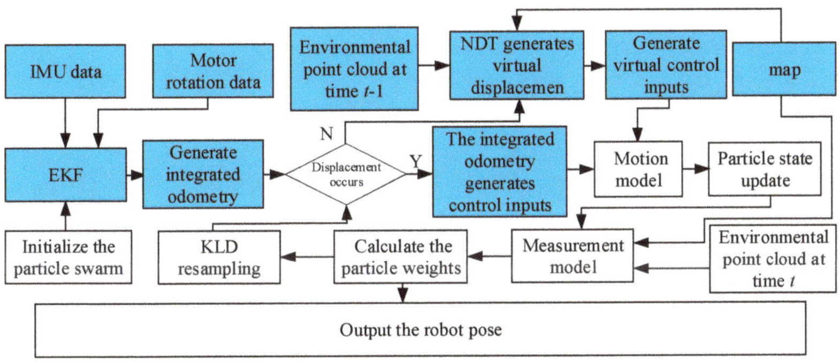

Figure 2. Framework of the improved AMCL algorithm.

3.1. Virtual Motion Model Based on NDT

As a probabilistic point cloud matching method, the NDT point cloud matching algorithm maps point cloud data to a smoothed surface, represented as a set of local probability density functions (PDF). By optimizing the objective function, it iteratively obtains the relative pose difference between two frames of point clouds. In this paper, the environmental map is converted into a reference point cloud. The NDT algorithm is used to obtain the relative pose difference between the reference point cloud and the laser point cloud. This relative pose difference generates virtual control $\hat{u}_t$ and virtual displacement, further completing the particle state update step. The NDT generation $\hat{u}_t$ process is as follows.

Step 1: Convert the map into a point cloud and divide it into multiple grid cells, calculating the PDF for each cell.

$$p(\vec{s}) = \frac{1}{(2\pi)^{D/2}\sqrt{|\Sigma|}} \exp\left(-\frac{(\vec{s}-\vec{\mu})^{\mathrm{T}}\Sigma^{-1}(\vec{s}-\vec{\mu})}{2}\right) \tag{7}$$

In the equation, $\vec{\mu}$ and Σ represent the mean vector and covariance matrix of the points on the grid cell within the cell, as shown in Equations (8) and (9).

$$\vec{\mu} = \frac{1}{m}\sum_{k=1}^{m} \vec{s}_k \tag{8}$$

$$\Sigma = \frac{1}{m-1}\sum_{k=1}^{m} (\vec{s}_k - \vec{\mu})(\vec{s}_k - \vec{\mu})^{\mathrm{T}} \tag{9}$$

In the equation, $\vec{s}_{k=1,\ldots,m}$ represents the position of the points within the grid cell.

Step 2: Set pose $\vec{pos} = [\hat{x}_t, \hat{y}_t, \hat{\theta}_t]^{\mathrm{T}}$ to map the scan point cloud $Z_{t-1} = \left\{\vec{Z}^{1}_{t-1}, \ldots, \vec{Z}^{n}_{t-1}\right\}$ at time $t-1$ to the map point cloud. The normal distribution of each grid cell after transformation is calculated using the following equation:

$$p(T(\vec{pos}, \vec{Z}^{k}_{t-1})) = -\exp\left(-\frac{(T(\vec{pos}, \vec{Z}^{k}_{t-1}) - \vec{\mu}_k)^{\mathrm{T}}\Sigma_k^{-1}(T(\vec{pos}, \vec{Z}^{k}_{t-1}) - \vec{\mu}_k)}{2}\right) \tag{10}$$

In the equation, the transformation T represents the relationship of transformation between the point cloud scanned by the LiDAR at time $t-1$ and the point cloud as converted to the map. Σ_k^{-1} represents the inverse of the covariance matrix for the k-th grid cell in Equation (9). $\vec{Z}^{k}_{t-1}$ represents the vector representation of the point cloud within the k-th grid cell scanned by the LiDAR, centered on the robot.

Step 3: Use Equation (11) to calculate the probability density for each grid cell, which serves as the objective function for determining the optimal pose $\vec{pos}$.

$$s(\vec{pos}) = -\sum_{k=1}^{n} \exp\left(-\frac{(T(\vec{pos}, \vec{Z}^{k}_{t-1}) - \vec{\mu}_k)^{\mathrm{T}}\Sigma_k^{-1}(T(\vec{pos}, \vec{Z}^{k}_{t-1}) - \vec{\mu}_k)}{2}\right) \tag{11}$$

Step 4: Obtain the objective function through Step 3 and use the Newton iteration algorithm to calculate the optimal pose $\vec{pos}$.

Step 5: Calculate the virtual control input $\hat{u}_t$ from the current robot pose $\vec{x}_{t-1} = [x_{t-1}, y_{t-1}, \theta_{t-1}]^{\mathrm{T}}$ to the optimal pose $\vec{pos}$.

$$\begin{cases} \hat{\Delta}x = \hat{x}_t - x_{t-1} \\ \hat{\Delta}y = \hat{y}_t - y_{t-1} \\ \hat{\Delta}\theta = \hat{\theta}_t - \theta_{t-1} \end{cases} \tag{12}$$

$$\hat{u}_t = [\frac{\hat{\Delta}x}{\Delta t}, \frac{\hat{\Delta}y}{\Delta t}, \frac{\hat{\Delta}\theta}{\Delta t}]^{\mathrm{T}} \tag{13}$$

Generate $\hat{u}_t$ as the virtual control for when the robot has not moved and input it into the motion model to optimize and adjust the particles of the AMCL algorithm. Then, use the improved motion model to predictively update the optimized particles.

3.2. Odometry Motion Model Based on EKF Integration

During the robot's movement, wheel speed information is typically collected through motor encoders to generate the odometry model. However, in practical applications, the odometry model may exhibit deviations due to encoder errors and wheel slippage. This inaccuracy can affect the results of the predictive update of the particles, causing them to deviate from the expected theoretical outcome, thereby reducing localization accuracy. Therefore, this paper uses the EKF algorithm to integrate the robot motion information measured by the IMU and the encoder as the robot odometry motion model. The specific steps are as follows.

Step 1: The robot pose prediction at time t is completed through Equation (14).

$$\hat{\vec{x}}_t = f(\vec{x}_{t-1}, u_{odo,t}) = \begin{bmatrix} x_{t-1} + \Delta x_{odo,t}\cos\theta_{t-1} - \Delta y_{odo,t}\sin\theta_{t-1} \\ y_{t-1} + \Delta y_{odo,t}\sin\theta_{t-1} + \Delta y_{odo,t}\cos\theta_{t-1} \\ \theta_{t-1} + \Delta\theta_{odo,t} \end{bmatrix} \tag{14}$$

In this equation, $x_{t-1}, y_{t-1}, \theta_{t-1}$ represents the position coordinates and yaw angle of the robot at time $t-1$, while $u_{odo,t} = [\Delta x_{odo,t}, \Delta y_{odo,t}, \Delta\theta_{odo,t}]^{\mathrm{T}}$ represents the incremental displacement of the x, y axes and the change in yaw angle measured by wheeled odometer.

Step 2: The term F_t denotes the Jacobian matrix of the function f at the $t-1$ instance. The derivative of function f with respect to $\vec{x}_{t-1}$ at $(\vec{x}_{t-1}, u_{odo,t})$ is given by

$$F_t = \frac{\partial f}{\partial \vec{x}} = \begin{bmatrix} 1 & 0 & -\Delta x_{odo,t}\sin\theta_{t-1} - \Delta y_{odo,t}\cos\theta_{t-1} \\ 0 & 1 & \Delta x_{odo,t}\cos\theta_{t-1} - \Delta y_{odo,t}\sin\theta_{t-1} \\ 0 & 0 & 1 \end{bmatrix} \tag{15}$$

Step 3: Wheeled odometer noise error can be expressed as

$$Q_t = \begin{bmatrix} \sigma^2\Delta x_{odo} & 0 & 0 \\ 0 & \sigma^2\Delta y_{odo} & 0 \\ 0 & 0 & \sigma^2\Delta\theta_{odo} \end{bmatrix} \tag{16}$$

In this equation $\sigma\Delta x_{odo}, \sigma\Delta y_{odo}, \sigma\Delta\theta_{odo}$, respectively, represent the standard deviation of the x and y-axis displacement noise of the wheeled odometer and the standard deviation of the yaw angle noise.

Step 4: A prediction error covariance matrix $\hat{P}_t$, used to represent the uncertainty error of predicted pose values, can be described as

$$\hat{P}_t = F_t P_{t-1} F_t^{\mathrm{T}} + Q_t \tag{17}$$

Step 5: The observation model is shown in the following formula, using IMU data as observational input.

$$h(\vec{x}_t, \vec{x}_{t-1}) = \begin{bmatrix} (\vec{x}_t - \vec{x}_{t-1})\cos\theta_{t-1} + (\vec{y}_t - \vec{y}_{t-1})\sin\theta_{t-1} \\ -(\vec{x}_t - \vec{x}_{t-1})\sin\theta_{t-1} + (\vec{y}_t - \vec{y}_{t-1})\cos\theta_{t-1} \\ \theta_t - \theta_{t-1} \end{bmatrix} \tag{18}$$

Step 6: The observation Jacobian matrix can be given by the following formula.

$$H_t = \frac{\partial h}{\partial \vec{x}} = \begin{bmatrix} \cos\theta_{t-1} & \sin\theta_{t-1} & (\vec{x}_t - \vec{x}_{t-1})\cos\theta_{t-1} + (\vec{y}_t - \vec{y}_{t-1})\sin\theta_{t-1} \\ -\sin\theta_{t-1} & \cos\theta_{t-1} & -(\vec{x}_t - \vec{x}_{t-1})\sin\theta_{t-1} + (\vec{y}_t - \vec{y}_{t-1})\cos\theta_{t-1} \\ 0 & 0 & 1 \end{bmatrix} \tag{19}$$

Step 7: The Kalman gain is calculated through Equation (20).

$$K_t = \hat{P}_t H_t^{\mathrm{T}} (H\hat{P}_t H_t^{\mathrm{T}} + R_t)^{-1} \tag{20}$$

In the equation, R_t represents the covariance of the IMU observation noise, given by the following formula:

$$R_t = \begin{bmatrix} \sigma^2_{\Delta x_{imu}} & 0 & 0 \\ 0 & \sigma^2_{\Delta y_{imu}} & 0 \\ 0 & 0 & \sigma^2_{\Delta \theta_{imu}} \end{bmatrix} \tag{21}$$

In the equation, $\sigma\Delta x_{imu}, \sigma\Delta y_{imu}, \sigma\Delta\theta_{imu}$, respectively, represent the standard deviation of the x and y-axis displacement noise of the IMU and the standard deviation of the yaw angle noise.

Step 8: The robot pose update at time t is completed through Equation (22).

$$\vec{x}_t = \hat{\vec{x}}_t + K_t(\vec{z}_t - h(\hat{\vec{x}}_t, \vec{x}_{t-1})) \tag{22}$$

In the equation, $\vec{z}_t = [\Delta x_{imu}, \Delta y_{imu}, \Delta\theta_{imu}]^{\mathrm{T}}$ represents the displacement increment of the robot's coordinate system, calculated by the IMU and the change in the yaw angle measured by the IMU.

Step 9: The estimated error covariance matrix at time t is calculated through Equation (23).

$$P_t = (I - K_t H_t)\hat{P}_t \tag{23}$$

Step 10: Calculate the actual control input u_t of the robot pose from time $t-1$ to time t.

$$\begin{cases} \Delta x = x_t - x_{t-1} \\ \Delta y = y_t - y_{t-1} \\ \Delta\theta = \theta_t - \theta_{t-1} \end{cases} \tag{24}$$

$$u_t = [\frac{\Delta x}{\Delta t}, \frac{\Delta y}{\Delta t}, \frac{\Delta\theta}{\Delta t}]^{\mathrm{T}} \tag{25}$$

In the traditional AMCL algorithm, a particle filter cycle can only be performed when the scale of the robot's movement reaches a certain threshold. If the scale of movement is small, or the robot undergoes small-scale pose changes, the algorithm does not update the robot's pose. The improved localization algorithm incorporates an NDT process and optimizes the motion model to enhance the speed of particle convergence and positioning accuracy while improving system stability. This allows for pose correction and updates of the robot using scan information, even when the robot is not moving, thereby increasing localization accuracy and reducing localization time. The pseudocode for the NDT-based improved AMCL algorithm is as in Algorithm 1.

Algorithm 1: Improved AMCL Algorithm Based on NDT

Input: The particle set from the previous time step X_{t-1};
Current lidar scan point cloud Z_t;
Environmental map m;
Previous time-step lidar scan point cloud Z_{t-1};
Total number of particles M;

Output: Particle set at the current time step X_t;

Static $\omega_{slow}, \omega_{fast}$;
$\overline{X}_t = X_t = \varnothing$;
if The robot has not undergone any displacement **then**
Calculate $\hat{u}_t$ according to Equation (13);
for i = 1 to M **do**
Update $x_t^{[i]}$ using $\hat{u}_t$ according to Equation (1);
Calculate the weight $\omega_t^{[i]}$ of each particle based on the map and lidar information;
end for
else
for i = 1 to M **do**
Update $x_t^{[i]}$ using u_t according to Equation (1);
Calculate the weight $\omega_t^{[i]}$ of each particle based on the map and lidar information;
end for
$\overline{X}_t = \overline{X}_t + < x_t^{[i]}, \omega_t^{[i]} >$;
Calculate $\omega_{slow}, \omega_{fast}$;
for i = 1 to M **do**
with probability $\max(0, 0, 1.0 - \omega_{fast}/\omega_{slow})$ **do**
add random pose to X_t;
else
draw $x_t^{[i]}$ from $\overline{X}_t$ with probability $\propto \omega_t^{[i]}$;
$X_t = X_t + < x_t^{[i]}, \omega_t^{[i]} >$;
end with
end for
return X_t;

4. Results

The traditional AMCL algorithm for localization consists of two stages. In the first stage, the robot's odometry trajectory information is used to obtain the pose transformation information of the robot between times t and $t-1$. By calculating the trajectory between these two moments, the control inputs needed to transition the robot's pose from $t-1$ to t are determined. These control inputs are then fed into the motion model to predict the robot's pose. In the second stage, point cloud data from the LIDAR scan at time t is used as posterior data to update the measurement of the first stage and pose prediction, completing the robot's pose estimation after multiple iterative cycles. However, in the first stage, if the robot does not move, there is no trajectory information for computation, causing the localization process to remain in the first stage and increasing the localization duration. The improved AMCL algorithm proposed in this paper addresses localization when the robot's initial pose is unknown. In the first stage, when the robot is stationary, point cloud data from the LIDAR scan at time $t-1$ is used as prior data. The NDT algorithm generates virtual movement data for the robot, creating virtual control inputs for the time interval between t and $t-1$, which are fed into the motion model to predict the robot's pose, advancing the localization to the second stage. When the robot moves, cumulative errors in the wheel odometry can cause data drift, making the obtained pose transformation information from t to $t-1$ inaccurate. This paper uses an EKF to fuse wheel odometry and IMU data to correct odometry errors. The corrected odometry is used to calculate the trajectory from time $t-1$ to t, generating the control inputs needed for the robot's pose transition from $t-1$ to t, which are then used in the motion model for pose prediction, allowing the localization process to reach the second stage.

The primary improvements of the proposed AMCL algorithm focus on localization time and accuracy when the robot's initial pose is unknown. To validate this, three sets of experiments were designed in both simulated and real environments, comparing the localization performance under the same conditions to the traditional AMCL, GMCL, and Cartographer algorithms.

4.1. Simulation Validation

This paper compares the localization time and accuracy of the traditional AMCL algorithm, the improved AMCL algorithm, the GMCL algorithm, and the cartographer algorithm in ROS Noetic on Ubuntu 20.04. A simulation experiment is designed in the Gazebo 11.x software, adding models to simulate real-world physical and collision properties. The simulation environment is shown in Figure 3a, which includes items such as tables and bookshelves to simulate the real world. The world coordinate system of the simulation environment is indicated by a red arrow, with the arrow's starting point representing the origin of the world coordinate system. The measurement range of the lidar is set to 6 m, with a resolution of 0.01 m and a scan frequency of 5.5 Hz. The update frequency of the IMU is 100 Hz. An environmental map is constructed using the Gmapping algorithm, as shown in Figure 3b. The map coordinate system is also indicated by a red arrow, with the arrow's starting point as the origin of the map coordinate system. In the map, the white areas represent obstacle-free regions, black indicates obstacles, and gray represents unknown areas.

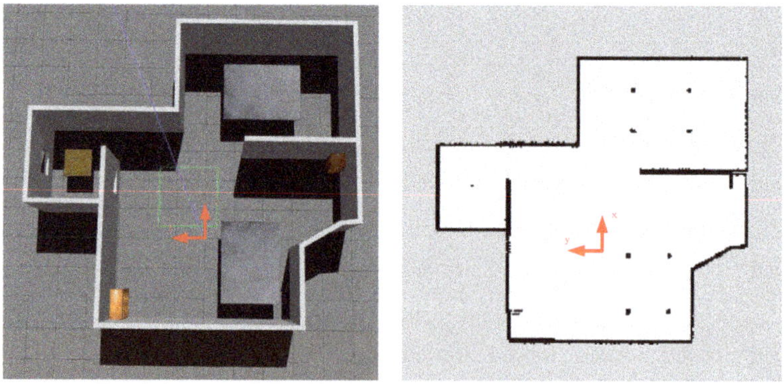

(**a**) Simulation environment (**b**) Environmental map

Figure 3. Simulation environment and map.

(1) Cold start localization time consumption test

Start localization was set with the robot in different poses in the simulated world. The robot's maximum linear velocity is 0.2 m/s, and the maximum angular velocity is

0.2 rad/s. The robot begins random motion from the initial position. Localization is considered successful if the robot's positioning error is within 0.3 m and the localization yaw angle is within 15 degrees. The robot starts localization from different poses in the environment. Timing begins when the robot starts localization and ends when the robot completes localization. A total of 15 experiments are conducted, and the robot's localization time is shown in Figure 4. The AMCL algorithm has an average localization time of 20.77 s, the GMCL algorithm has an average localization time of 31.88 s, the cartographer algorithm has an average localization time of 19.95 s, and the improved AMCL algorithm has an average localization time of 13.68 s.

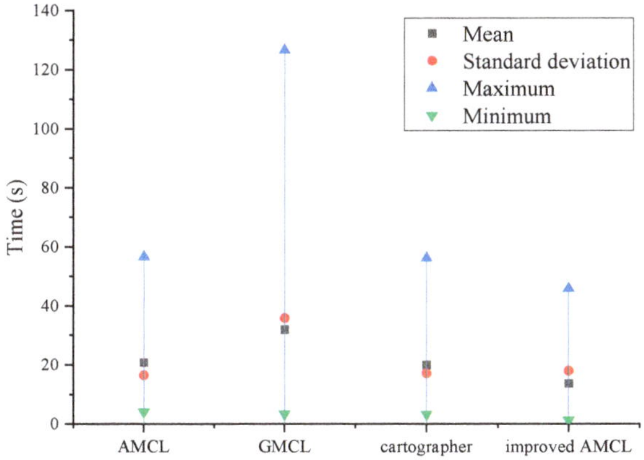

Figure 4. Comparison of robot localization time.

(2) Pose accuracy test

To further verify the localization effectiveness of the improved algorithm for the robot in dynamic scenarios, the robot could move randomly in the environment. The robot's localization information and the actual information in the simulated environment were read at regular intervals. Data were read 15 times, and the displacement of the robot's localized position relative to the actual position is shown in Figure 5. The coordinate point (0,0) indicates that the robot's localized position coincides with the actual position, with an error of 0. The further a point deviates from the coordinate point (0,0) in the figure, the larger the localization error. The gray square points represent the positional offsets produced by the traditional AMCL algorithm, the red circular points indicate the offsets produced by the GMCL algorithm, the blue upward-pointing triangles represent the offsets produced by the cartographer algorithm, and the green downward-pointing triangles represent the offsets produced by the improved AMCL algorithm.

The yaw angle errors generated during the robot's movement localization are shown in Figure 6. In the figure, points closer to a deviation of 0 indicate smaller localization errors. Taking the counterclockwise direction as the positive direction of the yaw angle, points above 0 indicate a positive deviation in the yaw angle of localization, while points below 0 indicate a negative deviation. The gray square points represent the yaw angle deviation produced by the traditional AMCL algorithm, the red circular points represent the deviation produced by the GMCL algorithm, the blue upward-pointing triangles represent the deviation produced by the cartographer algorithm, and the green downward-pointing triangles represent the deviation produced by the improved AMCL algorithm.

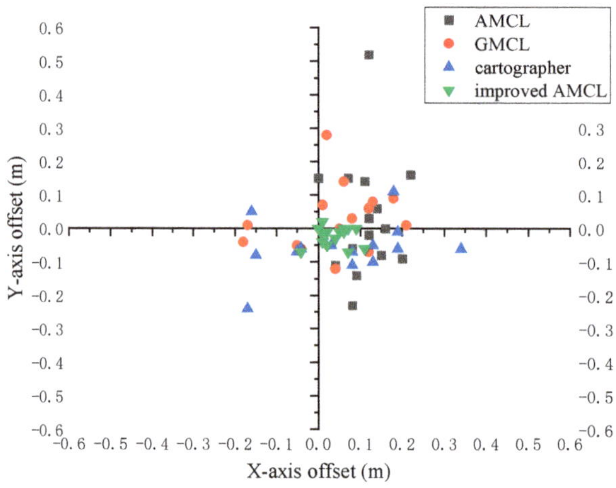

Figure 5. The offset of the localized position relative to the true position.

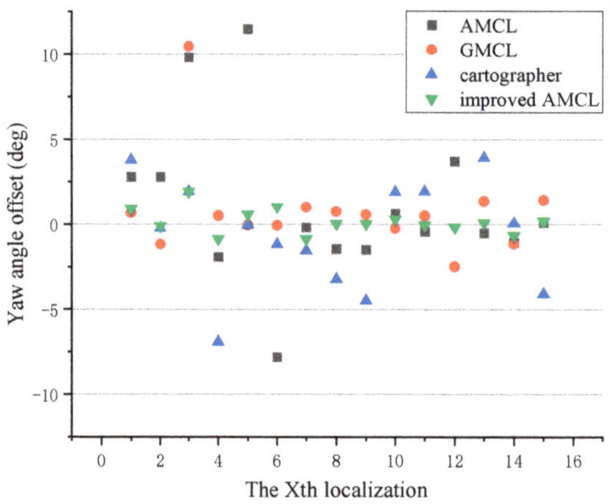

Figure 6. The offset of the localized yaw angle relative to the true yaw angle.

The data statistics of Figures 5 and 6 yield the results shown in Figures 7 and 8, with the position error and yaw angle error given by Equation (26).

$$\begin{aligned} \Delta &= \sqrt{\Delta x^2 + \Delta y^2} \\ \delta &= |\theta - \theta_*| \end{aligned} \quad (26)$$

In Equation (26), $\Delta x, \Delta y$ represents the displacement of the robot's localized position relative to the actual position using different algorithms and θ, θ_* represents the yaw angle of the robot localized by different algorithms and the actual yaw angle of the robot, respectively.

Through the data statistics of Figures 7 and 8, it was observed that both the position error and the yaw angle error of the improved AMCL algorithm were reduced. Specifically, the average position error decreased from 0.12 m to 0.08 m, and the average yaw angle error decreased from 3.08 degrees to 0.51 degrees.

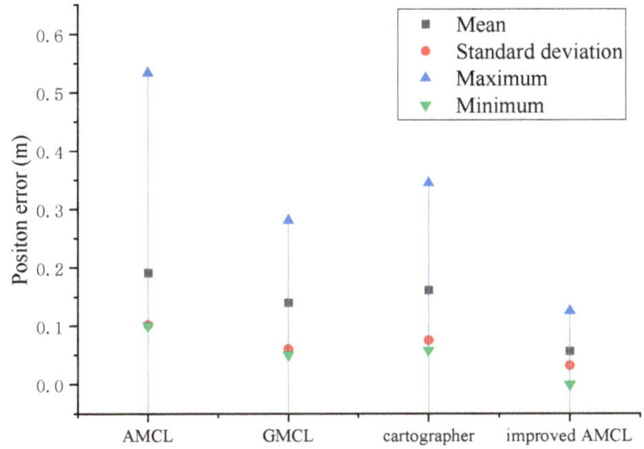

Figure 7. Position error statistics results.

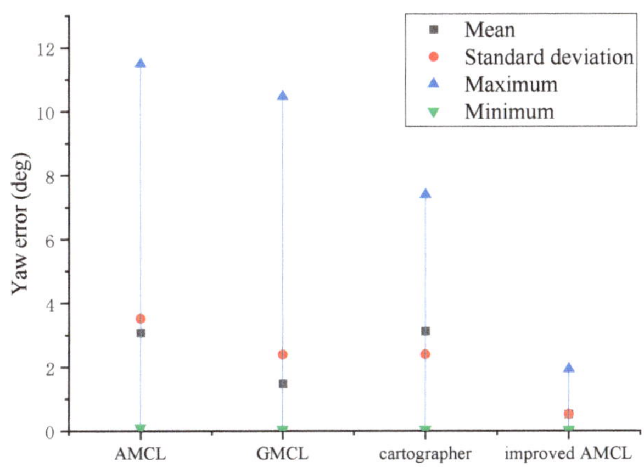

Figure 8. Yaw angle error statistics results.

(3) Preset trajectory localization test

The robot was guided to follow a preset trajectory in the environment, and different algorithms were used to generate corresponding trajectories based on the robot's localization data. The final trajectory curves are shown in Figure 9. The black curve represents the actual trajectory of the robot's movement, with the real trajectory data provided by Gazebo. The red curve represents the motion trajectory generated by the traditional AMCL algorithm based on localization data, the blue curve represents the trajectory generated by the GMCL algorithm, the green curve represents the trajectory generated by the cartographer algorithm, and the purple curve represents the trajectory generated by the improved AMCL algorithm.

By performing data statistics on the localization trajectories of different algorithms shown in Figure 9 and calculating the position error using Equation (26), the statistical results are shown in Figure 10. When the robot follows a specific trajectory, the average position error for the improved AMCL algorithm is 0.024 m, the average deviation for the cartographer algorithm is 0.070 m, the average deviation for the GMCL algorithm is 0.050 m, and the average position error for the traditional AMCL algorithm is 0.064 m.

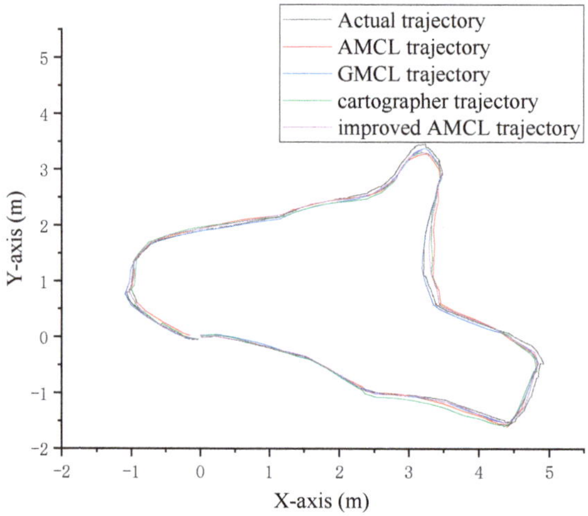

Figure 9. Robot motion trajectory and localization trajectory.

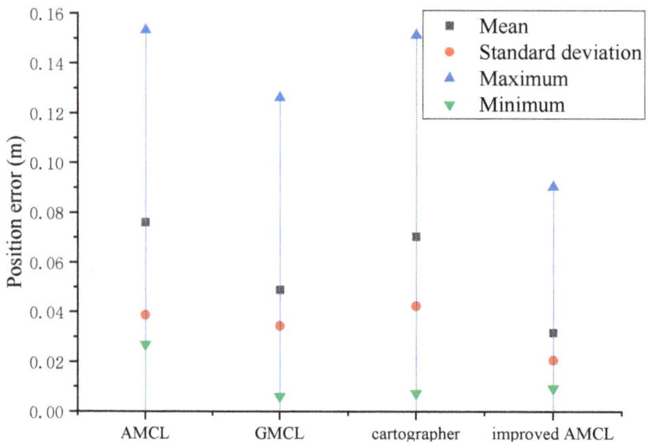

Figure 10. Trajectory error statistics results.

The three experiments described above demonstrate that the improved AMCL algorithm achieves reductions in both localization time and accuracy compared to the traditional AMCL algorithm, the GMCL algorithm, and the cartographer algorithm. Specifically, the minimum localization time has been reduced to 1.29 s, which is 2.81 s less than the original AMCL algorithm, 2.127 s less than the GMCL algorithm, and 1.95 s less than the cartographer algorithm. The average time has been reduced by 7 s, 18 s, and 6 s compared to the original AMCL, GMCL, and cartographer algorithms, respectively, and it has a smaller standard deviation, meaning localization times are more concentrated around the mean. The improved AMCL algorithm reduces the average positional error to below 0.057 m, which is an improvement of 0.133 m, 0.083 m, and 0.103 m compared to the original AMCL's 0.19 m, GMCL's 0.14 m, and cartographer's 0.16 m, respectively. The average yaw angle error is reduced to 0.519 degrees, which is 2.561 degrees less than the 3.08 degrees of the original AMCL algorithm, and the data are more concentrated around the mean. This indicates that in a simulation environment, the improved AMCL algorithm shows enhancements in localization time and pose accuracy.

4.2. Real-World Environment Experiment

To further test the localization performance of the improved algorithm, a robot, as shown in Figure 11, was constructed. The structural diagram below illustrates the distribution of the robot's key sensors and the position of the robot's body coordinate system (base_link) within the robot. The main sensor parameters of the robot are as follows. The LiDAR uses the SLAM-based RPLIDAR A1, which provides high-precision environmental sensing capabilities. The IMU uses the ICM20948, which can provide real-time motion status and directional information. Regarding the motors, the robot uses standard DC motors equipped with TB6112 driver chips and Hall encoders, ensuring stable power output and precise position control.

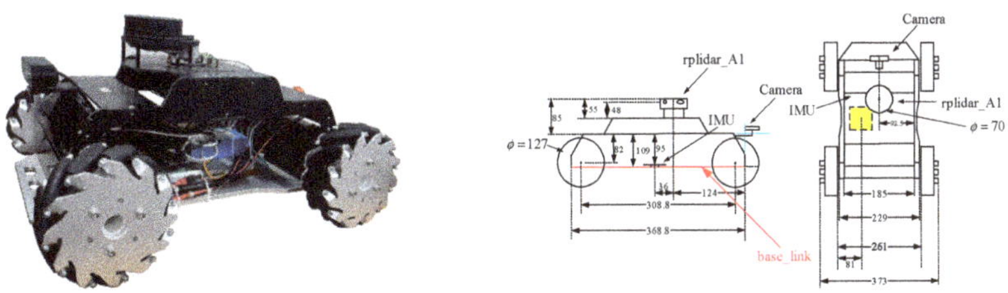

(**a**) Real object image of the robot (**b**) Structural Framework

Figure 11. Photographs and structural diagrams of the robot.

For ease of expression during the localization process, the pose information of the base_link coordinate system will be used as the robot's pose information. The positions of the robot's LiDAR, IMU, and camera will be provided based on their relative position information with respect to the base_link. The point cloud information z_t^k scanned by the LiDAR is transformed into the base_link coordinate system for processing, as shown in Equation (27).

$$\begin{bmatrix} x_{z_t^k} \\ y_{z_t^k} \end{bmatrix} = \begin{bmatrix} x_{base\_link} \\ y_{base\_link} \end{bmatrix} + \begin{bmatrix} \cos\theta & -\sin\theta \\ \sin\theta & \cos\theta \end{bmatrix} \begin{bmatrix} x_{k,rplidar} \\ y_{k,rplidar} \end{bmatrix} + z_t^k \begin{bmatrix} \cos(\theta + \theta_{k,rplidar}) \\ \sin(\theta + \theta_{k,rplidar}) \end{bmatrix} \tag{27}$$

The experimental environment is chosen as shown in Figure 12a, located in an irregular room with an area of approximately 35 square meters. The red coordinate system in the figure is defined as the map coordinate system and is used as the reference coordinate system for the robot's localization. The robot's pose is represented as pose information in the map coordinate system. An environmental map, as shown in Figure 12b, is constructed based on the real environment.

(**a**) Real experimental environment (**b**) map

Figure 12. Real experimental environment and map.

(1) Cold start localization time consumption test

In the experiment, the robot's initial pose was set to be unknown, with a maximum linear velocity of 0.2 m/s and a maximum angular velocity of 0.2 rad/s. The robot started localization from different poses in the environment. Localization was considered complete when the error between the robot's localized position and the actual position was less than 0.3 m, and the error between the localized yaw angle and the actual yaw angle was less than 15 degrees. The robot began localization from different poses in the environment, and timing started when the robot began localization and ended when localization was complete. Fifteen experiments were conducted, and the robot's localization time is shown in Figure 13. The traditional AMCL algorithm has an average localization time of 32.56 s, the GMCL algorithm has an average localization time of 43 s, the cartographer localization algorithm has an average localization time of 34.86 s, and the improved AMCL algorithm has an average localization time of 20.56 s.

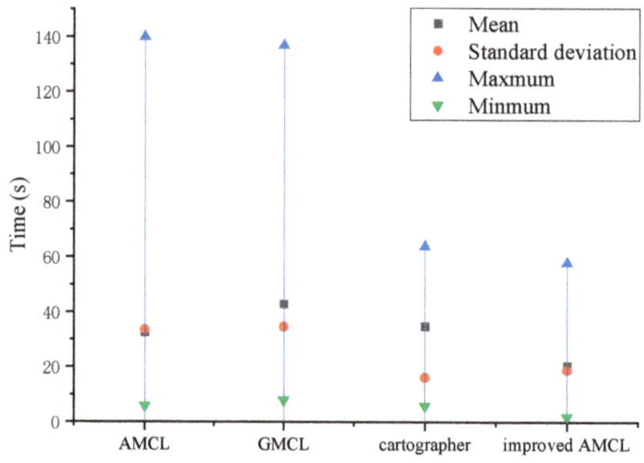

Figure 13. Comparison of robot localization time.

(2) Pose accuracy test

To verify the localization performance of the robot during movement, a mobile localization experiment was designed. The robot randomly moved in the environment, and after moving for a certain period, the robot's current actual pose was measured and compared with the pose information obtained through the localization algorithm. Through fifteen data readings, the displacement of the robot's localized position relative to the actual position is shown in Figure 14. In the figure, the center of the coordinate axis represents the actual position, and the farther away from the center, the greater the localization error. Gray squares represent the displacement when the robot uses the traditional AMCL for localization, red dots represent the displacement when using GMCL, the blue upward-pointing triangles represent the displacement when using cartographer, and the green downward-pointing triangles represent the displacement when using improved AMCL.

The yaw angle errors of different algorithms are shown in Figure 15, with the counterclockwise direction being positive and the clockwise direction being negative. A difference greater than zero indicates that the localized yaw angle is positively deviated from the actual yaw angle, while a difference less than zero indicates a negative deviation. In the figure, gray squares represent the difference between the yaw angle localized by the traditional AMCL algorithm and the robot's actual yaw angle, red dots represent the difference for the GMCL algorithm, blue upward-pointing triangles represent the difference for the cartographer algorithm, and green downward-pointing triangles represent the difference for the improved AMCL algorithm. The closer the difference in the figure is to zero, the more accurate the yaw angle localization of the algorithm.

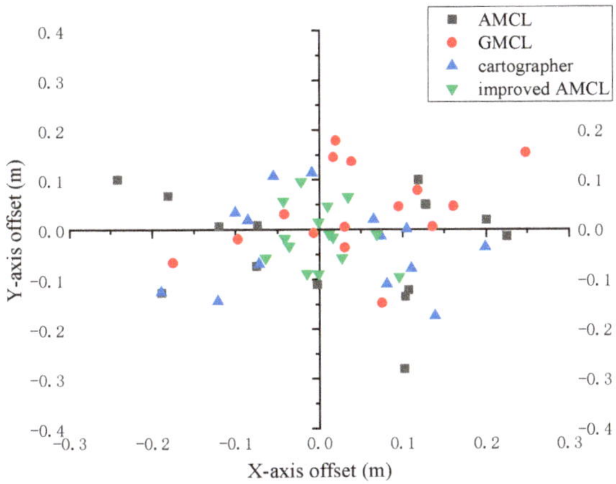

Figure 14. The offset of the localized position relative to the true position.

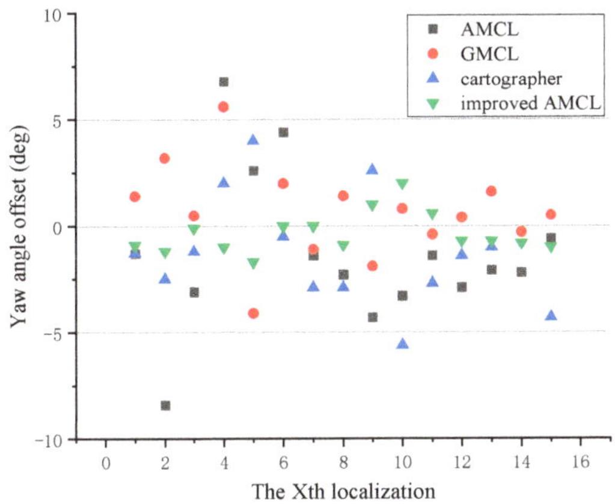

Figure 15. The offset of the localized yaw angle relative to the true yaw angle.

Based on the data statistics from Figures 14 and 15, the results shown in Figures 16 and 17 are obtained, with the position error and yaw angle error given by Equation (17). The improved AMCL algorithm has an average positioning error of 0.06 m and an average yaw angle error of 0.84 degrees. The cartographer algorithm has an average positioning error of 0.13 m and an average yaw angle error of 2.9 degrees. The GMCL algorithm's average positioning error is 0.12 m, with an average yaw angle error of 1.68 degrees. The original AMCL algorithm has an average positioning error of 0.17 m and an average yaw angle error of 3.14 degrees.

(3) Preset trajectory localization test

The robot walks along a preset trajectory in the real environment, and different algorithms generate corresponding trajectories based on the robot's localization data. The final trajectory curves are shown in Figure 18. In the figure, the black curve represents the reference preset trajectory during the robot's movement, the red curve represents the motion trajectory generated by the traditional AMCL algorithm based on localization data, the blue curve represents the motion trajectory generated by the GMCL algorithm, the green curve represents the motion trajectory generated by the cartographer algorithm, and the purple curve represents the motion trajectory generated by the improved AMCL algorithm.

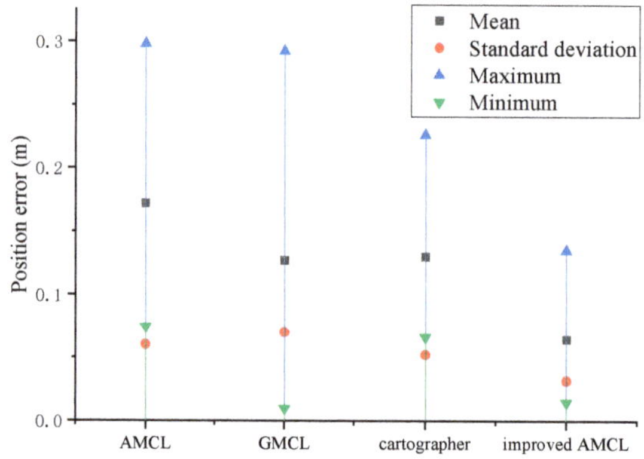

Figure 16. Position error statistics chart.

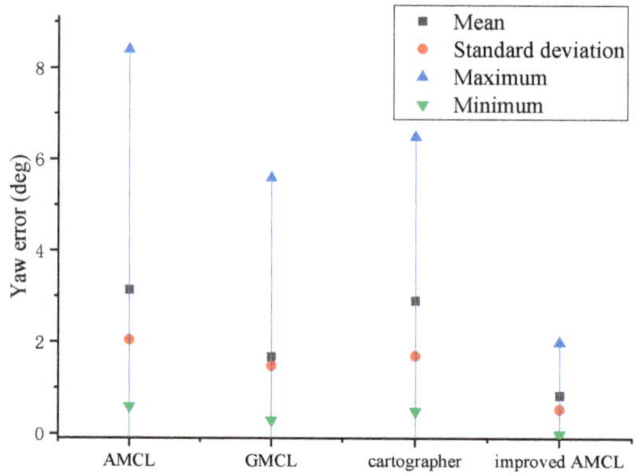

Figure 17. Yaw angle error statistics chart.

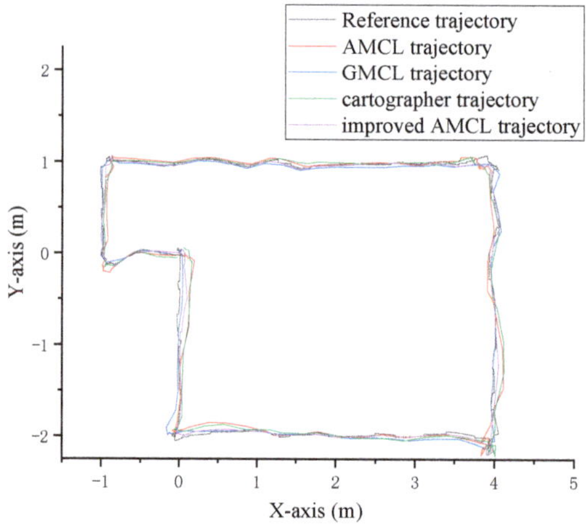

Figure 18. Robot motion trajectory and localization trajectory.

The data statistics of the different localization trajectory algorithms in Figure 18 were analyzed, and the localization position error was calculated using Equation (26). The statistical results are shown in Figure 19. When the robot walks along a specific trajectory, the average localization displacement is 0.031 m for the improved AMCL algorithm, 0.076 m for the traditional AMCL algorithm, 0.049 m for the GMCL algorithm, 0.046 m and for the cartographer algorithm.

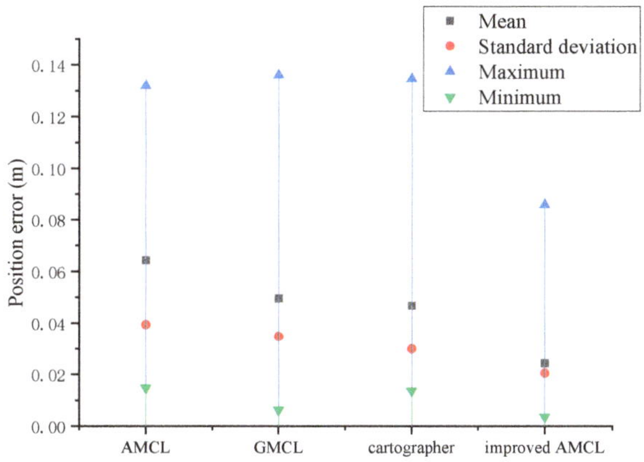

Figure 19. Trajectory error statistics chart.

The three experiments described above demonstrate that the improved AMCL algorithm achieves reductions in both localization time and accuracy compared to the traditional AMCL algorithm, the GMCL algorithm, and the cartographer algorithm. Specifically, the minimum localization time decreased to 2 s, which is 4 s less compared to the traditional AMCL algorithm, 6 s less compared to the GMCL algorithm, and 4 s less compared to cartographer algorithm. The average time decreased by 12 s, 23 s, and 14 s compared to the traditional AMCL, the GMCL algorithm, and cartographer algorithms, respectively, with a smaller standard deviation, indicating that the localization times are more concentrated around the mean value. The improved AMCL algorithm reduces the average positioning error to below 0.06 m, which is 0.11 m, 0.06 m, and 0.07 m less than the original AMCL algorithm's 0.17 m, the GMCL algorithm's 0.12 m, and cartographer's 0.13 m, respectively. The average yaw angle error is reduced to 0.84 degrees, decreasing by 2.3 degrees compared to the original AMCL algorithm's 3.14 degrees, by 0.84 degrees compared to the GMCL algorithm's 1.68 degrees, and by 2.06 degrees compared to cartographer's 2.90 degrees. Additionally, the data are more concentrated around the mean. This indicates that in real environments, the improved AMCL algorithm shows advancements in localization time and pose accuracy.

5. Conclusions

This paper investigates the issues of localization accuracy and efficiency of mobile robots under the ROS system. An improved AMCL algorithm is proposed to address the problem of excessive localization time of the traditional AMCL algorithm when the robot's initial pose is unknown. This algorithm, while inheriting the high robustness of the AMCL algorithm, incorporates NDT and EKF as enhancements. NDT uses the environment map as reference information by converting the environment map into a reference point cloud, and the LiDAR scan data as the source data, transforming the LiDAR scan data into an input point cloud and obtaining the transformation matrix between point clouds using the Newton iteration method. EKF enhances the system's ability to assess changes in the robot's pose by integrating motor encoder data and IMU data, resulting in a more accurate motion model. Experimental results show that in the Gazebo simulation environment, the improved algorithm reduces the average localization time by 7 s and increases the average localization accuracy by 0.133 m when the robot's initial pose is unknown. In a real-world scenario, the improved algorithm reduces the average localization time by 4 s and increases the average localization accuracy by 0.11 m under unknown initial pose conditions.

Author Contributions: Conceptualization, C.Z.; Methodology, C.Z.; Software, D.X.; Validation, D.X. and X.T.; Writing—original draft, C.Z. and D.X.; Writing—review & editing, C.Z., L.W. and H.Z.; Visualization, X.T.; Supervision, L.W. All authors have read and agreed to the published version of the manuscript.

Funding: The research is financially supported by National Key R&D Program of China (2023YFC3011100), the National Natural Science Foundation of China (No. 62373146), the Natural Science Foundation of Hunan Province (No. 2022JJ30265), the Young Talent of Lifting Engineering for Science and Technology in Hunan Province (No. 2022TJ-Q03), the Outstanding Youth Project of Education Department of Hunan Province (No. 22B0476), and the Project Supported by Scientific Research Fund of Hunan Provincial Education Department (23A0382).

Institutional Review Board Statement: Not applicable.

Informed Consent Statement: Not applicable.

Data Availability Statement: Data is unavailable due to privacy restriction.

Conflicts of Interest: The authors declare no conflict of interest.

References

1. Brahmi, B.; Ahmed, T.; El Bojairami, I.; Swapnil, A.A.Z.; Assad-Uz-Zaman, M.; Schultz, K.; McGonigle, E.; Rahman, M.H. Flatness based control of a novel smart exoskeleton robot. *IEEE/ASME Trans. Mechatron.* **2021**, *27*, 974–984. [CrossRef]
2. Pashna, M.; Yusof, R.; Ismail, Z.H.; Namerikawa, T.; Yazdani, S. Autonomous multi-robot tracking system for oil spills on sea surface based on hybrid fuzzy distribution and potential field approach. *Ocean. Eng.* **2020**, *207*, 107238. [CrossRef]
3. Keung, K.L.; Lee, C.K.; Ji, P. Industrial internet of things-driven storage location assignment and order picking in a resource synchronization and sharing-based robotic mobile fulfillment system. *Adv. Eng. Inform.* **2022**, *52*, 101540. [CrossRef]
4. Ren, Y.; Friderikos, V. Interference aware path planning for mobile robots under joint communication and sensing in mmWave networks. *Comput. Commun.* **2024**, *225*, 1–9. [CrossRef]
5. Ding, T.; Zhang, Y.; Wang, Z.; Hu, X. Velocity-based rack storage location assignment for the unidirectional robotic mobile fulfillment system. *Transp. Res. Part E Logist. Transp. Rev.* **2024**, *186*, 103533. [CrossRef]
6. Ullah, I.; Adhikari, D.; Khan, H.; Anwar, M.S.; Ahmad, S.; Bai, X. Mobile robot localization: Current challenges and future prospective. *Comput. Sci. Rev.* **2024**, *53*, 100651. [CrossRef]
7. Yang, X.; Huangfu, S.; Yan, S. Fusion positioning method of UWB/IMU/odometer based on improved UKF. *J. Chin. Inert. Technol.* **2023**, *31*, 462–471.
8. Kuncara, I.A.; Widyotriatmo, A.; Hasan, A.; Nazaruddin, Y.Y. Enhancing accuracy in field mobile robot state estimation with GNSS and encoders. *Measurement* **2024**, *235*, 114903. [CrossRef]
9. Ngamakeur, K.; Yongchareon, S.; Yu, J.; Islam, S. Passive infrared sensor dataset and deep learning models for device-free indoor localization and tracking. *Pervasive Mob. Comput.* **2023**, *88*, 101721. [CrossRef]
10. Yeh, S.C.; Wang, C.H.; Hsieh, C.H.; Chiou, Y.S.; Cheng, T.P. Cost-effective fitting model for indoor positioning systems based on bluetooth low energy. *Sensors* **2022**, *22*, 6007. [CrossRef]
11. Bi, J.; Zhao, M.; Yao, G.; Cao, H.; Feng, Y.; Jiang, H.; Chai, D. PSOSVRPos: WiFi indoor positioning using SVR optimized by PSO. *Expert Syst. Appl.* **2023**, *222*, 119778. [CrossRef]
12. Li, Y.; Gao, Z.; Xu, Q.; Yang, C. Comprehensive evaluations of NLOS and linearization errors on UWB positioning. *Appl. Sci.* **2023**, *13*, 6187. [CrossRef]
13. Zhang, H.; Pan, S. LSOS: An FG Position Method Based on Group Phase Ranging Ambiguity Estimation of BeiDou Pseudolite. *Remote Sens.* **2023**, *15*, 1924. [CrossRef]
14. Brena, R.F.; García-Vázquez, J.P.; Galván-Tejada, C.E.; Muñoz-Rodriguez, D.; Vargas-Rosales, C.; Fangmeyer, J., Jr. Evolution of indoor positioning technologies: A survey. *J. Sens.* **2017**, *2017*, 2630413. [CrossRef]
15. Nguyen DT, A.; Joung, J.; Kang, X. Deep gated recurrent unit-based 3D localization for UWB systems. *IEEE Access* **2021**, *9*, 68798–68813. [CrossRef]
16. Mendoza-Silva, G.M.; Torres-Sospedra, J.; Huerta, J. A meta-review of indoor positioning systems. *Sensors* **2019**, *19*, 4507. [CrossRef]

17. Piao, W.; Zhao, X.; Li, Z. MMGraphSLAM: Autonomous indoor positioning based on millimeter wave GraphSLAM. *Measurement* **2023**, *220*, 113300. [CrossRef]
18. Lee, S.; Song, J.B. Robust Mobile Robot Localization Using Optical Flow Sensors and Encoders. In Proceedings of the IEEE International Conference on Robotics and Automation, Proceedings ICRA'04, New Orleans, LA, USA, 26 April–1 May 2004; IEEE: Piscataway, NJ, USA, 2024; pp. 1039–1044.
19. Huang, L.; Li, H.; Li, W.; Wu, W.; Kang, X. Improvement of Pedestrian Dead Reckoning Algorithm for Indoor Positioning by using Step Length Estimation. *Int. Arch. Photogramm. Remote Sens. Spat. Inf. Sci.* **2022**, *48*, 19–24. [CrossRef]
20. Teng, X.; Shen, Z.; Huang, L.; Li, H.; Li, W. Multi-sensor fusion based wheeled robot research on indoor positioning method. *Results Eng.* **2024**, *22*, 102268. [CrossRef]
21. Li, L.; Schulze, L.; Kalavadia, K. Promising SLAM Methods for Automated Guided Vehicles and Autonomous Mobile Robots. *Procedia Comput. Sci.* **2024**, *232*, 2867–2874. [CrossRef]
22. Thrun, S.; Burgard, W.; Fox, D. *Probabilistic Robotics*; Cao, H.Y.; Tan, Z.; Shi, X.X., Translators; China Machine Press: Beijing, China, 2017; pp. 179–189.
23. Mohseni, A.; Duchaine, V.; Wong, T. Improvement in Monte Carlo localization using information theory and statistical approaches. *Eng. Appl. Artif. Intell.* **2024**, *131*, 107897. [CrossRef]
24. Guan, R.P.; Ristic, B.; Wang, L.; Palmer, J.L. KLD sampling with Gmapping proposal for Monte Carlo localization of mobile robots. *Inf. Fusion* **2019**, *49*, 79–88. [CrossRef]
25. Wang, N.; Wang, J.; Li, L.H. An improved adaptive Monte Carlo localization method for robot. *J. Navig. Position.* **2019**, *7*, 31–37.
26. Zhang, L.; Wu, X.; Gao, R.; Pan, L.; Zhang, Q. A multi-sensor fusion positioning approach for indoor mobile robot using factor graph. *Measurement* **2023**, *216*, 112926. [CrossRef]
27. Hatem, I.; Khalil MA, A. Development of a new technique in ROS for mobile robots localization in known-based 2D environments. *Tishreen Univ. J. Res. Sci. Stud. Eng. Sci. Ser.* **2021**, *43*, 1–19.
28. Friedland, B.; Bernstein, I. Estimation of the State of a Nonlinear Process in the Presence of Nongaussian Noise and Disturbances. *J. Frankl. Inst.* **1966**, *281*, 455–480. [CrossRef]
29. Peter, B. The Normal Distributions Transform: A New Approach to Laser Scan Matching. In Proceedings of the IROS2003, Las Vegas, NV, USA, 27–31 October 2003.
30. Li, Q.; Li, R.; Ji, K.; Dai, W. Kalman Filter and Its Application. In Proceedings of the 2015 8th International Conference on Intelligent Networks and Intelligent Systems (ICINIS), IEEE, Tianjin, China, 1–3 November 2015; pp. 74–77.

sensors

MDPI

Article

Three-Dimensional Extended Target Tracking and Shape Learning Based on Double Fourier Series and Expectation Maximization

Hongge Mao [1,2] and Xiaojun Yang [1,*]

[1] School of Information Engineering, Chang'an University, Xi'an 710064, China; maohongge@nynu.edu.cn
[2] School of Artificial Intelligence and Software Engineering, Nanyang Normal University, Nanyang 473061, China
* Correspondence: xjyang@chd.edu.cn

Abstract

This paper investigates the problem of tracking targets with unknown but fixed 3D star-convex shapes using point cloud measurements. While existing methods typically model shape parameters as random variables evolving according to predefined prior models, this evolution process is often unknown in practice. We propose a particular approach within the Expectation Conditional Maximization (ECM) framework that circumvents this limitation by treating shape-defining quantities as parameters estimated directly via optimization. The objective is the joint estimation of target kinematics, extent, and orientation in 3D space. Specifically, the 3D shape is modeled using a radial function estimated via double Fourier series (DFS) expansion, and orientation is represented using the compact, singularity-free axis-angle method. The ECM algorithm facilitates this joint estimation: an Unscented Kalman Smoother infers kinematics in the E-step, while the M-step estimates DFS shape parameters and rotation angles by minimizing regularized cost functions, promoting robustness and smoothness. The effectiveness of the proposed algorithm is substantiated through two experimental evaluations.

Keywords: target tracking; shape estimation; expectation conditional maximization; double Fourier series; axis-angle

1. Introduction

The accurate perception and tracking of targets in three-dimensional (3D) environments are a fundamental requirement for many modern intelligent systems, including autonomous vehicles, robotic platforms, and advanced surveillance systems [1–4]. While traditional point target tracking algorithms assume that one target only gives rise to one measurement, they are insufficient when dealing with objects possessing significant spatial extent relative to sensor resolution, known as extended target [5,6]. The effective tracking of these extended targets demands algorithms that can estimate not only their kinematic state (position, velocity) but also their physical characteristics, including shape and orientation, within a 3D space.

A substantial body of work considers the extended target tracking (ETT) problem, especially in 2D space [5,6]. Corresponding methods process 2D measurements based on diverse extent models. One common approach employs random matrix to model the target's shape as an ellipse [7,8]. These extent models often assume that the extent state follows a specific probability distribution, such as a Gaussian distribution. However, this underlying distribution is frequently unknow in practice. Addressing this limitation, paper [9] presents an alternative method where extent characteristics, including semi-axis lengths, are treated as parameters for direct estimation. For complex star-convex targets, more sophisticated shape modeling methods, such as B-splines [10] and Gaussian processes (GPs) [11,12], have been developed. Notably, the random hyper-surface model (RHM) is a typical approach for representing arbitrary star-convex shapes within the frequency domain [13,14]. This model defines the target's spatial extent using a Fourier series expansion, with its coefficients jointly estimated alongside the target's kinematics. RHM offers a parameterization particularly well-suited for star-convex extents, effectively balancing representational accuracy with computational complexity. Numerous adaptations of such Fourier-based approaches have subsequently been explored across diverse application settings [15–21].

When 3D surface measurements from devices like multilayer LiDAR sensor or depth cameras are available, it becomes possible to model the target's 3D shape. A few approaches to 3D extended target tracking involved measurement models using basic geometric shapes [22] or restricted shape estimation [23–25]. The methods are suitable for when the detailed shape information is not critical for the application. For instance, where the target's shape is unknown beforehand, random matrix models can represent the target using forms like ellipsoid, or rectangular cuboids. However, the representation power of RM models remains limited to these relatively simple geometric shapes. To address more complex geometries, Gaussian Processes have been developed for 3D space, allowing the modeling of arbitrary shaped star-convex targets [26,27]. While GPs effectively capture complex 3D shapes, their representation (often via a radius function) typically results in a high-dimensional state vector. This high dimensionality poses challenges for filtering algorithms and increases the computational burden. In contrast, the double Fourier series (DFS) shape model proposed in [28] utilizes a shape vector of significantly lower dimensionality compared to GP models, while also enabling the estimation of both shape and pose. It's worth noting that a common characteristic across many of these 3D shape measurement models is the modeling of the shape-characterizing state as a random variable, typically assumed to follow a Gaussian distribution.

Differing from traditional 3D methods that rely on prior shape evolution models, our approach models the parameters defining the extent state directly, treating them as quantities with clear physical meaning. Within the Expectation Conditional Maximization (ECM) framework, we propose an extended target tracking method for the joint estimation of kinematics, extent, and orientation in 3D space. The main contributions are listed as follows.

(1) The unknown but fixed 3D shape is represented via a radial function in spherical coordinates, with a Double Fourier Series (DFS) expansion employed for its modeling. This approach converts the continuous radial function into a parametric representation characterized by a finite set of coefficients.
(2) The axis-angle representation constitutes a compact and singularity-free approach for parameterizing 3D orientation. It describes a rotation by specifying a unit vector, which defines the axis of rotation, and a scalar angle, which denotes the magnitude of rotation about this axis. Furthermore, this method exhibits both geometric intuitiveness and notable computational efficiency.
(3) Joint estimation of the target's kinematic state and shape parameters is achieved via the Expectation Conditional Maximization (ECM) framework. The E-step infers the kinematic state using an unscented Kalman smoother filter, while the M-step estimates shape and rotation parameters by minimizing separate cost functions with added regularization for robustness and smoothness.

The remainder of this paper is organized as follows. Section 2 formulates the problem and presents the system, including the motion model, shape representation, and measurement model. Section 3 derives an ECM-based algorithm for 3D ETT problem. Section 4 presents simulation results to validate the proposed approach across two distinct scenarios. Finally, Section 5 summarizes the main findings and concludes the article.

Notation: Throughout this paper, the superscripts "-1" and "T" represent the inverse and transpose of a matrix, respectively; $\mathcal{N}(\mu, \Sigma)$ represents the Gaussian distribution with mean μ and covariance Σ; $E[\cdot]$ and $E[\cdot|\cdot]$ is mathematical expectation and conditional expectation. The notation $\otimes$ refers to the Kronecker product. I denote the identity matrix with proper dimensions; $|\cdot|$ and $Tr[\cdot]$ denote the determinant and trace of matrices, respectively; $||\cdot||_2$ denotes the Euclidean norm.

2. Problem Formulation

This section first introduces a model for the spatial extent of 3D target, which employs a radial function parameterized via a Double Fourier Series (DFS). The target's orientation is then described using the axis–angle representation, a method characterized by its compactness and singularity-free nature. Following this, two kinematic state evolution models specifically designed for 3D extended target tracking are presented. Finally, the paper details the measurement model that integrates this DFS-based shape representation with the selected axis–angle orientation.

2.1. DFS-Based Shape Model for 3D Target

Advanced sensors, such as depth cameras and LiDAR, can generate detailed three-dimensional (3D) point cloud data, extracting precise shape estimates from such data is crucial for object classification and predicting future behavior. To effectively learn shapes from these point clouds, it is essential to formulate a suitable object extent description method that possesses high representational power for diverse 3D shapes while also being sufficiently compact to support efficient online tracking algorithms.

This section details the modeling of the target shape in spherical coordinates via a radial function $f(\theta, \varphi)$. Such functions relate angular coordinate pairs (θ, φ) to a radial distance in each orientation and are adept at representing star-convex geometries. In our methodology, this radial function is modeled using a Double Fourier Series (DFS) expansion, the specific form of which is given by:

$$\begin{aligned} f(\zeta, \theta, \varphi) &= \zeta_0 + \sum_{n=1}^{N} \sum_{m=1}^{M} [\zeta_{nm}^{cc} \cos(n\theta)\cos(m\phi) + \zeta_{nm}^{ss} \sin(n\theta)\sin(m\phi) \\ &\quad + \zeta_{nm}^{sc} \sin(n\theta)\cos(m\phi) + \zeta_{nm}^{cs} \cos(n\theta)\sin(m\phi)] \\ &= S(\theta, \varphi) \cdot \zeta \end{aligned} \tag{1}$$

where

$$\zeta = [\zeta_0, \zeta_{11}^{cc}, \zeta_{11}^{ss}, \zeta_{11}^{sc}, \zeta_{11}^{cs}, \zeta_{12}^{cc}, \zeta_{12}^{ss}, \zeta_{12}^{sc}, \zeta_{12}^{cs}, \ldots, \zeta_{NM}^{cc}, \zeta_{NM}^{ss}, \zeta_{NM}^{sc}, \zeta_{NM}^{cs}]$$

$$\begin{aligned} S(\theta, \varphi) = [1,\ & \cos\theta\cos\phi, \sin\theta\sin\phi, \sin\theta\cos\phi, \\ & \cos\theta\sin\phi, \cos\theta\cos 2\phi, \sin\theta\sin 2\phi, \sin\theta\cos 2\phi, \cos\theta\sin 2\phi \ldots, \\ & \cos N\theta\cos M\phi, \sin N\theta\sin M\phi, \sin N\theta\cos M\phi, \cos N\theta\sin M\phi] \end{aligned}$$

Here, ζ_0 is the base radial parameter, $\zeta_{nm}^{cc}, \zeta_{nm}^{ss}, \zeta_{nm}^{sc}, \zeta_{nm}^{cs}$ are the Fourier coefficients, and N, M are the number of angle θ and ϕ. $\theta \in [-\pi, \pi]$ is azimuth and $\phi \in [-\pi/2, \pi/2]$ is elevation. The shape parameter ζ collects ζ_0 and all $4NM$ coefficients, resulting in a dimension of $d_\zeta = 1 + 4NM$.

2.2. Orientation Representation (Axis-Angle)

To represent the target's three-dimensional (3D) orientation, this work utilizes the axis-angle parameterization, often expressed compactly as an orientation vector $\omega = \gamma a$. In this formulation, a is a unit vector denoting the axis of orientation, and γ is the scalar angle defining the magnitude of rotation orientation according to the right-hand rule.

This choice of representation is underpinned by several compelling advantages pertinent to tracking applications. Firstly, the axis–angle representation (particularly in its three-parameter rotation vector form ω) provides a geometrically intuitive and unique mapping for any 3D rotation (up to 2π periodicity in angle), without requiring the satisfaction of complex inter-parameter constraints for its validity—unlike, for instance, the nine-parameter rotation matrix which must adhere to strict orthogonality and determinant conditions. Secondly, the practical application and conversion of this representation are significantly facilitated by Rodrigues' rotation formula. This formula not only provides a straightforward and computationally efficient method for determining the effect of the rotation on a vector (i.e., rotating a vector) but also, more fundamentally, offers an explicit algorithm to compute the exponential map from the Lie algebra so (3) (the space of 3×3 skew-symmetric matrices, representing infinitesimal rotations) to the Special Orthogonal group SO (3) (the Lie group of finite 3D rotations) [29]. The corresponding rotation matrix $R(\omega)$ is computed using Rodrigues' rotation formula:

$$R(\omega) = I_3 + \sin(\gamma)[b]_\times + (1 - \cos(\gamma))[b]_\times^2 \tag{2}$$

where

$$[b]_\times = \begin{pmatrix} 0 & -b_3 & b_2 \\ b_3 & 0 & -b_1 \\ -b_2 & b_1 & 0 \end{pmatrix} \tag{3}$$

$[b]_\times$ is the skew-symmetric cross-product matrix of b. The axis-angle coordinates $\omega = \gamma a$ are a natural and compact rotation representation in terms of its geometric building blocks.

2.3. Kinematic State Model

A. line kinematic model

The kinematic state $x_t = \left[c_t^\top\ v_t^\top\right]^\top$ at time t includes the center position $c_t = \left[c_{x,t}\ c_{y,t}\ c_{z,t}\right]$ and velocity components $v_t = \left[v_{x,t}\ v_{y,t}\ v_{z,t}\right]$. The dynamic model of extended target is modeled by

$$x_{t+1} = Fx_t + w_t \tag{4}$$

where F is the state transition matrix, the process noise $w_t \sim \mathcal{N}(0, C)$.

For constant velocity model (CV) [27]:

$$F = \begin{bmatrix} 1 & T \\ 0 & 1 \end{bmatrix} \otimes I_3 \quad C = \begin{bmatrix} \frac{T^3}{3} & \frac{T^2}{2} \\ \frac{T^2}{2} & T \end{bmatrix} \otimes (\sigma_c^2 I_3) \tag{5}$$

where T is sampling time, σ_c is the process noise variance for the centroid, $\otimes$ is the Kronecker product, and I_3 denotes an identity matrix.

B. nonlinear kinematic model

While the target moves with a constant acceleration (CA) between time step, the acceleration state vector $a_t = \left[a_{x,t}\ \mathrm{a}_{y,t}\ \mathrm{a}_{z,t}\right]$ is augmented the kinematic state. Combining these into the 9-dimensional state vector $x_t = \left[c_t^\top, v_t^\top, a_t^\top\right]^\top$, the state evolution model is given as follows [30],

$$x_{t+1} = F^{CA} x_t + w_t^{CA} \tag{6}$$

where F^{CA} is the state transition matrix, the process noise $w_t^{CA} \sim \mathcal{N}(0, Q^{CA})$.

$$F^{CA} = \begin{bmatrix} I_3 & TI_3 & \frac{1}{2}T^2 I_3 \\ 0_3 & I_3 & TI_3 \\ 0_3 & 0_3 & I_3 \end{bmatrix} \quad Q^{CA} = \sigma_a^2 \begin{bmatrix} \frac{T^5}{20} I_3 & \frac{T^4}{8} I_3 & \frac{T^3}{6} I_3 \\ \frac{T^4}{8} I_3 & \frac{T^3}{3} I_3 & \frac{T^2}{20} I_3 \\ \frac{T^3}{6} I_3 & \frac{T^2}{2} I_3 & TI_3 \end{bmatrix} \tag{7}$$

The scalar σ_a^2 represents the intensity of the underlying noise process and is a crucial tuning parameter for the CA model.

2.4. Measurement Model

Sensor data at time t consists of a set of N_t noisy 3D measurements $\{z_{j,t}\}_{j=1}^{N_t}$ originating from the target's surface. Each measurement $z_{j,t}$ is related to the state and parameters via the non-line measurement function plus noise. A single measurement can be expressed as

$$z_{j,t} = c_t + f_{local} + e_t \tag{8}$$

where c_t is the center of the target, f_{local} is the radial function representation which describes the target's shape in local coordinate. However, the measurement $z_{j,t}$ is obtained in global coordinates. Consequently, a global-to-local transformation is essential. This transformation yields the following local representation,

$$f_{local} = R_{global}^{local} f_{global} = R(\omega) \begin{bmatrix} f(\zeta, \theta_{j,t}, \phi_{j,t}) \cos(\phi_{j,t}) \cos(\theta_{j,t}) \\ f(\zeta, \theta_{j,t}, \phi_{j,t}) \cos(\phi_{j,t}) \sin(\theta_{j,t}) \\ f(\zeta, \theta_{j,t}, \phi_{j,t}) \sin(\phi_{j,t}) \end{bmatrix} \tag{9}$$

$f(\zeta, \theta_{j,t}, \phi_{j,t})$ represents the radial function form in global coordinate, which modeled using DFS parameterized by the coefficient ζ, and e_t is zero-mean Gaussian measurement noise with covariance R_t. $(\theta_{j,t}, \phi_{j,t})$ is the angle pair of the measurement source on the target surface that originates $z_{j,t}$.

$R(\omega)$ is the rotation matrix that converts the global coordinates to local coordinates for the corresponding orientation parameter ω. This parameters ω utilize the axis-angle representation, expressed as $\omega = \gamma a$, where $a = [a_x, a_y, a_z]$ is a unit vector ($||a||_2 = 1$) indicating the axis of rotation orientation, and γ is the angle of rotation about the axis a (typically in radial, following the right-hand rule). This parameter ω compactly stores both axis and angle information. Specifically, its magnitude $||\omega||_2 = \gamma$ give the rotation angle, while its direction $\omega\,/||\omega||_2 = a$ give the axis for rotation orientation (for $\gamma \neq 0$). This axis-angle representation uses only 3 parameters (the components of ω) and advantageously avoids the gimbal lock singularities associated with Euler angles.

The parameter ζ is the set of coefficients for the DFS that defines the entire 3D shape of the target. The core assumption of the model is that the target is a rigid body. Therefore, this single, unified shape parameter ζ is used to calculate the surface radius for any point (θ_t, ϕ_t) on the object's surface. When process the jth measurement, and we use its specific projected angles $(\theta_{j,t}, \phi_{j,t})$ as inputs to the universal shape function defined by ζ to get the predicted radius at that exact spot.

Let $Z^G = z_{j,t} - c_t$ and $(\theta_{j,t}, \phi_{j,t})$ be expressed as,

$$\theta_{j,t} = \arccos\left(\frac{x_t^G}{||z_{j,t} - c_t||_2}\right) \tag{10}$$

$$\phi_{j,t} = \text{atan2}(y_t^G, x_t^G) \tag{11}$$

where $Z_t^G \triangleq (x_t^G, y_t^G, z_t^G)$.

As the measurement $z_{j,t}$ cannot be isolated as an explicit function of the state vector and noise, Equation (8) defines an implicit measurement model. It can be rewritten as:

$$\begin{aligned} 0 = & \underbrace{-z_{j,t} + c_t + R(\omega)\begin{bmatrix} f(\zeta,\theta_{j,t},\phi_{j,t})\cos(\phi_{j,t})\cos(\theta_{j,t}) \\ f(\zeta,\theta_{j,t},\phi_{j,t})\cos(\phi_{j,t})\sin(\theta_{j,t}) \\ f(\zeta,\theta_{j,t},\phi_{j,t})\sin(\phi_{j,t}) \end{bmatrix} + e_t}_{h(x_t,z_{j,t},\zeta,\omega)} \\ & = h(x_t, z_{j,t}, \zeta, \omega) + e_t \end{aligned} \tag{12}$$

The zero vector appearing in Equation (12) can be interpreted as representing a pseudo-measurement. This pseudo-measurement is effectively a nonlinear function of the state and the original measurement, assumed to be corrupted by additive Gaussian noise.

3. ECM Tracking Algorithm Based on DFS

This section details the DFS-ECM framework for joint motion state, shape, and orientation estimation. Its core, the Expectation Conditional Maximization (ECM) algorithm, uses a batch method involving iterative Kalman smoothing over a sliding data window. This incorporates information from previous and subsequent scans. For a batch window from time 1 to T_b, the ECM framework defines these variables:

$$X = (x_1, \ldots, x_t \ldots, x_{T_b})$$

$$Z = (Z_1, \ldots, Z_t \ldots, Z_{T_b})$$

$$\Lambda = (\Lambda_1, \ldots, \Lambda_t \ldots, \Lambda_{T_b})$$

where X denotes the latent variables within the sliding window; Z represents the set of measurements within the sliding window, including $Z_t = (z_{1,t}, \ldots z_{j,t}, \ldots, z_{N_t,t})$; Λ denotes the complete sequence of unknown parameters, where each $\Lambda_t = (\omega, \zeta)$ of the sliding window at time step t. ECM is an iterative optimization method whose objective is to maximize the log-likelihood function of the incomplete data (latent variable) X with respect to the parameter Λ.

The main terms used in this algorithm can be found in Table 1.

Table 1. Summary of the main terms used in the derivation of the algorithm.

P_t	The error covariance of target state at time t
$P_{t\|t-1}$	The predict error covariance of target state at time t
$P_{j,zz,t\|t-1}$	The jth innovation covariance at time t
$P_{j,xz,t\|t-1}$	The jth cross-covariance at time t
$P_{j,t}$	The jth error covariance at time t
$P_{j,t\|T_b}$	The jth smoothed covariance at time t
$P_{t\|T_b}$	The smoothed error covariance at time t
$p(x_t\|x_{t-1})$	Transition Probability density of state x_t
$p(z_{j,t}\|x_t, \Lambda^{(i)})$	The likelihood function
$p(X^{(i+1)}, Z\|\Lambda^{(i)})$	The likelihood function of complete data
$p(x_0)$	The prior Probability density of state
$P_s(\zeta)$	The smoothness penalty term of the cost function
i	Iteration times of ECM
j	Index of measurement number

We define the complete-data log-likelihood function $L(\Lambda^{(i+1)})$ and its corresponding conditional expectation function $Q(\Lambda^{(i+1)};\Lambda^{(i)})$ of the $(i+1)$th iteration as follows:

$$L(\Lambda^{(i+1)}) = \ln p(X^{(i+1)}, Z \big| \Lambda^{(i+1)}) \tag{13}$$

$$Q(\Lambda^{(i+1)};\Lambda^{(i)}) = E[L(\Lambda^{(i+1)}) \big| Z, \Lambda^{(i)}] \tag{14}$$

As the Q function involves multiple unknown parameters, a direct application of the traditional maximization step is not feasible. Instead, we utilize multiple conditional maximization steps to iteratively estimate the parameters.

3.1. E-Step

The primary purpose of the E-step is to estimate the state x_t and its corresponding error covariance matrix P_t, a process that involves an Unscented Kalman forward filter (UKF) [31,32] and backward Rauch–Tung–Striebel Smoother smoother [33,34]. The forward propagation is performed as,

Let n be the dimension of the kinematic state x_t, we generate $2n+1$ sigma points $\mathcal{X}_{k,t-1}$ based on the previous state estimate x_{t-1} and error covariance P_{t-1} (state initialized as $x_0 = \mu_0$, $P_0 = \sum_0$):

$$\mathcal{X}_{k,t-1} = \begin{cases} x_{t-1} & k=0 \\ x_t + (\sqrt{(n+\lambda)P_{t-1}})_k & k=1,\ldots,n \\ x_t - (\sqrt{(n+\lambda)P_{t-1}})_k & k=n+1,\ldots,2n \end{cases} \tag{15}$$

The weights for mean $W_k^{(m)}$ and covariance $W_k^{(c)}$ are:

$$W_k^{(m)} = \begin{cases} \frac{\lambda}{n+\lambda} & k=0 \\ \frac{1}{2(n+\lambda)} & k \neq 0 \end{cases} \tag{16}$$

$$W_k^{(c)} = \begin{cases} \frac{\lambda}{n+\lambda} + (1-\alpha^2+\beta) & k=0 \\ \frac{1}{2(n+\lambda)} & k \neq 0 \end{cases} \tag{17}$$

where $\lambda = \alpha^2(n+\kappa) - n$ is a scaling parameter, α controls the distance of the Sigma points from mean, κ is a secondary parameter β is used to incorporate covariance information.

Prediction step of UKF:

(1) Propagate Sigma Points: Pass each state sigma point through the process model F,

$$\mathcal{X}_{k,t|t-1} = F\mathcal{X}_{k,t-1} \tag{18}$$

(2) Compute the predicted state $\hat{x}_{t|t-1}$ and predict error covariance $P_{t|t-1}$:

$$\hat{x}_{t|t-1} = \sum_{k=0}^{2n} W_k^{(m)} \mathcal{X}_{k,t|t-1} \tag{19}$$

$$P_{t|t-1} = \sum_{k=0}^{2n} W_k^{(c)} (\mathcal{X}_{k,t|t-1} - \hat{x}_{t|t-1})(\mathcal{X}_{k,t|t-1} - \hat{x}_{t|t-1})^\top + C \tag{20}$$

Update step of UKF: the update proceeds sequentially for each measurement. Initialize the updated state and error covariance for this time step with predicted values,

$$\hat{x}_{0,t} = \hat{x}_{t|t-1} \tag{21}$$

$$P_{0,t} = P_{t|t-1} \tag{22}$$

the measurements $\{z_{j,t}\}_{j=1}^{N_t}$ are processed using the following steps,

(1) Predict the jth measurement $\mathcal{Z}_{j,k,t|t-1}$ for each sigma point $\mathcal{X}_{k,t|t-1}$ using the parameter ζ, ω,

$$\mathcal{Z}_{j,k,t|t-1} = h(\mathcal{X}_{k,t|t-1}, \zeta, \omega) \tag{23}$$

(2) Compute the jth weighted average measurement $\hat{z}_{j,t|t-1}$ of the predicted sigma point,

$$\hat{z}_{j,t|t-1} = \sum_{k=0}^{2n} W_k^{(m)} \mathcal{Z}_{k,t|t-1} \tag{24}$$

(3) Estimate the jth innovation covariance $P_{j,zz,t|t-1}$ and the cross-covariance $P_{j,xz,t|t-1}$,

$$P_{j,zz,t|t-1} = \sum_{k=0}^{2n} W_k^{(c)} (\mathcal{Z}_{k,t|t-1} - \hat{z}_{j,t|t-1})(\mathcal{Z}_{k,t|t-1} - \hat{z}_{j,t|t-1})^\top + R_t \tag{25}$$

$$P_{j,xz,t|t-1} = \sum_{k=0}^{2n} W_k^{(c)} (\mathcal{X}_{k,t|t-1} - \hat{x}_{t|t-1})(\mathcal{Z}_{k,t|t-1} - \hat{z}_{j,t|t-1})^\top \tag{26}$$

(4) Calculate the jth Kalman Gain $K_{j,t}$,

$$K_{j,t} = P_{j,xz,t|t-1} P_{j,zz,t|t-1}^{-1} \tag{27}$$

(5) Update the jth state $\hat{x}_{j,t}$ and its error covariance $P_{j,t}$,

$$\hat{x}_{j,t} = \hat{x}_{j-1,t} + K_{j,t}(0 - \hat{z}_{j,t|t-1}) \tag{28}$$

$$P_{j,t} = P_{j-1,t} - K_{j,t} P_{j,zz,t|t-1} K_{j,t}^\top \tag{29}$$

When processing all N_t measurements, the UKF state $\hat{x}_t$ estimation and its error covariance P_t are:

$$\hat{x}_t = \hat{x}_{N_t,t} \tag{30}$$

$$P_t = P_{N_t,t} \tag{31}$$

After sequentially processing the data from time $t = 1$ to $t = T_b$ with an Unscented Kalman Filter (UKF), we obtain the series of filter states $\{\hat{x}_t\}_{t=1}^{T_b}$ and error covariances $\{P_t\}_{t=1}^{T_b}$. Subsequently, the Rauch–Tung–Striebel smoothing algorithm is applied to compute the smoothed state $\hat{x}_{t|T_b}$ and its error covariance $P_{t|T_b}$ as follows,

The backward pass begins at the final time step T_b. (1) Initialize the smoothed state $\hat{x}_{0,t|T_b}$ and error covariance $P_{0,t|T_b}$ for this time step with UKF estimation,

$$\hat{x}_{0,t|T_b} = \hat{x}_{T_b} \tag{32}$$

$$P_{0,t|T_b} = P_{T_b} \tag{33}$$

The algorithm then iterates backward in time, from $t = T_b - 1$ down to $t = 1$. For each time step t, the following computations with the measurements $\{z_{j,t}\}_{j=1}^{N_t}$ are performed:

(2) Compute the jth Smoothed Gain $G_{j,t}$,

$$G_{j,t} = P_{j,xz,t+1|t} P_{t+1|t}^{-1} \tag{34}$$

where $P_{j,xz,t+1|t}$ is the jth cross covariance and $P_{t+1|t}$ is the predict error covariance of forward UKF, both obtained from the forward UKF pass.

(3) Compute the jth Smoothed state $\hat{x}_{j,t|T_b}$,

$$\hat{x}_{j,t|T_b} = \hat{x}_{j-1,t|T_b} + G_t(\hat{x}_{t+1|T_b} - \hat{x}_{t+1|t}) \tag{35}$$

(4) Compute the jth Smoothed error Covariance $P_{j,t|T_b}$,

$$P_{j,t|T_b} = P_{j-1,t|T_b} + G_{j,t}(P_{t+1|T_b} - P_{t+1|t})G_{j,t}^{\top} \tag{36}$$

When processing all N_t measurements, the smoothed state $\hat{x}_{t|T_b}$ estimation and its error covariance $P_{t|T_b}$ are,

$$\hat{x}_{t|T_b} = \hat{x}_{N_t,t|T_b} \tag{37}$$

$$P_{t|T_b} = P_{N_t,t|T_b} \tag{38}$$

Considering the Markov property of state propagation and the independence of measurement sources, the log-likelihood function $L(\Lambda^{(i+1)})$ is written as,

$$L(\Lambda^{(i+1)}) = \ln p(x_0) + \sum_{t=1}^{T_b} \ln p(x_{t+1}|x_t) + \sum_{t=1}^{T_b}\sum_{j=1}^{N_t} \ln p(z_{j,t}\big|x_t, \Lambda^{(i+1)}) \tag{39}$$

So,

$$\begin{aligned} Q(\Lambda^{(i+1)};\Lambda^{(i)})) &= E_{Z,\Lambda^{(i)}}[\ln p(x_0)] + \sum_{t=1}^{T_b} E_{Z,\Lambda^{(i)}} \ln p(x_t|x_{t-1}) + \sum_{t=1}^{T_b}\sum_{j=1}^{N_t} E_{Z,\Lambda^{(i)}} \ln p(z_{j,t}\big|x_t, \Lambda^{(i+1)}) \\ &= \sum_{t=1}^{T_b}\sum_{j=1}^{N_t} E_{Z,\Lambda^{(i)}} \ln p(z_{j,t}\big|x_t, \Lambda^{(i+1)}) + Const \end{aligned} \tag{40}$$

Under the assumption of Gaussian measurement noise, the measurement likelihood becomes,

$$p(z_{j,t}|x_t, \omega, \zeta) = N(z_{j,t} - h(x_t, z_{j,t}, \omega, \zeta); 0, R_t) \tag{41}$$

Insert Equation (41) into (40), Q is,

$$\begin{aligned} &Q(\Lambda^{(i+1)};\Lambda^{(i)}) == \sum_{t=1}^{T_b}\sum_{j=1}^{N_t} E_{Z,\Lambda^{(i)}} \ln p(z_{j,t}\big|x_t, \Lambda_t^{(i+1)}) + Const \\ &= -\frac{1}{2}\sum_{j=1}^{N_t}\left\{\sum_{t=1}^{T_b} \ln|R_{j,t}| + \mathrm{Tr}\sum_{t=1}^{T_b}(R_{j,t})E_{Z,\Lambda^{(i)}}[(z_{j,t} - h(x_t, z_{j,t}, \omega, \zeta))(\bullet)^{\top}]\right\} + Const \end{aligned} \tag{42}$$

where $(\bullet)$ denotes the term same as the former one.

3.2. M-Step

After obtaining $Q(\Lambda^{(i+1)};\Lambda^{(i)})$ in the E-step of the algorithm, the task of the M-step is to find $\Lambda^{(i+1)}$, thereby updating the target's shape and axis-angle parameters from the previous iteration. Since $Q(\Lambda^{(i+1)};\Lambda^{(i)})$ is a non-linear function of the parameters, its analytical solution is generally difficult to obtain. Here, parameters are estimated by minimizing regularized cost function which equivalent to maxing $Q(\Lambda^{(i+1)};\Lambda^{(i)})$ estimation.

(1) Shape parameter optimization

The shape parameter ζ are estimated by minimizing the cost function $J(\zeta)$ over a sliding window of T_b time steps ($t = 1$ to T_b).

$$J_{original}(\zeta) = \Big(\frac{1}{\sum_{t=1}^{T_b} N_t}\sum_{t=1}^{T_b}\sum_{j=1}^{N_t}\Big|\Big|z_{j,t} - h(\hat{x}_t, z_{j,t}, \zeta, \hat{\omega})\Big|\Big|^2_{R_t^{-1}}\Big) + \eta_\zeta\Big|\Big|\zeta_{2:end}\Big|\Big|_2^2 + P_s(\zeta) \tag{43}$$

where, $\hat{x}_t$, $\hat{\omega}$ are the estimates of state and orientation at time t within the window. $||e||^2_{R^{-1}} = e^\top R^{-1} e$ denotes the squared Mahalanobis distance. The first term is the average squared Mahalanobis distance over all valid measurements in the window. η_ζ is the L2 regularization parameter applied to the coefficients (excluding η_0) to prevent overfitting and improving conditioning. $P_s(\zeta)$ is the smoothness penalty term.

Remark 1. *Using the original cost function $J_{original}(\zeta)$ can lead to artificial bulges or distortions near the poles ($\phi = \pm\, \pi/2$). This occurs because the spherical coordinate system used by the DFS is degenerate at the poles, and measurement near these regions can exert disproportionate influence during optimization, leading to overfitting localized at the poles.*

To mitigate these artifacts, we introduce a latitude-dependent weight factor $w_{pole}(\phi)$ that down-weights measurements near the poles. A common choice is

$$w_{pole}(\phi) = \cos^h\Big(\Big|\phi\Big|\Big) \tag{44}$$

where $h \geq 1$ (e.g., $h = 1$ or $h = 2$) is a tuning parameter controlling the strength of the suppression.

The modified cost function $J(\zeta)$ is,

$$J(\zeta) = \Big(\frac{1}{\sum_{t=1}^{T_b}\sum_{j=1}^{N_t} w_{pole}(\hat{\phi}_{j,t})}\sum_{t=1}^{T_b}\sum_{j=1}^{N_t} w_{pole}(\hat{\phi}_{j,t})\Big|\Big|z_{j,t} - h(\hat{x}_t, z_{j,t}, \zeta, \hat{\omega})\Big|\Big|^2_{R_t^{-1}}\Big) + \eta_\zeta\Big|\Big|\zeta\Big|\Big|_2^2 + P_s(\zeta) \tag{45}$$

Remark 2. *By multiplying the squared Mahalanobis distance of each measurement by $w_{pole}(\phi)$, we reduces the contribution of measurements where $|\phi|$ is close to $\pi/2$. This is mathematically equivalent to assuming a higher effective measurement noise variance ($R^*_{j,t} = R_t / w_{pole}(\hat{\phi}_{j,t})$) near the poles. It prevents the optimizer from overly fitting to potentially unreliable projections in these geometrically sensitive regions. Allowing the regularization terms to enforce a smoother, more plausible shape. The tuning parameter h control how aggressively measurements near the poles are down-weighted.*

Improved Smoothness Penalty: A physically meaningful smoothness penalty for the DFS on a sphere penalizes higher spatial frequencies more heavily. The indices n and m in the DFS correspond to the frequencies in the azimuth and elevation directions, respectively. Higher values of n and m represent finer details or oscillations on the surface. A suitable penalty terms $P_s(\zeta)$ is given by:

$$P_s(\zeta) = \eta_s \sum_{n=1}^{N}\sum_{m=1}^{M}(n^\nu + m^\nu)((\zeta^{cc}_{n,m})^2 + (\zeta^{ss}_{n,m})^2 + (\zeta^{sc}_{n,m})^2 + (\zeta^{cs}_{n,m})^2) \tag{46}$$

where η_s is the smoothness regularization strength parameter. The term $(n^\nu + m^\nu)$ acts as a weighting factor that increases with the frequency indices n and m. A common is

$v = 2$, making the penalty proportional to $(n^2 + m^2)$. The sum of squares involves all four coefficient types for the given (n, m) pair.

(2) Orientation parameter optimization

Minimize $J(\omega)$ at step t, regularizing towards the previous estimate $\hat{\omega}$:

$$J(\omega) = \frac{1}{N}\sum_{j=1}^{N_t}\left|\left|z_{j,t} - h(\hat{x}_t, z_{j,t}, \hat{\zeta}, \omega)\right|\right|^2_{R_t^{-1}} + \eta_\omega \left|\left|\omega - \hat{\omega}\right|\right|_2^2 \quad (47)$$

where $\hat{x}_t, \hat{\zeta}$ are the current estimates of state and shape. $\hat{\omega}$ is the estimated orientation parameter from the previous time step. η_ω is the regularization parameter controlling the penalty for deviating from the previous orientation estimation, promoting temporal smoothness.

(3) Optimization Constraints

Constraints can be applied during optimization:

A. Shape parameter constraints:

$\zeta_0 > \varepsilon$ (small positive value)

B. Rotation parameter constraints:

angle limit: $||\omega||_2 \leq \pi$.

Iteration of the EM algorithm ceases upon convergence or reaching a limit. Convergence is typically determined when the change in the likelihood function value between successive iterations is sufficiently small. Alternatively, termination occurs if the iteration count reaches a predefined upper bound.

The overall algorithm flow is summarized in Algorithm 1.

Algorithm 1. DFS-ECM

Initial parameters: Measurement set batch Z, State batch X, orientation and extent parameters Λ, Maximum Iterations I

begin
Setup $i = 0$
    while not converged **and** $i < \mathrm{I}$ **do**
        expectation:
            for $t = 1 : T_b$ **do**
                for $j = 1 : N_t$ **do**
                    Calculate $\hat{x}_{j,t}$ and $P_{j,t}$ according to (28–29)
                end for
            end for
            for $t = T_b : -1 : 1$ **do**
            Calculate smoothed $\hat{x}_{t|T_b}$ and $P_{t|T_b}$ according to (37–38)
            end for
        maximization:
            Calculate ζ, ω according to (45,47)
        $i++$;
    end while
end begin

4. Simulation Results

Our proposed method, referred to as DFS-ECM in this section, are evaluated in two distinct scenarios. The simulations are run on an Intel Core™ i7-7700 processor at

3.6 GHz with 32 GB of RAM using MATLAB 2021a. Root Mean Square Error (RMSE) and the Intersection-Over-Union (IOU) are used to assess overall performance.

4.1. Evaluation Index

Root-Mean-Square Error (RMSE) is defined below:

$$RMSE_c = \sqrt{\frac{1}{M}\sum_{i=1}^{M}(C_i - c_i)^2} \tag{48}$$

where $RMSE_c$ represents the RMSE of the parameter c, C_i represents the true and c_i represents the estimated value.

The shape estimation performance is evaluated by Intersection-Over-Union (IOU). IOU is defined as follows:

$$IOU(v_{true}, \hat{v}) = \frac{volume(v_{true} \cap \hat{v})}{volume(v_{true} \cup \hat{v})} \tag{49}$$

where volumes v_{true} is the true target shape, and $\hat{v}$ represents the estimate. $IOU(v_{true}, \hat{v}) \in [0, 1]$, 1 corresponds to a perfect match, while 0 indicates no intersection between the true and the estimated extent.

4.2. Scenario I

In this scenario, we test the DFS-ECM algorithm with ellipsoid and cubic target. Every target moves in a linear trajectory with the constant velocity (CV) model, and with a sensor located at the origin. The trajectory of target motion is presented in Figure 1. The initial kinematic state is set to [0 m, 0 m, 0 m, 10 m/s, 10 m/s, 0 m/s]. At each instant, 100 points measurements are originated from random sources which are sample from a uniform distribution defined over the target surface. The sampling time of all algorithms is set to 1 s. The orientation is represented by the axis angle parameter $\omega = (0, 0, 0)$.

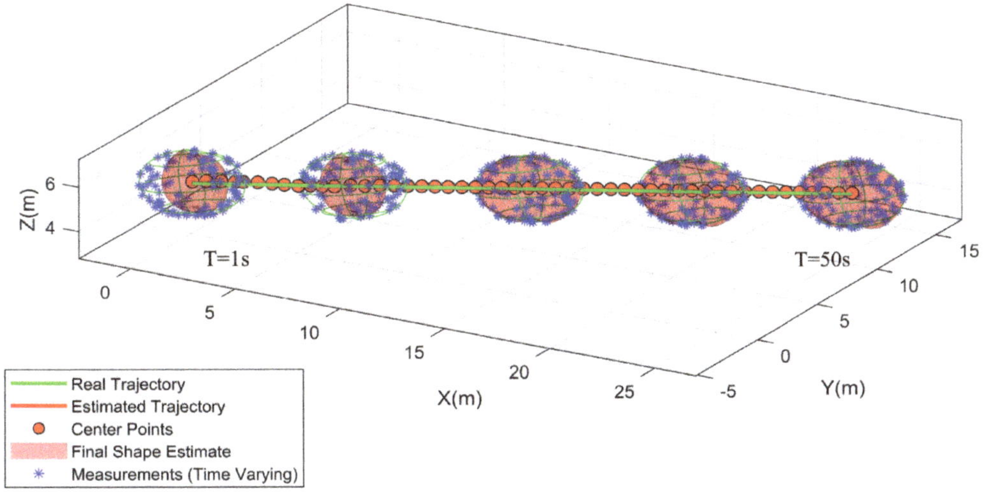

(a)

Figure 1. *Cont.*

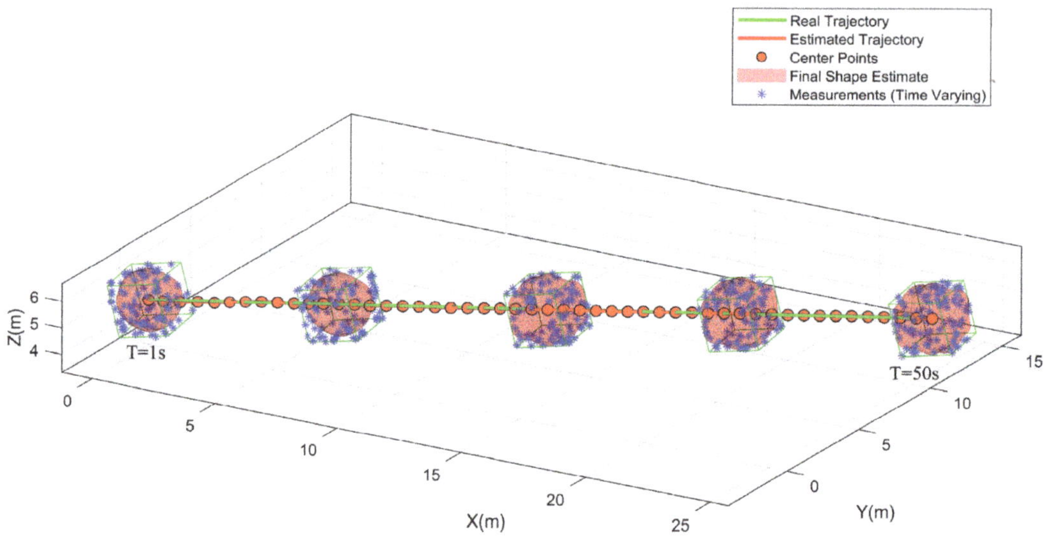

(**b**)

Figure 1. (**a**) ellipsoid target tracking; (**b**) cubic target tracking.

The process noise standard deviations are set to $\sigma_c = 0.1$, For the sigma points, a scaling parameter $\alpha = 1$, $\kappa = 0$, $\beta = 2$. For the extent parameter, $\zeta_0 = 1.0$ and $\zeta_{2,\ldots,d_\zeta} = 0$, the covariance matrix is set to $diag(0.1, \ldots, 0.1) \in \mathbb{R}^{d_\zeta \times d_\zeta}$. The degree of double Fourier series is chosen as $M = N = 2$.

Figure 2 illustrates the typical results and tracking trajectories in one Monte Carlo run. Figure 2a,b depict the tracking scene for the ellipsoid, cube, respectively. For enhanced clarity regarding the reconstruction outcomes, Figure 3 illustrates the reconstruction results for 3D shapes obtained without incorporating the weight. A comparison between Figures 2 and 3 reveals that optimizing shape reconstruction with the weight yields superior accuracy.

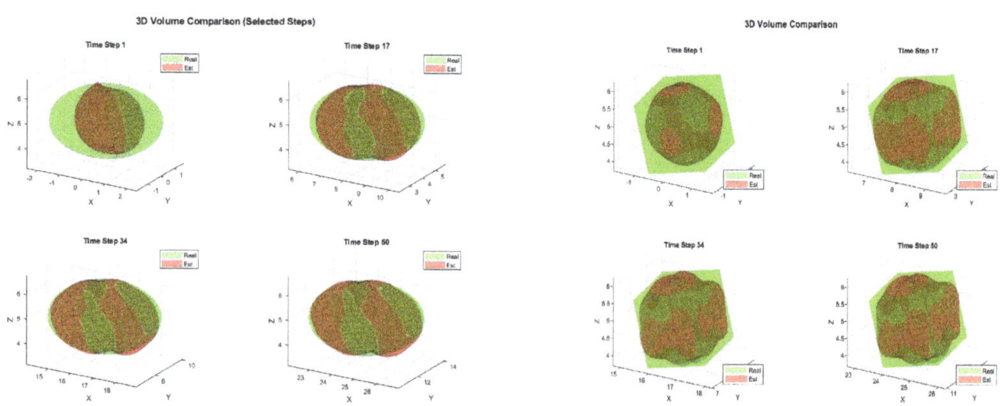

(a) (b)

Figure 2. Target construction, (**a**) ellipsoid target, (**b**) cube target.

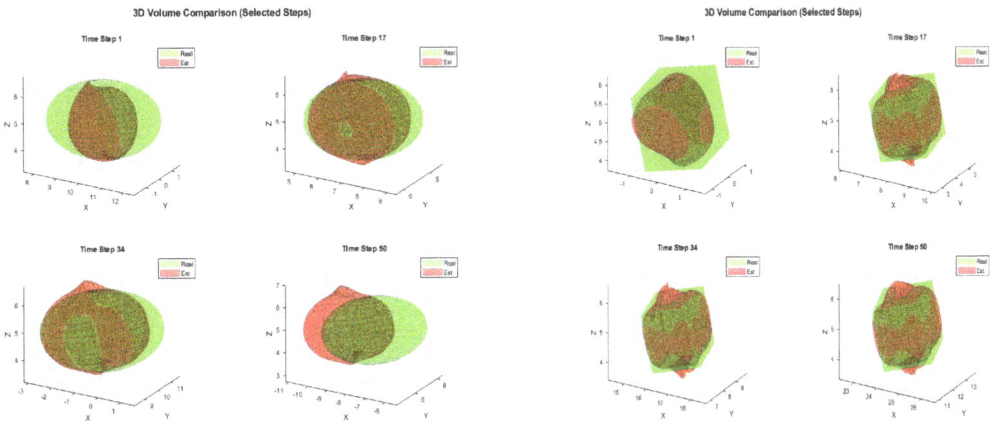

(a) (b)

Figure 3. Target construction without weight (**a**) ellipsoid target (**b**) ellipsoid target.

We evaluated centroid estimation for linearly moving ellipsoidal and cubic targets using 100 Monte Carlo trials (randomizing noise and sample points). Figure 4 shows the centroid RMSE for sliding window lengths T_b = 5, 10, and 15. For the ellipsoid (Figure 4a), $T_b = 5$ yields the minimum RMSE; performance worsens for $T_b = 10$ and $T_b = 15$, suggesting potential overfitting. For the cube (Figure 4b), $T_b = 15$ is optimal. Generally, errors decrease over the simulation (indicating stabilization), and the relative performance of different window sizes remains consistent.

Figure 5 illustrates the RMSE values of the velocity estimates with the sliding window length T_b = 5, 10, and 15. This indicates that while there's a large initial uncertainty in velocity, the estimation becomes highly accurate over time, and the choice of window size (within this range) has minimal impact on the long-term velocity estimation accuracy.

Figure 6 presents the Intersection over Union (IOU) values for the shape estimates, obtained using sliding window lengths of T_b = 5, 10, and 15. For both the ellipsoidal and cubic targets, the analysis indicates that the highest IOU score (representing the best shape estimate accuracy) is achieved with the window length $T_b = 15$.

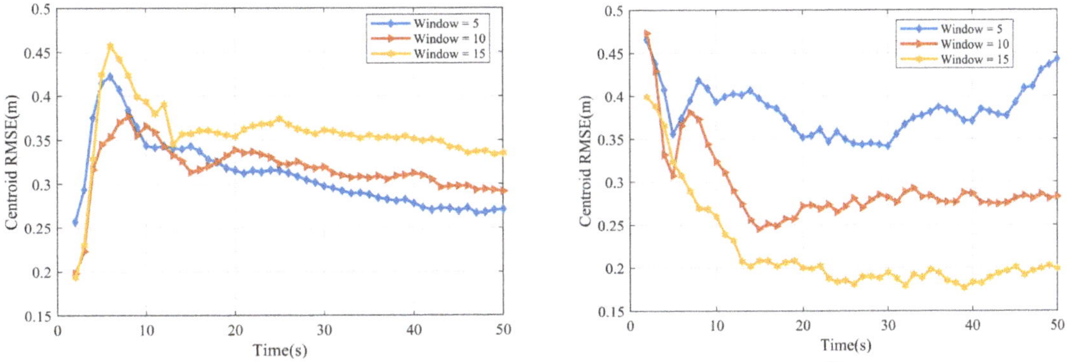

(a) (b)

Figure 4. Centroid RMSE (**a**) ellipsoid target RMSE (**b**) cube target RMSE.

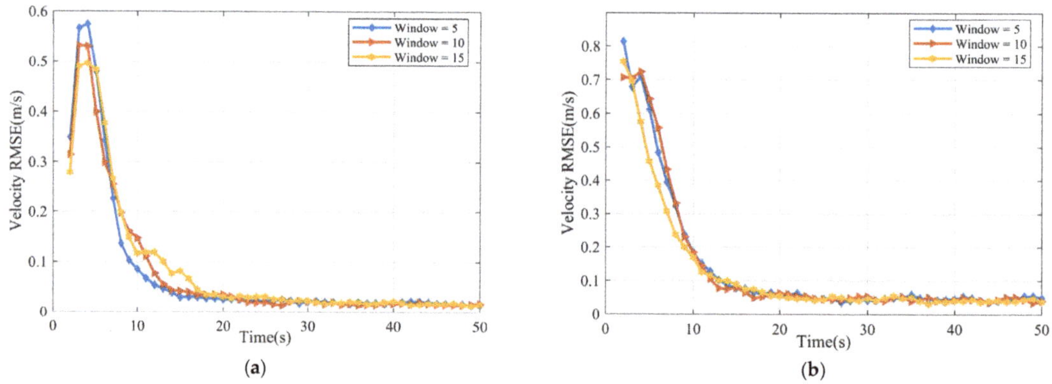

Figure 5. Velocity RMSE (**a**) ellipsoid target RMSE (**b**) cube target RMSE.

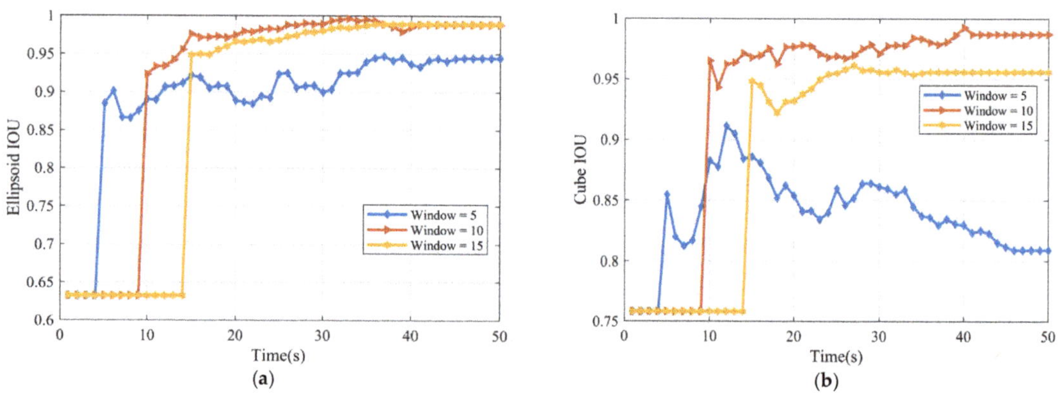

Figure 6. IOU (**a**) ellipsoid target IOU (**b**) cube target IOU.

As detailed in Table 2, the errors associated with the estimated orientataion parameter components remain consistently below 0.1 for both ellipsoidal and cubic target types, indicating a high degree of estimation accuracy.

Table 2. Orientation parameters estimation.

Target	Angle	T = 5 (rad)	T = 10 (rad)	T = 15 (rad)
Ellipsoid	ω_x	0.020	0.030	0.030
	ω_y	0.080	0.050	0.060
	ω_z	0.050	0.050	0.050
Cube	ω_x	0.043	0.042	0.043
	ω_y	0.004	0.004	0.003
	ω_z	0.015	0.012	0.016

For comparison, we use a standard random matrix-based extended target tracker [35], henceforce denoted as RM. Table 3 summarizes the comparative performance of the RM and DFS-ECM approaches. The results consistently show that the DFS-ECM method outperforms the RM model in terms of centroid estimation accuracy, velocity estimation accuracy, and shape estimation quality (as measured by IOU) across all tested scenarios.

Table 3. Performance comparation of two diffenernt algorithm.

Target	Algorithm	Centroid RMSE (m)	Velocity RMSE (m/s)	IOU
Ellipsoid	RM	0.27	0.10	0.90
	DFS-ECM	0.26	0.07	0.96
Cube	RM	0.27	0.10	0.70
	DFS-ECM	0.20	0.07	0.95

As presented in Table 4, increasing the number of measurements positively impacts overall tracking performance. Specifically, a higher measurement count leads to a lower Centroid RMSE, indicating improved positioning accuracy. Concurrently, the shape estimation quality, measured by IOU, steadily improves for both the ellipsoid and the cube, confirming that more data yields more accurate reconstructions.

Table 4. Performance comparation of different muber of measurements.

Target	Number of Measurement	Centroid RMSE (m)	IOU
Ellipsoid	10	0.29	0.91
	20	0.28	0.92
	40	0.27	0.96
Cube	10	0.30	0.90
	20	0.28	0.92
	40	0.288	0.95

4.3. Scenario II

The performance of the DFS-ECM algorithm is further evaluated using a nonlinear scenario. In this experiment, a cubic target moves along a curved trajectory in three-dimensional space, as depicted in Figure 7. The target's motion is governed by the constant acceleration (CA) model, which is defined in Equation (6). This simulation utilizes the same DFS model hyperparameters and EM parameters as those employed in Scenario I. Within the CA model, the scalar parameter σ_a^2 is set to 0.001. At each instant, 100 measurement points are sampled uniformly from the target's surface. The orientation is represented by the axis angle parameter $\omega = (0.1, -0.2, 0.05)$.

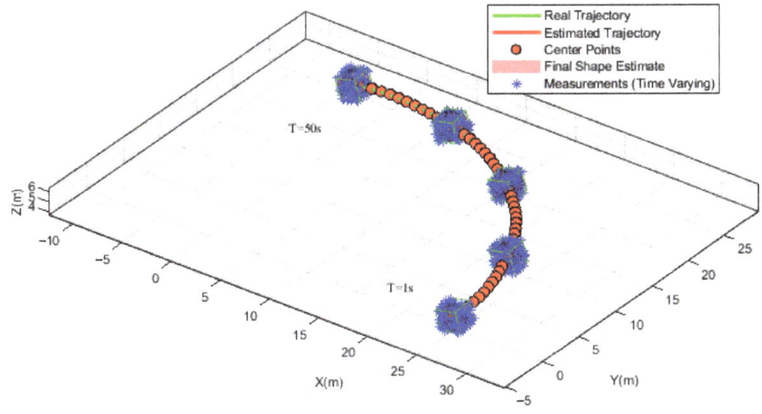

Figure 7. Cube targe tracking.

To better illustrate the reconstruction quality, Figure 8 presents magnified views of typical reconstruction examples for the cubic target. These examples demonstrate satisfactory reconstruction performance.

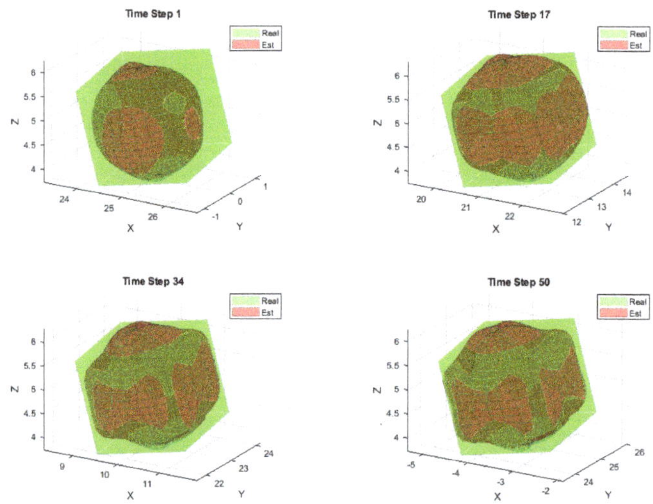

Figure 8. Cube target.

Figure 9 presents the centroid RMSE over time for the cubic target, comparing window lengths $T_b = 5$, 10, and 15. Optimal performance (lowest RMSE) is achieved with the shortest window, $T_b = 5$. Conversely, the larger window lengths ($T_b = 10$ and $T_b = 15$) yield higher errors, potentially indicating underfitting in this nonlinear scenario. However, all three RMSE curves exhibit convergence towards stable values over time.

The velocity RMSE for the cubic target is presented in Figure 10. Notably, while the larger window sizes ($T_b = 10$ and $T_b = 15$) show similar performance with lower peak errors and better long-term stability (lower RMSE), the smallest window ($T_b = 5$) suffers from a higher peak error (occurring later) and maintains a higher error level beyond approximately 18 s.

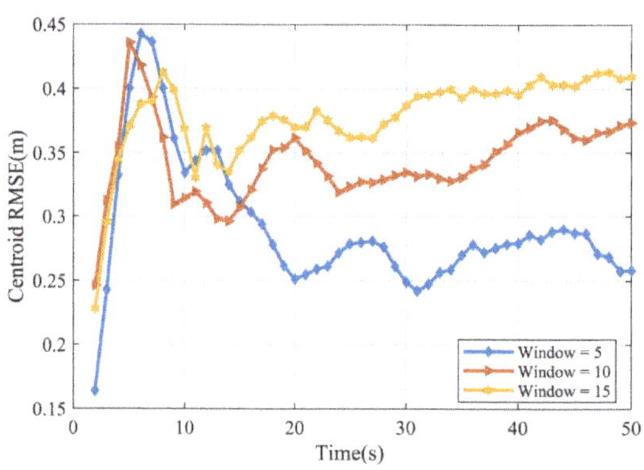

Figure 9. Centroid RMSE of cube.

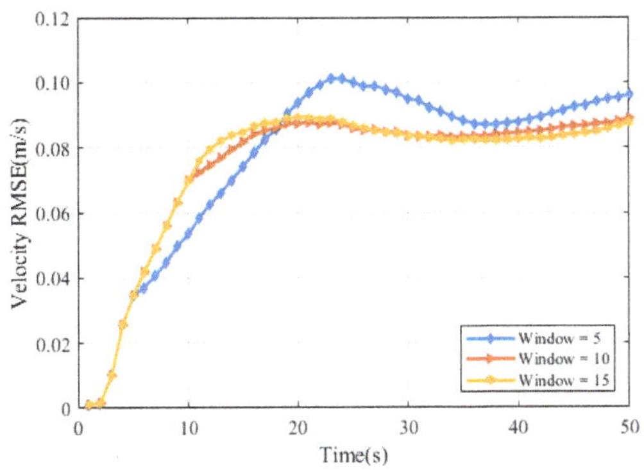

Figure 10. Velocity RMSE of cube.

Figure 11 plots the volume Intersection over Union (IOU), representing shape estimation accuracy, over time for three different window sizes. Towards the end of the observation period (after 40 s), the performance stabilizes for all window sizes. They converge to similar and relatively high IOU values, approximately in the range of 0.86–0.88. While these results are strong, the accuracy is slightly below that achieved for the cube in the linear scenario. We attribute this minor performance difference to the inherent challenges of the nonlinear model. The nonlinearity likely introduces small errors into the centroid position estimate, which subsequently impacts the overall accuracy of the shape reconstruction.

The performance of orientation parameter estimation is detailed in Figure 12. The error dynamics for individual rotation components within a single Monte Carlo run are illustrated in Figure 12a–c. The overall accuracy, depicted by the average Root Mean Square Error (RMSE) across all Monte Carlo runs, is shown in Figure 12d. The low error values demonstrate the algorithm's capability for accurate angle estimation.

Figure 13 illustrates the computational cost, specifically the average processing time per step required by the algorithm, as a function of different window lengths ($T_b = 5$, 10, and 15). A clear positive correlation is evident: longer window lengths necessitate incNreased computation time per step.

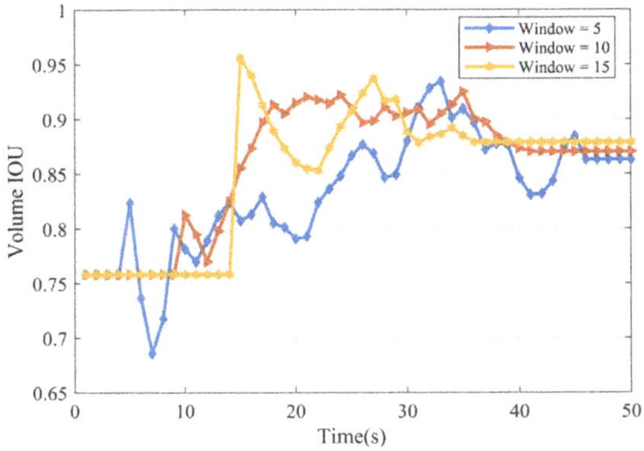

Figure 11. IOU of cube.

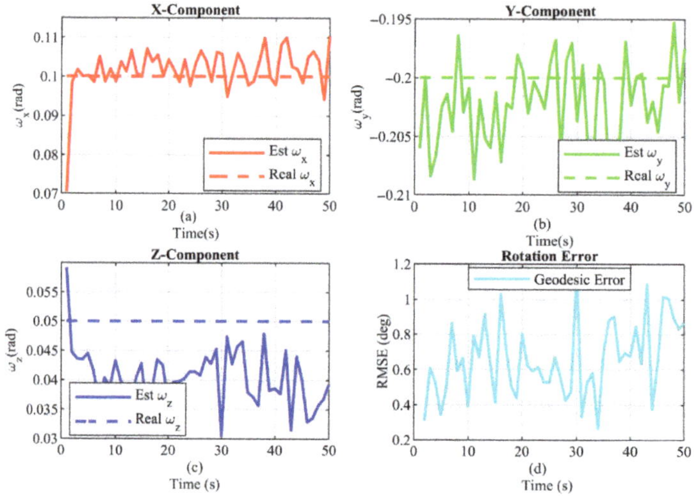

Figure 12. Rotation parameters estimate (**a**) rotation about the x-axis (**b**) rotation about the y-axis (**c**) rotation about the z-axis (**d**) orientation RMSE.

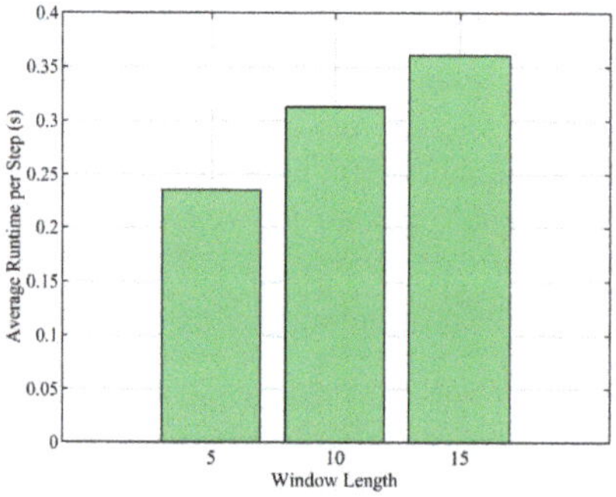

Figure 13. Average runtime.

5. Conclusions

This paper proposed a method for joint 3D extended object tracking with unknown but fixed shape in the framework of Expectation Conditional Maximization (ECM). By introducing a shape representation based on a Double Fourier Series (DFS) radial function and utilizing the axis-angle representation for orientation, this method enables the direct estimation of target shape and orientation parameters, overcoming the reliance on prior shape evolution models. Leveraging the ECM algorithm, kinematic state inference is performed using an Unscented Kalman Filter (UKF) in the E-step, while shape and orientation parameters are estimated via minimization of regularized cost functions in the M-step. This ultimately achieves the joint, robust, and smooth estimation of kinematics, shape, and orientation. The methods provide a full expression for the target's shape, this detailed information can be used in 3D applications such as navigation, identification and classification. Future work will focus on adaptive tuning and joint estimation schemes to further advance the capability of 3D target tracking.

Author Contributions: Methodology, X.Y.; Software, H.M.; Investigation, X.Y.; Writing—original draft, H.M.; Writing—review & editing, X.Y. All authors have read and agreed to the published version of the manuscript.

Funding: This research received no external funding.

Institutional Review Board Statement: Not applicable.

Informed Consent Statement: Not applicable.

Data Availability Statement: The data presented in this study are available upon request from the corresponding authors.

Conflicts of Interest: The authors declare no conflicts of interest.

References

1. Faion, F.; Zea, A.; Steinbring, J.; Baum, M.; Hanebeck, U.D. Recursive Bayesian Pose and Shape Estimation of 3D Objects Using Transformed Plane Curves. In Proceedings of the 2015 Sensor Data Fusion: Trends, Solutions, Applications (SDF), Bonn, Germany, 6–8 October 2015; pp. 1–6.
2. Moosmann, F.; Stiller, C. Joint self-localization and tracking of generic objects in 3D range data. In Proceedings of the 2013 IEEE International Conference on Robots and Automation (ICRA), Karlsruhe, Germany, 16–10 May 2013; pp. 1146–1152.
3. Held, D.; Levinson, J.; Thrun, S.; Savarese, S. Robust Real-time Tracking Combining 3D Shape, Color, and motion. *Int. J. Robot. Res.* **2016**, *35*, 30–49. [CrossRef]
4. Kraemer, S.; Bouzouraa, M.E.; Stiller, C. Simultaneous Tracking and Shape Estimation Using a Multi-layer Laserscanner. In Proceedings of the 20th IEEE International Conference on Intelligent Transportation Systems (ITSC), Yokohama, Japan, 16–19 October 2017; pp. 1–7.
5. Granström, K.; Baum, M.; Reuter, M. Extended Object Tracking: Introduction, Overview and Applications. *J. Adv. Inf. Fusion.* **2017**, *2*, 139–174.
6. Mihaylova, L.; Carmi, A.Y.; Septier, F.; Gning, F.; Pang, S.K.; Godsill, S. Overview of Bayesian Sequential Monte Carlo Methods for Group and Extended Object Tracking. *Digit. Signal Process.* **2014**, *25*, 1–16. [CrossRef]
7. Thormann, K.; Yang, S.; Baum, S. A Comparison of Kalman filter Based Approaches for Elliptic Extended Object Tracking. In Proceedings of the IEEE 23rd International Conference on Information Fusion (FUSION), Rustenburg, South Africa, 6–9 July 2020; pp. 1–8.
8. Thormann, K.; Baum, K. Fusion of Elliptical Extended Object Estimates Parameterized with Orientation and Axes Lengths. *IEEE Trans. Aerosp. Electron. Syst.* **2021**, *4*, 2369–2382. [CrossRef]
9. Liu, S.; Liang, Y.; Xu, L.; Li, T.; Hao, X. EM-based Extended Object Tracking Without a Prior Extension Evolution Model. *Signal Process.* **2021**, *188*, 108181. [CrossRef]
10. Dahlén, K.M.; Lindberg, C.; Yoneda, M.; Ogawa, T. An Improved B-spline Extended Object Tracking Model Using the Iterative Closest Point Method. In Proceedings of the 25 the International Conference on Information Fusion (FUSION), Linköping, Sweden, 4–7 July 2022; pp. 1–8.
11. Wahlström, N.; Özkan, E. Extended Target Tracking Using Gaussian Processes. *IEEE Trans. Signal Process.* **2015**, *16*, 4165–4178. [CrossRef]
12. Thormann, K.; Baum, M.; Honer, J. Extended Target Tracking Using Gaussian Processes with High-resolution Automotive Radar. In Proceedings of the 21st International Conference on Information Fusion (FUSION), Cambridge, UK, 10–13 July 2018; pp. 1764–1770.
13. Baum, M.; Hanebeck, U.D. Shape Tracking of Extended Objects and Group Targets with Star-convex RHMs. In Proceedings of the 14th International Conference on Information Fusion(FUSION), Chicago, IL, USA, 5–8 July 2011; pp. 1–8.
14. Baum, M.; Hanebeck, U.D. Extended Object Tracking with Random Hypersurface Models. *IEEE Trans. Aerosp. Electron. Syst.* **2014**, *50*, 149–159. [CrossRef]
15. Baum, M.; Feldmann, M.; Fränken, D.; Hanebeck, U.D.; Koch, W. Extended Object and Group Tracking: A Comparison of Random Matrices and Random Hypersurface Models. In Proceedings of the Informatik 2010 Service Science—Neue Perspektiven für Die Informatik, Beitrage Der 40 Jahrestagung Der Gesellschaft Fur Inform. e.V., Leipzig, Germany, 1 January 2010; pp. 904–906.
16. Zea, A.; Faion, F.; Baum, M.; Hanebeck, U.D. Level-set Random Hypersurface Models for Tracking Nonconvex Extended Objects. *IEEE Trans. Aerosp. Electron. Syst.* **2016**, *52*, 2990–3007. [CrossRef]
17. Liu, Y.; Ji, H.; Zhang, Y. Measurement Transformation Algorithm for Extended Target Tracking. *Signal Process.* **2021**, *186*, 1–12. [CrossRef]
18. Steinbring, J.; Baum, M.; Zea, A.; Faion, F.; Hanebeck, U.D. A Closed-Form Likelihood for Particle Filters to Track Extended Objects with Star-Convex RHMs. In Proceedings of the IEEE International Conference on Multisensor Fusion and Integration for Intelligent Systems (MFI), San Diego, CA, USA, 14–16 September 2015; pp. 1–6.
19. Kaulbersch, H.; Baum, M.; Willett, P. EM Approach for Tracking Star-convex Extended Objects. In Proceedings of the 20th International Conference on Information Fusion, Xi'an, China, 10–13 July 2017; pp. 1–8.

20. Liu, Y.; Ji, H.; Zhang, Y. Gaussian-like Measurement Likelihood Based Particle Filter for Extended Target Tracking. *IET Radar Sonar Navig.* **2022**, *12*, 1–15. [CrossRef]
21. Özkan, E.; Wahlström, N.; Godsill, S.J. Rao-Blackwellised Particle Filter for Star-convex Extended Target Tracking Models. In Proceedings of the International Conference on Information Fusion, Heidelberg, Germany, 5–8 July 2016; pp. 1193–1199.
22. Faion, F.; Baum, M.; Hanebeck, U.D. Tracking 3D Shapes in Noisy Point Clouds with Random Hyper Surface Models. In Proceedings of the 15 th International Conference on Information Fusion (FUSION), Singapore, 9–12 July 2012; pp. 2230–2235.
23. Zea, A.; Faion, F.; Hanebeck, U.D. Tracking Extended Objects Using Extrusion Random Hypersurface Models. In Proceedings of the Sensor Data Fusion: Trends, Solutions, Applications (SDF), Bonn, Germany, 8–10 October 2014; pp. 1–6.
24. Hoher, P.; Baur, T.; Reuter, J.; Griesser, D.; Govaers, F.; Koch, W. 3D Extended Object Tracking and Shape Classification with a Lidar Sensor using Random Matrices and Virtual Measurement Models. In Proceedings of the 27th International Conference on Information Fusion (FUSION), Venice, Italy, 8–11 July 2024; pp. 1–6.
25. Baur, T.; Reuter, J.; Zea, A.; Hanebeck, U.D. Extent Estimation of Sailing Boats Applying Elliptic Cones to 3D Lidar Data. In Proceedings of the 25th International Conference on Information Fusion (FUSION), Linköping, Sweden, 4–7 July 2022; pp. 1–8.
26. Kumru, M.; Özkan, E. 3D Extended Object Tracking Using Recursive Gaussian Processes. In Proceedings of the International Conference on Information Fusion (FUSION), Cambridge, UK, 10–13 July 2018; pp. 1–8.
27. Kumru, M.; Özkan, E. Three-Dimensional Extended Object Tracking and Shape Learning Using Gaussian Processes. *IEEE Trans. Aerosp. Electron. Syst.* **2021**, *5*, 2795–2814. [CrossRef]
28. Baur, T.; Reuter, J.; Zea, A.; Hanebeck, U.D. Shape Estimation and Tracking using Spherical Double Fourier Series for Three-Dimensional Range Sensors. In Proceedings of the IEEE International Conference on Multisensor Fusion and Integration for Intelligent Systems (MFI), Karlsruhe, Germany, 23–25 September 2021; pp. 1–6.
29. Hu, H.; Beck, J.; Lauer, M.; Stiller, C. Continuous Fusion of IMU and Pose Data using Uniform B-Spline. In Proceedings of the Multisensor Fusion and Integration for Intelligent System(MFI), Karlsruhe, Germany, 14–16 September 2020; pp. 1–6.
30. Li, X.R.; Jinkov, V. Survey of Maneuvering Target Tracking. Part I: Dynamic Models. *IEEE Trans. Aerosp. Electron. Syst.* **2003**, *4*, 1333–1364.
31. Wan, E.; Merwe, R. The Unscented Kalman Filter for Nonlinear Estimation. In Proceedings of the IEEE 2000 Adaptive Systems for Signal Processing, Communications, and Control Symposium, Lake Louise, AB, Canada, 4 October 2000; pp. 1–6.
32. Julier, S.J.; Uhlmann, J.K. Unscented filtering and nonlinear estimation. *Proc. IEEE* **2004**, *3*, 401–422. [CrossRef]
33. Särkkä, S. Unscented Rauch-Tung-Striebel smoother. *IEEE Trans. Autom. Control.* **2008**, *3*, 845–849. [CrossRef]
34. Huang, Y.; Zhang, Y.; Zhao, Y.; Mihaylova, L.; Chambers, J.A. Robust Rauch-Tung-Striebel smoothing framework for heavy-tailed and/or skew noises. *IEEE Trans. Aerosp. Electron. Syst.* **2020**, *1*, 415–441. [CrossRef]
35. Feldman, M.; Franker, D.; Koch, W. Tracking of extended objects and group targets using random matrices. *IEEE Trans. Signal Process.* **2011**, *4*, 1409–1420. [CrossRef]

MDPI AG
Grosspeteranlage 5
4052 Basel
Switzerland
Tel.: +41 61 683 77 34

Sensors Editorial Office
E-mail: sensors@mdpi.com
www.mdpi.com/journal/sensors

www.ingramcontent.com/pod-product-compliance
Lightning Source LLC
Chambersburg PA
CBHW061600220926
51850CB00005B/381
* 9 7 8 3 7 2 5 8 4 9 7 9 6 *